Innovative Trends in Civil Engineering for Sustainable Development

Patron

Prof. N. V. Ramana Rao

Director, National Institute of Technology, Warangal

Chairman

Prof. M. Chandrasekhar

Head of the Department

Organising Secretaries

Prof. N. V. Umamahesh

Dr. K. Venkata Reddy

Dr. Venkaiah Chowdary

Dr. P. Hari Prasad Reddy

Dr. S. Venkateswara Rao

Joint Secretaries

Dr. P. Venkateswara Rao

Dr. S. Shankar

Dr. D. Ravi Prasad

Treasurer

Prof. T.D. Gunneswara Rao

Sponsored by

TEQIP III

Organized by

Department of Civil Engineering
National Institute of Technology
Warangal – 506 004, Telangana State India

Disclaimer

The Publisher and Organisers do not claim any responsibility for the accuracy of the data, statements made, and opinions expressed by authors. The authors are solely responsible for the contents published in the paper.

Published by

A unit of **BSP Books Pvt. Ltd.**

4-4-309/316, Giriraj Lane, Sultan Bazar,
Hyderabad - 500 095
Phone : 040 - 23445600, 23445688
e-mail : info@bspbooks.net
www.bspbooks.net

ISBN: 978-93-89354-57-7

Organizing Committee

Fax : 0870-2459119
E-mail : director@nitw.ac.in

Phone : 0870 - 2459216
0870 - 2462000

राष्ट्रीय प्रौद्योगिकी संस्थान

वरंगल - ५०६ ००४, (टि.एस.) भारत

NATIONAL INSTITUTE OF TECHNOLOGY

WARANGAL - 506 004, (T.S.) INDIA

web: www.nitw.ac.in

डॉ. एन वी. रामना राव
निदेशक
Dr. N.V. Ramana Rao
B.E.(OU),M.Tech (IITD),PGDCS(UOH), Ph.D (UK)Post Doc. (UK)
Director

Director's Message

I am happy to understand that Department of Civil Engineering has taken up a challenging task of dealing with a very sensitive and most vital issue of sustainable development during the International conference on Innovative Trends in Civil engineering for Sustainable Development (ITCSD) to be held on 13-15 September, 2019. Sustainability is an issue of global importance in the 21st century. All the stakeholders in all the nations have to plunge into action to achieve sustainable Development goals. The civil engineering discipline is no exception, as it uses an enormous amount of resources and energy in various sectors. However, civil engineering is also essential for creation of infrastructure that forms the basis of socio economic activities. Hence, it is necessary that we need to utilize these technologies by implementing the sustainable practices for a greener tomorrow. I understand that the department of civil engineering is already doing its share of contributing to conservation and sustainable development. The deliberations of the conference will address the cause and solution for challenges faced in various sectors of Civil Engineering. I am confident that this will be a forum to play a dominant role in exchange of information and professional interaction among those who are involved in civil engineering profession. This conference is an opportunity to consolidate progress in civil engineering research for sustainable development.

Once again, I congratulate the conference organizers for identifying the important research themes for the conference and I wish them, all the best.

N.V. 7/9/19
DIRECTOR

Message from Head, CED

The department of Civil Engineering is one of the oldest branches established in the year 1959. The department offers one UG and seven PG programs. The department is well equipped to carry out research in thrust areas of civil engineering. The department offers consulting and testing services also. Civil engineering is a broad field of engineering that deals with planning, construction and maintenance of fixed structures, or public works, as they are related to earth, water or civilization and their processes. This is a profession that makes the world a more agreeable place to live in. Civil Engineering is one of the core and most fundamental branches of engineering. In the modern world of construction where we witness revolutionary growth in construction methods, materials and skills, developing sustainable materials and technologies plays a key role.

The three day International conference on Innovative Trends in Civil Engineering for Sustainable Development (ITCSD) provides a platform to the researchers, faculty and practitioners to showcase their ideas and innovations related to Sustainable Development. I wish all the participants good luck and have a nice time three day stay in NIT Warangal.

I welcome all the delegates to the three day International Conference on Innovative Trends in Civil Engineering for Sustainable Development on behalf of the department of Civil Engineering, National Institute of Technology Warangal.

(Prof M CHANDRA SEKHAR)

Message from Organising Secretary

Civil Engineers are the builders of the nation. They always have played a significant role in providing the infrastructure required for the society. Ever increasing population, rapid urbanization and industrialization, global warming and climate change, increasing incidents of natural and manmade hazards and environmental degradation have become major concerns of the society today and these concerns pose a challenge to the Civil Engineers. There has been a noticeable change in the environment within which civil engineers carry out their work. There is a growing concern on the adverse consequences of the development activities. There is a need for Civil Engineers to rise up to face these challenges and contribute to the sustainable development. A multitude of threats confront us in our effort to achieve sustainability. Civil engineers have always been looking for innovative tools and techniques for solving the challenging problems that they face during planning, analysis, design and construction of civil engineering structures. It is in this context that the Department of Civil Engineering, National Institute of Technology, Warangal has come out with the proposal of organising an International Conference on "Innovative Trends in Civil Engineering for Sustainable Development". This conference is aimed to provide a forum to all Civil Engineers to deliberate on the innovative tools and techniques that are needed to face the challenge of sustainable development.

We received an overwhelming response to our call for submission of abstracts to this conference. We received more than 200 abstracts from across the country and abroad. It is heartening to note that several of the contributions that we received are from young researchers/ faculty. Through our review process we ensured that only quality papers were selected for presentation during the conference. About 150 abstracts were selected for presentation during the conference. We are proposing to bring out an Edited Volume consisting of selected papers from the papers that are presented during the conference.

I hope that this three day conference will provide a platform, especially for young budding researchers to share their ideas and vision on new and innovative technologies that are needed for Civil Engineers to face the challenges of Sustainable Development. I also hope that this conference will help in building a network of researchers, faculty and practitioners that will work towards the goal of achieving sustainable development.

N. V UMAMAHESH

TEQIP Coordinator's Message

I feel extremely elated to learn that the coordinators have selected a very important area for dissemination of their knowledge from their rich experience and bringing researchers from different corners of the world to participate in vital discussions. Today, we have come a long way on sustainable development around the globe. As our world has become smaller and technology has grown leaps and bounds, we cannot ignore what is happening in the world and think only of one country as in the past. When we consider sustainable development in a broad sense global, we must see what is taking place in various parts of the world. I am sure the technical sessions will provide learning and networking opportunities on the latest advances, technical knowledge, continuing research, tools and solutions for innovative trends in civil engineering. I can understand how important civil engineering technologies are for economy, social responsibility and environmental aspects. These are the three pillars of sustainable development. Every sector in civil engineering in some form or another is responsible for our civilized wellbeing. However, the technologies used also require sustainable energy use, reduced carbon dioxide emissions. We all have a duty to behave responsibly and the related wasteful and polluting activities need to be dealt with. To do this in a committed and balanced way, it requires both knowledge and experience.

I once again appreciate the coordinators for ticking a very important research area, which can go a long way for a better living.

(Prof L KRISHNANAND)

Contents

Engineering Structures

Shear Strength of Monolithic Geopolymer Concrete by ACI 318

Sumanth Kumar B[1*], Prof. D Rama Seshu[2]
[1, 2] *Department of Civil Engineering, National Institute of Technology, Warangal, India*
**bsk109@rediffmail.com*

Introduction

The making of Ordinary Portland Cement involves enormous size of energy consumption, leading to a mammoth discharge of carbon di-oxide to the air, which is being a great task to the sustainable advance. Efforts are required to grow an environmental sociable construction material to reduce release of green-house gases to the atmosphere. One of the effort to reduce the carbon foot print, waste by products are used as alternative binders to the cement such as fly ash, ground granulated blast furnace slag (GGBS) etc., along with alkaline activated solution forms a matrix called "Geopolymer concrete" (GPC). Generally, Structural members fail in flexural and shear. However, Flexural can be avoided by providing tension / flexural reinforcement and elements with shear reinforcement will carry tension as tie and compression by concrete struts. To envisage this shear strength, shear friction model is adopted.

Shear friction started as a theory for a design of concrete connections. From basic mechanics theory, shear transferred $Vu = N \tan \phi$, where Vu is maximum shear force transferred and N is normal force acting at interface and $\tan \phi$ is contact friction coefficient. This paper presents a study on the shear strength of monolithic GPC interface. 18 push-off specimens with and without transverse reinforcement at interface were cast and tested. The test shear strength of GPC is compared with the shear strength assessed by different editions of ACI 318 and are conservative in evaluating the shear strength of GPC.

Materials and Methods

Shear friction theory along layers is resisted by Cohesion and after crack Cohesion is lost and transfer is in combination of shear – friction and dowel action. ACI Committee 711 proposed shear resistance of unreinforced interfaces are equal to allowable shear stress of unreinforced beam. Research based on shear friction started since 1960's because of accurate test data fits this analogy number of test specimens. Mast developed first shear friction chapter for American Concrete Institute ACI 318 – 1971. Based on the research Cohesion term, which is dependent on type of aggregate was included along with friction term and with minimum clamping force in ACI 318-1983 was introduced. The same expression was considered for shear strength capacity as per shear friction model till 2014. Only upper limit for shear capacity is adopted based on type of interfaces in ACI 318 – 2008. Shear friction equation omits cohesion term and only friction term is considered as per latest edition i.e., ACI 318 – 2014.

Fly ash and GGBS are considered as binders, Fine Aggregate of river sand conforming to Zone-2 of IS: 383-2016. Coarse Aggregate is well graded aggregate conforming to IS: 383-2016 with 20mm nominal size of granite. Potable water was used in the experimental work. Alkaline Solution consists of Sodium Silicate to Sodium Hydroxide (8 Molarity) with ratio 2.5:1 and stored at room temperature (25±2°C) for 24 hours and relative humidity of 65% before using it in the casting of GPC push off specimens.

Mix proportion for GPC push off specimens was adopted from procedure given by G Mallikarjuna Rao et al and mix quantity shown in Table.1 after making different trials having different strengths.

Table 1. Materials used in GPC (per Cu.m)

S. No	Grade of GPC	Materials						
		Coarse Agg. (kg)	Fine Agg. (kg)	Fly Ash (kg)	GGBS (kg)	NaOH Sol. 8 Molarity (kg)	Sodium Silicate (kg)	SP* (kg)
1	A20	965	812	294	126	66	165	4.2
2	B30	965	812	252	168	66	165	4.2
3	C40	965	812	210	210	66	150	4.2

*SP: Super plasticizer (SP 430, Make: Fosroc Chemicals).

Sizes of the push-off specimens considered for investigation are shown in Figure 1. The samples were cast with and without reinforcement through the shear interface as shown in Figure 2. The samples were loaded axially till failure. The Push-off models with and without reinforcement across the slip plane, tested and failed by developing a crack along the interface.

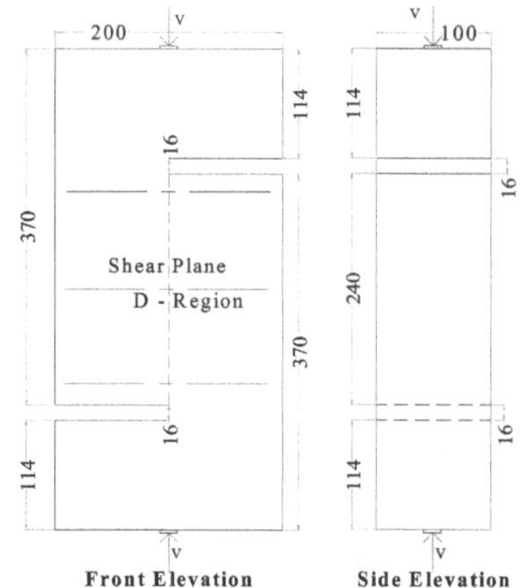

Figure 1. Push off Specimen

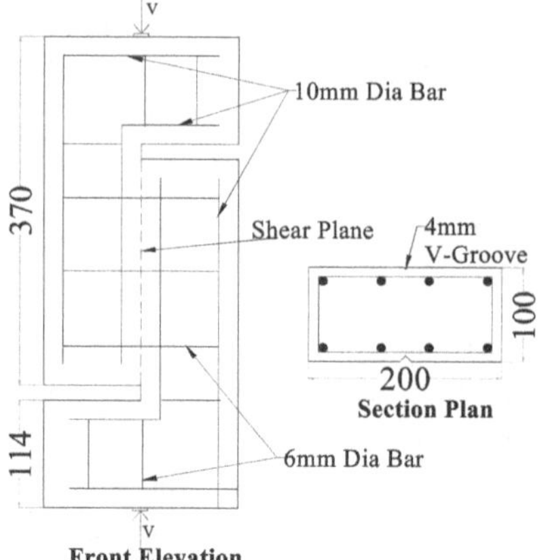

Figure 2. Reinforcement Details for Push -Off specimen

Results and Concluding Remarks

The push-off specimens has ruptured along the interface. In case of specimens without reinforcement across the interface, failed suddenly and reinforced specimens showed visible cracking along the shear plane at about 70 to 80 percent of the ultimate loads. Due to the provision of suitable reinforcement in the L shape of specimen, early flexural failure was not seen in none of the specimens in vertical or horizontal arms of specimens In the event of reinforced shear interfaces the shear strength has increased about 28% of the corresponding compression strength.

The comparison of shear strength of GPC obtained with the shear strength of normal concrete predicted by different editions of ACI 318 shear-friction equations. The comparative study indicates that the available normal concrete shear strength prediction models are highly conservative in estimating the shear strength of unreinforced and reinforced monolithic shear interfaces in GPC. ACI 318 – 1983 to ACI 318 – 2008 seems to give better forecast of shear strength of GPC in the case of unreinforced and reinforced monolithic interfaces respectively

References

ACI Committee 711. (1953). Minimum Standard Requirements for Precast Concrete Floor Units (ACI 711-53). ACI Journal Proceedings, 50(9), 1-1

ACI Committee, 1971. Building code requirements for reinforced concrete.

ACI Committee, 1983. 318 Building Code Requirements for Reinforced Concrete (ACI-318-83).

ACI Committee, American Concrete Institute and International Organization for Standardization, 2008. Building code requirements for structural concrete (ACI 318-08) and commentary. American Concrete Institute.

ACI (American Concrete Institute), 2014. Building code requirements for reinforced concrete. ACI 318-14.

Davidovits, J., 1991. Geopolymers: inorganic polymeric new materials. Journal of Thermal Analysis and calorimetry, 37(8), pp.1633-1656.

IS: 383, 2016. Specification for coarse and fine aggregates from natural sources for concrete.

Mast, R.F., 1968. Auxiliary reinforcement in concrete connections. Journal of the Structural Division.

Rao, G.M., Rao, T.D., Seshu, R.D. and Venkatesh, A., 2016. Mix proportioning of geopolymer concrete. Cement Wapno Beton, 21(4), pp.274-285.

Standard, A.A., 2011, August. Building Code Requirements for Structural Concrete (ACI 318-11). In American Concrete Institute.

Experimental Investigation on Strength Characteristics of Self-Curing Concrete: A Sustainable Approach

Mahesh Navnath Patil[1], Shailendrakumar Dubey[2]
[1] *Research Scholar, K.B.C.N.M.U., Jalgaon, MS, India*
[2] *Professor, Department of Civil Engineering, S.S.V.P.S College of Engineering Dhule, MS, India*
** e-mail: m.patil123@gmail.com, dubey.dhule@gmail.com*

Introduction

Self-curing (also recognized as internal curing concrete) represents the method of generating the source of water inside the body of the concrete itself which can be made available for the hydration process subsequently (Dhir, Hewlett et al. 1995). In the traditional method of curing, the conditions are so made up that, water should not be lost from the surface i.e. curing from outside to inside while in self-curing concrete, curing is done from inside to outside of the concrete through the reservoir of water developed inside the body of the concrete. This paper aims to review the existing literature of self-curing concrete and some value addition from experimentation done. Self-curing can be an admixture based concrete. It can be generated with courteous addition of Super Absorbent Polymer (SAP), Polyethylene Glycol (PEG) and with the surface coating of Curing Compound (CC). In the present study, all the mentioned admixtures and curing compound were tried.

The intention of the paper is to provide significant data and mix design guideline for the researcher and professional engineers in the places with water scarcity or the structure where the curing process is very difficult and tedious.

Materials and Methods

All the experimental work mentioned in this paper is carried out well-equipped concrete technology laboratory in the Civil Engineering Department of RCPIT, Shirpur and SSVPS College of Engineering, Dhule.

The material used, designing of the test specimen and actual testing procedure were discussed in the following sections.

Materials

The cement used is the Portland Pozzolana Cement (PPC) from Ambuja Cement Ltd. (Unit: Maratha Cement Works) confirming to requirements as per IS: 1489 (Part1):2015 with 32 % of fly ash in the given Portland cement and specific gravity 2.76. Chemical and physical requirements of cement as supplied by provided is given in Table 1. The fine aggregate used is natural sand which passes over 4.75 mm IS sieve, permitted as per IS 383: 2016. It is clean and free from the impurities with specific gravity 2.52. Coarse aggregate used is a natural stone retaining on 4.75 mm IS sieve and contains so much of fine material which is permitted as per IS 383: 2016 with specific gravity 2.88. The mechanical properties of fine aggregate and coarse aggregate are as per Table 2 and Table 3 respectively as given below.

Potable water free from impurities as per Indian Standard Code of practice was used mixing and curing of concrete specimens. High range water reducing admixture i.e. super plasticizer under the Brand name Auramix 200 by Fosroc was used to increase workability without affecting the strength of the concrete and without the use of additional water.

Four types of self-curing regimes were performed in this study. The first was done by using superabsorbent polymer (SAP or hydrogel) also known as Sodium Poly-acrylate, second is using chemical curing agents Polyethylene glycol (PEG 400 and PEG 600) produced by Loba Chemie Pvt. LTD in India as a chemical self-curing agent in the liquid form for self-curing of concrete, third with a surface coating on a concrete surface using the curing compound to create the adiabatic condition. Curing compound helps to prevent loss of moisture from concrete due to evaporation and ultimately results in the development of full strength without using water for curing. The curing compound used is FARECURE WC white pigmented concrete curing aid confirming to ASTM C309-06 standard.

Method

The experimental program is divided into four steps. In the initial stage effect of the water-cement ratio on the strength of the concrete is studied. In this experimental program, four different water-cement ratio are tried i.e. 0.3, 0.36, 0.4 and 0.45 (Aruntas, Yilmaz et al. 2008). M40 grade concrete is designed as per IS 10262:2009 and IS 10262:2017. Total of 72 cubes was cast. 36 with superplasticizer, 36 with controlled concrete without superplasticizer. The compressive strength of these cubes is tested at 3 days, 7 days and at 28 days of curing. The water-cement ratio which gives the best result under compression and split tensile strength is chosen for further study to generate self-curing concrete.

In the second step self-curing concrete is generated, again this second step is further divided into 3 more stages.

Stage I- Super Absorbent Polymer (SAP) was used as a self-curing agent as 0.05%, 0.15%, 0.20%, 0.25%, 0.3% of the cement content (Justs, Wyrzykowski et al. 2015).

Stage II- Chemical curing agent's Polyethylene glycol 400 (PEG 400) (Kamal, Safan et al. 2018) was used as a self-curing agent. This chemical was added as 1%, 2%, 3%, 4% and 5% of the weight of the cement.

Stage III- Chemical curing agent's Polyethylene glycol 600 (PEG 600) (Kamal, Safan et al. 2018) was used as a self-curing agent. This chemical was added as 1%, 2%, 3%, 4% and 5% of the weight of the cement.

Stage IV- The curing compound FARECURE WC white pigmented concrete curing aid was used for surface coating of the cubes with plasticizer and controlled mix.

The specimens used in this test were cubes of dimensions 150 mm x 150 mm x 150 mm to determine compressive strength, cylinders of 150mm diameter and 300 mm length to determine split tensile strength.

Results and Concluding Remarks

In this research paper, vast experimentation was carried out to investigate the strength behavior of concrete under different circumstances. Based on this experimental work following conclusion can be drawn.

1. Out of the four different water-cement 0.3, 0.36, 0.4 and 0.45, water cement ratio 0.36 giving best results in compression as well as in the split tensile test.
2. Workability of the self-curing concrete directly depends on the type of the curing agent used.
3. Curing compound based self-curing concrete shows best results followed by SAP, PEG600 and PEG 400.
4. The optimum value was obtained at 0.15% SAP, 2% PEG 400 and 2% of PEG 600. The externally applied curing compound gives the best result.

References

Aruntas, H. Yilmaz et al. 2008. "Effects of Super Plasticizer and Curing Conditions on Properties of Concrete with and without Fiber." Materials Letters 62(19): 3441–43. doi:10.1016/j.matlet.2008.02.064

Dhir, R.K., P.C. Hewlett, and T.D. Dyer. 1995. "Durability of 'Self-Cure' Concrete." Cement and Concrete Research 25(6): 1153–58.

El-Dieb, A. S. 2007. "Self-Curing Concrete: Water Retention, Hydration and Moisture Transport." Construction and Building Materials 21(6): 1282–87. http://dx.doi.org/10.1016/j.conbuildmat.2015.08.042.

Justs, J., M. Wyrzykowski, D. Bajare, and P. Lura. 2015. "Internal Curing by Superabsorbent Polymers in Ultra-High Performance Concrete." Cement and Concrete Research 76: 82–90. https://doi.org/10.1016/j.cemconres.2015.05.005

Kamal, M. M., M. A. Safan, A. A. Bashandy, and A. M. Khalil. 2018. "Experimental Investigation on the Behavior of Normal Strength and High Strength Self-Curing Self-Compacting Concrete." Journal of Building Engineering 16(July 2017): 79–93. https://doi.org/10.1016/j.jobe.2017.12.012

Study of behaviour of Uniaxial column bonded with Basalt FRP wraps

Toufeeq Anwar[1*], Shaik Abdul Rafi[2]
[1] *Associate professor,* [2]*PG scholars, Dept. Of Civil Engg., Muffakham Jah College of Engineering and Technology, Hyderabad,India.*
** e-mail: toufeeqanwar@mjcollege.ac.in*

Introduction

The use of externally bonded fiber reinforced polymer (FRP) reinforcement to strengthen reinforced concrete (RC) structures is becoming an increasingly popular strengthening technique. The light weight and formability of FRP reinforcement makes these systems easy to install. Because the materials used in these systems are noncorrosive, nonmagnetic, and generally resistant to chemicals, they are an excellent option for external reinforcement.

R. Sudhakar et al. 2017 has conducted test on columns strengthened with Glass FRP fabrics and concluded that axial compressive strength of RCC column wrapped with single layered and double layer GFRP is increased by 15.31% and 31.35% compare with control column. Jibi Sam et al 2017 has done comparative study on strengthening of RC column using stainless steel wire mesh and Glass FRP and concluded that the columns wrapped with SSWM showed higher strength than those columns wrapped with GFRP. Present study investigated the behaviour of uniaxial RC columns externally bonded basalt FRP. Since it has advantages as low cost, excellent resistance towards corrosion, sound mechanical properties but scanty studies using BFRP as retrofitting material are found in literature

Materials and Methods

In order to carry out the experiment, 9 C shape columns of cross section and length as shown in table 1 were casted. Longitudinal reinforcement having 4# 10mmϕ and 2-legged 5mm ties at 125mm c/c is provided.

Table 1. Details of RC members that were cast in the present study.

Columns	Type	No. of Specimen	Size of specimen
Set 1	Control Specimen	3	900mm x 100mm x 150mm
Set 2	Full Tension Zone	3	900mm x 100mm x 150mm
Set 3	Full Tension Zone +Top & Bottom face of C	3	900mm x 100mm x 150mm

The method used to retrofit is externally bonding technique in which basalt FRP is externally bonded to RC specimens using epoxy adhesive. Basalt FRP having density 300gsm is externally bonded to full tension region of columns for set 2 whereas for set 3 it is bonded on tension face as well as critical zone of columns. Ordinary Portland cement of grade 53 having specific gravity of 2.9 is used, the maximum aggregate size was about 20 mm, and the cement:sand:gravel ratio was 1:1.45:2.62 by weight. Cubes and cylinders were cast and tested along with RC specimens in order to determine concrete compressive strength and split tensile strength. To get efficient bond between FRP and concrete surface the following steps were taken: (a) removal of laitance on the concrete surface where the FRP has to be installed (b) blowing the concrete surface with air; (c) applying lapox ultra epoxy-adhesive; (d) adhesion of BFRP fabrics; (e) applying second layer of resin; and (f) removing excessive resin using a plastic roller.

To provide an eccentricity of 225mm, columns are casted in the form of C shape and they were tested under UTM and under each load increment, deflection is noted. Type of failure and maximum load for each specimen is recorded.

Results and Concluding Remarks

Columns strengthened with Basalt FRP shows increase load carrying capacity. Table 2 shows the load carrying capacity of RC columns and the average percentage increment.

Table 2. Load and percentage increment.

Column	Specimen number	Ultimate Load	Average	% Increment
Set 1	A	56.2kN		
	B	49 kN	54.4 kN	Control Specimen
	C	58.2 kN		
Set 2	A	59.2kN		
	B	71 kN	65 kN	19.48%
	C	64.5 kN		
Set 3	A	63kN		
	B	67 kN	68.51 kN	25.9%
	C	73.55 kN		

The load carrying capacity of RC columns using Basalt FRP was found to be greater than that of control specimens. No debonding of Basalt FRP sheet was observed.Hence it is concluded that Basalt FRP has played a significant role in enhancing the load carrying capacity of Rc columns.

Acknowledgments: The R&D cell MJCET provided financial assistance for this research, to which writers are grateful.

References

R. Sudhakar and Dr. P. Partheeban (2017) Strengthening of RCC Column Using Glass Fibre Reinforced Polymer (GFRP). International Journal of Applied Engineering Research ISSN 0973-4562 Volume 12, Number 14 (2017) pp. 4478-4483 . http://www.ripublication.com

Jibi Sam.S, Bennet A Ipe(2017) Comparative study on strengthening of Rc short columns using SSWM and GFRP wraps.International Research Journal of Engineering and Technology (IRJET) e-ISSN: 2395 -0056 Volume: 04 http://www.irjet.net

Concrete Mix Proportioning-Guidelines IS-10262:2009

Shear performance of RC members retrofitted with Basalt FRP

Toufeeq Anwar[1*],Syed Jawwad Ahmed[2],Mohammed Rehan Ali[3] ,Syed Jafer Hussaini[3]
[1] Associate professor, [2] Assistant professor and [3] PG scholars, Dept. Of Civil Engg., Muffakham Jah College of Engineering and Technology, Hyderabad, India
** e-mail: toufeeqanwar@mjcollege.ac.in*

Introduction

Retrofitting of RC structures, nowadays have become a major part of the construction industry in many countries. Need for retrofitting arises due to committing mistakes in design or construction, exposure to unpredicted loads, changing the usage pattern of the structure, deterioration of concrete structures mainly due to corrosion. Conventional retrofitting methods such as Steel jacketing, Shotcrete, Reinforced plaster, Grout and epoxy injection etc., do not always offer the most appropriate solutions. Use of fibre reinforced polymers (FRP) as retrofitting method provides a more economical and technically better option to the conventional methods in many cases. The FRPs have higher strength-to-weight ratios, more durable than conventional materials such as steel, and requires less manpower and less equipment-intensive retrofitting work.

Riza Secer et al. 2017 has conducted test on beams strengthened with Carbon FRP fabrics and observed significant increase in shear strength and change of failure mode from shear to flexural. J.H. Gonzalez-Libreros et al. 2017 investigate the behavior of RC beams strengthened in shear with externally bonded composites and observed an increase in shear strength. Bassam Q. Abdulrahman et al. 2017 investigated the effectiveness of strengthening flat slab to column corner connections by using Carbon FRP fabrics. From series of slabs tested, it was concluded that strengthening of slabs increased the ultimate punching capacity by 11%. M. Gherdaoui et al. 2018 investigated the punching behaviour of reinforced concrete slabs strengthened with carbon fiber reinforced polymer (CFRP). Test results showed an increase of 23-65% compared to unstrengthened slabs. Present study investigated the behaviour of RC beams in shear and RC slabs in punching shear, retrofitted with externally bonded basalt FRP fabric. Basalt fiber reinforced polymer is relatively newcomer to the FRP line-up and has advantages as low cost , excellent resistance towards corrosion, sound mechanical properties but scanty studies using BFRP as retrofitting material are found in literature.

Materials and Methods

In order to increase the shear strength of RC beams and punching capacity of RC slabs using FRP, series of tests were carried out.Table 1 shows details of RC members that were cast to carry out the study.

Table 1. Details of RC members that were cast in the present study.

RC Member	No. of Specimen	Size of specimen	Reinforcement provided
Beam	3	100mm x 175mm x 1000mm	2# 8mm ϕ on top & 2#12mm ϕ at bottom
Slab	3	500mm x 500mm x 75mm	8mm ϕ @ 115mm c/c in both directions
Slab	3	300mm x 300mm x 75mm	8mm ϕ @ 115mm c/c in both directions

After casting the specimens mentioned in the above table, they were cured for 28 days and then tested for their capacities in the UTM.After failure, all the specimen were retrofitted and again tested to examine their retrofitted capacities. The method used to retrofit is externally bonding technique in which basalt FRP is externally bonded to RC specimens using epoxy adhesive. Basalt FRP is externally bonded to full shear span region of beams whereas for larger slabs it is bonded on tension face of slabs in a zone critical for punching shear and for smaller slabs it is bonded on four side faces of the slabs.

Basalt FRP fabrics used for both slabs and beams have density 200 gsm, thickness 0.158mm, and orientation as bi-directional according to the data provided by the manufacturer. Ordinary portland cement of grade 53 having specific gravity of 2.9 is used, the maximum aggregate size was about 20 mm, and the cement:sand:gravel ratio was 1:1.45:2.62 by weight. Cubes and cylinders were cast and tested along with

RC specimens in order to determine concrete compressive strength and split tensile strength. Results showed average cube compressive strength of 39.7 N/mm^2 and an average split tensile strength of 3.72 N/mm^2 for 28 days of curing. To get efficient bond between FRP and concrete surface the following steps were taken: (a) removal of laitance on the concrete surface where the FRP has to be installed (b) blowing the concrete surface with air; (c) applying lapox ultra epoxy-adhesive; (d) adhesion of BFRP fabrics; (e) applying second layer of resin; and (f) removing excessive resin using a plastic roller.

A four-point bending system was adopted for the beam testing and slabs were tested for punching shear strength. At the end of each load increment, deflection is noted. Ultimate load, type of failure were carefully observed and recorded. After failure , RC specimens were retrofitted using basalt FRP and similarly tested again for their ultimate capacities and also the deflections were recorded.

Results and Concluding Remarks

Using Basalt FRP for retrofitting improved the shear resistance of RC beams and punching capacity of RC slabs in a very significant manner. Table 2 shows the average ultimate load carrying capacities of tested RC members before and after retrofitting and the average percentage increment in the load capacities of retrofitted members.

Table 2. Average Load at failure of control and retrofitted specimen and percentage increment of retrofitted members .

RC Member	Size of specimen	Control specimen	Retrofitted specimen	% Increment
Beam	100mm x 175mm x 1000mm	94.4 kN	105.83 kN	12.1%
Slab	500mm x 500mm x 75mm	103.3 kN	109.85 kN	6.34%
Slab	300mm x 300mm x 75mm	92.4 kN	99.3 kN	7.46%

The shear strength of retrofitted RC beams and punching capacity of retrofitted RC slabs was found to be greater than that of control specimen. Hence, it can be concluded that Basalt FRP fabrics are noteworthy for increasing the shear strength of beams and punching strength of slabs.

Acknowledgments: The R&D cell MJCET provided financial assistance for this research, to which writers are grateful.

References

Bassam Q.Abdulrahman, Zhangjian Wu, Lee S.Cunningham (2017) Experimental and numerical investigation into strengthening flat slabs at corner columns with externally bonded CFRP. Journal of Construction and building materials 139(2017): 132-147. https://doi.org/10.1016/j.conbuildmat.2017.02.056

J.H. Gonzalez-Libreros, L.H. Sneed, T. D'Antino, C. Pellegrino (2017) Behavior of RC beams strengthened in shear with FRP and FRCM composites. Journal of Engineering Structures 150(2017): 830-842. https://doi.org/10.1016/j.engstruct.2017.07.084

M. Gherdaoui, M. Guenfoud, R. Madi (2018) Punching behavior of strengthened and repaired RC slabs with CFRP. Journal of Construction and building materials 170(2018): 272-278. https://doi.org/10.1016/j.conbuildmat.2018.03.093

Riza Secer Orkun Keskin, Guray Arslan, Kadir Sengun (2017) Influence of CFRP on the shear strength of RC and SFRC beams. Journal of Construction and building materials 153(2017): 16-24. https://doi.org/10.1016/j.conbuildmat.2017.06.170

Utilization of Industrial Wastes GGBS in Regular Concrete by Material Replacement Technique

Ramansh Bajpai[1], Deepesh Singh[2]
[1]Department of Civil Engineering, HBTU, Kanpur, India.
[2]Department of Civil Engineering, HBTU, Kanpur, India.
*email: ramanshbajpai786@gmail.com, dr.deepeshsingh@gmail.com

Introduction

Dumping of Industrial waste is the major problem that we are facing today, Cement industry is one of the largest carbon emission industries in the world which is responsible for the most concerned problem for our planet that is global warming. Concrete is the second most consumed material in the world after water and the waste produced from iron and steel industries is in tones, that can be partially utilized in the regular concrete in place of cement and thus it will reduce CO_2 emission at large scale. The experimental study has been done by partially replacing the OPC (Ordinary Portland Cement) by GGBS (Ground granulated blast furnace slag) in different percentages. This paper focuses on the technical specification and utilization of GGBS and small portion of RHA in concrete and producing a combined material of similar technical properties as the regular concrete. M. Dina et. al. (2017) had investigated the OPC replaced by High volume GGBS Pastes modifies with micro sizes metakaolin subjected to elevated temperature and found out that Compressive Strength before and after being exposed to elevated temperatures increased with micro metakaolin Content. *P.S Joanna et. al. (2014)* had conducted the experimental study on flexural behavior of partially replaced concrete by GGBS and concluded that the deflections under the service loads for the concrete beams with 40% GGBS were same as that of the controlled beams at 28 days testing. *Mathew et. al.* (2013) inferred that low strength of bottom ash GGBS based concrete is due to large particle size. *Arivalagan (2014)* had experimented on material replacement techniques. He replaced ordinary Portland cement with GGBS of 20% 30% and 40% in proportion with mix design of M35 at 7 and 28 days and came to conclusion that strength increases for 20% replacement. P.J. Wainwright et. al. (2000) had studied the influence of GGBS additions and time delay on the bleeding of concrete and concluded that the partial replacement of cement upto 55% GGBS increased the bleed capacity by 30%.

Experimental investigation on GGBS (ground Granulated Blast Furnace Slag) has carried out which is by product of iron industry and also can be used as a replacement for ordinary Portland cement in concrete. Use of GGBS as Cementitious binder reduces the cost of concrete and help to reduce the rate of cement consumption. The investigation is done for economical utilization of wastes from iron and steel industries and its applicability in concrete structures.

Materials and Methods

Ground Granulated Blast Furnace Slag (GGBS)is the byproduct of iron and steel industries. Iron ore, coke and lime stone are fed into the furnace and the resulting molten slag floats above the molten iron at a temperature of about 1500°C to 1600°C. The molten slag has composition close to chemical composition of Portland cement. It is the primary material that has been used for experimental purpose with small quantity of rice husk. Locally available OPC 43 grade cement and river Ganges sand of Kanpur is used for concrete mixtures. GGBS is delivered by Guru Corporation which deals with disposal of iron and steel industries wastes in Gujarat. Rice husk is locally available near kannauj city rice mill.

Eight different mixtures were prepared G0 to G8 that is 0 to 80 % GGBS replacement with 5 % of rice husk in every mixture. W/c is taken as 0.45 for M20 mix design and 0.45 for M40 (1:1.88:2.76) mix. Material is mixed in mechanical mixer at speed of 75 rpm and after mixing material is poured in Cylindrical and cube moulds. 6 samples for each mix is prepared for testing (7 and 28 days) with 3 and 5 layer compaction and vibrated for 40 seconds to remove air bubbles. The test is done after 7 and 28 days for compressive strength and split tensile strength also durability test is performed under different conditions to check the suitability of geopolymer (GGBS based) concrete use.

Results and Concluding Remarks

The prepared cubes and cylindrical samples are tested in compression testing machine after 7 and 28 days which gave the results as follows.

The compressive strength increases up to 50% replacement of cement with GGBS (45%) and rice husk ash (5%) after that it starts decreasing. The average compressive strength for 50% replacement of GGBS and RHA for M20 mix for 7 and 28 days respectively is 16.78 N/mm^2 and 25.13 N/mm^2 and for M40 mix it is 34.05 N/mm^2 and 45.87 N/mm^2. Geopolymer is tested for sulphate attacks and for its durability under severe conditions.

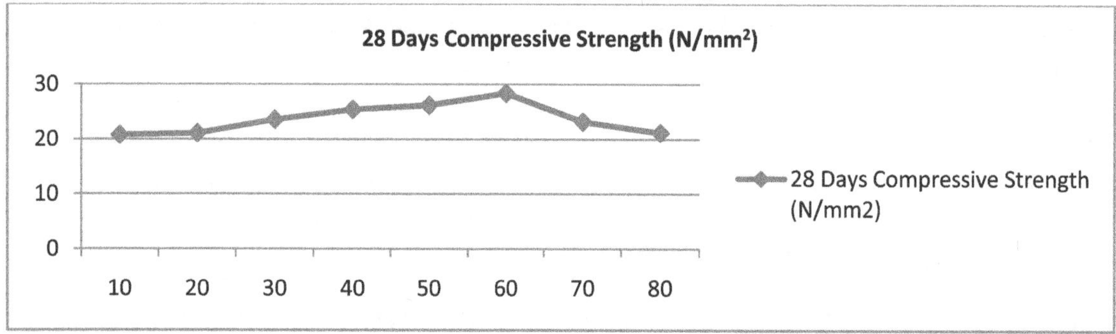

Fig, 1. Variation of Compressive strength for partially replaced cement by GGBS and RHA (M20 grade)

Durability test carried out by the investigation through acid attack test and chloride test with 1% Sulphuric acid and 3% hydrochloric revealed that when cement is replaced with 45% GGBS and 5% RHA in concrete is more durable in terms of durability factors than control mix as regular concrete losses nearly 7% compressive strength.

Average loss in weight of cube and compressive strength for acid resistant test with 1% Sulphuric acid by wt. of water is 2.21% and 5.70% respectively and average loss in weight of cube and compressive strength for Chloride attack test with 3% HCL by wt. of water is 2.97% and 6.53 % respectively.
The inference after testing is that nearly 50% of cement can be replaced by GGBS and rice husk ash which can be used for structural purpose and have good workability, strength and durability. Use of GGBS (45%) and RHA (5%) up to 50% is economical and can decrease the carbon emission in large scale and helpful in decreasing the global CO_2 emission.

Table 1. Use of GGBS replaced concrete for different types of structures

Type of Structures	Usable GGBS Percentage
For normal ground concrete structures	GGBS replacement is 30-35%
For Underground concrete structures with average strength	GGBS replacement is 35-45%
For marine structures and sewerage treatment plants	GGBS replacement is 50-65%

Table 1 describes the possible replacement of GGBS with cement for different types of structures. Concrete with 50% to 65 % GGBS is more durable for sulphate action and hence used for marine structures. For normal ground structures such as buildings, Offices etc. cement can be replaced up to 35% by GGBS, which has similar properties and more compressive strength than regular OPC based concrete.

Future scopes of GGBS are positive due to benefits in Durability, Sustainability, Appearance and strength obtains by partial replacement of GGBS with cement in concrete.

References

1. Alaa M. Rashad and Dina M. Sadek (2017),"An investigation on Portland cement replaced by high-volume GGBS pastes modified with micro-sized metakaolin subjected to elevated temperatures" , Elsevier-IJSBE, Vol. 6, pp. 91-101, 2017.
2. Arivalagan. S (2014), "Sustainable Studies on Concrete with GGBS as a Replacement Material in Cement", Jordan Journal of Civil Engineering, Vol 8, No. 3, pp 263-270.
3. W. C. Leung. Peter. and Wong H. D., "Final Report on Durability and Strength Development of Ground Granulated Blast furnace Slag Concrete ", Geo Report No. 258, no. SPR 1/2010, July 2010.
4. G. Singh, S.Das, A.A. Ahmed, S. Saha and S. Karmarkar (2015), "study of Granulated Blast Furnace Slag as Fine Aggregate in Concrete for Sustainable Infrastructure". Procedia- Social and Behavioral Sciences, Vol. 195, pp. 2272-2279, July 2015.
5. P.J. Wainwright and N. Rey (2000), "The Influence of ground granulated blast furnace slag (GGBS) additions and time delay on the bleeding of concrete" , Elsevier-Cement & Concrete Composites, Vol. 22, pp. 253-257, 2000.

Experimental investigation on the use of waste (PET) fibre as a supplement of fine aggregate in concrete

A. Monika Thokchom[1], B.Chitra Shijagurumaum[2*], C.Thokchom Suresh[1]

[1]*Department of Civil Engineering, Manipur Institute of Technology, Takyelpat, Imphal, 795004, Manipur*
[2] *Research Scholar*
* *e-mail:chitras10shij@gmail.com*

Introduction

Plastic waste is now a serious environmental threat caused by the modern way of living. According to the latest study of 2018, 70% of total plastic consumption is discarded as waste in India. The improper disposal of plastic bottle has been a major concern to the environment as it is not easily degradable. Therefore, one possible solution to this problem is to recycle polyethylene terephthalate (PET) in the construction industry as a reinforcing fibre in concrete. Prabhu et al. 2014 suggested that the maximum percentage increase in compressive strength and split tensile strength was observed at 1% of fibre content. Prabha et al. 2017 concluded that Fibre reinforced concrete (FRC) having pet fibre aspect ratio 4 up to a level of 2% volume in M25 grade concrete may be utilized in infrastructural applications like concrete structures.

The objective of this paper is to determine the mechanical properties and the optimum percentages of recycled PET fibres in conventional concrete. In this study, straight and irregular recycled polyethylene terephthalate (PET) fibres were used. The fibres were simply cut from waste plastic bottles, water bottles, beverage bottles in dimension 50mm X 5mm. The percentage of recycled PET fibres added in the concrete mix were 1.0%, 2.0%, 3.0% and 4.0% respectively according to the volume of concrete. A water cement ratio of 0.45 was accepted for all ranges. The test conducted include the slump test, compressive strength test, splitting tensile strength test, ultrasonic pulse velocity(UPV) test ,rebound hammer test, bulk density and water absorption test. The specimens were tested after 28 days after the concrete was mixed.

Materials and Methods

The following materials were used while conducting the different experimentsWaste plastic bottle fibres or polyethylene terephthalate (PET) fibre, Ordinary Portland Cement (OPC), Fine aggregate (passing through IS sieve 4.75mm), Coarse aggregate (12.5mm and 20mm), Water (Tap Water).
PET is a hard, stiff, strong and dimensionally stable material that is usually used as packaging for carbonated beverages, water and many food products. Its crystalline structure varies from amorphous to fairly high crystalline. PET is found to be effective in replacing aggregates in concrete because of its versatile behaviour. It is lightweight, flexible, strong, moisture- resistant and cheap. PET plastic bottles were collected, cleaned and dried before being cut into fibre form to get rid of any impurities. They were cut into smaller pieces into the desired shape and size of 50mm in length and 5mm in width. Three cubes of dimension (150*150*150mm) and three cylinders of 150mm diameter and 300mm length for each mix were casted and tested for compression and split tensile test.The specimens were tested for their strength and other properties after curing for 28 days.

Results and Concluding Remarks

The specimens reinforced with PET fibres (PC1, PC2, PC3 and PC4) were tested after curing for 28 days. The compressive strength and split tensile strength of the conventional concrete (CC) was found to be 30MPa and 5MPa respectively. The compressive strength increases up to 33MPa at 2% replacement and it gradually decreases to 31Mpa at 4% replacement i.e. the compressive strength increases till 2% replacement and decreases with increase in the percentage of fibre addition.

The split tensile strength of the concrete reached its optimum value of 5.4MPa at 3% replacement but in case of 4% replacement the split tensile strength decreased up to 5.3MPa. Slump test was also performed to check the workability of the concrete.

The non – destructive test like Ultrasonic Pulse Velocity (UPV) test and Rebound Hammer test were also conducted on the specimens. The Ultrasonic Pulse Velocity (UPV) test gave no satisfactory result due to the presence of fibre in the concrete.

In case of water absorption test, the specimens PC1, PC2, PC3 and PC4 absorbed less amount of water than CC (Conventional Concrete) since the PET fibres are moisture resistant in nature. Bulk Density Test was also performed to check the ability of concrete to function for structural support, water and solute movement, and durability.

The concrete with PET fibres reduced the weight of concrete and thus the concrete can be used as light weight concrete. From the present experiment, it can be concluded that the replacement of fine aggregate up to 2% of PET fibre is preferable for obtaining high values of compressive strength. But, in case of split tensile strength test, the tensile strength increases with increase in the percentage of PET fibre added and reaching its peak value at 3% replacement.

References

1. P.G. Prabhu, C.A. Kumar, R.Pandiyarj, P.Rajesh & L.S. Kumar (2014) " Study on Utilization of Waste PET Bottle Fibre in Concrete". IMPACT: International Journal of Research in Engineering and Technology (IMPACT:IJRET) ISSN(E):2321-8843; ISSN(P):2347-4599 Vol.2, Issue 5,May 2014, 233-240.
2. D. Prabha V M, Dr. S. George (2017) " Shear Behavior of Reinforced Concrete Using PET Bottle Fibre". International Journal of New Technology and Research (IJNTR) ISSN:2454-4116, Volume-3, Issue 5, May 2017 Pages 54-56.
3. S.Shahidan, N.A.Ranle, S.S. M. Zuki, F. S. Khalid, A.R.M. Ridzuan, F.M. Nazri (2018) " Concrete Incorporated with Optimum Percentage of Recycled Polyethylene terephthalate (PET) Bottle Fibre." International Journal of Integrated Engineering Vol.10 No.1(2018)p.1-8. DOI: https://doi.org/10.30880/ijie.2018.10.01.001
4. Ms. K. Ramadevi, Ms.R.Manju (2012) "Experimental Investigation on the properties of Concrete With Plastic PET (Bottle) Fibres as Fine Aggregates". International Journal of Emerging Technology and Advanced EngineeringWebsite: www.ijetae.com (ISSN 2250-2459, Volume 2, Issue 6, June 2012)
5. C. Marthong , D. K. Sarma (2017) " Mechanical Behaviour of Recycled PET Fiber Reinforced Concrete Matrix " World Academy of Science, Engineeing and Technology International Journal of Civil and Environmental Engineering Vol:9, No:7,2015.
6. M. Mokhtar, M. Kaamin, S. Sahat and N. B. Hamid (2017) " The Utilization of Shredded PET as Aggregate Replacement for Interlocking Concrete Block" E3S Web of Conferences 34, 01006 (2018) CENVIRON 2017 https://doi.org/10.1051/e3sconf/20183401006
7. N. L. Holland, J. M. Nichols and A. B. Nichols(2012) " Fibre Reinforced Concrete Using Polyethylene Strips" BEFIB2012- Fibre reinforced concrete Joaquim Barros et al (Eds)

Flexural Strength of Steel-Concrete Composite Beams with Flexible Shear Connector: Numerical Studies

A. Sougata Chattopadhyay[1], B. Umamaheswari.N[2*]
[1] *Department of Civil Engineering, SRM Institute of Science and Technology, Kattankulathur-603203, Tamil Nadu, India*
[2] *Department of Civil Engineering, SRM Institute of Science and Technology, Kattankulathur-603203, Tamil Nadu, India*
* *e-mail: njuma_sus@yahoo.com*

Introduction

Steel-concrete composite beam is a typical construction which is made up of two dissimilar materials such as steel and concrete. It is used mainly for construction of commercial buildings and bridges. Due to the combination of steel and concrete, strength and ductility of composite section is highly enhanced. Besides that many other advantages are found out. Shear connector is a basic component through which steel and concrete are mechanically connected to transfer the load from concrete slab to steel beam. In the present numerical study, two common flexible shear connectors such as stud connector (Lam and Ehab 2005; Xue et al. 2008; Chung and Chan 2011; Balasubramanian and Baskar 2018; Ling et al. 2018) and channel connector (Maleki and Bagheri 2008; Eray and Cem 2012) are used to study the flexural behaviour of steel-concrete composite section. The normal strength as well as high strength steel sections are preferred (Zhao and Yuan 2010; Chung and Chan 2011; Huiyong and Bradford 2013). The normal strength concrete is used (Jingquan et al. 2019). Geometry as well as layout of shear connectors is considered as key factors involved in enhancement of strength and stability of the structure.

Performance of stud connector and channel connector under push out test has been analysed by the researchers for the last few decades but enough research data is not available in literature to understand the structural behaviour of steel-concrete composite beam under flexure. Thus the response of steel, concrete and shear connector which are used in steel-concrete composite beam subjected to bending is considered as important parameters for numerical approach used in this research work.

Materials and Methods

Steel-concrete composite structure is typically made up of steel and concrete which is mechanically connected by shear connector to transfer the load from concrete slab to steel beam. In this current study, two common type of shear connectors such as stud and channel are taken into consideration to analyse the strength and stability of beam. Normal as well as high strength steel sections are considered for steel beam section and normal strength concrete is taken into account for concrete section.

Finite element analysis is a cost-effective method to understand the performance of materials as well as structures. In the present research work, numerical model is made using finite element software package ANSYS, validated with the test results available in literature and, further numerical models are developed to demonstrate the flexural behaviour of steel-concrete composite beam connected with stud as well as channel connector subjected to four point loading condition. The variables considered for the current study are layout and geometry of shear connectors, steel and concrete strength. Meshing, boundary condition and interaction between the element are the important factors for the solution of finite element problem with accuracy. Bottom face of shank of the stud/ bottom face of flange of the channel connector is considered to be connected with the flange of steel beam section through welding. To analyse the flexural performance of steel-concrete composite beam, hinge at one support and roller at other support is chosen as boundary condition to ensure the simply supported beam criteria.

Results and Concluding Remarks

Effect of using normal as well as high strength steel, geometry and spacing of flexible connectors are taken as main parameters for this present numerical study to identify the flexural performance of steel-concrete composite beam section. The observations made are load-deflection behaviour and ultimate load capacity. Increasing diameter along with length of stud connector is found to enhance the load carrying capacity of the stud. In the case of channel connector, ultimate load capacity as well as stiffness is found to be enhancing with minimum deflection when the arrangement of channel section is considered with minimal spacing of connector. Also provision of higher strength steel section is capable of enhancing the load carrying capacity as well as ductility of the composite beams.

References

Eray B, Cem T (2012) An experimental study on channel type shear connectors. Journal of Constructional Steel Research 74: 108-117. http://doi.org/10.1016/j.jcsr.2012.02.015.

Chung KF, Chan CK (2011) Advanced Finite Element Modelling of Composite Beams with High Strength Materials and Deformable Shear Connectors. Procedia Engineering 14: 1114-1122. http://doi.org/10.1016/j.proeng.2011.07.140.

Jingquan W, Jianan Q, Teng T, Qizhi X, Hongliang X (2019) Static behavior of large stud connectors in steel-UHPC Composite structures. Engineering Structures 178: 534–542. http://doi.org/10.1016/j.engstruct.2018.07.058.

Xue W, Ding M, Wang H, Luo Z (2008) Static Behavior and Theoretical Model of Stud Shear Connectors. Journal of Bridge Engineering 13(6): 623-634. http://doi.org/10.1061/(ASCE)1084-0702(2008)13:6(623).

Ling Y, Zhaoyl Z, Yang TY, Hongwel M (2018) Behaviour and modeling of the bearing capacity of shear stud connectors. International Journal of Steel Structures. https://doi.org/10.1007/s13296-018-0154-3.

Maleki S, Bagheri S (2008) Behavior of channel shear connectors, Part II: Analytical study. Journal of Constructional Steel Research 64: 1341–1348. http:// doi.org/10.1016/j.jcsr.2008.01.006.

Huiyong B, Bradford MA (2013) Flexural behaviour of composite beams with high strength steel. Engineering Structures 56: 1130-1141. http://dx.doi.org/10.1016/j.engstruct.2013.06.040.

Lam D, Ehab EL (2005) Behavior of Headed Stud Shear Connectors in Composite Beam. Journal of Structural Engineering 131(1): 96-107. http://doi.org/10.1061/(ASCE)0733-9445(2005)131:1(96).

Zhao H, Yuan Y (2010) Experimental studies on Composite beams with high strength steel and concrete. Steel and Composite Structures 10(4): 297-307.

Balasubramanian BR, Baskar R (2018) Performance study of steel-concrete composite beam involving flexible shear connector. Int. Journal of Structural Engineering 9(1): 70-80.

Numerical Investigation on Pullout response of Steel Fibers in Concrete

A. Thiagarajan. K[1], B.Umamaheswari. N[2*]
[1]*Deparment of Civil Engineering/ SRM Institute of Science and Technology, Kattankulathur-603203,Tamilnadu, India*
[2]*Deparment of Civil Engineering/ SRM Institute of Science and Technology, Kattankulathur-603203,Tamilnadu, India*
e-mail: njuma_sus@yahoo.com

Introduction

Cementitious concrete is known for its brittle failure because of low tensile strength. Research since last four decades on steel fiber reinforced concrete (SFRC) reveals that it is one of the effective materials to overcome the brittleness of concrete. Mechanical properties of SFRC depend on the development of fiber/ matrix interfacial zone since it plays an important role in load transfer. Bond shear stress and the slip between the interface is effectively identified by the fiber pullout tests. Naaman et al. 1991 conducted an experimental investigation on fiber pullout behaviour to study the maximum pullout load and slip. This topic of research was further extended by Zile and Zile 2013 by varying the types of fibers and embedment length of fiber in the matrix. Experimental investigations on fiber pullout focussed on peak load and displacement of fibers from the matrix, Soulioti et al.2013. Ellis et al. 2014 carried out an analytical investigation to predict the energy dissipation, plastic strain and stress in the fiber and matrix and showed that fibers with uneven shapes provided better resistance than the flat straight fiber. A comparative analysis by Breitenbucher et al. 2014 on both experimental and analytical study revealed an effective validation between the results obtained. Inclination of fibers in the matrix is found to increase the ultimate load and decrease the fiber efficiency and pullout work. Pullout behaviour of twisted fiber studied by Ye and Liu 2019 showed better resistance due to the untwisting phenomenon.

The primary objective of this research is to study the bond between fiber and matrix by conducting numerical investigation. A single fiber placed at the centre of cubical matrix is studied using Finite Element Analysis (FEA) software. Analysis of fiber pullout is carried out in straight, crimped and hooked end fibers at different embedment lengths in the matrix. This numerical study provides information on effective stress and strain in the fiber along with the deformation from matrix. This improved analysis of pullout response will help in apprehending the bond mechanism between the fiber/ matrix which is very effective to improve the bond quality and choice of steel fiber.

Numerical Models and Methods

A three-dimensional cubical model is designed using FEA software with a dimension of 7.5cm x 2.5cm x2.5cm. Zile et al. 2013 has been referred for validation of numerical models. Aspect ratio of steel fiber considered is 80. Matrix and fiber is assembled together and fiber is placed exactly at the centre of the matrix. Fiber is placed at various embedment lengths in the matrix for a comparative analysis. The boundary conditions were applied based on the previous experimental investigation by Soulioti et al.2013, bottom and the top faces of the fibers are fixed and tensile force is applied at the free end of the fiber with the smooth step function. Mesh is generated using the software for matrix and fiber separately. The results obtained are pullout force versus slip behaviour, stress-strain behaviour of the elements and deformation of the fiber. The flowchart showing the steps involved in the numerical analysis is shown in Figure 1.

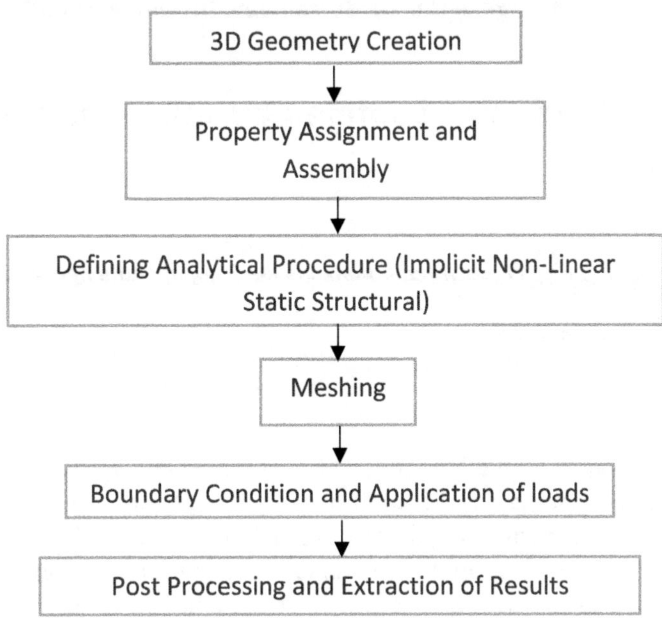

Figure 1. Procedure for numerical analysis

Results and Concluding Remarks

In this numerical investigation, Finite Element Analysis of pullout response of various types of fibers were performed with different embedment lengths. Pullout load versus slip behaviour was obtained and maximum load was obtained for the crimped fiber because of the uneven shape. The crimp in the fiber provide better bond to hold in the matrix rather than the straight and hooked end fiber. Fiber slips away from the matrix as the pullout load increases with significant change in shape of the fiber. Von Mises Stress response due to the pullout force increases as the fiber is pulled out. Stress values for crimped fibers are higher than that of straight and hooked end fiber. Damage patterns from the current analysis confirms that mechanical interlocking in hooked end and crimped fiber requires additional force to pullout from the matrix. This causes the additional damage to the matrix and fiber. The stress versus strain values also confirm that uneven shaped fibers have higher resistance to pullout force than the straight fibers. The bond between the fiber and matrix is effectively explained in this numerical study. Further research could be performed to obtain detailed observations and discussions about pullout response of steel fibers with changes in the matrix and fiber configuration.

References

Naaman A E, George G, Namur, Jamil M, Ahwan, Najm H M (1991)Fiber pullout and bond slip. II:Experimental validation. Journal of Structural Engineering 117(9): 2791-2800.

Zile E, Zile O (2013) Effect of the fiber geometry on the pullout response of mechanically deformed steel fibers. Cement and Concrete Research 44: 18-24.

Soulioti D V, Barkoula N M, Koutsianopoulos F, Charalambakis N, MatikasT E (2013) The effect of fibre chemical treatment on the steel fibre/cementitious matrix interface. Construction and Building Materials 40: 77-83.

Ellis B D, McDowell D L, Zhou M (2014) Simulation of single fiber pullout response with account of fiber morphology. Cement and Concrete Composites 48: 42-52.

Breitenbucher R, Meschke G, Song F, Zhan Y (2014) Experimental, analytical and numerical analysis of the pullout behaviour of steel fibres considering different fibre types, inclinations and concrete strengths. Structural Concrete 15: 126-135.

Ye J, Liu G (2019) Pullout Response of Ultra-High-Performance Concrete with Twisted Steel Fibers. Applied Sciences 9:658.

Behavior of concrete columns confined by using various FRP composite strips under axial loads

G.N. Narule[1*], V.P. Kare[2]
[1,2] *Department of Civil Engineering, Vidya Pratishthan's Kamalnayan Bajaj Institute of Engineering and Technoogy, Baramati,Pune,India.*
** e-mail: corresponding.gnnarule@gmail.com*

Introduction

In recent years, the concrete construction industry has faced a very significant challenge in view of the deterioration of infrastructure. A large number of bridges, buildings and other structural elements require rehabilitation and repair. There is currently a range of techniques available for extending the useful life of structurally deficient and functionally obsolete structures. Fiber reinforced polymer composites (FRPCs) as external reinforcement is one of the very effective techniques to resolve issues associated with deterioration and retrofitting of concrete structures in the construction industry. Various FRPCs have been used to retrofit concrete members like columns, slabs beams and girders in structures such as bridges, parking decks and buildings. Among these the application of FRPCs to strengthen the concrete columns has perhaps received the most attention from the research community.
Highly aggressive environmental condition effects on the durability and structural integrity of steel reinforced concrete piles, piers and column. Also the corrosion of steel rod damages the structure. Dealing with the problem, steel wrapping is used to repair; rehabilitation and strengthening of reinforced concrete column but it suffers from problem of steel rebar, corrosion, poor durability and its weight.

To overcome this problem, the combination of GFRP and CFRP strips can be used. It improves the ultimate stress, ultimate lateral strain, ultimate axial strain and ductility. It also produces good lateral confinement. This technique provides significant contribution and increase the compressive ultimate strength.

Materials and Methods

Eighteen RC circular column specimens were cast as per test matrix of concrete grade M20. Nine specimens were wrapped with one layer of unidirectional FRP strips. Three specimens wrapped by alternate strips of glass fiber and three column specimens confined by carbon fiber. The remaining three specimens strengthened by alternate strips of glass fiber and carbon fibers. Resin epoxy was used as matrix and had two components (base and hardener) mixed in the ratio of 100:35 as suggested by the manufacturer. All specimens were pre-treated with resin primer-11 in order to achieve a smooth surface and good bond between FRP and the specimen surface. Strain gauges were installed on concrete and FRP surface to record the strain during application of axial compressive loading. Deformator were used to trap the axial deformation of control as well as strengthened specimens while axial testing. Specimens were tested on a 200ton compression testing machine with loading rate of 1.5 kN/sec. The strain gauge readings were taken and established relation between load vs. deformation and stress vs. strain with the effect of variable confinements.

Figure 1. Unwrapped and FRP wrapped columns with CFRP, GFRP and combination of GFRP and CFRP strips

Table 1: Test Matrix for column specimens

Items	Circular column
Dimension(mm)	230mm diameter,450mm
Unwrapped columns	09
CFRPC strips strengthened columns	03
GFRPC strips strengthened columns	03
CFRPC and GFRPC strips strengthened columns	03

Results and Concluding Remarks

The following experimental results are obtained for unwrapped and FRP strip wrapped specimens as shown in table 2 and Fig.2.

Table 2. Comparison of experimental results for circular concrete column specimens.

column specimens/ Type of FRP strips used for strengthening	Average experimental values (MPa)		% increase in strength after wrapping with strips
	Ultimate strength of unwrapped specimen	Ultimate strength of wrapped specimen by strips	
Circular-230mm Dia.450mmHt.(GFRPC strips)	22.25	33.10	48.37
Circular-230mm Dia. 450mm Ht.(CFRPC strips)	21.10	39.01	84.89
Circular-230mm Dia. 450mm Ht. (CFRPC and GFRPC strips)	22.12	41.32	86.81

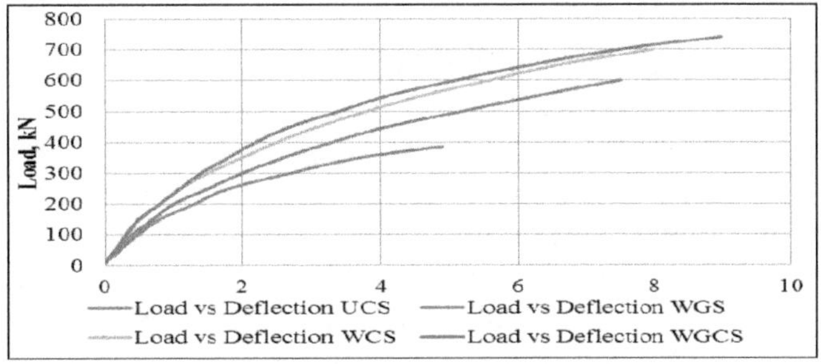

Figure 2. Combined load vs. deflection behavior for unwrapped and wrapped cylinder.

As per experimental and modelling study, the following conclusions are drawn:

1 The enhancement in axial compressive strength achieved by columns wrapping with combination of GFRP and CFRP strip was remarkably higher than columns with columns separately strengthened by GFRP and CFRP strips.

2 Axial stress vs. axial strain diagrams revealed the close relationship between compressive strength, effect of shape and ductility of the column.

References

1. Firmo J P, Correia J R, Pitta D, Tiago C, Arruda M R (2015) Experimental characterization of bond between externally bonded reinforcement CFPP strips and concrete at elevated temperature, Cement & Concrete Composites 60(2): 44–54
2 Zhang S S, Teng J G(2013) Bond slip model for CFRP strips near surface mounted to concrete, Engineering Structures 56(3): 945–953
3. Sun W, Ghannoum W M (2015) Modelling of anchored CFRP strips bonded to concrete, Construction and Building Materials 85(2):144–156.
4. Peng H, Hao H, Zhang J, Liu Y(2015) Experimental investigation of behaviour of interface between near surface CFRP strips and concrete, Construction and Building Materials 96(4):11–19
5. Esfahani M R, Kianoush M R (2005) Axial Compressive Strength of Reinforced Concrete Columns Wrapped with Fibre Reinforced Polymers (FRP), International Journal of Engineering Transactions B: Applications, 18(1): 9-19

Durability Studies on Pumice Lightweight Aggregate Self Compacting Concrete"

N. Ramanjaneyulu[1], Dr. M. V. Seshagiri Rao[2] and Dr. V.Bhaskar Desai[3]

[1]*Research Scholar, JNTUA College of Engineering,Anantapuram and Asst. Professor, Civil Engineering Department, CVR College of Engineering, Hyderabad, India*

[2]*Professor, Civil Engineering Department, CVR College of Engineering, Hyderabad, India*

[3]*Professor, Civil Engineering Department, JNTU Anantapur, Anantapuramu, India*

* *e-mail:* rams.613@gmail.com

Abstract

The present paper presents some of the durability properties of Normal self-compacting concrete and pumice light weight aggregate Self-compacting concrete of standard grade made with pumice aggregate partially replacing conventional granite aggregate of 10 mm nominal size of different grades of concrete i.e., M20, M30, M40 and M60. Viscosity Modifying Admixtures (VMA) is used to eliminate possible segregation. Fine powdered materials fly ash is used for eliminating the possible segregation. Superplasticizer is used to enhance the flow of mix. The mixes were designed using rational mix design procedure for SCC and satisfies the EFNARC (2005) guidelines. The mixes were designed using the rational mix design procedure for self-compacting concrete. The present paper consists of two phases. In the first phase, SCC mixes for different grades are developed without pumice and with pumice. The mechanical properties like compressive strength of the different grades were studied. In the second phase, Durability studies like Acid attack factors, Acid-Durability factors, Sulphate attack factors, were discussed. Detailed studies have revealed that the pumice Aggregate Self Compacting Concrete made with the pumice and it displays a better performance with reduced density.

1. Introduction

Concrete with low density is termed lightweight concrete. The unit weight of such concretes is about two-thirds of normal concrete. Most structural lightweight concretes weigh between 1600 and 1760 kg/m3. Design strengths of 20 – 35 MPa are common. The lightweight nature of these concretes is usually obtained either by using lightweight cellular aggregates. LWC can be produced by using lightweight materials like Lightweight Expanded Clay Aggregate, Pumice stone, expanded shale, Perlite etc. Structural lightweight aggregate can be produced naturally or from environmental waste. use of these aggregates can reduce the density of concrete, the self-weight of the structure and it helps to construct larger precast unit. Lightweight concrete (LWC) is an excellent solution in terms of decreasing the dead load of the structure, while self-compacting concrete (SCC) eases the pouring and removes construction difficulties.

KEYWORDS: Pumice lightweight aggregate, superplasticizers, VMA, SCC, flyash, etc.

2. Materials and Methods

Mix design: A rational mix design procedure for self-compacting concrete by S.V.Rao etal,

Cement

Cement is used Ordinary Portland Cement of Grade 53 is used conforming to various specifications of IS 12269-1987.

Fine aggregate

Locally available Fine aggregate used is confirming to the requirements of IS: 383-1970.

Coarse aggregate

Locally available crushed granite coarse aggregates used.

Pumice Lightweight aggregate

Pumice Lightweight aggregate (PLA) is used as the lightweight aggregate which is of size 8 to 12mm.

Chemical Admixture

CONPLAST SP 430 is used to improve the workability of the concrete.

Viscosity Modifying Agent: A Viscosity modified admixture Roof Plast2 is used for Rheodynamic Concrete.

Water

Locally available Clean potable water is used for concrete mixes to bind the cement content and aggregates.

3. Results and Concluding Remarks

Table 1. **Acid Durability Loss Factor of NWSCC and LWSCC for 56 Days**

Type of Concrete	Grade of Concrete	Percentage of pumice replacement	Chemical Used (5 %)	Acid Durability loss factor for 56 days
NWSCC	M20		NA2SO4	26.47
			HCL	1376.32
			H2SO4	24874.37
LWSCC		10%	NA2SO4	**6.53**
		20%	**HCL**	3311.66
		30%	H2SO4	41879.10
NWSCC	M30		NA2SO4	14.49
			HCL	929.98
			H2SO4	49134.16
LWSCC		10%	NA2SO4	**28.50**
		20%	**HCL**	1008.71
		30%	H2SO4	53855.61
NWSCC	M40		NA2SO4	40.68
			HCL	1346.8
			H2SO4	34504.76
LWSCC		10%	NA2SO4	**9.97**
		20%	**HCL**	999.35
		30%	H2SO4	37453.79

Fresh concrete test results Average values of slump flow (D_{ave}), T500 slump flow time, V-funnel time and 5 min delayed V-funnel time (V5min), L-box (h1/h2) and air content of the mixtures are given in Table 6. The fresh concrete data were classified in accordance with European guidelines for SCC (Table 7).[17]

The EFNARC guideline[2] recommends the minimum slump flow (D_{ave}) value as 650 mm. However, the European guidelines for SCC classify SCC having slump flow values between 550 and 650 mm

The following conclusions are drawn from the Experimental Investigation:

The density of the present light weight aggregate self-compacting concrete has found to be varying from 1800 to 1950 Kg/m3 depending upon the percentage replacement. Due to the less density PUMICE was found to be segregating and floating on the surface of self-compacting concrete, which needs to be corrected with appropriate dosage of super plasticizer and viscosity modifying admixture. With the increase in duration of exposure to the acidic environment the AAF found to be increase in both LWSCC and NWSCC in both acids and sulphate. However, the Dimension Loss is less in LWSCC when compared to NWSCC. With the increase in duration of exposure to the acidic environment the AWLF and ASLF are found to be increase in both LWSCC and NWSCC in both acids and sulphate. However, the loss of weight and loss of strength is more in LWSCC when compared to NWSCC. When compare to the plain SCC the LWSCC was found to be less durable in both Acids and Sulphate.

References

1. European Federation for Specialist Construction Chemicals and Concrete Systems. Specifications and Guidelines for Self-Compacting Concrete. EFNARC, Farnham, 2002.
2. K.Devi Pratyusha, S. Ram Lal. (2017) "Strength and Workability Investigations on M40 Grade Concrete with Partial replacement of Aggregate and With Pumice". International Journal of Advanced Technology and Innovative Research Volume. 09, IssueNo.12, November-2017, Pages: 1974-1979.
3. N.Sivalingarao and N.Manju. "A Brief study on Mechanical Properties of Silica Fume Light Weight Aggregate(pumice) Concrete." IOSR Journals of Mechanical and Civil Engineering (IOSR-JMCE). DOI: 10.9790/1684-16053016671.PP 66-71.

4. T. Divya Bhavana, Senior Assistant Professor "Study of Lightweight Concrete" International Journal of Civil Engineering and Technology (IJCIET), Volume 8, issue 4, April 2017, pp.1223-1230.

With reference to the suggestions given for 1,2,3,4 and 5 the explanation is as fallows

Suggession-1: Mix design methodology for SCC is not clear

Remark: A Rational mix design procedure for self-compacting concrete by S.V.Rao etal, as fallowed the same is incorporated in the paper.

Suggession-2: Mention the percentage replacement of Pumice

Remark: 10%, 20% and 30%, and same is incorporated in paper

Suggession-3: Check the durability factors. With the increase of dosage of pumice, loss factors are increased. It's controversial

Remark: The pumice aggregate is porous in nature, so with increase in the percentage of pumice aggregate the porous nature of the concrete is increasing. Hence the durability loss factors are found to be increasing with the increase in pumice aggregate.

Suggession-4: strength the conclusions

I. Remark: The density of the present light weight aggregate self-compacting concrete has found to be varying from 1800 to 1950 Kg/m3 depending upon the percentage replacement.
II. Due to the less density, PUMICE was found to be segregating and floating on the surface of self-compacting concrete, which needs to be corrected with appropriate dosage of super plasticizer and viscosity modifying admixture.
III. With the increase in duration of exposure to the acidic environment the AAF found to be increase in both LWSCC and NWSCC in both acids and sulphate. However, the Dimension Loss is less in LWSCC when compared to NWSCC.
IV. With the increase in duration of exposure to the acidic environment the AWLF and ASLF are found to be increase in both LWSCC and NWSCC in both acids and sulphate. However, the loss of weight and loss of strength is more in LWSCC when compared to NWSCC.
V. When compare to the plain SCC the LWSCC was found to be less durable in both Acids and Sulphate.

Strengthening properties of concrete using industrial waste like coal bottom ash and marble dust and its microstructural investigation

A. Sunil Kumar[1]*, B. Dr. Shobha Ram[2], C. Gaurav Chand[3]
[1] *Department of Civil Engineering, Gautam Buddha University, Greater Noida, India*
[2] *Assistant Professor*
**sunilchauhan0097@gmail.com*

Introduction

The rate of production of industrial waste is increasing at an alarming rate. These industrial wastes are primarily source of environment and water pollution. Industrial waste like coal bottom ash obtained from coal based thermal power plants and marble dust from marble industry has a serious issue of disposal. Hence, it is creating a problem of solid waste management. However, these uses are of low technical use hence productivity is less. Some researchers such as (Archana *et al.* 2014) have also suggested that bottom ash based artificial lightweight aggregate have higher potential for large scale utilisation in the construction work. This application also reduces the dependency on exploitation of natural fine aggregates. Moreover, the same method can be applied for marble dust also for replacing fine aggregates because of their identical properties. Some researchers such as (Naik *et al.* 2010) have suggested that cement and fine aggregate can be replaced by marble dust due to its chemical properties. These both methods solve the environmental problem and prove to be sustainable in nature also. In the present experimental investigation, authors have examined the comparative analysis of replacing fine aggregate with coal bottom ash and marble dust individually. Moreover, a hybrid concrete is also produced in which both industrial wastes are used altogether with silica fumes and its mechanical and durability properties are analysed. Furthermore, to check the filling of remaining pores in the concrete specimens is also examine using Scanning electron microscopy (SEM) study.

Materials and Methods

In the present experiment, authors have casted four different M30 grades of concrete i.e. CS, MDC, BAC and hybrid concrete (HC) are prepared and the material used includes OPC grade 43 cement, fine aggregates, natural coarse aggregates, coal bottom ash, marble dust, silica fumes, superplasticizer (sika 5202) and water. In the preparation of CBA specimens, firstly, 10%, 20%, 30% replacement of natural fine aggregates with coal bottom ash is adopted. Secondly, MDC specimens are prepared using 10%, 20%, 30% replacement of natural fine aggregates with marble dust. Lastly, a hybrid concrete HC specimen is prepared using 10%, 20%, 30% replacement of natural fine aggregates with marble dust and coal bottom ash together in equal proportion. Meanwhile, 7 % of cement is partially replaced by silica fume to further strengthen the mechanical and durability properties of concrete.

In the present investigation total 210 specimens are casted of cubes (150mmx150mmx150mm) and cylinders (300mmx150mm) to study the mechanical and durability properties of concrete. For mechanical properties, compressive strength test, split tensile strength test are carried out at 7, 14, and 28 days. For durability properties, acid attack test was carried by immersing cubes in 2% sulphuric acid (0.2N H_2SO_4) solution for 28 days and loss in compressive strength and weight is calculated. In figure 1 and 2, test of compressive strength and split tensile strength is shown.

Figure 1. compressive strength

Figure 2. Splitting tensile strength

Figure 3. SEM images of CS showing pores

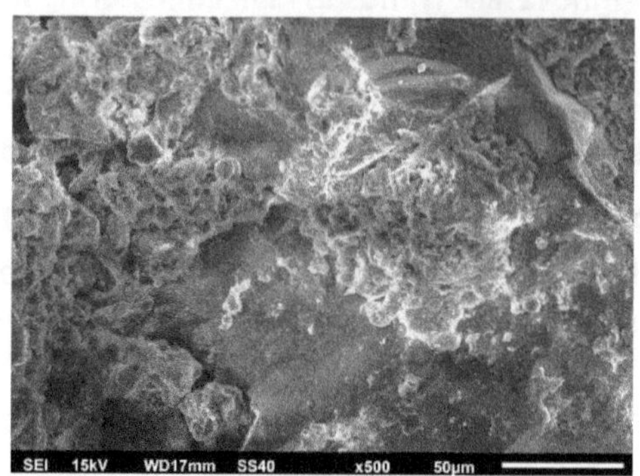

Figure 4. SEM images of HC showing pores

Results and Concluding Remarks

The concrete prepared using fractional replacement of coal bottom ash and marble dust separately found to be porous in nature in SEM images. However, on using them together, the marble dust helps in filling the pores. Meanwhile, using silica fume as a partial replacement with cement in the hybrid concrete filled the remaining pores of concrete.

Table 1. Results of mechanical properties at 28 days.

Grade	Compressive strength (MPa)	% of increment in strength	Splitting tensile strength (MPa)	% of increment in tensile strength
CS	37.6	0	3.1	0
MDC	41.2	9.6	3.4	9.6
BAC	38.6	2.6	3.3	6.4
HC	43.9	16.7	3.6	16.1

In the HC specimen's porosity is removed by 70-80%. This is shown in SEM images figure 3 and 4. The compressive strength of MD, and HC are obtained to be increased by 9.6%, and 16.7% of control specimen at 20% and 30% replacement of fine aggregate respectively. And compressive strength of coal bottom ash is obtained increase by 2.6% of control specimen at 20% replacement of fine aggregate. The established results from the test are shown in Table 1.

References

1. *Archana, B., & Porchejian, C. (2014). replacement of fine aggregate by bottom ash in concrete.* http://www.ijscer.com/uploadfile/2015/0421/20150421032942506.pdf
2. *Singh, M., & Siddique, R. (2013). Effect of coal bottom ash as partial replacement of sand on properties of concrete. Resources, conservation and recycling, 72, 20-32.* https://doi.org/10.1016/j.resconrec.2012.12.006
3. Singh, M., & Siddique, R. (2013). Effect of coal bottom ash as partial replacement of sand on properties of concrete. *Resources, conservation and recycling, 72*, 20-32. https://doi.org/10.1016/j.resconrec.2012.12.006
4. Singh, M., & Siddique, R. (2016). Effect of coal bottom ash as partial replacement of sand on workability and strength properties of concrete. *Journal of cleaner production, 112*, 620-630. https://doi.org/10.1016/j.jclepro.2015.08.001
5. Rafieizonooz, M., Mirza, J., Salim, M. R., Hussin, M. W., & Khankhaje, E. (2016). Investigation of coal bottom ash and fly ash in concrete as replacement for sand and cement. *Construction and Building Materials, 116*, 15-24. https://doi.org/10.1016/j.conbuildmat.2016.04.080
6. Siddique, R. (2014). Utilization of industrial by-products in concrete. *Procedia Engineering, 95*, 335-347. https://doi.org/10.1016/j.proeng.2014.12.192
7. Corinaldesi, V., Moriconi, G., & Naik, T. R. (2010). Characterization of marble powder for its use in mortar and concrete. *Construction and building materials, 24*(1), 113-117. https://doi.org/10.1016/j.conbuildmat.2009.08.013

Developing bio-concrete in laboratory and investigating its mechanical, structural and self-healing properties

S. Nirala[1], P. Kumar[1], J. Kumar[1], K. Paul[2*], D. Prasad[3]

[1] *Under-Graduate student, Department of Civil & Enviromental Engineering, BIT Mesra, Ranchi, India*

[2] *Asst. Professor, Department of Civil & Enviromental Engineering, BIT Mesra, Ranchi, India*

[3] *Asst. Professor, Department ofBio-Engineering, BIT Mesra, Ranchi, India*

* *e-mail: koushik_p77@yahoo.co.in*

Introduction

Concrete has a very high resistance towards compressive force but fails when the applied tensile stress exceeds the tensile strength of concrete. These tensile forces lead to crack formation and reduce the strength and durability of concrete. Self-healing concrete or bio-concrete has been developed in recent years to overcome the problem of maintenance of these cracks. It is a novel technique in which bacteria are introduced in the concrete either directly or indirectly which helps in the healing of cracks without human intervention.

Ghosh et al. (2005) had successfully increased the compressive strength of cement-sand mortar by 25% by addition of about 10^5cell of thermophilic anaerobic micro-organisms per ml of mixing water. The bacteria produced calcium carbonate which helped in crack filling. Jonkers et al. (2010) selected a specific group of alkali-resistant spore-forming bacteria related to the genus *Bacillus* to act as self-healing agent in concrete. Tittelboom et al. (2010) showed that ureolytic bacteria (such as *Bacillus sphaericus*) were able to produce calcium carbonate by converting urea into ammonium and carbonate which helps in repair of cracks in concrete.

The objective of this work is to develop bio-concrete in laboratory, compare the compressive and flexural strength of both types of concrete and investigate the self-healing property of the bio-concrete. Various samples of M25 grade concrete have been cast (keeping water/cement ratio fixed at 0.45) with varying the amount of bacterial solution and tested in the laboratory.

Materials and Methods

There are various carbonate producing bacteria such as *Bacillus subtilis, Bacillus pasteurii, Bacillus licheniformis, Bacillus pseudofirmus, Escherichia coli* which has been used by previous researchers in their work. In this current work, the authors have selected *Bacillus subtilis* which is a gram positive, rod-shaped bacterium naturally found in upper crust of soil, harmless and can tolerate extreme environment condition. *Bacillus subtilis* strains were obtained from Microbial Type Culture Collection & Gene Bank (MTCC), Chandigarh, India.

Procedure for cultivation of bacteria

Preparation of nutrient agar

Nutrient agar medium was used for supporting growth of bacteria. The pure culture of *Bacillus subtilis* is then preserved on nutrient agar slants. The bacteria formed irregular colonies in the agar slants.

Preparation for bacterial solution

For preparation of nutrient broth, Luria Bertani powder is mixed in distilled water at a rate of 25 gm/litre and then autoclaved at 121°C and 15 PSI to sterilize the mixture. Bacteria are inoculated from nutrient agar slants to this nutrient broth to make the bacterial solution. This nutrient broth is then incubated at 37°C in an orbital shaker incubator at 130 RPM for 24 hours. An opaque suspension is observed showing the growth of bacterial cells. Afterwards, for harvesting the bacterial cells, the mixture is centrifuged at 5000 RPM for 15 mins. The concentration of bacterial solution was adjusted to (Optical Density) OD 1 at 600 nm (approximately $8x10^8$ cells/ml) for further use. These bacteria are then washed twice in normal saline solution to make a protective layer of urea and calcium chloride for survival of the bacterial cells.

Cement and aggregates

Ordinary Portland Cement of 43 Grade was used in the present study; sand conformed to zone III of IS: 383-1970 (Reaffirmed 2002) and locally available coarse aggregate passing through 20 mm IS sieve and retaining on 12.5 mm IS sieve was used in casting of concrete. M25 grade concrete was cast in each case using concrete mix design (water: 195 kg; cement: 433.33 kg; sand: 595.7 kg and coarse aggregate: 1097.5 kg per cum of concrete), keeping a constant water/cement ratio of 0.45.

Results and Concluding Remarks

Results indicate that the use of *Bacillus subtilis* in concrete can improve the 28 days compressive strength of concrete by almost 30% and flexural strength by 10%. Compressive strength of concrete increases with increase the amount of bacteria in concrete until 90 ml of bacterial solution, while maximum flexural strength is achieved against incorporation of 30 ml bacterial solution while mixing concrete. Similarly bacteria are found to initiate self-healing of cracks in about 2 weeks.

Healing of cracks

The addition of bacteria into the concrete resulted in precipitation of bio-minerals, helped it to auto-heal and filling up of pores (Figure 1). The process of healing could be visible after 7-14 days and complete healing of hairline shrinkage cracks (limited to 1 mm) took more than one month. It is presumed that, when cracks are formed, water or moisture helps to activate the bacteria for producing the limestone which fills the gaps in the concrete.

Scanning Electron Microscopy (SEM)

The morphology and mineralogical composition of the deposited crystals were investigated using scanning electron microscope. Encapsulated bacterium (red arrow) are lodged inside the bio-concrete (Figure 2), which on favourable conditions germinate and starts growing inside the hairline cracks, simultaneously precipitating bio-minerals to fill the cracks. SEM micrographs were obtained using a JEOL JSM 5600 LV model Philips XL 30 attached with EDX unit, with accelerating voltage 20 KV, magnification 10x upto 2000x and resolution upto 1μm.

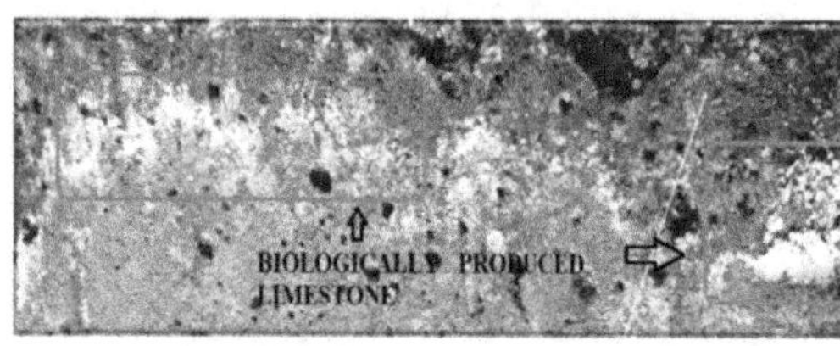

Figure 1. Healing of pores and cracks.

Figure 2. Bio-concrete structure under Scanning Electron Microscopy.

Acknowledgments: The author/s thankfully acknowledges the financial support provided by BIRLA INSTITUTE OF TECHNOLOGY for carrying out B.E. project work in this subject.

References

Ghosh P, Mandal S, Chattopadhyay BD, Pal S (2005) Use of microorganism to improve the strength of cement mortar. Cement and Concrete Research 35(10): 1980 –1983.

Jonkers HM, Thijssen A, Muyzer G, Copuroglu O, Schlangen E (2010) Application of bacteria as self-healing agent for the development of sustainable concrete. Ecological Engineering 36 (2): 230–235.

Van Tittelboom K, De Belie N, DeMuynck W, Verstraete W (2010) Use of bacteria to repair cracks in concrete. Cement and Concrete Research 40(1): 157-166.

REVIEW ON APPLICATIONS OF FIBER REINFORCED CONCRETE (FRC) IN STRUCTURES

Kapil Prabhat Verma[1*], Dr. Shobha Ram[1 2], Gaurav Chand[1]
[1]Department of Civil Engineering, Gautam Buddha university, Greater Noida, India.
[2]Assistant Professor
*kvvkapil@gmail.com

Introduction

The conventional plain concrete possesses a very low tensile strength, limited ductility and little resistance to cracking. Plain concrete having poor strength due to cracks and it is leading to brittle failure of concrete. Development of such micro cracks is the main cause of inelastic deformation in concrete. It has been observed that the addition of small, closely spaced and uniformly dispersed fibres to concrete would act as a crack arrester and would significantly improve its static and dynamic tensile strength, energy absorbing characteristic and better fatigue strength.

Literature review

In the past years number of researches have been conducted to verify the whether normal concrete can be substituted with the FRC.Aim of all these research is to make structures more efficient, economic and sustainable. In 2016, Xing-wen Liang conducted an experiment on beam column joint, in that FRC is poured into different potential plastic hinge zone depth of beam – column joint and specimen is subjected to seismic loading. In 2013, Eva Azhra latifa conducted an experiment on Hardened concrete to find various properties of normal concrete by adding different quantity of steel fiber. In 2015, E. Garcia-Taengua conducted a pull out test on different specimens with different percentage of fiber content that shows variations in the bond strength. In 2014, Nemkumar Banthia conducted a comprehensive test program carried out to fully understand the benefits of fiber reinforcement in cement-based repair materials. In 2014, Eshan Ahmed conducted an experiment to show that the steel fiber is used as a replacement of minimum shear reinforcement for one-way thick bridge slab.

Review Analysis

RESEARCH PAPER TITLE	
(1) Seismic performance of fiber-reinforced Concrete Interior beam column joints	In this research polyvinyl alcohol fiber (PVA) is used to improve the beam-column joint seismic performance. two types of Beam-column joint specimens are made Some of regular concrete and some of regular concrete Poured with FRC into joint core zone, potential plastic hinge zone of joint and it is subjected to seismic loading. In results, it is found that the joint improved its load Carrying capacity and energy dissipation capacity.
(2) Bond of reinforcing bars to steel fiber Reinforced Concrete	In this research steel fibers are used to improve the bond strength between reinforcement and concrete. Authors have prepared three different types of concrete with various fiber content. The prepared specimens are subjected to loading pull out test. It is established from the result that the bond strength is improved in shorter fiber as compared to longer fiber at same fiber content.

(3) Performance of steel fiber concrete as Rigid Pavement	In this research hook shaped steel fibers are used to improve performance of concrete used as rigid pavement. In this work, samples are made with various percentage of Fibers in regular concrete. Samples are tested in lab on fresh and hardened concrete behaviour. In results, it is found that the best performance was achieved at 9% of Steel fiber at 0.5 w/c ratio.
(4) Sustainable fiber reinforced concrete for Repair applications	In this research cellulose and polyvinyl alcohol(PVA) fibers used as repair materials in regular concrete. In this, two types of Specimen are made with different volume of fiber. Cellulose specimens are subjected to permeability test to detect resistance against cracking and PVA speci- -mens are subjected to corrosion test to detect resistance against climatic condition. In results, it is found that it is durable and strong.
(5) Steel fiber as replacement of minimum Reinforcement for one-way thick bridge Slab.	In this research steel fibers are used as a replacement of minimum shear reinforcement for one-way thick bridge slab. In this experimental program, it consist of three variables (i) presence of min. shear reinforcement (ii) Use of steel fiber reinforced concrete (iii) pre stress or no pre stress. These different types of specimens are subjected to loading and their behaviour is observed. In results, it is found that the load bearing capacity and ductility of FRC members is relatively higher than the regular concrete member which is reinforced with minimum shear reinforcement.

Conclusion

For the sustainable development, in the field of civil engineering structures, the continuous exploitation of natural resources used in the production of reinforcement can be significantly reduced, since the tensile strength property of reinforcement is substituted by FRC. From the established results it is obtained that concrete prepared using FRC is efficient, economic and sustainable as well. In this review analysis it is observed that the characteristics of normal concrete is improved because of adding fibers but standard dosage of fibers for particular concrete construction work is still in experimental zone.

References

1. Liang, X. W., Wang, Y. J., Tao, Y., & Deng, M. K. (2016). Seismic performance of fiber-reinforced concrete interior beam-column joints. *Engineering Structures*, *126*, 432-445. http://dx.doi.org/10.1016/j.engstruct.2016.08.001

2. Garcia-Taengua, E., Martí-Vargas, J. R., & Serna, P. (2016). Bond of reinforcing bars to steel fiber reinforced concrete. *Construction and Building Materials*, *105*, 275-284. http://dx.doi.org/10.1016/j.conbuildmat.2015.12.044

3. Latifa, E. A., Aguswari, R., & Wardoyo, P. H. (2013). Performance of Steel Fiber Concrete as Rigid Pavement. In *Advanced Materials Research* (Vol. 723, pp. 452-458). Trans Tech Publications https://doi.org/10.4028/www.scientific.net/AMR.723.452

4. Banthia, N., Zanotti, C., & Sappakittipakorn, M. (2014). Sustainable fiber reinforced concrete for repair applications. *Construction and Building Materials*, *67*, 405-412.http://dx.doi.org/10.1016/j.conbuildmat.2013.12.073

5. Ahmed, E., Legeron, F., & Ouahla, M. (2015). Steel fiber as replacement of minimum shear reinforcement for one-way thick bridge slab. *Construction and Building Materials*, *78*, 303-314.http://dx.doi.org/10.1016/j.conbuildmat.2014.12.095

Experimental Investigation on Manufactured Sand CFST Column

Dhiraj D. Ahiwale[1*], Rushikesh R. Khartode[1], Giridhar N. Narule[1], Kamalkishor M. Sharma[2]
[1]*Assistant Professor, Department of Civil Engineering, VPKBIET, Baramati, Pune.*
[2]*P.G.Student, Department of Civil Engineering, VPKBIET, Baramati, Pune.*
E-mail: dhirajahiwale1990@gmail.com[1], sharmakamal8586@gmail.com[2]

Introduction:

Steel members have the advantages of high ductility and tensile strength, while concrete members have the advantages of high compressive strength and fire resistance. Composite members made from steel and concrete, have the beneficial qualities of both materials. Concrete filled steel tube column is a recent concept. Concrete filled steel tube columns provide higher stiffness due to confined effect as compare to hollow steel tube column. The concrete filled aluminium stub column test strength was compared with the design strengths predicted using the American specifications and Australian/New Zealand standards for aluminium and concrete structures (Feng Zhou et al. 2009). Very few research works are conducted on concrete filled steel tube short columns. It is observed that this type column has more load carrying capacity.

In this paper, experimentation is carried out on three hollow specimens and twelve specimens of concrete filled steel tube with variable ratio of diameter to wall thickness. All the composite column specimens tests are carried out under Universal Testing Machine. The axial behaviour of hollow tube and concrete filled steel tube columns have been observed in terms of load carrying capacity, strains and buckling effect and also describes the how much contribution of concrete filled in steel tube as compare to hollow section. As an experimental work, it is observed that the ultimate load carrying capacity of concrete filled steel tube columns in axial compression are more than ultimate load calculated by Euro code 4 as well as AISC 360-10.

Parametric study

In this experimentation two type of core concrete, including natural river sand concrete and manufactured sand concrete were used together with steel tubes in the specimens for the sake of comparison. Secondly, the structural performances of the tested Concrete Filled Steel Tubular specimens using these two types of core concrete were compared and analyzed based on the testing data and observation. The structural behaviors of stub columns are focused in the current study. The influences of important parameters are studied, including cross-sectional shape, steel tubular thickness, and material properties of steel and concrete. The comparisons of variation of load verses lateral deflection are as shown in Figure 1 respectively.

Table 1. Information of the concrete filled steel tube columns.

Sr. No.	Section profile	Specimen labelling	Diameter (mm)	Thickness (mm)	Fcu (Mpa)	Type of fine aggregate
1		C-3-RS-1	88.7	3.2	76.9	River sand
2		C-4-RS-1	88.7	4.0		
3	Circular	C-5-RS-1	88.7	4.8		
4		C-3-MS-1	88.7	3.2	78.4	Manufactured sand
5		C-4-MS-1	88.7	4.0		
6		C-5-MS-1	88.7	4.8		

*Corresponding author E-Mail: dhirajahiwale1990@gmail.com

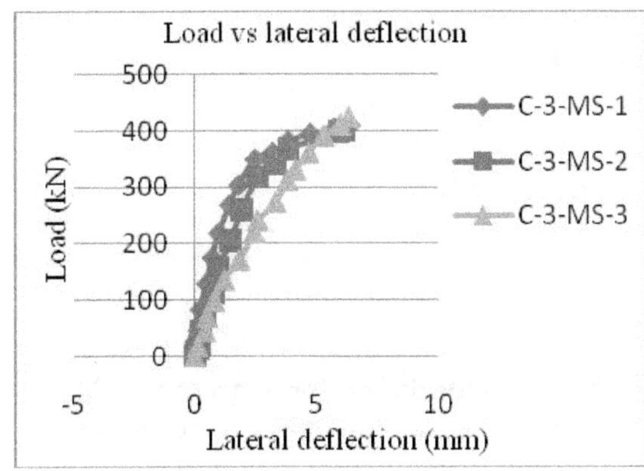

Fig.1 Lateral deflection versues Load curve for manufactured sand CFST column

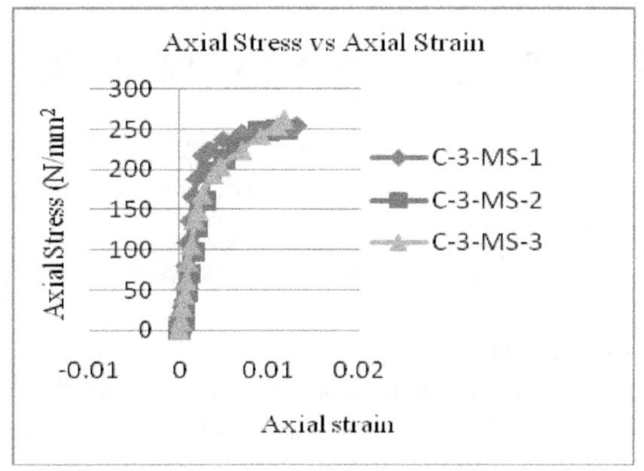

Fig.2 Axial Strain versues Axial Stress curve for manufactured sand CFST column

Concluding Remarks

In the present study, three hollow specimens and twelve specimens of concrete filled steel tube with variable ratio of diameter to wall thickness were experimentally investigated. The following conclusion is drawn based on the test results and analysis.

The experimental load carrying capacity of hollow tube column specimens is more than 30% and 35% as compared to Euro Code-4 and AISC respectively. Similarly, the experimental load carrying capacity of concrete filled steel tube column specimens is more than 15% and 10% as compared to Euro Code-4 and AISC respectively. Generally, manufactured sand concrete filled steel tube column specimens showed excellent ductility during the axial compression tests. The ultimate strength of manufactured sand concrete filled steel tube specimens is always more than river sand concrete filled steel tube column specimens. Also, the failure modes of such type of composite columns specimens were outward buckling at mid height.

References

1. Fei-Yu Liao, Chao Hou, et.al (2019) "Experimental investigation on sea sand concrete-filled stainless steel tubular stub columns", Journal of construction steel research, Vol. 155, Pp: 46-61.

2. Feng Zhou, Ben Young, (2009) "Concrete-filled aluminum circular hollow section column tests", Thin walled structures, Vol. 47, Pp: 1272-1280.

3. Feng Zhou, Ben Young, *(2012)* "Numerical analysis and design of concrete-filled aluminium circular hollow section columns", *Thin-Walled Structures, Vol: 50, Pp: 45–55.*

4. Dalin Liu, *(2004),* "Behaviors of high strength rectangular concrete-filled steel hollow section columns under eccentric loading", *Thin-Walled Structures, Vol: 42, Pp: 1631–1644.*

Durability studies on high strength concrete

K. Ramesh[1], J. Karthikeyan[2], S. Srikanth[1*]
[1] *Assistant professor, Civil engineering department, Malla reddy engineering college (A)*
[2] *Associate professor, NIT Trichy*
* *e-mail: ssseelamsrikanth@gmail.com*

Introduction:

High Performance Concrete (HPC) is that concrete which meets special performance and uniformity requirements that cannot always be achieved by conventional materials, normal mixing, placing and curing practices. Special performance requirements using conventional materials can be achieved only by adopting low water binder ratio, which necessitate the use of high cement content. But the addition of chemical and mineral admixtures can reduce the cement content and these results in the economical HPC.

The use of mineral admixtures in concrete production improves the compressive strength, pore structure, and permeability of the concrete; this is attributed to the pozzolanic reaction. This approach will have the potential to reduce costs, conserve energy, and waste minimization. Use of various mineral admixtures in producing mortar decreases the porosity and capillary, thus improves the durability against water and aggressive solutions. From the literature review, the properties of concrete depend on ingredients used. The optimum % replacement may vary based on the properties of GGBS and ingredients used. (Gaurav et al. 2015, Nagender and Meena 2017, Wei-Hao Lee et al. 2019).

The objectives of this research are to study the durability of HPC for different replacement levels of mineral admixtures by alternate wetting and drying phenomenon which includes; acid attack, sulphate resistance and marine Environment and also to determine the Compressive strength of HPC concrete as a partial replacement of cement with mineral admixture (GGBFS) for 28 days.

Materials and Methods:

The following materials were used for the study are: cement (OPC 53), fine aggregate, coarse aggregate, mineral admixtures (GGBFS), chemical admixtures (GLENIUM B233) and acids (HCL, H_2SO_4, Ammonium Sulphate, Marine solution)

The following methodology is used for this study:

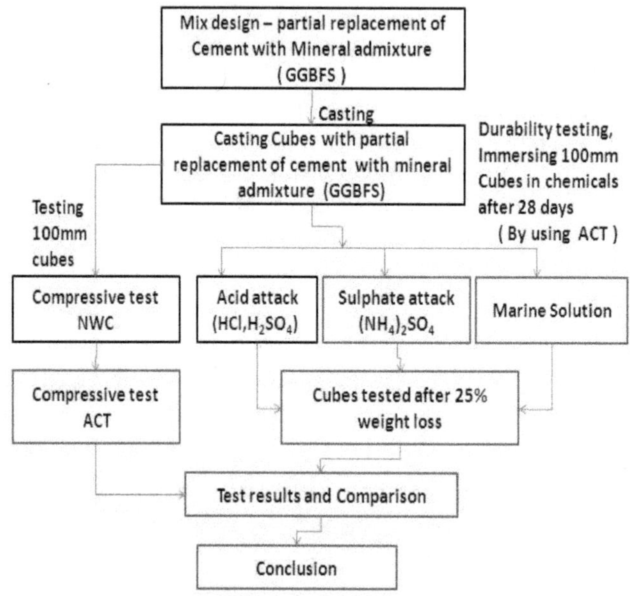

Results and Concluding Remarks:

Compressive strength

The 100 mm cube specimens were casted to use for compression test of concrete. Four specimens were casted for each test. For the 28th day compressive strength was found by testing. The average of best 3 results for each test was taken. The compressive strength of HPC concrete compared with normal water curing (NWC) and accelerated curing test (ACT) was given in Figure 1.

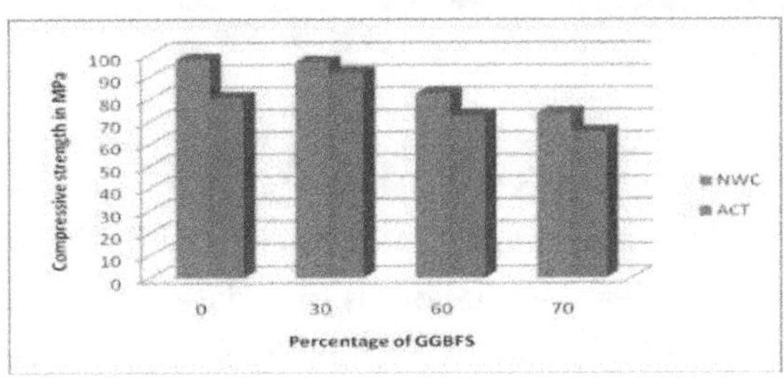

Figure. 1 Comparison of compressive strength of HPC concrete with NWC & ACT

Compressive Strength:

- Maximum Compressive Strength of 96.4MPa at 28days is found for the HPC specimens at a partial replacement of cement with 30% of GGBFS with Normal Water Curing.
- Maximum Compressive Strength of 91.6MPa at 28days is found for the HPC specimens at a partial replacement of cement with 30% of GGBFS with Accelerated Curing Tank.

Durability Studies:

- The specimens (100mm size) kept in a different prepared solutions for alternate wetting and drying (one month, 15days each wetting and drying), for a period of 150days (i.e., 1-30, 30-60, 60-90, 90-120, 120-150days).

Acid Attack using 1% HCL

- In hydrochloric acid, 30% of GGBFS replaced with cement concrete shows more weight loss when compared to 0%, 60%, and 70% of GGBFS replaced with cement.

Acid Attack using 1% H_2SO_4

- In sulphuric acid, 60% of GGBFS replaced with cement concrete shows more weight loss when compared to 0%, 30%, and 70% of GGBFS replaced with cement.

Sulphate Attack using $(NH_4)_2SO_4$

- The Weight loss of HPC concrete was found to be more in ammonium sulphate solution $(NH_4)_2SO_4$ when compared to acid solutions (HCl, H_2SO_4). In sulphate solution, 60% of GGBFS replaced with cement concrete shows more weight loss when compared to 0%, 30%, and 70% of replacement.

Marine solution

- The specimens immersed in marine solution were found to have less weight loss when compared to acids and sulphate solution.
- In marine water, 60% of GGBFS replaced with cement concrete shows more weight loss when compared to 0%, 30%, and 70% of GGBFS concretes.

References

Singh G, Das S, Ahmed AA, Saha S, Karmakar S (2015) Study of granulated blast furnace slag as fine aggregates in concrete for sustainable infrastructure. Procedia-Social and Behavioral Sciences 195:2272-2279. https://doi.org/10.1016/j.sbspro.2015.06.316

Narender reddy A, Meena T (2017) Study on Compressive Strength of Concrete Incorporating Alccofine and Ground Granulated Blast Furnace Slag (GGBS). International conference on materials, manufacturing and modeling, Vellore.

Lee WH, Wang, JH, Ding YC, Cheng TW (2019) A study on the characteristics and microstructures of GGBS/FA based geopolymer paste and concrete. Construction and Building Materials 211:807-813. https://doi.org/10.1016/j.conbuildmat.2019.03.291

CROSS SECTION OPTIMIZATION OF A TRUSS SUBJECTED TO VARIOUS GROUND MOTIONS USING SALP SWARM ALGORITHM

Sistla Saiteja[1*], Kalyana Rama J. S [2]
[1] *Graduate student, Department of Civil Engineering, BITS Pilani Hyderabad Campus*
[2] *Assistant professor, Department of Civil Engineering, BITS Pilani Hyderabad Campus*
* *e-mail: h20181430050@hyderabad.bits-pilani.ac.in*

Introduction

In the modern era accuracy is one of the most important factors in the field of design. As we witnessed a lot of earthquakes in the recent times, the need of the hour is to update the spectral quantities which are vital for estimating the dynamic response of any given structure. This current study involves 7 places namely **Bhuj (2001), Uttarkashi (1991), Chamoli (1999), Chamba (1995), India-Burma border (1997), Nepal (2015), Ummulong(1986)** respectively. Previously many researchers have generated their own site-specific spectrum for nuclear power plant design, capital city of Agartala etc. [5] [6] [7] The ground acceleration data of these 7 locations was obtained from ww.strongmotion.org. These locations were specifically chosen since they all fall under the category of zone-5 according to the seismic micro zonation as per IS:1893 part 1 2016 [2].With the help of Newmark's β method the respective spectral quantities were estimated and effectively utilized to estimate the Lateral loading on the truss, which is further used as an input for the SALP SWARM ALGORITHM . The site-specific curve is then compared with design spectrum of IS:1893 2016[2] in terms of the weight of the truss. The pseudo code and the understanding of the SALP SWARM ALGORITHM can be obtained from [9].

Materials and Methods

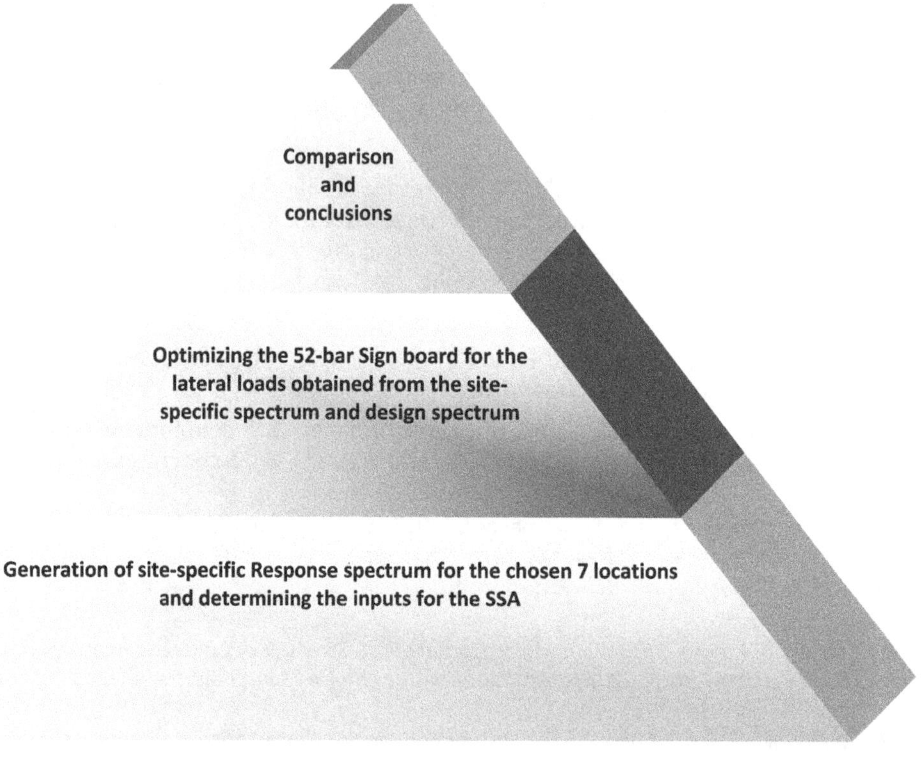

Comparison and conclusions

Optimizing the 52-bar Sign board for the lateral loads obtained from the site-specific spectrum and design spectrum

Generation of site-specific Response spectrum for the chosen 7 locations and determining the inputs for the SSA

Results and Concluding Remarks

With the main agenda to observe the trend of the latest earthquakes and their implication on the existing buildings and its comparison with existing design spectrum IS:1893 part 1 the following conclusions can be drawn

1) Robust classification of soil is mandatory. There is no variation in the peak response for different types of soils and hence the criteria for the existing design spectrum of IS:1893 2016 should be altered to make the responses more precise and accurate like that of EURO code 8.
2) The building response has drastic variations for different ground acceleration time histories. Hence the base shear and the total truss weight have significant variation for all the site-specific spectrums.
3) It is seen that some earthquakes fall below the design base shear and some above. This clearly shows the uncertainty involved with the current code and this would have a drastic effect on life and property.
4) The zone factor i.e. peak ground acceleration has some specific values, but nothing has been mentioned about the P.G.A which falls in between.
5) It is better to use Site-specific response spectrum for a site since it yields better and accurate values for design and in case of the absence of the Site-specific spectrum, the design spectrum maybe used.

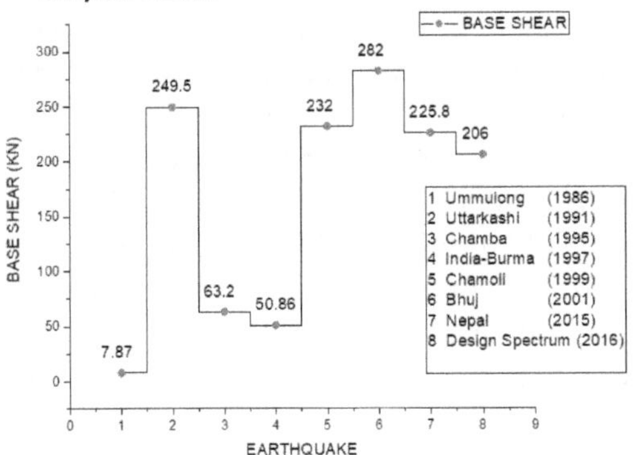

Fig 1: Variation in Base shear

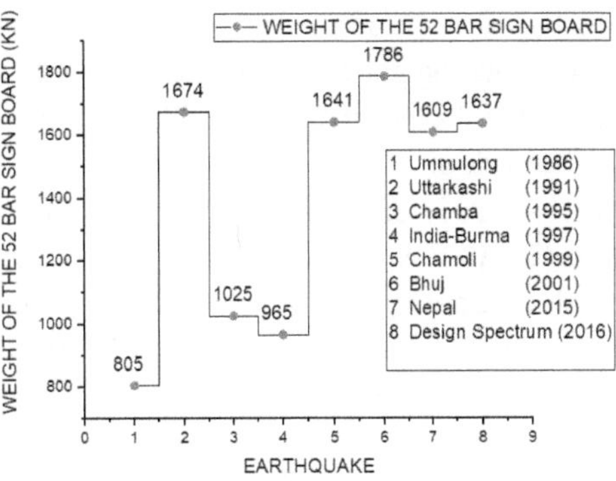

Fig 2: Variation in Base shear

References

1. Katona, M. C., & Zienkiewicz, O. C. (1985). A unified set of single step algorithms part 3: The beta-m method, a generalization of the Newmark scheme. International Journal for Numerical Methods in Engineering, 21(7), 1345-1359.
2. IS-1893 Part-1 2016 Criteria for earthquake resistant design of structures.
3. Chopra AK. Dynamics of structures: theory and applications to earthquake engineering. Prentice-Hall; 2001.
4. Jain AK. Dynamics of structures with MATLAB® applications. Pearson Education India; 2017.
5. Boominathan A, Vikshalakshie MG, Subramanian RM. Site specific seismic study for a power plant site at Samalkot, Godavari rift basin in Peninsular India. 2012
6. Poggi V, Edwards B, Fäh D. A comparative analysis of site-specific response spectral amplification models. Physics and Chemistry of the Earth, Parts A/B/C. 2017 Apr 1;98:16-26.
7. Sil A, Sitharam TG. Site specific design response spectrum proposed for the capital city of Agartala, Tripura. Geomatics, Natural Hazards and Risk. 2016 Sep 2;7(5):1610-30.
8. ABHIJIT MOLAWADE RC, Ramancharla VK. Generation of Site-Specific Response Spectrum near Nuclear Power Plants: A Preliminary Case Study at Rawatbhata Station. 2015
9. Mirjalili, S., Gandomi, A. H., Mirjalili, S. Z., Saremi, S., Faris, H., & Mirjalili, S. M. (2017). Salp Swarm Algorithm: A bio-inspired optimizer for engineering design problems. Advances in Engineering Software, 114, 163-191.Rajeev, S., & Krishnamoorthy, C. S. (1992). Discrete optimization of structures using genetic algorithms. Journal of structural engineering, 118(5), 1233-1250.

A Study on Punching Shear Strength of Fiber Reinforced High Volume Fly Ash (HVFA) Concrete Slabs

P.V.Hari Krishna[1],Dr. Ch.N.Sathish Kumar[2*],
[1]*Asst. Professor,Department of CE,QIS Institute of Technology, Ongole,A. P,India. e-mail: krishna1450@gmail.com.*
[2] *Professor,Department of CE,Bapatla Engineering College, Bapatla,A.P,India. e-mail : nagasathish123@gmail.com*

Introduction

Structural systems with reinforced concrete flat slabs are used commonly in practice. In such Systems, the slabs are supported directly by columns without beams that help to reduce considerably building height and increase used space. However, there is an important problem existing in this system that is punching shear failure of the slabs due to high concentration of stress in vicinity of slab- column connections. This failure type is very dangerous because of its brittle nature. Once, the punching shear failure occurs, resistance of the structure is significantly reduced, which causes separation of the column and slab, and then lead to collapse of the whole structure. To increase punching shear capacity of flat slab, a variety of methods have been proposed such as: *i*) Traditional shear reinforcing method using stirrups but this method is inapplicable to slabs with shallow depth less than 150 mm (ACI 318-2002); *ii*) New method using headed-studs but this one need much time for construction. Recently, new technique using fibers to improve the punching shear resistance and cracking control of slab-column connections has been proven to give good results. Some experimental research studies have been done in the last 30 years regarding the sequential behaviour of slab-column connections using fiber reinforced concrete (Harajli MH, Maalouf D, and Khatib H (1995, Swamy RN and Ali SAR (1982). De Hanai and Holanda (2008), Cheng and Parra-Montesinos (2010).Results have clearly shown that the addition of steel fibres has a significant effect in increasing slab capacity and its ductility, when subjected to monotonically increased concentrated load. The utilization of high volume fly ash (HVFA) concrete chooses in very well with feasible advancement. HVFA concrete will have lesser amount of cement and more volume of fly ash (up to 60%) Indrajit Patel, Modhera, C.D(2010).

The main scope of this project is to investigate the ultimate Punching shear strength and normal shear stress of slab – column specimens prepared by fiber reinforced *High volume fly ash*(HVFA) concrete of M30 grade in which the cement is replaced by 60% of fly ash. The obtained values of ultimate punching shear strength and normal shear stress are further compared with the codal provisions.

Materials and Methods

The materials used in this project are cement (OPC 53 Grade),locally available fine aggregate, Crushed coarse aggregates of size < 10mm and < 20mm are used in this mix with 60%-40% ratio. Fly Ash used in this project was taken from the VTPS (Vijayawada Thermal Power Plant), Vijayawada. The crimped steel fibers having aspect ratio of 70 are used. A 24mm length of fibrillated polypropylene fibers is used. Roof plast SP 45, a high range water reducing agent was added to the concrete mix in order to impart the high strength and desired workability for high volume fly ash concrete. Mix design for M30 grade of concrete is done as per IS 10262-2009.

The experimental program consisted of casting and testing of the circular slab- Column specimens of two different sizes namely large (L) and medium (M) of normal strength concrete with 60% of fly ash as replacement of cement. Cubes and cylinders were also casted and tested for compressive and tensile strength of HVFA concrete. In this procedure, for each series there are two different sizes of the slab column specimen's with 0%, 0.5% and 1% addition of steel and polypropylene fibers were casted and cured for 28 days and latter tested on Universal Testing machine(UTM) having capacity of 1000KN under punching shear force. The diagrammatic representation of test setup is shown in (Figure 1).The punching shear strength values obtained from the experimental results are compared with the values that are calculated by using the formulas available for the evaluation of punching shear strength in the ACI 318-08 and IS 456-2000 code provisions. The nominal shear stress values are calculated as per the ACI code and IS code provisions and also for the experimental values obtained. The load-deflection graphs of different sizes of slab-column specimens and the graphs of compressive and Split tensile strength are also drawn.

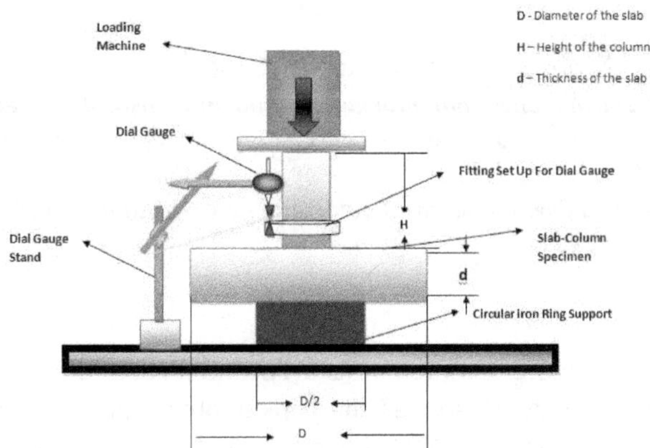

Figure 1 : Diagramatic representation of Test setup

Results and Concluding Remarks

Based on the investigations and from test results that are obtained, the following concluding remarks are stated:

1. By using the Steel Fibers of 0.5% and 1% volume fractions, the punching shear strength of the specimen was significantly increased from 23.6% to 56.3% for Large size Specimens and 58% to 85% for Medium size Specimens than the Plain Slab – Column specimens.
2. When the Polypropylene Fibers are used with volume fractions of 0.5% and 1%, a notable increase in punching shear strength was observed when compared with Plain specimens. The increase is from 7.17% to 15.5% for large size specimens and 10.66% to 34.98% for Medium size specimens.
3. When the experimental punching shear strength values are compared with the values calculated from the codal provisions, the obtained experimental results are very much higher than the predicted values.
4. From the comparison of Nominal shear stress values between theoretical and the experimental results it was observed that the nominal shear stress decreases with the increase in slab thickness as well as with the increase in % addition of fibers.
5. The increasing in slab thickness cause increasing in punching shear strength for slab but decreases the nominal shear stress.
6. The compressive and split tensile strength of plain HVFA concrete specimens shows a less strength values than those casted with Steel and polypropylene fibers.
7. In the plain HVFAC sample it is observed that the samples gain the targeted strength at 28 days. And there is a notable increase in strength than the Plain concrete. The HVFA based concrete performed well at 28 days than at the early stage of days.

References

1. Harajli MH, Maalouf D, and Khatib H (1995). Effect of fibers on the punching shear strength of slab–column connections. Cement & Concrete Composites. 17, pp.161–170.
2. Swamy RN and Ali SAR (1982). Punching shear behavior of reinforced slab–column connections made with steel fiber concrete. ACI Structural Journal. 79(6), pp. 392–406.
3. Indrajit Patel, Modhera, C.D., 'Study basic properties of fiber reinforced high volume fly ash concrete', Journal of Engineering and Science/Vol. I/Issue I/July-Sept. 2010/60-70.
4. ACI Committee 318. (2001). ACI 318-08: "Building Code Requirements For Structural Concrete". American Concrete Institute.
5. IS 456-2000, Indian standard " Plain And Reinforced Concrete - Code Of Practice" (*Fourth Revision*).
6. IS 10262-2009 Indian standard "Concrete Mix Proportioning- Guide Lines", Bureau of Indian Standard July 2009.

Pozzolanic Efficiency of Local Crushed Bricks for use as Supplementary Cementitious Material

Mrigakshee Sarmah[1], Mani Mohan[2], Birendra Kumar Singh[3*]
[1] *M.E. Student, Department of Civil and Environmental Engineering, Birla Institute of Technology, Mesra, Ranchi, India*
[2] *Research Scholar, Assistant Professor, Department of Civil and Environmental Engineering, Birla Institute of Technology, Mesra, Ranchi, India*
[3] *Professor, Department of Civil and Environmental Engineering, Birla Institute of Technology, Mesra, Ranchi, India*
** e-mail: birendrasingh@bitmesra.ac.in*

Introduction

Supplementary cementitious materials (SCMs) as a potential replacement for cement has gained wide acceptance in the past few years as it reduces the clinker production and makes use of industrial by – products. However, the usage of common SCMs becomes limited as they are not locally available in many places and hence not accessible to all. As inefficient production techniques of Indian brick industry lead to an annual massive wastage of bricks, these waste bricks from kilns in crushed form (Crushed Brick Powder or CBP) can be studied for use as potential SCMs. Use of CBP to reduce carbon dioxide emissions (Filho et. al., 2007), effect of ground clay brick powder on the fresh and hardened properties of both plain concrete (Ge et. al., 2015) and self compacting concrete (Heikal et. al., 2013) has been widely studied by researchers. Microstructural behaviour of cement paste replaced with CBP has been studied by Bediako et. al., 2018; Ortega et. al., 2018. Use of red clay waste brick for manufacture of masonry units (Robayo – Salazar et. al., 2017) and for use as a decorative plaster (Li et. al., 2016) has also been reported.

A common problem encountered in developing countries is the lack of availability of brick powder conforming to prescribed standards of pozzolanicity. Outdated manufacturing technologies leading to production of underburnt or overburnt bricks inhibit the pozzolanicity of CBP. This brings about a scarcity of standard supplementary materials to be used in manufacture of cement clinker in these countries. Limited research has been done on the utilization potential of such non – conforming waste materials in cement production. The present work analyzes the capability of one such locally obtained CBP to be utilized as a cement supplement despite not satisfying standard chemical composition of calcined clay given in IS: 1344 - 1981.

Materials and Methods

Ordinary Portland cement of grade 43 conforming to standard code IS: 8112 – 2013 was used in this experiment. Discarded clay bricks from a local brick kiln in Mesra, Ranchi were used. They were crushed in a concrete crusher and further milled in a pulveriser to get a fine powder. After that, this powder was passed through two sieves: 150 µm and 75 µm. Hence two different size fractions, i.e. passing through 150 µm but retained on 75 µm; and passing through 75µm were obtained. EDS data of CBP showed that net percentage of oxides of Si, Al and Fe is 57.33%, which is less than the minimum percentage required by IS 1344: 1981. Mortar specimens for testing of compressive strength were prepared and tested according to standard IS codes after 3, 7, 28 and 90 days respectively. Paste samples for TGA and XRD tests were prepared and tested for similar hydration periods according to standard procedures described in literature (Schöler et. al., 2015; Scrinever et. al., 2018).

Results and Concluding Remarks

Compressive strength of blended samples did not exceed that of control at any stage of hydration upto 90 days. But, the 90 days compressive strength of blended mortars was found to be significantly higher than their corresponding 28 days strength, whereas the compressive strength of control mortar did not increase adequately from 28 to 90 days. Hence, blended mortars showed weak pozzolanicity, observable only at advanced curing ages. This is due to presence of crystalline phases (detected from XRD) inhibiting pozzolanicity like cristobalite. However, presence of additional phases like AFm and early appearance of common phases like monosulfoaluminate indicate that the volume of hydrates produced in blended samples is higher as compared to that of OPC sample, which maybe responsible for higher compressive strength of blended mortars at 28-90 days. Intensity of portlandite peaks observed from XRD graphs and quantification of portlandite from TGA data both indicate that OPC has lesser CH content than CBP blended samples at 90 days, confirming the weak pozzolanic action of CBP. 10 – 15% replacement of OPC by CBP was found to be optimum for mortar samples.

As it has been proved extensively in literature that CBP has pozzolanic characteristics which can be easily detected after 28 to 90 days of hydration, it can only be concluded that the chemical characteristics of CBP varies from place to place and depends upon the geological characteristics of the raw clay as well as the manufacturing process adopted in producing the brick. Presence of kaolinite, followed by montmorillonite, which show the highest pozzolanic activity (Ambroise et. al., 1985; Tironi et. al., 2013) were not observed from the XRD scans and only illite was detected. Due to these reasons, the CBP obtained from a local brick kiln in Mesra, Ranchi, India did not show adequate pozzolanic characteristics. Based on the above findings, it has been concluded that CBP selected from a proper site can be replaced upto 15% in OPC for attainment of long term mechanical strength and for secondary construction works like plastering of walls.

Acknowledgments: This study was funded by Department of Civil end Environmental Engineering, Birla Institute of Technology, Mesra, Ranchi, India. The authors would like to thank Central Instrumentation Facility, BIT Mesra, in which a part of the experimental work was carried out and Emami Cement Limited for supply of OPC 43.

References

Ambroise J, Murat M, Pera J (1985). Hydration reaction and hardening of calcined clays and related minerals V. Extension of the research and general conclusions. Cement and Concrete Research 15(2): 261-268. https://doi.org/10.1016/0008-8846(85)90037-7

Bediako M (2018). Pozzolanic potentials and hydration behavior of ground waste clay brick obtained from clamp-firing technology. Case Studies in Construction Materials 8: 1-7. https://doi.org/10.1016/j.cscm.2017.11.003

Ge Z, Wang Y, Sun R, Wu X, Guan Y (2015). Influence of ground waste clay brick on properties of fresh and hardened concrete. Construction and Building Materials 98: 128-136. https://doi.org/10.1016/j.conbuildmat.2015.08.100

Heikal M, Zohdy K M, Abdelkreem M (2013). Mechanical, microstructure and rheological characteristics of high performance self-compacting cement pastes and concrete containing ground clay bricks. Construction and Building Materials 38: 101-109. https://doi.org/10.1016/j.conbuildmat.2012.07.114

IS 1344 (1981): Specification for calcined clay pozzolana [CED 2: Cement and Concrete]

IS 8112 – 2013: Ordinary Portland Cement, 43 Grade Specification.

Li H, Dong L, Jiang Z, Yang X, Yang Z. (2016). Study on utilization of red brick waste powder in the production of cement-based red decorative plaster for walls. Journal of Cleaner Production 133: 1017-1026. https://doi.org/10.1016/j.jclepro.2016.05.149

Ortega JM, Letelier V, Solas C, Moriconi G, Climent MÁ, Sánchez I (2018). Long-term effects of waste brick powder addition in the microstructure and service properties of mortars. Construction and Building Materials 182: 691-702. https://doi.org/10.1016/j.conbuildmat.2018.06.161

Robayo-Salazar RA, Rivera JF, de Gutiérrez RM (2017). Alkali-activated building materials made with recycled construction and demolition wastes. Construction and Building Materials 149: 130-138. https://doi.org/10.1016/j.conbuildmat.2017.05.122

Schöler A, Lothenbach B, Winnefeld F, Zajac M. (2015). Hydration of quaternary Portland cement blends containing blast-furnace slag, siliceous fly ash and limestone powder. Cement and Concrete Composites 55: 374-382. https://doi.org/10.1016/j.cemconcomp.2014.10.001

Scrivener K, Snellings R, Lothenbach B. (Eds.). (2018). A practical guide to microstructural analysis of cementitious materials. Crc Press.

Tironi A, Trezza MA, Scian AN, Irassar EF (2013). Assessment of pozzolanic activity of different calcined clays. Cement and Concrete Composites 37: 319-327. https://doi.org/10.1016/j.cemconcomp.2013.01.002

Toledo Filho RD, Gonçalves JP, Americano BB, Fairbairn EMR (2007). Potential for use of crushed waste calcined-clay brick as a supplementary cementitious material in Brazil. Cement and Concrete Research 37(9): 1357-1365. https://doi.org/10.1016/j.cemconres.2007.06.005

Study on Optimization and Efficiency of Outrigger system as Virtual Outriggers

B. Venkat Rao[1*], T. Manasa Lakshmi[2], Mallika Alapati[3], G.K. Viswanath[4]
[1]*Assistant Professor of Civil Engineering, V.R. Siddhartha Engineering College, Vijayawada, Andhra Pradesh*
[2]*P.G student of Civil Engineering, V.R. Siddhartha Engineering College, Vijayawada, Andhra Pradesh*
[3]*Professor of Civil Engineering, VNR Vignana Jyothi Institute of Engineering & Technology, Hyderabad, Telangana*
[4]*Professor of Civil Engineering & Director UGC-HRDC, JNTU, Hyderabad, Telangana*
[*]*e-mail : bvrvrsec@gmail.com*

Introduction

Today by the advanced approaches in structural design and structural systems, building weight is reduced and slenderness is increased which entail lateral loads into consideration. Due to lateral loads the structure undergoes heavy lateral drifts and overturning moments. To reduce these effects use of outriggers in building systems are widely used. Babaei (MBabaei,2017) considered tall steel structures with core and outrigger belt-truss system for 20 to 30 stories and carried out with pareto-front optimal solutions for minimum displacement and weight as objective functions.Some authors ((B. Smith, 1981)conducted parametric studies by considering the flexibility and the core moment distributions for outrigger-braced tall building structures.optimum location of outriggers for minimum top drift was arrived.Alexetal(Alex Coull, Otto Lau W.H, 1989)conducted a study on multi outrigger-braced structure based on continuum approach and also presented design curves for assessing the lateral drift and core base moments for any structural configuration.Some researchers (Shivacharan k, Chandrakala S, Narayana G, Karthik N.M, 2015) studied about the qptimum position of outrigger for vertically irregular tall structures.

Methodology

In the present study, many steel frame models with different number and different locations of belt trusses are considered and analyzed. Linear static and response spectrum analysis are performed. The selected topology is that two outriggers, however, one is fixed at top storey and second one is optimized for location such that where, maximum reduction in displacement occurs. Moment resisting frames with bay length of 6 m and storey height of 3.5 m are considered for the study.

Different sections for columns and different bracing system for outrigger positions are used. Structural steel with yield strength of 250 N/m^2 is used for all the structural members including columns, beams and belt trusses. Dead load and live loads of intensity 2 k N/m^2 and 3 k N/m^2 are considered on the floors. Wind and seismic loads are calculated as per Indian standards.

Results and Conclusions

Building without outrigger and core element results 88.103mm and 0.00134 values as displacement and maximum storey drift ratio respectively. Building having outrigger system along with core element shows 73.874mm and 0.00115 resulted displacement and maximum drift ratio respectively. This gives the reduction of 16.15% displacement and 14.17% maximum storey drift. The position of first outrigger is fixed at top storey and the second outrigger position is varying along the total height of the building. When the second outrigger is at 2/3 rd height of the building maximum reduction in displacements and storey drift ratios are observed as 60.089mm and 0.000871mm.

It was concluded that maximum reduction of 31.79% displacement and 35.07% maximum storey drift ratio was observed at 2/3rd height of building when outrigger is placed at that position along with the top storey. Also by optimizing different sections for columns and as well as for outriggers, the position of outrigger with maximum reduction is seen between 1/2th to 2/3rd height of the building. While using different bracing for outrigger an inverted "v" shape bracing system is effective. It is observed that there is considerable reduction of forces in core, bending moment in particular when outrigger is added to the structure.

References

B. Smith (1981) Parameter study of outrigger braced tall building structures, Journal of the Structural Division, Vol-107, No.10

M Babaei (2017) Multi-objective optimal number and location for steel outrigger-belt truss system. Journal of Engineering, Science and Technology, Vol-12: 2599-2612.

Alex Coull and Otto Lau W.H (1989) Multi outrigger braced structures, Journal of Structural Engineering, ASCE, Vol-115, No.7

Shivacharan K, Chandrakala S, Narayana G, Karthik N.M (2015) Analysis of outrigger system for tall vertical irregularities structures subjected to lateral loads, International Journal of Research in Engineering and Technology, Vol-04, Issue-05, pp.84-88

Taranath B.S (2012) Structural Analysis and Design of Tall Buildings: Steel and Composite Construction CRC, e Book ISBN: 978-1-4398-5090-9

Krunal Z Mistry, Dhruti J Dhyani (2015) Optimum outrigger location in Outrigger structural system for high rise building, International Journal of Advance Engineering and Research Development, Vol-02, ISSN(p): 2348-6406

R. Shankar Nair (1998) Belt trusses and basements as virtual outriggers for tall buildings, American Journal of Steel Construction, 4th Quarter

Choi HS, Ho G Joseph L, Math N (2014) Outrigger design for high-rise buildings, Available from: http://store.ctbuh.org/PDF_Previews/Books/2012_CTBUH_Outrigger_Guide-preview.pdf [cited 16.01.14]

N. Herath, N. Haritos, T. Ngo & P. Mendis (2009) Behaviour of Outrigger Beams in High rise Buildings under Earthquake loads, Australian Earthquake Engineering Society 2009 Conference

J. C. D. Hoenderkamp and M. C. M. Bakker (2003) Analysis of high-raised braced frames with outriggers, The structural Design of tall and special Buildings, Struct. Design Tall Spec. Build. **12**, 335–350

REVIEW ON SUSTAINABLE UTILIZATION OF DIFFERENT SOLID WASTE MATERIALS IN CONSTRUCTION INDUSTRY

Priyanka[1,*], Dr. Shobha Ram[1,2], Gaurav Chand[1]

[1] *Department of Civil Engineering, Gautam Buddha Unversity, Greater Noida, India*

[2] *Assistant professor*

* *priyankap780@gmail.com*

INTRODUCTION

Industrialized nations are grappling with the problem of expeditious and safe waste disposal. Though the waste disposal was a problem from decades but now it has increased due to population growth and industrial growth. The environment is polluted by the lot of types of wastes and due to diminution of natural resources it is now very important to consume these wastes. This waste can be used in two ways, one of the ways is to reuse and other is to recycle it. Only disposing waste is not sufficient now and with the help of new researches the alternate use of waste in construction industry is proven possible. The aim of the study is to determine the effect of selective additives on the properties of cement mortar. These selective additives include C&D, commercial plastics, glass fibre and HIPS.

A low cost concrete can be prepared by using C&D waste as aggregates replacement and unsaturated polymer resin as cement replacement which is a composite material known as polymer concrete (PC). Unsaturated polyester resins have good chemical resistance characteristics. Foam glass improves sound absorption and thermal insulation properties due to their porous structure and its spherical shape improves the workability of mortar. HIPS is a rigid and tough plastic material which has high impact strength. Due to its high impact strength at low temperatures it is widely used for toys, packaging, furniture etc. HIPS is basically a modified bitumen. Plastic bottles, bags, wrappers, and other commercial plastics can also be used in construction.

LITERATURE REVIEW

Hameed, A. M., & Hamza, M. T. (2019) has done an experimental research and investigated the effect on mechanical and physical properties on polymer concrete with the changed percentage of unsaturated polymer resin (20, 25 and 30%). Dachowski, R., & Kostrzewa, P. (2016) has discussed the results of some experimental research and concluded that the increase in substitution level HIPS and glass fibre influence the properties of cement mortar. Appiah, J. K., Berko-Boateng, V. N., & Tagbor, T. A. (2017) has discussed that the use of high density polythene and polypropylene to enhance the overall dynamic and absolute viscosities of the binder. Kamaruddin, M. A., Abdullah, M. M. A., Zawawi, M. H., & Zainol, M. R. R. A. (2017, November) had investigated that under appropriate mix design the properties of concrete can improve with the addition of plastic aggregate and plastic fibre.

REVIEW ANALYSIS

Various researchers conducted test with these parameters to study their effect on properties of concrete. This paper presents a critical review of comparison of the effect between various additives on the properties of cement mortar. The finding of the analysis also shows that the mechanical and physical properties of the cement increases by using these additives. The use of waste materials can meaningfully contribute towards more sustainable development. The table 1 below signifies the effect on various properties of cement.

Table 1. Effect of various additives on properties of cement

Parameters	Source	Effect on properties of cement
Construction and demolition waste	Waste of concrete debris. (CO) Waste of ceramic tiles. (CR) Waste of building blocks. (BL) Natural sand. (NS) River sand. (RS)	Bulk density decreases, Compressive strength increase, splitting tensile strength increase with the increase in percentage of polymer resins as it binds the aggregate with each other.
Glass fibre	Glass Cullet	The impact on the decrease in density is greater than HIPS. With the increase in amount of glass fibre compressive and bending strength decreases.
HIPS	Plastic	Shows greater decrease in bulk density than glass fibre With the increase in HIPS granulates compressive strength and bending strength increases upto 15%.
Plastic aggregate (PA)	Plastic	PA considerably decreases bulk density. On increasing the amount of PA compressive strength decreases due to low bond strength between the plastic and aggregate. With the increases in percentage of PA tensile strength decreases.
Plastic fibre (PF)	Plastic	If using PF decreases little amount of bulk density. With the increase in amount of PF compressive strength decreases. On increasing amount of PF increases tensile strength

CONCLUSION

Overall it can be concluded that by using these materials properties of concrete can be improved if mix under appropriate composition. With the use of these materials amount of solid waste can be reduced which contributes towards sustainable development. As a result of studies it can be seen that different additives have different effect on properties of concrete. It can also be seen that it produces low cost mortars with required properties. The mortar prepared using C&D waste and UP resins produces mortar which can be used as precast. On comparing the results the results of HIPS, PA & PF, HIPS shows better properties in terms of compressive strength.

REFERENCES

1. Hameed, A. M., & Hamza, M. T. (2019). Characteristics of polymer concrete produced from wasted construction materials. *Energy Procedia, 157*, 43-50.
2. Kamaruddin, M. A., Abdullah, M. M. A., Zawawi, M. H., & Zainol, M. R. R. A. (2017, November). Potential use of plastic waste as construction materials: Recent progress and future prospect. In *Materials Science and Engineering Conference Series* (Vol. 267, No. 1, p. 012011).
3. Dachowski, R., & Kostrzewa, P. (2016). The use of waste materials in the construction industry. *Procedia engineering, 161*, 754-758.
4. Appiah, J. K., Berko-Boateng, V. N., & Tagbor, T. A. (2017). Use of waste plastic materials for road construction in Ghana. *Case studies in construction materials, 6*, 1-7.
5. Ansari, M., & Ehrampoush, M. H. (2018). Quantitative and qualitative analysis of construction and demolition waste in Yazd city, Iran. *Data in brief, 21*, 2622-2626.
6. Saikia, N., & De Brito, J. (2012). Use of plastic waste as aggregate in cement mortar and concrete preparation: A review. *Construction and Building Materials, 34*, 385-401.

A comparative study of GFRP and steel stiffened plates with rectangular opening

Subramanian Anitha Priyadharshani[1*], Anumolu Meher Prasad[2], Ranganathan Sundaravadivelu[2]
[1] *Assistant Professor, NIT Warangal, Warangal, Telengana, India*
[2] *Professor, IIT Madras, Chennai, Tamilnadu, India*
* *e-mail: priyadharshanianitha@nitw.ac.in*

Introduction

Glass Fiber Reinforced Polymer (GFRP) composites with high performance, light weight, high corrosion and fatigue resistance has replaced the conventional steel stiffened plates in marine and aircraft structures. Rectangular openings are required for access and maintenance in marine structures. In ship hulls, stiffened plates undergo axial compression due to hogging and sagging bending moments. Steel stiffened plates are heavier and stronger than GFRP stiffened plates. The axial load carrying capacity of steel and GFRP stiffened plates with and without opening are different for various plate and column slenderness. Hence, optimum cross-sections with light weight and high strength need to be determined for stiffened plates with and without rectangular opening. A brief review of work carried out by some of the researchers on parametric studies of stiffened plates is presented. Kumar et al. 2009 studied steel stiffened panels with central circular opening subjected to in-plane, out-of-plane and combined loading for various plate widths, plate and column slenderness ratios. Lauterbach et al. 2010 analyzed the effect of length, width and location of crack on the crack growth and collapse behavior of stringer-stiffened CFRP panels. Wadee and Farsi 2015 carried out parametric studies to determine the critical combination of local and global slenderness which causes interactive buckling. It is observed from existing literature that, no/little work was carried out to study the effect of rectangular opening which cuts the intermediate stiffeners in stiffened plates. In the present study, GFRP and steel stiffened plates with and without rectangular opening subjected to axial compression are analyzed using the finite element software ABAQUS. The influence of rectangular opening, plate and column slenderness on the behavior of GFRP and steel stiffened plates are studied.

Methodology

The influence of plate slenderness (PS) and column slenderness (CS) ratios on the behavior of GFRP and steel stiffened plates with and without rectangular opening is studied. PS and CS ratios are varied by varying the plate thickness and stiffener height respectively. PS and CS ratios considered in the present study are shown in Table. 1.

Table 1. Details of numerical study

Property	Plate thickness (t_p) (mm)	Plate slenderness (β)	Stiffener height (h_s) (mm)	Column slenderness (λ)
GFRP	9	3.3	75	1.65
	6.35	4.73	50	2.70
	4.5	6.6	25	6.20
Steel	9	1.04	75	0.52
	6.35	1.48	50	085
	4.5	2.07	25	1.94

For different PS and CS ratios, GFRP stiffened plates without opening are designated as GFRP1, GFRP2, GFRP3, GFRP4, GFRP5, GFRP6, GFRP7, GFRP8 and GFRP9 and with opening are designated as GFRPO1, GFRPO2, GFRPO3, GFRPO4, GFRPO5, GFRPO6, GFRPO7, GFRPO8 and GFRPO9. For different PS and CS ratios, steel stiffened plates without opening are designated as ST1, ST2, ST3, ST4, ST5, ST6, ST7, ST8 and ST9 and with opening are designated as STO1, STO2, STO3, STO4, STO5, STO6, STO7, STO8 and STO9. The behavior of GFRP and steel stiffened plates with and without rectangular opening for various PS and CS ratios is investigated using ABAQUS .

Results

GFRP and steel stiffened plates with and without rectangular opening are analyzed using ABAQUS under axial compression. The axial load carrying capacities (P_f) of GFRP and steel stiffened plates with and without rectangular opening are compared for different PS and CS ratios in Table. 2. The weight of GFRP and steel stiffened plates considered in this study are 5.5 t and 1.3 t respectively. Hence, steel stiffened plates are 4 times heavier than GFRP stiffened plates. Whereas, the axial load carrying capacity of steel stiffened plates (with and without opening) are 3 to 8 times that of GFRP stiffened plates (with and without opening) for different PS and CS ratios considered (Table. 2). For β_{GFRP} = 6.60 & βsteel = 2.07 and λ_{GFRP} = 6.20 & λ_{steel} = 1.94, steel stiffened plates without opening are found to be 8 times stronger and 4 times denser than GFRP stiffened plates under axial compression. For β_{GFRP} = 4.73, 6.60 & β_{steel} = 1.48, 2.07 and λ_{GFRP} = 6.20 & λ_{steel} = 1.94, steel stiffened plates with rectangular opening are found to be 7 times stronger and 4 times denser than GFRP stiffened plates with rectangular opening under axial compression.

Table 2. Comparison of GFRP and steel stiffened plates

Specimen type	λ_{GFRP}	λ_{steel}	$(P_f)_{steel}$ / $(P_f)_{GFRP}$		
			β_{GFRP} = 3.3 & βsteel = 1.04	β_{GFRP} = 4.73 & βsteel = 1.48	β_{GFRP} = 6.6 & βsteel = 2.07
Without opening	6.2	1.94	6.57	7.05	7.92
	2.7	0.85	5.36	4.90	4.55
	1.65	0.52	4.00	2.65	4.44
With rectangular opening	6.2	1.94	6.44	6.67	6.61
	2.7	0.85	5.56	5.89	5.44
	1.65	0.52	5.13	4.13	5.75

Conclusions

The steel stiffened plates without opening are 4 times heavier and 8 times stronger than GFRP stiffened plates without opening for β_{GFRP} = 6.6 & βsteel = 2.07 and λ_{GFRP} = 6.2 & λ_{steel}= 1.94. The steel stiffened plates with rectangular opening are 4 times heavier and 7 times stronger than GFRP stiffened plates with rectangular opening for β_{GFRP} = 4.73, 6.6 & βsteel = 1.48, 2.07 and λ_{GFRP} = 6.2 & λ_{steel} = 1.94. Hence, the optimum cross-sections are found to be steel stiffened plates with stiffener height of 25 mm and plate thickness of 4.5 and 6.35 mm. The reduction in load carrying capacity of GFRP stiffened plates due to the presence of rectangular opening is minimum (2%) for β_{GFRP} = 3.3 and λ_{GFRP} = 6.20. The reduction in load carrying capacity of steel stiffened plates due to the presence of rectangular opening is minimum (4%) for β_{steel} = 1.04 and λ_{steel} = 1.94. Hence, the least effect of rectangular opening in GFRP and steel stiffened plates subjected to axial compression can be obtained for a plate thickness of 9 mm and stiffener height of 25 mm. With the increase in β, the reduction in load carrying capacity of GFRP and steel stiffened plates without opening is minimum for λ_{GFRP} = 2.70 and λ_{steel} = 1.94 respectively. With the increase in β, the reduction in load carrying capacity of GFRP and steel stiffened plates with rectangular opening is minimum for λ_{GFRP} = 1.65 and λ_{steel} = 0.52 respectively. With the increase in λ, the reduction in load carrying capacity of GFRP and steel stiffened plates without opening is minimum for β_{GFRP} = 6.60 and β_{steel} = 2.07 respectively. With the increase in λ, the reduction in load carrying capacity of steel stiffened plates with rectangular opening is minimum for β_{GFRP} = 3.3 and 1.04 respectively.

References

Kumar MS, Alagusundaramoorthy P, Sundaravadivelu R (2009) Interaction curves for stiffened panel with circular opening under axial and lateral loads. Ships and Offshore Structures 4 : 133-143.

Lauterbach S, Orifici AC, Wagner W, Balzani C, Abramovich H, Thomson R (2010) Damage sensitivity of axially loaded stringer-stiffened curved CFRP panels. Composites Science and Technology 70 : 240–248.

Wadee MA, Farsi M (2015) Imperfection sensitivity and geometric effects in stiffened plates susceptible to cellular buckling. Structures 3 : 172–186.

A STUDY ON THE EFFECTIVENESS OF HELICALLY WOUND REINFORCING BARS AS ANCHORAGES

Eshwar sai Battu[1*], Ashok Amara[2], Sai deep Samanthula[3], Rakesh Pilli[4], Dr. C B Kameswara Rao[5]
[1,2,3,4] Civil Engineering Department, National Institute of Technology, Warangal.
[5] Professor of Civil Engineering, National Institute of Technology, Warangal.
** eshwarsaibattu98@gmail.com*

Introduction

Steel and concrete are two different components which make the reinforced concrete. But these two components have unequal behaviour and physical features. For the satisfactory operation of reinforced concrete as construction material, composite action of concrete and embedded reinforcement is required. For the composite action of reinforced concrete, the bond between steel and concrete is very important and essential for no slip condition in a loaded structure. The load transfer is achieved by means of bond/and/ or anchorage. To avoid the bond failure between reinforcing bar and concrete, sufficient development length is to be provided. When the sufficient space is not available to provide the development length bends, hooks and anchorages are used. In this investigation, an attempt has been made to study the anchorage provided by helically wound reinforcing bars and to see if development length can be provided with less length of steel which can be accommodated in small space.

Materials and Methods

Twenty-seven 150 × 300 mm cylindrical specimens with centrally embedded helically wound reinforcing bars were cast, using ordinary Portland cement with water – cement ratio of 0.42; cement 530.55 Kg/m^3; Fine aggregate 730.82 Kg/m^3 (Zone II) and coarse aggregate 840.84 Kg/m^3 with a nominal size of 12.5mm conforming to IS 383. Plain steel bars of diameters 6mm, 8mm and 10mm were used. Nine specimens of each diameter of steel bars were used viz. three one turn, three one and a half turn and three two turn. Outer diameter of helix was approximately 4 times the diameter of the bar and the pitch of helix was approximately ten times the diameter. An embedded length of five times the diameter was maintained beyond the helical portion of bar. After adjusting for embedded length it is centred in the mould and then cast. A test rig was fabricated (as shown in fig 2) to carry out the pull-out test using UTM. Specimens were cured for 28 days. Then the specimen was setup in the test rig as shown in fig 2. A dial gauge was also set up to the bar to record the relative slip. Then this entire setup was fixed in the UTM (as shown in figure 3). Load was applied till either the concrete failed or the bar failed. Dial gauge readings were noted down for every one kilonewton increase in the load. Few specimens were tested with clamps.

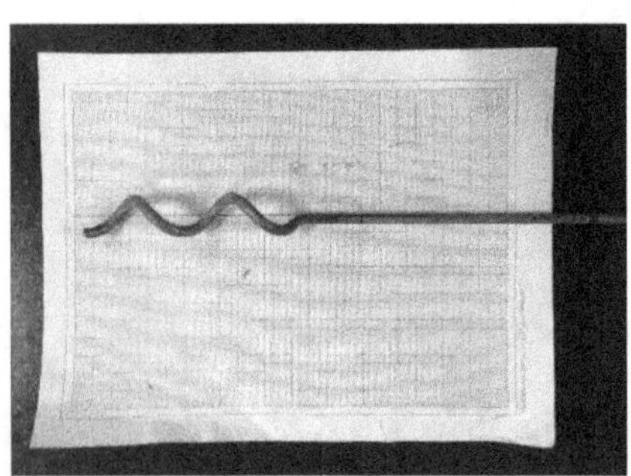

Figure 1. Helical reinforcing bar

Figure 2. Specimen with test rig

Figure 3. Testing Setup

Results and Concluding Remarks

A total of 27 concrete cylinders of grade M35 with centrally embedded helically wound reinforcing bars were tested to study their pull-out behaviour. An examination of these specimens showed that 6mm specimens and 8mm specimens have not shown any cracks and the reinforcing bar has failed under the load. In case of specimens with 10mm bar, concrete cracked but the bars were not pulled out. Further examination of these specimens revealed that the 10mm bars showed a tendency to unwind and to attain the normal straight bar shape. In this process the reinforcing bar exerted a hoop force creating tension in the concrete. Thus, the specimens developed cracks from the top of the specimens up to the helix and the helix was still found to be well embedded in the concrete. So, this clearly shows that helical form of reinforcement is providing sufficient anchorage. In order to examine if the reinforcing bar fails when the concrete cracking is arrested, a steel collar was fixed to the cylinder and the specimens were tested. It was found that the concrete developed crack but the wire also snapped. Results of few samples are shown in table 1.

Table 1. Results of the specimens

Specimen	Total Embedded length (Straight + Helical) (mm)	Development length required (mm)	Peak load (KN)	Comments
A-1	138	540	13.5	Bar failed
A-1.5	170	540	11.68	Bar failed
A-2	212	540	17.3	Bar failed
B-1	185	852	33.15	Bar failed
B-1.5	262	852	27.4	Bar failed
B-2	342	852	32.6	Bar failed
C-1 [C]	230	982	45.78	Concrete and bar, both failed.
C-1.5	288	982	33.5	Concrete failed.
C-2	336	982	37.5	Concrete failed.

The specimen designation A, B, C represents the diameters of the bar 6, 8, 10mm respectively, the numerical beside the alphabet represents the number of turns of helix and C in the bracket beside the designation indicates the usage of clamp. An examination of the total embedded length and the development length required clearly shows that the full strength can be obtained by providing one helix itself for 6mm and 8mm reinforcing bars. This shows that 25% of development length is sufficient to develop the full strength of bar if helical form is used. This will save the reinforcement and also can be best used where there is insufficient development length. However, at the kink of the helix the concrete needs to be provided with necessary hoops to resist the tensile forces developed in concrete due to opening of the kink. Hence from this study it can be concluded that helically wound reinforcing bars possess good anchorage capacity and also an effective means for providing anchorage.

References

Muhammad Saleem, Walid A. Al-Kutti (2015) Non-destructive Testing Procedure to Evaluate the Load-Carrying Capacity of Concrete Anchors. Journal of Construction Engineering and Management Vol. 142 Issue 5 DOI: https://doi.org/10.1061/(ASCE)CO.1943-7862.0001105.

Mohamad Hairi Osman1, Mohamad Nur Mustaqim Abd Shukor, Suraya Hani Adnan, Mohamad Luthfi Ahmad Jeni, Mohd Sufyan Abdullah, Salman Salim, Hannifah Tami, and Nor Azira Abdul Rahman (2018) Comparison of Pigtail with J Anchor Bolt in Normal Concrete. MATEC Web of Conferences DOI: https://doi.org/10.1051/matecconf/201815003001.

Homayoun H. Abrishami and Denis Mitchell (1996) Analysis of bond stress distributions in pull-out specimens. Journal of Structural Engineering Vol. 122, Issue 3 DOI: https://doi.org/10.1061/(ASCE)0733-9445(1996)122:3(255).

Study on behaviour of welded and bolted T-joints in circular hollow sections

A. Biju V[1*], B. Dr. Bindhu KR[1], C. Dr. Lalumangal[2]
[1*, 1] *Department of Civil Engineering, College of Engineering Trivandrum, Kerala, India*
[2] *Formerly Professor, Department of Civil Engineering, TKM College of Engineering Kollam, Kerala, India*
** e-mail: bijurit@gmail.com*

Introduction

Circular hollow sections (CHS) are commonly used in steel construction industry and many types of joints like T, K, X etc., are preferred during construction. T-Joint consists of a chord and a brace member connected at right angle to chord. T-joint in CHS can be connected by using welded and bolted technique. Orientation of gusset plate is an important parameter while dealing with bolted connection. In the present study, the behaviour of welded and bolted T-joints with longitudinal and transverse orientation of gusset plate when subjected to quasi static cyclic loading is evaluated.

Several research are there depicting the behaviour of welded connections. However studies dealing with behaviour of bolted connection to CHS are scarce. Andrew and Jeffrey, (2012), conducted numerical study on T-type branch plate to CHS connection and suggested correctness to the present design provisions. Wei and Yi-Yi, (2007), performed quasi static experimental study on eight T-joint specimens under cyclic axial loading. The authors found that, failure of axially loaded members were under weld cracking in tension and chord plastification in compression.

The Indian codes IS 806: 1968 (reaffirmed 1998) and IS 800: 2007 does not give valuable suggestions regarding joint strength for CHS. The experimental study on CHS T-joints based on Indian standards is scanty and present study deals with the application of CIDECT guidelines in Indian Standard sections.

Materials and Methods

The main objective of the work is to study the behaviour of T-joint on CHS considering various connection techniques using weld and bolt assembly. The longitudinal and transverse orientation of gusset plate in bolted connection is also studied.

The methodology adopted is as follows: A typical T-joint was chosen for the study and the tubular sections used for the T-joints were selected as per IS 1161:1998 and the joint design was carried out as per IS 800:2007 and IS 806:1998. Weld and bolt connections were also designed. For all the specimens, chord diameter and thickness were kept constant while brace diameter has been varied. Three specimens for welded and six specimens for bolted with different orientation of gusset plate (longitudinal and transverse) are fabricated and tested under quasi static cyclic loading. For welded connections, the brace members were cut to fit the contour of chord members and were connected using 6 mm fillet weld. Two HSFG 10 mm bolts were used for connecting the bolt to gusset plate. The gusset plate is of 10 mm thick and is welded to chord member either transverse or longitudinal to chord member axis.

The tests were displacement controlled and the load was applied using 20 t double acting hydraulic jack. Displacement was applied at an increment of 1 mm as positive and negative cycles and load obtained for each increment of tension and compression cycle is noted. The variation of parameter, diameter ratio β (ratio of brace diameter to chord diameter for welded and ratio of gusset plate width to chord diameter for bolted assembly) and orientation of gusset plate for bolted connections are studied. Based on the results, ultimate failure load under tension and compression, mode of failure, hysteresis curve, load deflection envelope and ductility ratio are determined and compared for the two cases. The strength obtained is compared with design strength as predicted by CIDECT guidelines.

The test setup used for the study of welded and bolted specimens is as shown in figure 1.

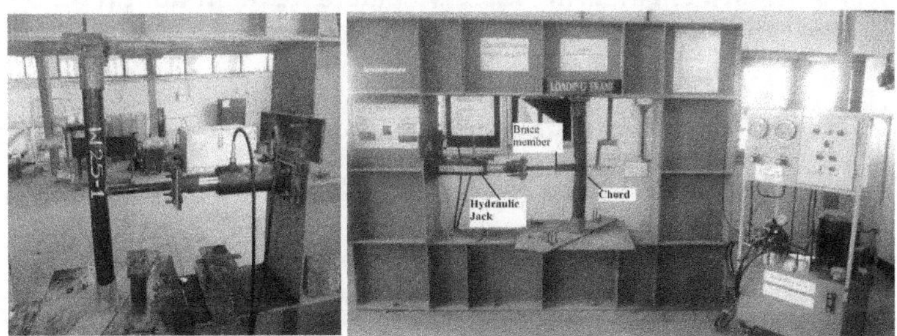

Figure 1.Test setup adopted for welded and bolted specimen.

Results and Concluding Remarks

The dimensions, strength predicted by CIDECT guidelines and experimental observations like strength and mode of failure for specimens tested are tabulated in table 1.

Table 1. Dimensions and experimental observations of tested specimens.

Specimen designation	Geometric parameter, β	Design Strength(CIDECT), kN	Experimental Observations		
			Cycles to failure	Ultimate Load, kN	Mode of failure
WS1(80MX25M)	0.38	30.18	10	54.72	Chord plastification
WS2(80MX40M)	0.54	46.59	13	63.20	Chord plastification
WS3 (80MX65M)	0.86	100.06	15	69.36	Shear failure of chord
BS1(80MX25M) L	0.38	22.45	7	35.45	Chord plastification
BS2(80MX40M) L	0.54	23.7	10	41.11	Chord plastification
BS3 (80MX65M) L	0.86	26.19	12	48.86	Chord plastification
BS4(80MX25M) T	0.38	27.52	9	41.45	Chord punching shear
BS5(80MX40M) T	0.54	41.39	11	46.24	Chord plastification
BS6(80MX65M) T	0.86	83.68	12	52.41	Chord punching shear
(WS- welded specimen, BS - Bolted specimen, L- longitudinal, T- Transverse, M- medium thickness pipe)					

From the test results, it is observed that the joint strength increases as the β value increases. For lower value of β (25 and 40 diameter brace members), strength value obtained from experimental studies is quite higher than that predicted by CIDECT. For higher value of β (60mm diameter brace), design guides overestimate the ultimate strength which may lead to unsafe design. Orientation of gusset plate transverse to chord axis gives more strength than that of longitudinal orientation. Based on the load to each cycle of displacement, hysteresis curve, load displacement envelope and ductility ratio were analysed. Ductility ratio gets increased with increase in β value for all specimens. Performance of transverse connection is more than that of longitudinal connection.

From the present study, it can be concluded that CIDECT guidelines are applicable for welded and bolted connections. Welded connections perform better than that of bolted connections. It can be inferred that in bolted connections, the designers can opt for transverse plate orientation of gusset plate, as it performs better than that of longitudinal orientation.

Acknowledgments: Research seed fund under Centre for Engineering Research & Development (CERD), College of Engineering Trivandrum, Kerala.

References

Andrew P Voth, Jeffrey A Packer (2012) Numerical study and design of T-type branch plate to circular hollow section connections. Engineering Structures (41): 477-489.

Wei Wang, Yi-Yi Chen, (2007) Hysteretic behaviour of tubular joints under cyclic loading. Journal of Constructional Steel Research 63: 1384-1395.

CIDECT DESIGN GUIDE 1 (2008), Design guide for circular hollow section joints under predominantly static loading.

IS 800:2000, General construction in steel – Code of Practice. Bureau of Indian Standards, New Delhi.

IS 806:1968 (Reaffirmed 1998), Code of practice for use of steel tubes in general building construction. Bureau of Indian Standards, New Delhi.

IS 1161:1998, Indian standard steel tube for structural purpose. Bureau of Indian Standards, New Delhi.

Compressive strength assessment of concrete mixed with binary blends of GGBS and RHA as partial replacement in cement

G. Kishore Kumar[1*], Dr. R. Dayakar Babu[2], L N V N Vara Prasad[3]
[1] *Ph.D. Scholer, Civil Engineering, NIT Trichy, Tiruchirapalli, Tamilnad, India,* kishore7142@gmail.com
[2] *Prof in Dept. of Civil Engineering, KITS-Divili, East Godavari, Andhra Pradesh India,* rdbgkk25@gmail.com
[3]P.G Student, JNTUK Kakinada East Godavari, Andhra Pradesh India, prasadcivil109@gmail.com

Introduction

The huge demand for the concrete using the ordinary Portland cement has resulted in the emission of more volume of CO_2 in to the atmosphere. Also it leads to creation of imbalance in the ecological and continuous depletion of natural resources. The contribution of OPC production worldwide to greenhouse gas emission is estimated to be 6% of the total greenhouse gas emission (Robert McCaffrey 2002). On the other hand lot of industrial waste was generated every year from lot of major industries like agricultural, steel, thermal power, etc. Our environment gets polluted because of those waste materials increasing day by day. In the present paper, an attempt was made to use different waste materials replace the cement in the production of concrete. The waste materials considered were Rice Husk Ash (RHA), a potential agro industrial waste and Ground Granulated Blast furnace Slag (GGBS), a waste from steel industry, with different proportions as a replacement to the cement.

Review of Literature

The partial replacement of OPC in concrete by GGBS, not only provides the economy in the construction but it also facilitates environmental friendly disposal of the waste slag which is generated in huge quantities from the steel industries (Yogendra O. Patil et al. 2013; Yogendra O. Patil, Prof.P.N.Patil and Dr. Arun Kumar 2013). GGBS is very useful in the design and development of high quality cement paste/mortar and concrete (Rafat Siddique et al. 2012; Rafat Siddique and Rachid Bennacer 2012). Test results obtained in this study indicate that up to 30% of RHA could be advantageously blended with cement without adversely affecting the strength and permeability properties of concrete (K.Ganesan et. al. 2008; K.Ganesan, K.Rajagopal and K.Thangavel 2008)

Materials

From the past few decades, lot of work has been going on the supplement materials for the cement to produce the high strength and durable concrete in order to mitigate the waste disposal problem. **Cement:** A 53 Grade Ordinary Portland Cement (OPC) confining to IS: 12269-1987 was bought from Market. Initial and final setting times were 38 min and 360 min respectively. **Fine Aggregate:** Locally available Godavari river sand was bought as a fine aggregate. From the various properties we get from laboratory tests, the sand is belongs to zone-II. **Coarse Aggregate:** For the present study coarse aggregate was bought from quarry available in the nearby village called Prathipadu in East Godavari district of Andhra Pradesh. **Ground Granulated Blast furnace Slag (GGBS):** GGBS was brought from the JSW store, chemical composition of Cao - 37.34%, Al_2O_3 – 14.42%, Fe_2O_3 – 1.11%, SiO_2 – 37.73%, MgO – 8.71%. Properties of considered GGBS were density – 2.87 g/cc, Fineness 348 m^2/kg, 28 days slag activity index – 97%. **Rice Husk Ash (RHA):** For this study locally available Rice Husk Ash (RHA) was collected from a rice mill near Peddapuram.

Methodology

The following steps were give the details in the methodology, Casting of concrete cubes for M40 Grade with various proportions of RHA & GGBS as a replacement to the cement. Different variables in materials are given in below table,

Table 1. Different variables studied

S. No.	Material	Variables Studied
1.	GGBS	0%, 10%, 20% & 30%.
2.	RHA	0%, 3% & 6%.

- Laboratory tests were conducted on fresh and hardened concrete prepared with different mix proportions of % replacement of RHA & GGBS in cement. The specimens were cured for 7 Days, 14 Days and 28 days and then tested to determine the compressive strength & split tensile strength.

Mix Proportion: Based on the guidelines given in IS: 10262-2012, the design mix proportion of M40 grade of concrete considered in the present work was **1:0.8:2.6 with w/c ratio of 0.36.**

Results and Concluding Remarks

From the workability test, it was observed that there is decrease in the slump value with increase in both the RHA & GGBS quantities. The slump value was decreased by 68% at 64% Cement + 6% RHA + 30% GGBS compared with conventional concrete. Results (Figure 1) show that Compressive strength of M40 grade concrete gradually increased with the increase in % RHA and GGBS up to 20%, but for 30% GGBS it decreased for 6% RHA. The compressive strength at 64% Cement + 6% RHA + 20% GGBS was improved by about 11% when compared with conventional concrete. This improvement in strength can be attributed to the enhanced gel formation due to the availability of reactive silica present in GGBS & RHA. The Split tensile strength of M40 grade concrete had shown a similar trend as of the compressive strength. From the above discussions, it can be concluded that usage of proposed industrial waste materials in concrete as a replacement to the cement will give us a better strength concrete along with a solution for disposal of such an industrial waste.

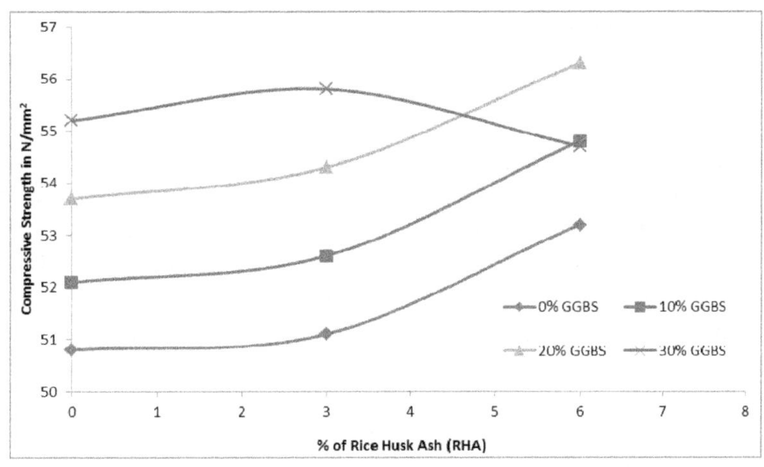

Figure 1. Variation of compressive strength at 28 days curing

References

Robert McCaffrey (2002.). Climate Change and the Cement Industry, Environmental overview. GCL Magazine, environmental special issue. London, UK: Cement Trends.

Yogendra O. Patil, Prof.P.N.Patil and Dr. Arun Kumar (2013) GGBS as Partial Replacement of OPC in Cement Concrete – An Experimental Study; Volume : 2 | Issue : 11

RachidBennacer, RafatSiddique (2012) Use of iron and steel industry by-product (GGBS) in cement paste and mortar. Volume 69, Pages 29-34

K.Ganesan, K.Rajagopal and K.Thangavel (2008) Rice husk ash blended cement: Assessment of optimal level of replacement for strength and permeability properties of concrete; Volume 22, Issue 8,Pages 1675-1683.

Performance evaluation studies on the efficiency of WTC as partial replacement in coarse aggregate in ternary blended concrete

B. Siva Ganesh Raja[1*], Dr. R Dayakar Babu[2], Ch. Bhaskara Rao[3]
[1] *Asst. Prof in Dept. of Civil Engg., AEC(A)-Surampalem, E.G Dist., Andhra Pradesh India, sivaganeshr9@gmail.com*
[2] *Prof in Dept. of Civil Engineering, KITS-Divili, East Godavari, Andhra Pradesh India, rdbgkk25@gmail.com*
[3] *Asst. Prof in Dept. of Civil Engg., ACET-Surampalem, E.G Dist., A.P, India, bhaskarnaidu.mahesh1212@gmail.com*

Introduction

The growing demand for concretes with higher performance, lower cost and reduced environmental impact when compared to those produced with conventional Portland cements has promoted the development of clinker-free alternative cementitious materials. CO_2 reduction by minimizing the use of Portland cement has recently realized the use of cementitious materials such as fly ash (FA), silica fume (SF), ground granulated blast furnace slag (GGBS) and metakaolin as partial replacements for cement. The combinations of relatively small levels of silica fume (e.g., 3 to 6%) and moderate levels of high CaO fly ash (20 to 30%) were very effective in reducing expansion due to ASR and also produced a high level of sulphate resistance (M.D.A.Thomas et al. 1999; M.D.A.Thomas, M.H.Shehata and K.Cail). Concretes made with these proportions generally show excellent fresh and hardened properties since the combination of silica fume and fly ash is somewhat synergistic. On the other side the vehicle tyres which are disposed to landfills constitute one important part of solid waste. Stockpiled tyres also present many types of health, environmental and economic risks through air, water and soil pollution.

The present paper show cases the work to assess the performance of concrete prepared Waste Tyre Chips (WTC) as a partial replacement to coarse aggregate in the ternary blended mix proportion which contains 64% of Cement + 30% of Flyash + 6% of Silica fume.

Literature Review

Environmental issues are due to disposal problems associated with FA which has led researchers to utilize it in the production of Portland cement. Further, many researchers have focused their attention on the use of FA as a source material for the development concrete (Ahmaruzzaman M.; 2010). An industrial waste is generally used as pozzolanic materials to partially replace Portland cement for a high performance concrete (good mechanical and durability properties) (Chalee W et al. 2013; Chalee W, Sasakul T, Suwanmaneechot P and Jaturapitakkul C 2013) . However, they cannot totally replace Portland cement since silica (SiO_2) and alumina (Al2O3) in pozzolanic materials still need $Ca(OH)_2$ from hydration process for its pozzolanic reaction to produce calcium silicate hydrate (CSH) and calcium aluminate hydrate (CAH) which is primarily responsible for the concrete strength. Reduction in unit weight of 14.33 % was observed in rubberized concrete mixes and thus give a viable alternative where there is requirement of normal loads, low unit weight, Medium strength, high toughness etc. (Mohammed Mudabheer Ahmed Siddiqui; 2016).

Materials and Methodology

For this present study 53 grade cement was brought from market, locally available fine aggregate from Godavari River at Rajahmundry and coarse aggregate from a stone quarry in the Prathipadu village of East Godavari district of Andhra Pradesh. Silica fume for the present study was purchased through India mart. Fly Ash was collected from Dr. Narla Tata Rao Power Plant, Vijayawada of Andhra Pradesh. For the present study Waste Tyre Chips were collected from the scrap rubber and replaced as coarse aggregate with 15%, 30% & 45% to the ternary blended mix proportion which contains 64% of Cement + 30% of Flyash + 6% of Silica fume.

Mix Proportion

The concrete mix design was done for the present study based on the guidelines given by the IS code i.e 10262-2012. The design mix proportions are given in the table 1.

Table 1. Mix Proportions for various grades of concrete considered

S. No.	Grade of Concrete	Mix Proportion	Water Cement Ratio (w/c)
1.	M20	1:1.40:2.78	0.42
2.	M30	1:1.31:2.52	0.40
3.	M40	1:1.19:2.45	0.37

Results and Concluding Remarks

From the results (Figure 1), the strength of ternary blended concrete was more at any curing period compared with conventional concrete. The workability of concrete with various % of Waste Tyre Chips (WTC) as a replacement to the coarse aggregate was observed to decrease with increase in content of WTC. Optimum % of WTC for M20 & M30 is 30% & 15% respectively as a replacement in coarse aggregate. This is due to the reduced resistance offered in the concrete matrix imparted by the toughness of coarse aggregate as the tough HBG aggregate is partially replaced with relatively softer WTC. As the grade of concrete increases, the % improvement in strength decreases for all percentages of WTC as replacement. The above said reason is clearly visible from the results obtained for M40 grade at 45% replacement of WTC, where the strength at 28 days was the lowest but still it was higher than the corresponding characteristic compressive strength. Split tensile strength also improved considerably following the same trend as that of compressive strength. Concrete production by using these industrial wastes give us a dual advantage in decreasing the pollution in our environment and at the same time imparts better strength for concrete than the conventional concrete.

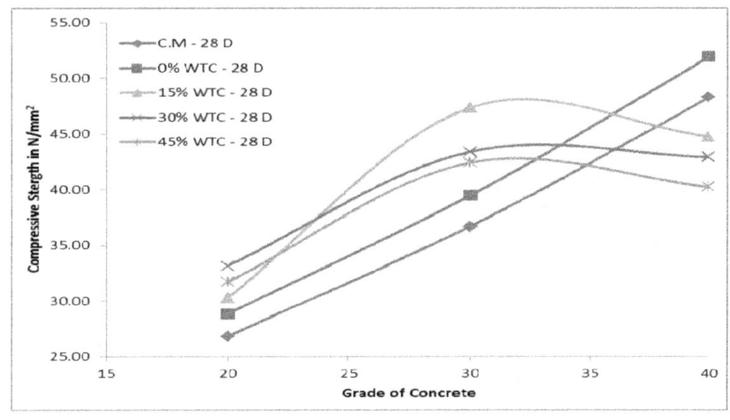

Figure 1. Variation of compressive strength

References

M.D.A.Thomas, M.H.Shehata, S.G.Shashiprakash, D.S.Hopkins and K.Cail. 1999. Use of ternary cementitious systems containing silica fume and fly ash in concrete. Volume 29, Issue 8, Pages 1207-1214

Ahmaruzzaman M. A review on the utilization of fly ash. Prog Energy Combust Sci 2010;36:327–63.

Chalee W, Sasakul T, Suwanmaneechot P, Jaturapitakkul C. Utilization of rice husk-bark ash to improve the corrosion resistance of concrete under 5-year exposure in a marine environment. Cem Concr Compos 2013;37:47–53.

Mohammed Mudabheer Ahmed Siddiqui; Study of Rubber Aggregates in Concretean Experimental Investigation; IJLRET; 2016; Volume 02 -Issue 12; PP. 36-57.

Concrete Carbonation in Blended Cement

Paritosh Kumar Singh[1], Mani Mohan[2*], Birendra Kumar Singh[1]
Birla Institute of Technology, Mesra, Ranchi, India
** mani.mohan@bitmesra.ac.in*

Introduction

The environmental impact of carbon dioxide (CO_2) emission due to production of Portland cement can be reduced by partial replacement of Portland cement with supplementary cementitious materials (SCMs). In the last few decades there is an increase in the demand of SCMs as the demand of cement is increasing, therefore large number of studies have shown that natural pozzolan such as Calcined clay (CC), which are locally available in large quantities in the Earth's crust can be used as an alternative source of SCMs as the high demand of SCMs cannot be met only by fly ashes, granulated blast furnace slag's and silica fume (Juenger and Siddique, 2015; Schneider et al., 2011). Metakaolin is one of the form of CC used as SCM and it has been reported by (Antoni et al., 2012) that 45% substitution of Portland cement by 30% Calcined clay in the form of metakaolin and 15% limestone gives better mechanical properties. Crushed clay brick has also been used as SCMs and it was seen that 25% replacement of Portland cement by crushed clay brick concrete achieved considerable compressive strength as compared to control specimen but above 25% replacement reduction in compressive strength was noticed (Aliabdo et al., 2014). This study focus on a "green" way to utilize under burnt bricks classified as third class bricks according to (BIS, 1971) and are used only for temporary structures due to low strength, as a SCM by crushing and pulverizing it into brick powder. The pozzolanic property of this material is studied by utilizing tools including SEM and particle size analyzer as well as the compressive strength and the durability (carbonation) of concrete containing Calcined clay in the form of under burnt brick powder, limestone powder and ordinary Portland cement in three types of binder system, binary binder comprising of OPC/CC45 (B1) and ternary binders OPC/CC35/LP10 (T1), OPC/CC25/LP20 (T2), OPC/CC15/LP30 (T3) and control blend OPC (C1) has been investigated.

Materials and Methods

A commercial 43 grade ordinary Portland cement (OPC) manufactured by Emami cement limited conforming to the requirements specified in (BIS, 2013) was used for all mixes. Locally available clean and dry river sand with fineness modulus of 2.76 was used as fine aggregate; approximately 90% were finer than 4.75 mm. Natural crushed stone with a maximum size of 20 mm was used as coarse aggregate. Under burnt bricks (Third Class Bricks) were procured from a brick kiln, Rodiya, Ranchi, India and locally available limestone powder (LS) was used. The bricks were crushed and pulverized into brick powder and the brick powder used in all mixes is of size less than 75μm.

Pozzolanicity of Calcined Clay

Strength activity index test (SAI) which involves measurement of compressive strength was used and the standard procedure according to (ASTM, 2019) was followed.

Compressive strength of concrete

M30 grade concrete mix proportion was calculated based on the guidelines of (BIS, 2009) and the compressive strength of concrete cube specimens (150×150×150 mm^3) containing cement replaced by Calcined clay and limestone powder in different blend composition (C1, B1, T1, T2 and T3) by weight was measured after 7, 28 and 56 days of curing using Compressive testing machine (CTM) by applying a uniformly rated compressive load of 140 kg/cm^2 per minute until failure.

Carbonation depth measurement

The concrete cubes (100×100×100 mm^3) with different blend composition (C1, B1, T1, T2 and T3) at a constant water to binder ratio of 0.42, were exposed to accelerated carbonation in a sealed chamber at CO_2 concentration of 4 % ± 0.5 %, temperature of 20°C ± 2°C and 55 ± 5 % relative humidity volume for 8 weeks after 28 days of curing. Traditional method was used to measure carbonation depth using the phenolphthalein solution, a 1% ethanol with 1 gm phenolphthalein and 90 ml 95.0 V/V% ethanol diluted in water to 100ml. Purple colour was seen where concrete was highly alkaline and the carbonated part where the alkalinity has reduced no coloration was seen. The average depth (X) in mm of the colourless part was measured from three points, perpendicular to the two edges of the split face (Chang and Chen, 2006).

Results and Concluding Remarks

For the Calcined clay in the form of brick powder bears Pozzolanicity as the strength activity index was about 0.756 when water binder ratio is 0.5 after 28 days and thus satisfies the ASTM specifications. Replacement of cement by Calcined clay and limestone powder reduces the strength of concrete. It is worth noting that the increase in percentage replacement from 0% to 30% of cement by limestone powder leads to reduction of strength up to 43% after 28 days of curing. The decrease may be because limestone powder does not bear Pozzolanicity and hence does not produce C–S–H gel also due to dilution effect there is a decrease in strength (Ramezanianpour et al., 2009). Results show that blend B1 containing Calcined clay 45% by weight of cement achieves 77% strength to that of control specimen after 56 days curing.

After 8 weeks of accelerated carbonation, as compared to control concrete other blend compositions showed higher carbonation depth as cement replacement increases the carbonation depth and the pozzolanic reaction of SCM leads to a decrease of portlandite reserve(Bucher et al., 2017).

The main function for inclusion of Calcined clay and limestone powder in concrete is to improve the compressive properties and durability of concrete. Result of present study may aid in selecting proportion of Calcined clay in the form of under burnt brick powder and limestone powder achieves desired compressive strength and durability capacity.

Acknowledgment: I gratefully acknowledge Civil & Environmental Engineering Department, Birla Institute of Technology, Mesra, Ranchi, India for funding my project to undertake my Masters in Engineering.

References

Aliabdo, A.A., Abd-Elmoaty, A.-E.M., Hassan, H.H., 2014. Utilization of crushed clay brick in concrete industry. Alex. Eng. J. 53, 151–168. https://doi.org/10.1016/j.aej.2013.12.003.

Antoni, M., Rossen, J., Martirena, F., Scrivener, K., 2012. Cement substitution by a combination of metakaolin and limestone. Cem. Concr. Res. 42, 1579–1589. https://doi.org/10.1016/j.cemconres.2012.09.006.

ASTM, 2019. C618: Standard Specification for Coal Fly Ash and Raw or Calcined Natural Pozzolan for Use in Concrete.

BIS, 2013. IS 8112: Specification for 43 grade ordinary portland cement, Bureau of Indian Standards.

BIS, 2009. IS 10262: Guidelines for concrete mix design proportioning, Bureau of Indian Standards.

BIS, 1971. IS 3102: Classification of Burnt Clay Solid Bricks, Buraeu of Indian Standards.

Bucher, R., Diederich, P., Escadeillas, G., Cyr, M., 2017. Service life of metakaolin-based concrete exposed to carbonation: Comparison with blended cement containing fly ash, blast furnace slag and limestone filler. Cem. Concr. Res. 99, 18–29. https://doi.org/10.1016/j.cemconres.2017.04.013.

Chang, C.-F., Chen, J.-W., 2006. The experimental investigation of concrete carbonation depth. Cem. Concr. Res. 36, 1760–1767. https://doi.org/10.1016/j.cemconres.2004.07.025.

Juenger, M.C., Siddique, R., 2015. Recent advances in understanding the role of supplementary cementitious materials in concrete. Cem. Concr. Res. 78, 71–80. https://doi.org/10.1016/j.cemconres.2015.03.018.

Ramezanianpour, A.A., Ghiasvand, E., Nickseresht, I., Mahdikhani, M., Moodi, F., 2009. Influence of various amounts of limestone powder on performance of Portland limestone cement concretes. Cem. Concr. Compos. 31, 715–720. https://doi.org/10.1016/j.cemconcomp.2009.08.003.

Schneider, M., Romer, M., Tschudin, M., Bolio, H., 2011. Sustainable cement production—present and future. Cem. Concr. Res. 41, 642–650. https://doi.org/10.1016/j.cemconres.2011.03.019.

Numerical Analysis of Three Legged Articulated Tower Supporting 5MW Offshore Wind Turbine

Chinsu Mereena Joy[1*], Anitha Joseph[2], Lalu Mangal[3]
[1] *Department of Civil Engineering, TKM College of Engineering, Kollam, India*
[2] *Department of Civil Engineering, Saintgits College of Engineering, Kottayam, India*
[3] *Structural Consultant, Manibil, Thiruvananthapuram, India*
** e-mail: chinsu@tkmce.ac.in*

Introduction

Offshore wind energy is widely recognized as a promising renewable energy capable to satisfy the increasing energy need. Compared to the other renewable energy resources wind energy resource exploitation is becoming more reliable due to technological advancements in the recent past. The offshore wind industry has proved to be a matured one, using bottom fixed support structures. Floating support structure based OWT is becoming more and more technically viable solution for 5MW and higher offshore wind turbine. In contrast to the fixed structures, complaint structures must provide enough buoyancy to withstand the wind turbine weight and also the environmental forces without any substantial displacement. Articulated tower is a compliant structure capable of displacing from its original position when subjected to environmental loads there by reducing the internal response. Developing further on the single leg articulated tower led to the realisation of multi legged articulated tower. The utilization of the universal joint ensures that that the columns always remain parallel to one another and the tower remains in vertical position. This paper highlights with the numerical analysis of a three legged articulated tower which supports an offshore wind turbine to study the motion response of the structure in a given sea state.

Numerical modeling and free vibration studies

The wind turbine and the sea state characteristics are considered as input data. The National Renewable Energy Laboratory (NREL, USA) has made a detailed design of a 5 MW offshore wind turbine (Jonkman et al. 2009). This turbine is assumed to be mounted over three legged articulated tower. The gross properties of the structure are given in Table 1.

Table 1. Properties of the structure.

Total mass of the structure	11232000 kg	Diameter of tower legs	7.2 m
Weight of nacelle assembly	350000 kg	Diameter of upper tower	4.8 m
Depth	144 m	Diameter of supporting beam	3 m
Draft	111 m	Total ballast weight	5175360 kg
Displacement	17947440 kg		

The numerical model of the three legged articulated tower is developed using ANSYS Aqwa. As the dimension of the legs are small compared to the wave length (i.e., for the considered waves diameter of leg is lesser than 0.5 times the wave length), tube elements are used to model the Morison elements in the design modeller of ANSYS Aqwa. Articulated joints are modeled as the joint, which transfers all translational responses. The turbine weight is given as a point mass at the top of the tower. Generated model is shown in Figure 1. The free vibration analysis is carried out and the natural periods of surge, sway and yaw are found to be 37.4 s, 37.4 s and 34.1s respectively. The natural periods of motion for these modes do not lie in the range of typical wave periods, these being from approximately 5 to 17 s.

Numerical investigations

Analysis is carried out in time domain under regular wave in Ansys Aqwa. Equation of motion is solved at each time step. The water particle kinematics is evaluated using Airy's wave theory for the estimation of hydrodynamic forces. Response Amplitude Operator (RAO) of the tower under regular wave loading is obtained in the active degree-of- freedom (surge) and is shown in Figure 2. Analysis under random waves will give the exact behaviour of the structure since the ocean waves are a combination of waves with different frequency and direction. Also the low natural frequency of the structure can be excited by wind loads. So to obtain the actual motion responses of the system, numerical simulations under wind-wave conditions are to be done. Time history analyses are performed under random waves alone and correlated wind and waves are investigated. The wind load on the tower supporting the turbine mass is given by considering a constant wind velocity at 10m height above sea level, which is assumed to be unidirectional and uniform with height. The horizontal aerodynamic thrust of NREL 5 MW wind turbine for operating condition and stop generation power condition are calculated (Zhang et al. 2013) and given as a point force at the top of the tower. The rated speed and cut out speed of NREL 5MW wind turbine are 11.4 m/s and 25 m/s respectively. When at the rated operation point, the wind thrust force reaches the maximum so that the wind-induced loads become most significant under this situation. So the performance of the articulated offshore wind tower subjected to environmental condition corresponding to rated wind speed is investigated. Environmental conditions for the study are Hs=4.55 m, Tp=9 s, U=11. 4m/s; Hs=6 m, Tp=10.28 s, U=25m/s (Oguz et al. 2018). Random waves are represented by JONSWAP spectrum. The wind and the wave are assumed to point along the same direction. All of the simulations conducted were run for 3000 seconds in order to allow sufficient time for the motion behaviour to be captured. For the analysis a time step of 0.1 seconds is used. Responses are compared to realize the effects induced by the wave loads only and the combination of wind and wave loads and are shown in Figure 3 and Figure 4. It is seen that the wind displaces the tower off the mean position where it vibrates.

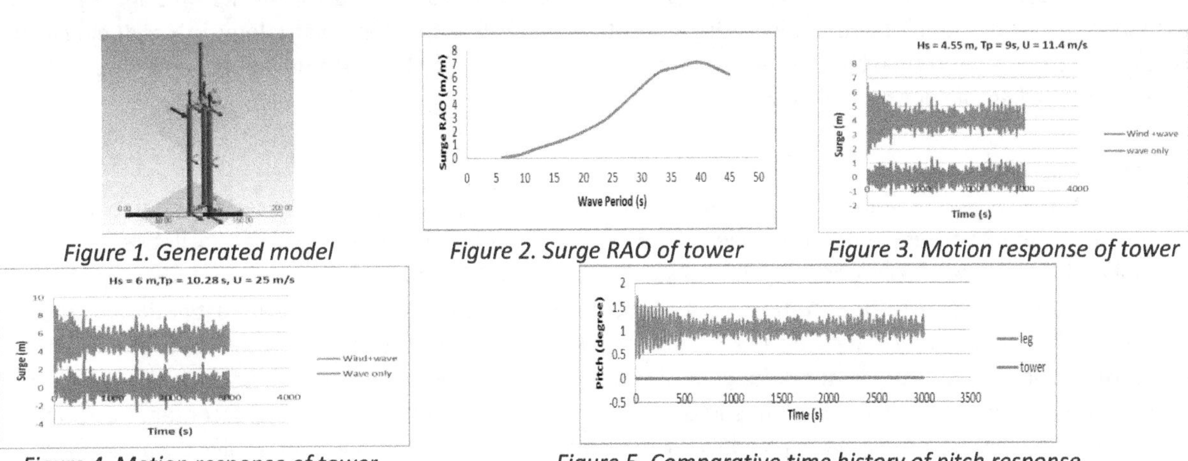

Figure 1. Generated model

Figure 2. Surge RAO of tower

Figure 3. Motion response of tower

Figure 4. Motion response of tower

Figure 5. Comparative time history of pitch response

Concluding Remarks

From the analysis it is shown that surge RAO reaches its peak at the corresponding natural frequency indicating less coupling with other modes. The structure suffers only minimum motion response. The pitch response of the tower and the leg (Figure 5) show that response of the tower is lesser than that of the legs. This guarantees that the tower remains vertical under the action of external lateral loads. Reduction in the tower response shall be attributed to the presence of the universal joint between the beam and the legs. The joints confine the transfer of the rotational displacements from the legs to the tower. So the proposed articulated tower is found to be satisfactory for supporting wind turbine.

References

Jonkman J., Butterfield S., Musial W. and Scott G. (2009) Definition of a 5-MW reference wind turbine for offshore system development, National Renewable Energy Laboratory Technical Report. NREL/TP-500–38060. USA

Zhang R., Tang Y., Hu J., Ruan S., Chen C. (2013) Dynamic response in frequency and time domains of a floating foundation for offshore wind turbines. Ocean Eng. 60: 115–123.http://dx.doi.org/10.1016/j.oceaneng.2012.12.015

Oguz E., Clelland D., Day A.H., Incecik A., Lopez J. A., Sanchez G. and Almeria G. G. (2018) Experimental and numerical analysis of a TLP floating offshore wind turbine. Ocean Eng. 147: 591–605. https://doi.org/10.1016/j.oceaneng.2017.10.052

Optimization of Clay Filled Flood Soil as Fine Aggregate and Source Material in Fly Ash Based Geopolymer Mortars

Sreedevi Lekshmi[1*], J. Sudhakumar[2], Reesha Bharath[3], Sunitha K Nayar[4]
[1,2,3] *Department of Civil Engineering,National Institute of Technology,Calicut,Kerala,India*
[4]*Department of Civil Engineering,Indian Institute of Technology,Palakkad,Kerala,India*
* *e-mail: sreedevi_p170033ce@nitc.ac.in*

Introduction

The boundless demand for cement over the globe results in excessive climatic change due to global warming. Thus, environmental protection and its prevention has become major concern for mankind. Many researches were conducted to find an alternate material that could be used by partial or full replacement of cement so as to reduce the carbon dioxide emission due to cement production. This has led to the invention of an alternate binder by Davidovits which make use of industrial or agricultural waste and are of rich in aluminates and silicates. The source material used is activated by an alkaline medium which leads to the formation of non-cementitious binder called geopolymer. This study focus on the development of fly ash based geopolymer mortar in which fly ash is partially replaced with treated clay filled sandy soil accumulated during floods, along with which an attempt is made to use the former as fine aggregate. In this paper, an effort is made to explore the potential of alkali activation technology in producing eco-friendly sustainable mortars which make use of waste soil such as clay filled sandy soil collected from flood affected areas of Kerala during the 2018 outburst.

Materials and Methods

Materials used in the present study includes class F fly ash, fine aggregate used for control mix was river sand , treated and untreated clay filled sandy soil accumulated during 2018 Kerala floods as source material and fine aggregate respectively, NaOH solution and Na_2SiO_3 solution.

Methodology of the study contains preliminary studies such as geotechnical investigations and SEM analysis of clay filled sandy flood soil (FS). A set of design trials were obtained using central composite design of response surface methodology in design expert software. Each individual trials gave mix proportion of molarity of NaOH solution, sodium silicate to sodium hydroxide ratio, percentage of FS to be added and temperature of curing. Each trials were performed experimentally and analysis was conducted for compressive strength, dry density, water absorption and shrinkage using design of experiment software. For all individual trials, treated FS was used to partially or fully replace fly ash and untreated or raw FS was used as fine aggregate. Lime treatment was performed on FS to be used as source material. Optimization was performed experimentally and analytically for compressive strength.

Results and Concluding Remarks

A set of 13 trials were obtained using central composite design of response surface methodology in design expert as shown in table 1 by keeping temperature of curing as constant at 60^{0}C and percentage of FS as source material as 50%. Also molarity of sodium hydroxide solution is taken as 6M to 10M and ratio of sodium silicate to sodium hydroxide (S/N) is taken as 0.5 to 2.5 for generating the design trials. Since the flood soil was of sandy in nature, it was unable to perform the geotechnical investigation before and after treatment using lime. Variation of responses such as dry density, water absorption, initial and final setting time and shrinkage were studied experimentally for all the generated trials and are reported in figure 1 to 5. The optimum percentage of lime to be added to the FS was estimated as 4%. Compressive strength was obtained between 7 MPa to 15MPa. The optimum mix obtained was with molarity of NaOH as 7M and ratio of sodium silicate to sodium hydroxide as 2 at 60^{0}C. The compressive strength of optimum mix was obtained as 13 MPa. Hence it is concluded that the optimum mix can be effectively used for plastering works, as crack fillers and generating eco-friendly adhesives for wrapping fibre reinforced polymers such as CFRP and GFRP for rehabilitation purpose of RC structures.

Table 1: Design trials generated using design expert

RUN	Molarity of NaOH (M)	Na_2SiO_3/NaOH (Ratio)
1	8	1.5
2	8	1.5
3	8	2.5
4	8	1.5
5	9	2
6	8	0.5
7	8	1.5
8	7	2
9	7	1
10	8	1.5
11	9	1
12	6	1.5
13	10	1.5

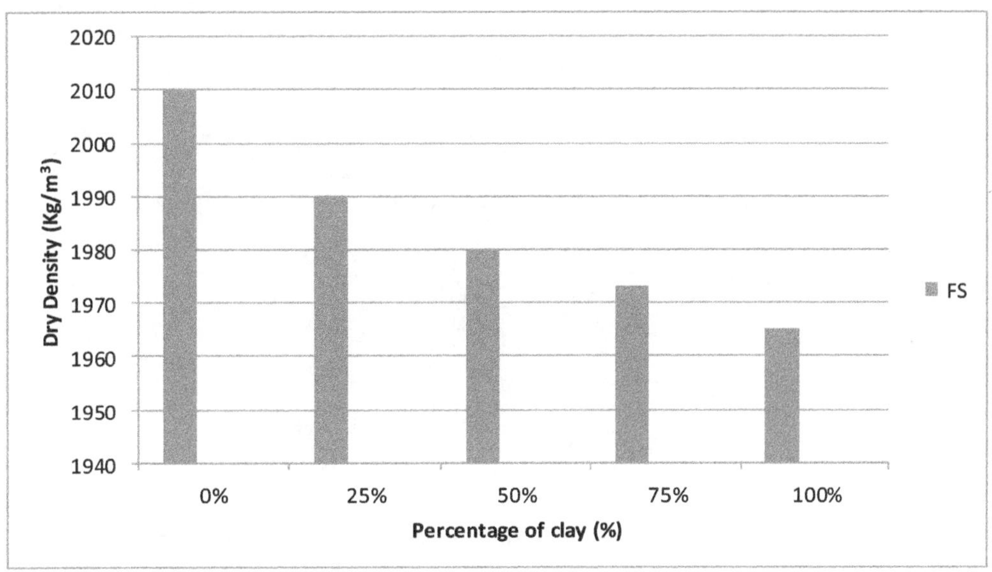

Fig 1: Variation of dry density w.r.t percentage of clay

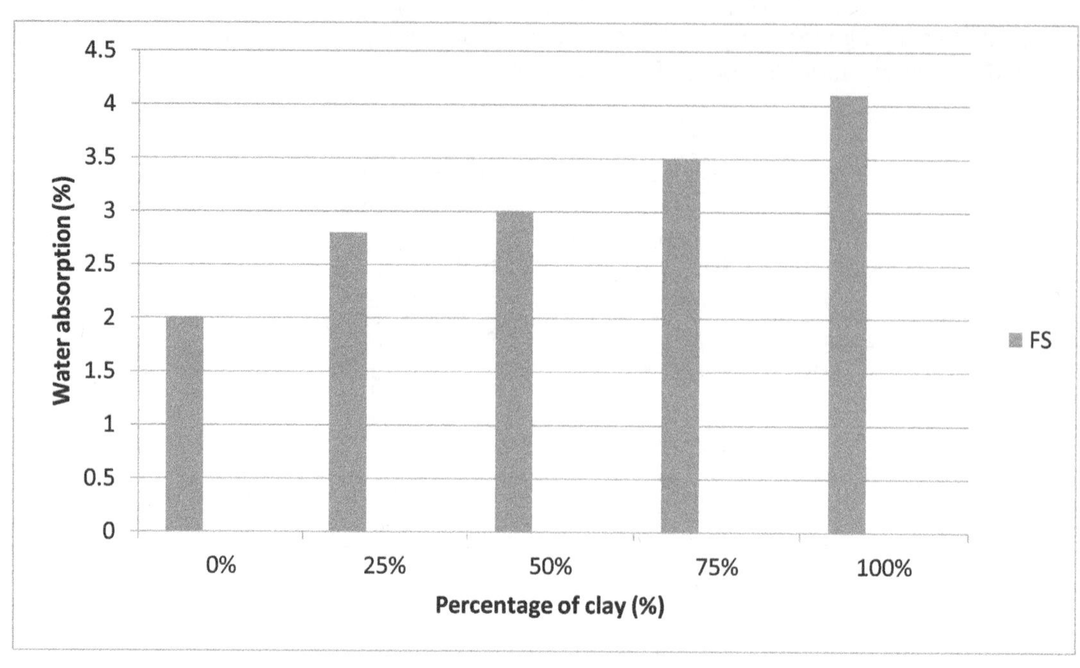

Fig 2: Variation of water absorption w.r.t percentage of clay

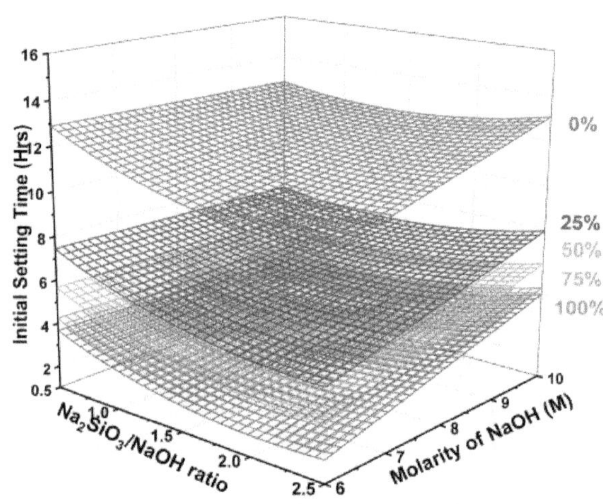

Fig 3: Variation of initial setting time w.r.t molarity of NaOH and S/N ratio

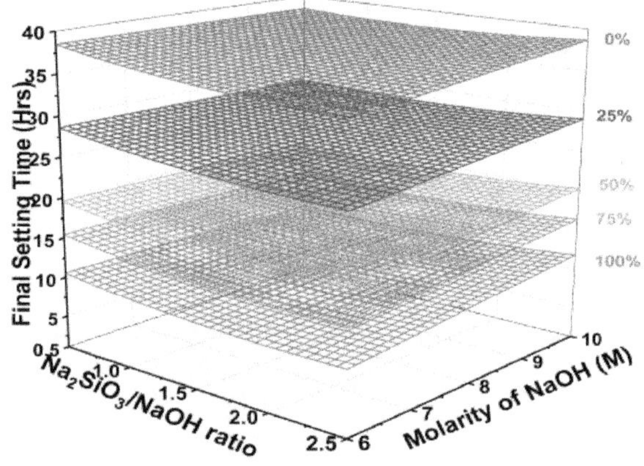

Fig 4: Variation of final setting time w.r.t molarity of NaOH and S/N ratio

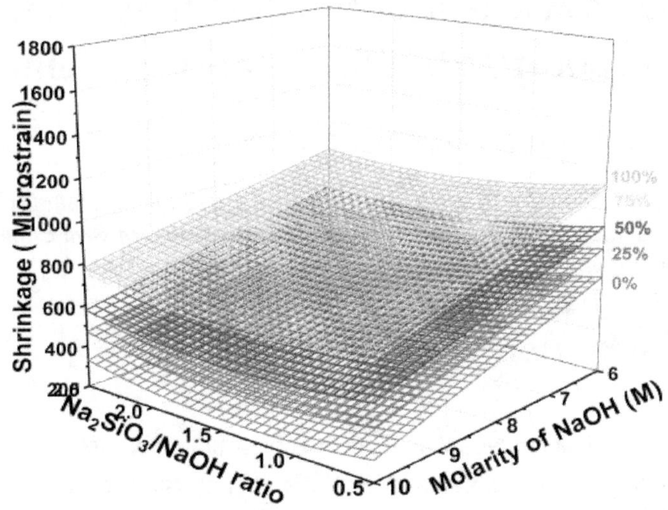

Fig 5: Variation of final setting time w.r.t molarity of NaOH and S/N ratio

References

1. Ogundiran et al.: Synthesis and Characterization of Geopolymer from Nigerian Clay, Applied Clay Science, Vol.108, pp. 173-181, 2015.
2. Priyadarshini P., Ramamurthy K., Robinson R. G.: Excavated Soil Waste as Fine Aggregate in Fly Ash based Geopolymer Mortar, Applied Clay Science, Vol.146, pp.81-91, 2017.

A Comparative Study on Mechanical and Durability behavior of Ternary Blended Mortars incorporating GGBS, Alccofine and Silica Fume.

A. Arnab Mondal[1], B.Dr Kushal Ghosh[2*], C.Dr Partha Ghosh[3]

[1] *Student, M.E in Construction Engineering with Specialization in Structural repair and retrofit Engg, Jadavpur University.*
[2] *Assistant Professor, National Institute of Technology,Sikkim, Department of Civil Engineering*
[3] *Associate Professor, Jadavpur university, Department of Construction Engineering*
* *e-mail: kushalghosh100@gmail.com*

Introduction

There is a push towards sustainability in recent times in the construction industry. As large scale production of cement causes emission of greenhouse gases and depletion of natural resources. Thus importance is being given on optimum utilization of OPC by partially replacing it with various industrial by products. The main objective of the research is to develop an eco friendly sustainable mortar mix by using varying percentages of GGBS along with Alccofine and Silica fume as partial replacement of OPC and to make a comparative study between the control mix and different supplementary cementitious materials (SCM's) mixes in terms of mechanical properties and durability exposures. This experimental study discusses the effect of different supplementary cementitious properties (SCM's) on the durability behavior of cement based mortar specimens under different exposure conditions. The specimens consist of ordinary cement mortars and mortars containing ternary mixtures of supplementary cementitious materials(SCM's) like Ground Granulated Blast Furnace Slag(GGBS), Alccofine and silica fume (SF) along with cement. Two series of mixes were prepared by replacing OPC with 30% and 50% of GGBS along with 5%,10% and 15% of Alccofine and Silica fume respectively . Compressive strength, ultrasonic pulse velocity test were performed on different samples to observe SCMs behavior in hardened conditions as well as different exposure conditions (acid, sulphate and thermal)after 7,28 and 56 days respectively. Microstructural studies using scanning electron microscopy(SEM) and electron dispersive X-ray(EDX) analysis were performed on samples before and after different exposure conditions.

Materials and Methods

Mix samples were prepared conforming to ASTM C 109/C109M-02 having constant w/b ratio of 0.36 and casted in 50x50x50 cube molds. Ternary mixes replacing OPC with 30% and 50% GGBS and 5%,10% , 15% by alccofine and Silica fume were prepared in two separate series including control mix. Compressive strength test were performed on samples under normal condition as well as different exposure conditions (acid, sulphate and thermal)after 7,28 and 56 days respectively. Microstructure studies using scanning electron microscopy(SEM) and electron dispersive X-ray(EDX) analysis were performed on samples before and after different exposure conditions.

Results and Concluding remarks

As illustrated in fig 1 control mix performed better than the other ternary mixes containing SCM's after 7 days water curing. Ternary mix series incorporating GGBS and alccofine offered better strength gain

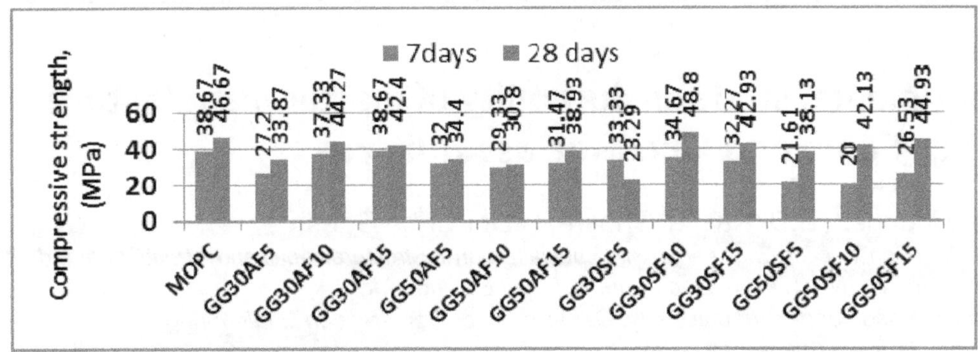

Fig 1: comparison of compressive strength of different mixes after 7 and 28 days respectively.

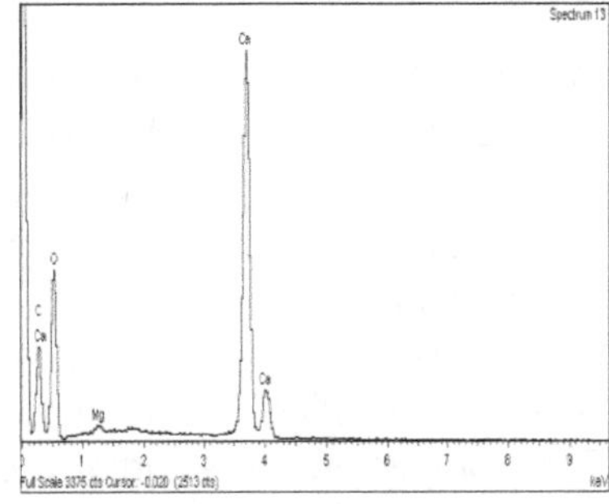

Fig 2: SEM image and EDX profile of ternary specimen GG30SF15 (surface) after H_2SO_4 exposure .

compared to GGBS and silica fume mix combination after 7 days. Further inclusion of GGBS shows a decreasing trend in compressive strength. Significant strength gain occurred for mixes blended with GGBS and silica fume compare to GGBS alccofine mixes after 28 days and even surpassed the strength value of control mix. Results signify the addition of SCM's responsible for significant change in compressive strength .Due to pozzolanic activity at later age accompanied with pore refinement enables silica fume to perform better after 28 days .

References

[1] A. Narender Reddy *, T. Meena,- A Study on Compressive Behavior of Ternary Blended Concrete Incorporating Alccofine Materials Today: Proceedings 5 (2018) 11356–11363.

[2] Saurabh Gupta1, Dr. Sanjay Sharma2, Er. Devinder Sharma3- A Review on Alccofine : A supplementary cementitous material, International Journal of Modern Trends in Engineering and Research.

[3] R. Sri Ravindrarajah-Acids attack on silica fume high-strength concrete

[4] Caijun Shi , Dehui Wang, Linmei Wu, Zemei Wu-The hydration and microstructure of ultra high-strength concrete with cement–silica fume–slag binder.

An Experimental study on durability of mortar manufactured using Fly Ash, Silica Fume and Alccofine as Additives.

A. Santanu Mondal[1], B.Dr Kushal Ghosh[2], C.Dr Partha Ghosh[3]
[1] *Student, M.E in Construction Engineering with Specialization in Structural repair and retrofit Engg, Jadavpur University.*
[2] *Assistant Professor, NIT Sikkim, Department of Civil Engineering*
[3] *Associate Professor, Jadavpur university, Department of Construction Engineering*
* *e-mail: kushalghosh100@gmail.com*

1. Introduction

From the context of sustainability, the need for alternative building materials has been increasing. In this research work the cement content being used in a mix has been tried to be reduced without compromising on the strength and durability characteristics. Effort has been made to study the various fresh and hardened properties of the mortar prepared with flyash, alccofine and silicafume as a partial replacement of the ordinary Portland cement.

The main objective of the paper is to study the performance of a binary and ternary mix prepared with 30% Flyash and 5 to 15 % alccofine, 30% Flyash and 5 to 15 % Silicafume, 50% Flyash and 5 to15% alccofine , 50 % Flyash and 5 to15% Silicafume as partial replacement of ordinary Portland cement of 43 Grade exposed to 4% H_2SO_4 for 56 days and 4% $MgSO_4$ for 56 days also performance of the under thermal exposure at 800°c for 4 hours. Also hardened property of the mix under unexposed conditions like 7 days and 28 days Compressive Strength, water absorption, apparent porosity, bulk density, sorptivity and their morphological changes by micro structural study like SEM & EDX will also be studied.

2. Materials and Methods

Materials:-

- **Cement:-** Ordinary Portland Cement of 43 grade ultra tech brand conforming to IS:8112:2013.
- **Sand:-** River sand of grading zone II conforming to IS:383:1970
- **Water:-** Normal tapwater
- **Flyash:-** Class F Flyash
- **Alccofine:-** Alccofine 1203 is a pozzolanic material prepared by the method of controlled granulation having sp gravity-2.9, fineness> 12000cm^2/gm.
- **Silicafume:-** is a product of silicon metal, is a very reactive pozzolan, conforming to ASTMC1240, having sp gravity-2.2 to 2.5 and fineness-15000-35000 m^2/kg

Test methods:-

- **Compressive strength:-** Test method conforming to ASTMC 109
- **Fresh property-** Sorptivity test method conforming to ASTMC1585, water absorption, bulk density, apparent porosity.
- **Durability Study:-** Sample exposed to H_2SO_4 and $MgSO_4$ for 56 days and sample exposed to elevated temperature up to 800°C for 4 hours.

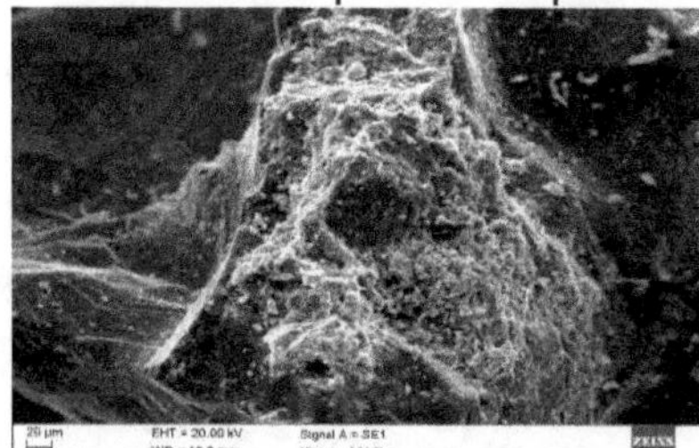

Figure 1. SEM image of the 28 days best *Performing sample C70FA30SF10*

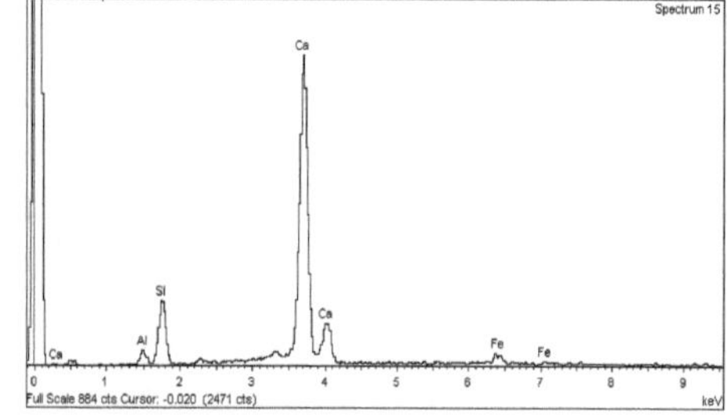

Figure 2. EDX image of the 28 days best *Performing sample C70FA30SF10.*

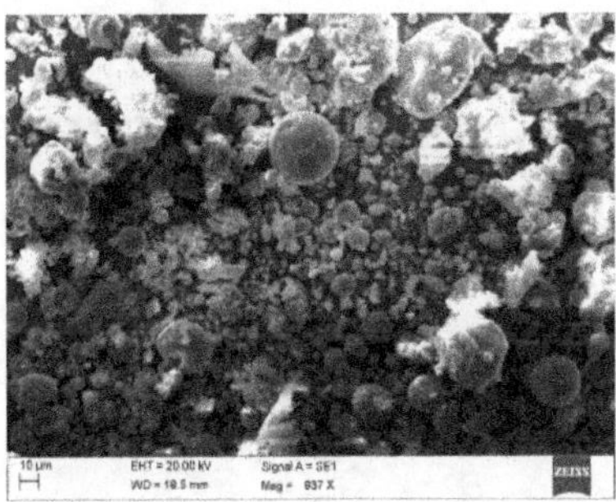

Figure 3. SEM image of the Flyash as raw material.

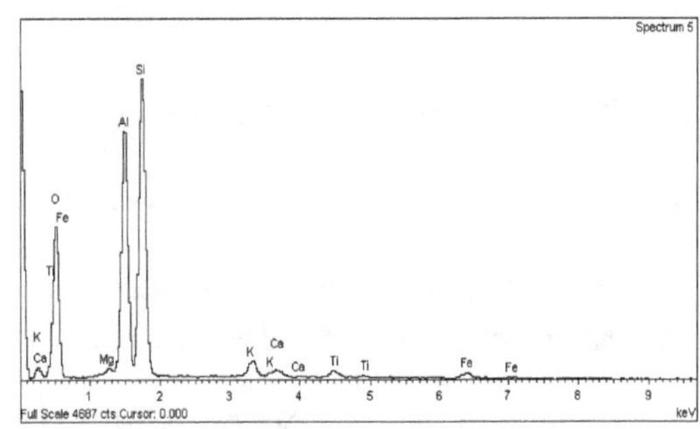

Figure 4. EDX image of the Flyash as raw material.

Results and Concluding Remarks

The following Table-1 contains the results of the various best performing samples under different exposure conditions and comparison with the control mix.

Table 1. Test results and comparisons.

Experiment	Best performing sample/results	Control mix results
7 days compressive strength	C70FA30AF5/30.13 MPa C70FA30SF15/30.1 MPa	C100/38.67 MPa
28 days compressive strength	C50FA50AF10/48 MPa C70FA30SF10/54 MPa	C100/46.67 MPa
Acid(H_2SO_4)exposure	C70FA30AF15/52.73% Residual Strength C50FA50SF15/45.14% Residual Strength	C100/45.12%Residual Strength
Sulphate($MgSO_4$)exposure	C70FA30AF10/ Surplus Strength 1.48% C70FA30SF15/ Surplus Strength 35.67%	C100/Surplus strength 5.12%

[1]It may be concluded from the various results obtained that the Flyash has an later age strength gaining property,[2]addition of slag,silica fume and alccofine leads to high early strength. Its observed that the compressive strength and bulk density increased and water absorption, apparent porosity decreased due to the presence of Flyash up to 30% and alccofine, Silicafume 5% to 15 %. [3]Flyash and Silicafume contain high amount of Al and Si compounds, In reaction with the $Ca(OH)_2$ attribute to more stable C-A-S-H with a higher filler effect in the concrete pores resulting into lesser weight loss, in this study it has been found that mix with 30% Flyash with alccofine 5 to 10 % retains more residual strength and mix with 50% flyash and 5 to 15% Silicafume retains more residual strength, which is closer to the ordinary Portland cement. It has been observed for $MgSO_4$ exposure that mix containing 30% flyash and 10 to 15 % alccofine and Silicafume increased strength substantially compared to control mix which attributes to the presence of more stable and dense C-S-H gel formation in the mix.

References

1. Study on effect of Alccofine & Fly ash addition on the Mechanical properties of High performance Concrete Mr. Sunil Suthar1 Dr. (Smt.) B. K. Shah2 Prof. P. J. Patel

2.S. Kumar, R. Kumar, and S. P. Mehrotra. Journal of Materials Science, 45(3),607 (2010), DOI: 10.1007/s10853-009-3934-5

3. Durability of fly ash based geopolymer concrete in the presence of silica fume,Francis N. Okoye a, Satya Prakash a, Nakshatra B. Singh.

4. Strength and resistance to sulfate and sulfuric acid of ground fluidized bed combustion fly ash–silica fume alkali-activated composite Prinya Chindaprasirt a, Pattanapong Paisitsrisawat b, Ubolluk Rattanasak.

PREDICTING THE STRENGTH AND PERMEABILITY OF POROUS CONCRETE PAVEMENT USING FEED-FORWARD NEURAL NETWORKS STRATEGY

T. Srinivas[1*], D.Tarangini[2],Dr. P Sravana[3]
[1]*M. Tech-Scholar, Department of Civil Engineering, JNTUH, Hyderabad, India;*[2]*Research Scholar,Department of Civil Engineering,JNTUH,Hyderabad,India;*
[3]*Professor &Head, Department of Civil Engineering,JNTUH, Hyderabad, India;*
** e-mail: suraj303677@gmail.com*

Introduction

Porous concrete (PC) pavement is a sustainable type of concrete pavement that can protect and restore natural ecosystem. The permeability is the most important characteristic of PC for infiltration of rain water. The purpose of this experimental study is to predict the effect of water-to-cement ratio (W/C) and size of aggregate on the permeability and compressive strength of PC using artificial neural networks(ANN). Based on the lowest root mean squared error (RMSE), the ANN model was chosen.This approach can reduce the number of trial mixes to get target performance of the samples. The 28-days compressive strength for pervious concrete ranges from 5.6 to 21.0 MPa, porosity ranging from 14 to 31%, and permeability coefficient varies from 0.25 to 6.1 mm/s (Schaefer et al. 2006). The effects of aggregate-cement ratio, aggregate sizes and type of binder material on strength of No-fine concrete have been reported in the past (Tennis et al. 2004; Malhotra 1976; Meininger 1988; Otani et al. 2005). The proposed network is intended to represent a reliable functional relationship between the input independent variables accounting for the variability of permeability and compressive strength of a pervious concrete as two output dependent variables.

Materials and Methods

Cement

Locally available 53 grade of Ordinary Portland Cement (Ultra Tech)confirming to IS: 12269 and also properties like specific gravity, initial and final setting time, fineness wereconducted.

Aggregates

Fine Aggregates

The locally available sand is used as fine aggregate in the present investigation. The sand is free from clayey matter, salt and organic impurities. The sand is tested for various properties like specific gravity,bulk density and water absorption. In this present study coarse aggregate is partially replaced with 0% and 10% fine aggregates.

Coarse Aggregates

Machine crushed angular granite metal of 10mm and 20mm nominal size from the local source is used as coarse aggregate. Different proportions of (20mm:10mm) in ratios of 50:50,60:40 & 70:30 were used. It is free from impurities such as dust, clay particles and organic matter etc.

Water:

Locally available water used for mixing and curing which is potable, shall be clean and free from injurious amounts of oils, acids, alkalis, salts, sugar, organic materials or other substances that may be deleterious to concrete or steel.

Testing Specimen:

Concrete specimens of size 150mm X 150mm X 150mm cubes were used to determine the compressive strength of Pervious Concrete and permeability.

Mix Design

As there is no specific code forpervious concrete, in the present work we have used ACI 522 R10 for preparing the samples. A total of 144 cubes were casted. The mixes used were having W/C ratio of 0.38 and 0.40, total aggregate to cement ratio (TA/C) of 6:1 and 5:1,the coarse aggregate with sizes of

20mm:10mm in ranges of 50:50, 60:40 and 70:30,fine aggregates as 0% and 10% replacement with coarse aggregate. The compressive strength for 7 and 28 days, porosity and permeability for 28 days .

Results and Concluding Remarks

It is observed that the average permeability coefficient of Pervious Concrete(PC)produced from various mixes were approximately in the range of 0.25cm/s to 0.543 cm/s.The average compressive strength of PC wasin between 11.92N/mm^2 and 17.71N/mm^2.Hence from study we can say that if the size of aggregates increases the porosity and permeability coefficient increases but compressive strength decreases. Also with the increase in percentage of fine aggregates the compressive strength of PC increasedbut reduced the permeability.However the strength and permeability were in the acceptablerange as per ACI code for pavement applications.The Root Mean Square Error(RMSE)of the modeling results is less than 2%, it can be concluded that the ANN can be used as fast tool for modeling the permeability coefficient and compressive strength.The goodness of fit factor also shows that there is an acceptable error in training(0.024%) and testing(0.017%) models of the networks.Hence it is suggested that application of ANN would be the better option to predict the required output with proper trained models. The following figures shows the fitness graphs of the trained model and predicted output for a known sample input.

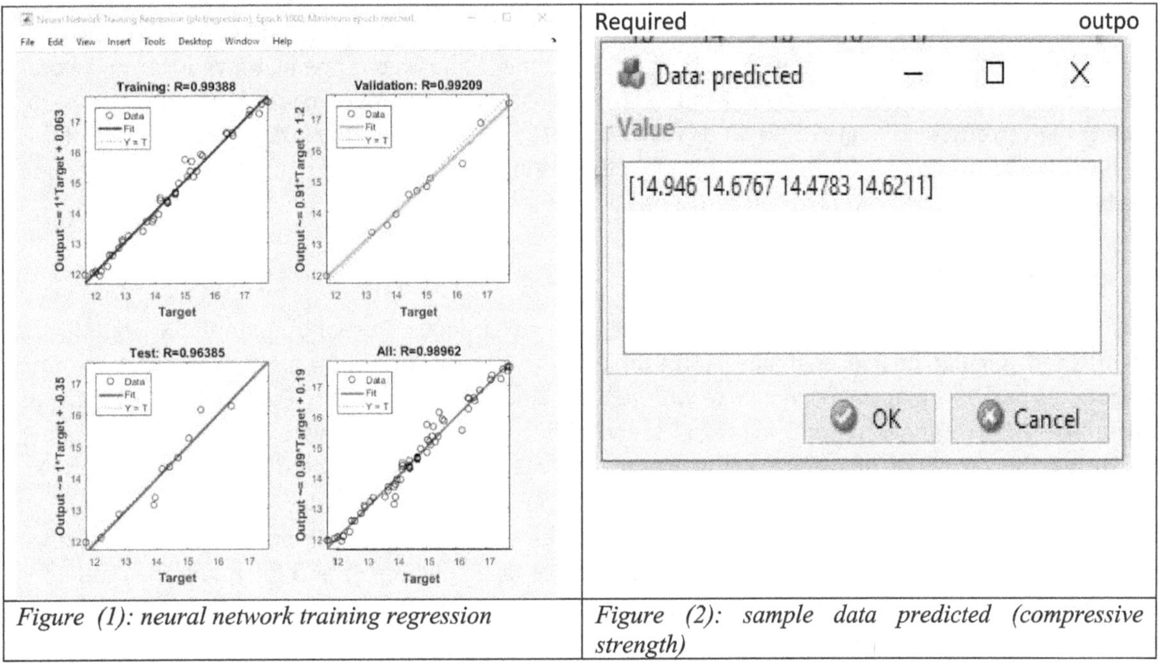

Figure (1): neural network training regression

Figure (2): sample data predicted (compressive strength)

References

1. Patil, Praveenkumar, and Santosh M. Murnal. "Study on the Properties of Pervious Concrete." *International Journal of Engineering Research & Technology (IJERT) IJERT ISSN*(2014): 2278-0181.
2. ASTM C 192 (2003). Standard practice for making and curing concrete test specimens. Annual book of ASTM standards 4.02. West Conshohocken, PA: ASTM international.
3. Consolazio GR (2000). Iterative equation solver for bridge analysis using neural networks. Comput.-Aided Civ. Infrastruct. Eng. 15(2):107-119.
4. Hajela P, Berke L (1991). Neurobiological computational models in structural analysis and design. Comput. Struct. 41(4):657-667.
5. Hola J, Schabowicz K (2005). Application of artificial neural networks to determine concrete compressive strength based on non-destructive test. J. Civ. Eng. Manage. 11(1):23-32.
6. Lian C, Zhuge Y (2010). Optimum mix design of enhanced permeable concrete-An experimental investigation. Constr. Build. Mater. 24(1):2664-2671. Mansour MY, Dicleli M, lee YJ, Zhang J (2004). Predicting the shear strength of reinforced concrete beam using artificial neural networks. Eng. Struct. 26(4):781-799.

OPTIMIZING PERVIOUS CONCRETE MIX USING RESPONSE SURFACE METHODOLOGY

Y.Swathi[1*], D.Tarangini[2],Dr.P Sravana[3]

[1] *M. Tech-Scholar, Department of Civil Engineering, JNTUH, Hyderabad, India;*
[2] *Research Scholar,Department of Civil Engineering, JNTUH,Hyderabad,India;*
[3] *Professor & Head, Department of Civil Engineering, JNTUH, Hyderabad, India;*
* *e-mail:saraswathi.yashoda@gmail.com*

Introduction

Current climatic changes are occurring due to various human and industrial activities. In particular, the effects of urbanisation & growing threat of Global warming caused increasing precipitation in many areas. To encounter these,Porous concrete can help us to minimize flooding risks, recharging ground water etc.The optimal share of the coarse aggregate turn to be 40.21%, the share of fine aggregate is 49.79% for achieving required compressive strength of 25 MPa, flexural strength of 4.31 MPa and porosity of 21.66%(**Ivana Barisic et al.2015)**.Response Surface Methodology (RSM) is employed to optimize a four-component concrete containing fly-ash,cement and high range water reducer and water binder ratio. A combination of 7.5% SF and 2.5% SP was the optimum dosage of admixtures for obtaining high strength concrete of 90 MPa **(T.M. Murali1, and S. Kandasamy 2015)**. The purpose of this study is to determine the mix design that will yield the optimum compressive strength, permeability and porosity by response surface methodology (RSM).

Materials and Methods:

In this study, the main concrete mix constituents such as Ordinary Portland cement 53 Grade(Ultra Tech),crushed angular coarse aggregate (60% of 20mm, 40% of 10mm),natural river sand as fine aggregate as partial replacement of coarse with (0%,5% &10%) and W/C ratio (0.38, 0.39 & 0.40).Cubes with different percentages of fine aggregates and different W/C ratios were prepared. This paper investigates the potential of using design of experiments (DOE-RSM) to predict the properties like compressive strength, permeability and porosity of pervious concrete at 28th day.

The response surface design used in this study is face-centred central composite response surface design. A face-centred central composite response surface design with α=1 and full quadratic model for each response was used. The statistical software "Minitab 2018 version" and DESIGN-EXPERT software is used to analyse the experimental design.

Results and concluding Remarks:

From the results we can say that with increase in W/C ratio and TA/C ratio Compressive Strength values decreases but increase in % of Fine aggregate by 10%, strength increased by 14.42%.Linear regression equations and significant R^2 values were obtained as 0.8941,0.7882 and 0.9539 for 0%,5%and 10% fine aggregate. Of all the mixes, the combination of 0.38 W/C ratio, T.A./C of 5 and 10% fine aggregate gives the maximum compressive strength of 16.8Mpa with the composite desirability of 0.94.Optimum combination of 0.39 of w/c ratio, T.A./C of 6 and 0% fine aggregate gives the maximum permeability of 0.525 cm/sec with the composite desirability value as 1.Optimum combination of 0.4 of w/c ratio, T.A./C of 6 and 0% fine aggregate gives the maximum porosity of 32.74% with the composite Desirability value as 0.95.By adopting Central Composite Design using Response Surface Methodology there is decrease in experimental cost and time for selected optimum mix. Full quadratic model is employed by using ANOVA method and all models are found to be significant.

Figure1 shows the Contour Plots and Surface Plots of Compressive Strength Vs W/C ratio and TA/C ratio. From these plots we can conclude that with the increase in W/C ratio and T.A/C ratio Compressive Strength values decreases.

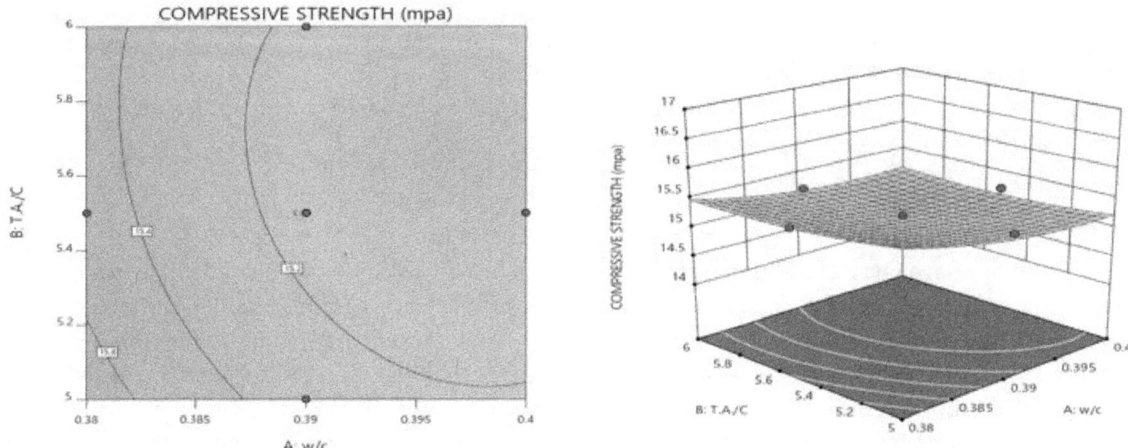

Figure 1: Contour Plot and Surface Plot of Compressive Strength Vs W/C ratio, T.A. to cement ratio

Likewise the remaining graphs can also be drawn for compressive strength ,Permeability and Porosity vs W/C ratio,Total Aggregate to cement ratio and % of Fine aggregate.

References:

[1]P.T. Weiss, M. Kayhanian, L. Khazanovich, J.S. Gulliver, Permeable Pavements in Cold Climates: State of t he Art and Cold Climate Case Studies, Final Report Published by Minnesota Department of Transportation Research Services & Library, June 2015.

[2]P.D., Tennis, M.L. Leming, DJ. Akers, Pervious Concrete Pavements, EB302.02, Portland Cement Associati on, Skokie, IL, and National Ready Mixed Concrete Association, Silver Spring, MD, USA, 2004.

[3]M. Scholz, P. Grabowwiecki, Review of permeable pavement systems, Build Env (2007) 42(11): 2972-297 9. https://doi.org/10.1016/j.buildenv.2006.11.016

[4] Hasan. M and Kabir, " Prediction of compressive strength of concrete from early age test result",4th Annual Paper Meet and 1st Civil Engineering Congress, 2011, 22-24, Dhaka, Bangladesh ISBN: 978-98433-4363-5.

[5] M. Sayed-Ahmed1, "Statistical Modeling and Prediction of Compressive Strength of Concrete" Concrete Research Letters, 2012, Vol. 3(2).

DYNAMIC ANALYSIS OF RC TALL STRUCTURAL SYSTEMS AS PER IS 16700: 2017

S.B. Umesh[1], B.O. Naveen[2*], C. Suraj[1], L.S. Manjunath[1], A.B. Suhas[1]
[1] *UG Student Department of Civil Engineering, The National Instituite of Engineering Mysore, India*
[2] *Assistant Professor, Department of Civil Engineering, The National Instituite of Engineering Mysore, India*
* *e-mail: naveenbo@nie.ac.in*

Introduction

Tall buildings have fascinated humans from the beginning of civilization as evidenced by the pyramids of Giza, Egypt; Mayan temples of Tikal, Guatemala; and Kutub Minar of Delhi, India. The motivation behind their construction was primarily for creating monumental rather than human habitats. By contrast, contemporary tall buildings are primarily a response to the demand by commercial activities, often developed for corporate organizations as prestige symbols in city centers. The feasibility of tall buildings has always depended upon the available materials and the development of the vertical transportation necessary for moving people up and down the building.

Analysis of free vibrations in tall buildings such as frame, shear wall, and frame shear wall structures was carried out and the solutions of mode shapes of framed structures are treated as shear beam with varying cross sections are expressed in Bessel's functions and parameters of fundamental period are determined (Li et al., 1994). The structural efficiency of tall buildings mainly dependent on lateral stiffness and resistance capacity. Among those structural systems for tall buildings outrigger system is one of the most common and efficient systems especially for those with relatively regular floor plan. In the past outrigger systems is only used to provide additional stiffness to reduce drift and deflection. New application for outrigger systems now move to provide additional damping to reduce wind load and acceleration, and also could be used as structural fuse to protect the building under a severe earthquake condition (Goman, 2016). Tall building developments have been rapidly increasing worldwide. The evolution of tall building's structural systems and the technological driving force behind tall building developments are reviewed and for the primary structural systems, a new classification interior structures and exterior structures – is presented (Ali and Moon, 2011). While most representative structural systems for tall buildings are discussed, the emphasis in this review paper is on current trends such as outrigger systems and diagrid structures. Auxiliary damping systems controlling building motion are also discussed. Further, contemporary "out-of-the-box" architectural design trends, such as aerodynamic and twisted forms, which directly or indirectly affect the structural performance of tall buildings, are reviewed. Finally, the future of structural developments in tall buildings is envisioned briefly.

Methodology

In this paper, Reinforced concrete (RC) tall structural systems as per IS 16700:2017 has been considered. Structural systems considered for the present study are conventional RC Moment Frame System (Model 1), Structural wall system with flat slab (Model 2), Flat slab with single tower system (Model 3), Shear wall flat slab with single tower system (Model 4), and Shear wall flat slab floor with single tower outrigger system (Model 5). In this study modal analysis, equivalent static method of analysis and dynamic analysis is carried out as per IS 1893 – 2016. The multi-storey building is Reinforced Concrete (RC) structure with 180 m height, consisting of 60 floors, each of height 3 meter in zone V. All the structural systems are modeled in ETABS and key responses are extracted. The performance of the RC structural system based on the specifications and requirements of IS 16700:2017 are evaluated. Main objectives of the study is to study the behaviour of tall structural systems under lateral loads, performing dynamic time history analysis of tall structural system under considerations and evaluation of structural safety of tall buildings for structural responses. Finally, the conclusion are drawn based on results and discussion made.

Results and Concluding Remarks

From the, analysis of different tall structural systems considered, Table 1. presents the behaviour of structural systems under modal, equivalent static and dynamic time history analysis.

Table 1. Comparison of models with different parameters of analysis.

Model No./Parameter	Model 01	Model 02	Model 03	Model 04	Model 05
Time period	Very high	high	moderate	low	Very low
Frequency	low	Very low	moderate	High	Very high
Base shear	low	Very low	moderate	High	Very high
Lateral drift	moderate	high	Very high	low	Very low
Stiffness	low	Very low	moderate	high	Very high
Displacement	moderate	high	Very high	Low	Very low
Peak displacement	high	very high	moderate	low	Very low
Peak acceleration	low	moderate	very low	high	Very high

From the present study it can be concluded that, particularly in tall structural systems, the seismic demand of the structure will be more with respect to story displacement and storey drifts. Hence selecting suitable structural systems will be challenging for a structural design engineer. In the present study the Single tower shear wall with outrigger system is found to be more effective in terms of lesser time period, lesser drift in both the directions, lower joint displacement in both the directions, higher base shear in both the directions, higher stiffness in both the directions, more tendency to vibrate with high natural frequency than other systems, The other structural systems can also be used with the provision of additional structural elements like outrigger, belt truss, shear wall in order to limit the lateral load responses within the limits of code provisions, This study is mainly concentrated on preliminary investigation of tall structure.

References

Li Q, Cao H, Li G (1994) Analysis of free vibrations of tall buildings. Journal of Engineering Mechanics 120(9): 1861-1876. https://doi.org/10.1061/(ASCE)0733-9399(1994)120:9(1861)

Ali M M, Moon K S (2011) Structural developments in tall buildings: current trends and future prospects. Journal of Architectural Science Review 50(3): 205-223. https://doi.org/10.3763/asre.2007.5027

Goman W. M. Ho (2016) The evolution of outrigger system in tall buildings. International Journal of High-Rise Buildings 5(1): 21-30. https://doi.org/10.21022/IJHRB.2016.5.1.21

Bungale S., Taranath A, (2010) Reinforced concrete design of tall buildings. Taylor and francis, New York.

Experimental Investigation on Deflection and Strength Characteristics of Reinforced Concrete Slab with Recycled Plastic Fillers

Swethaa. B[1], Rekha. M[1], Ramya. S[1]
[1] *Department of Civil Engineering, Sei Lakshmi Ammal Engineering College, Chennai , India*

*swethaabaskaran@gmail.com, jasminerekha19@gmail.com, ramya.chml@gmail.com

Introduction

Plastic refuse and its disposal have been a matter of great concern since it's evolution and with the eternal increase of it's production and usage, the problem of safe disposal has gained Himalayan importance. With a view of recycling the locally available plastic refuse, and incorporating them in concrete, the existing concept of bubble deck slabs were given a new filler material, polypropelene boxes. Some researches on Bubble deck technology has been pursued by K. R. Deepan, et. Al (2017) who has experimented with polypropelene balls, Daniela Mackowa (2015), who has compared various lightweight filling methods, Apurv.S Bokil (2013), who has experimented with elliptical plastic fillers and Tina Hai (2012) , who has presented a study on the structural behaviour of RC slab with hollow spherical spheres.

The prime objective of this study is to incorporate locally available plastic refuse in concrete as cuboidal fillers for reinforced slabs as a means of recycling the non degradable waste. This study emphasises the use of simple filler blocks of rectangular cross section rather than spherical, elliptical or specially moulded hollow blocks since the manufacture of such sections is relatively easier.

Materials and Methods

The study is conducted on M20 grade of concrete, suitably designed using the stipulations of IS 10262:2009, with the slab being designed as per IS 456:2000. Polypropelene refuse are collected, sorted and cast into filler boxes of suitable sizes. The slabs are moist cured for 7 and 14 days and are tested for flexural strength and deflection characteristics using a loading setup.

Results and Concluding Remarks

From the experimental study, it was seen that the flexural strength of slabs that were cast with the fillers was 9.3% lesser than that of the conventional slab and this difference increased to almost 25% at the 14 day mark. On the other hand, the ultimate deflection at failure of the slab with the fillers is found to be 3.7% lesser than the conventional slab and the deflection is seen to further reduce to a difference of 14.3% at 14 days. Thus, it is evident from the study that, the slab with fillers shows greater resistance to deflection as compared to the conventional slab, but exhibits lesser flexural strength. Hence, such structures may be considered suitable where the loads be comparatively lesser , or in places where the dead weight has to be kept under check, with lesser importance towards it's load carrying capabilities.

References

K.R.Dheepan, S.Saranya, S.Aswini (2017) Experimental Study on Bubble Deck Slab using Polypropylene balls International Journal of Engineering Development and Research, Volume 5, Issue 4 | ISSN: 2321-9939

Arati Shetkar and Nagesh Hanche An Experimental Study on Bubble Deck Slab System with Elliptical Balls Proceeding of NCRIET-2015 & Indian J.Sci.Res. 12(1):021-027, 2015 ISSN: 2250-0138 (Online)

EFFECT OF SHEAR WALL ON SESIMIC PERFORMANCE OF HIGH-RISE BUILDINGS

V.SuryaPrathap[1*], B.Rohini[2],G.Sreenivasulu[3],Chenna Rajaram[4]

[1]Graduate Student,School of Civil Engineering,RajeevGandhiMemorial College of Engineering and Technology,Nandyal

[2,4]Assistant Professor of Civil Engineering, RajeevGandhiMemorial College of Engineering and Technology,Nandyal

[3]H.O.D, School of Civil Engineering,RajeevGandhiMemorial College of Engineering and Technology,Nandyal

email:drchenna78@gmail.com

INTRODUCTION

Emporis Standards defines a high-rise as "A multi-story structure between 35–100 meters tall, or a building of unknown height from 12–39 floors". There are different structural components unique to a high-rise structure one of them is shear wall. Shear wall is a vertical structural member in a reinforced concrete framed structure to resist lateral forces and used in high-rise buildings. In building construction, a rigid vertical diaphragm capable of transferring lateral forces from exterior walls, floors, and roofs to the ground foundation in a direction parallel to their planes. They also provide adequate strength and stiffness to control lateral displacements.

Past researchers have demonstrated different research methods for the analysis of High-rise building, like Comparison of concrete tall building behaviour using an intermittent shear walls form in one frame with continuous shear walls (Davaran et.al.,(2006)), Seismic Performance of Shear-Wall and Shear-Wall Core Buildings Designed for Indian Codes(Surana M et.al., (2015)).Also an investigation on dynamic behaviour of shear walls on flexible foundation(Fard D.M.Y et.al., (2006)) is done.

The paper discusses about different ways in which a high-rise structure exhibits its structural behaviour in terms of its resistance to lateral loads. The analysis is carried out four different structures types open frame structure, building with shear walls at sides, building with shear walls at corners, and building with infill walls. The analysis considers a G+30 storey and designed as per IS456:2000. The structure is modelled in SAP2000, a Finite Element Method software. The structure is subjected to KTP ground motion of the 2015 Nepal earthquake. Further, the analysis is carried out using push-over analysis. The analysis results will be discussed in the following section.

METHODOLOGY

Time history analysis is a step-by- step analysis of the dynamic response of a structure to a specified loading that may vary according to the specified time function. It is used to determine the seismic response of a structure under dynamic loading of representative earthquake.

The data for Nepal Earthquake 2015 is obtained from four different stations and time history analysis is done. The output of this analysis will give us displacements of the structure for each earthquake mode. These displacement values are incorporated as inputs for displacement-controlled pushover analysis. Now, Pushover is a static-nonlinear analysis method where a structure is subjected to gravity loading and a monotonic displacement-controlled lateral load pattern which continuously increases through elastic and inelastic behaviour until an ultimate condition is reached.

The analysis is carried out in displacement control with 4.0 % drift. These drifts are applied in SAP 2000 on a point on top storey. The max displacement is then taken from the results of time history analysis and applied to the structures in push over analysis to obtain the pushover curve. This analysis will also give us data about plastic hinges and its location. Now the damage can be obtained from a set of elastic and inelastic equations through which the obtain results are subjected to. Once the damage assessment is done we go to retrofitting phase where we decide whether the building requires retrofitting or not and to how much extent. Once the retrofitting is done the analysis is repeated to check the percentage of improvement after repair.

RESULTS

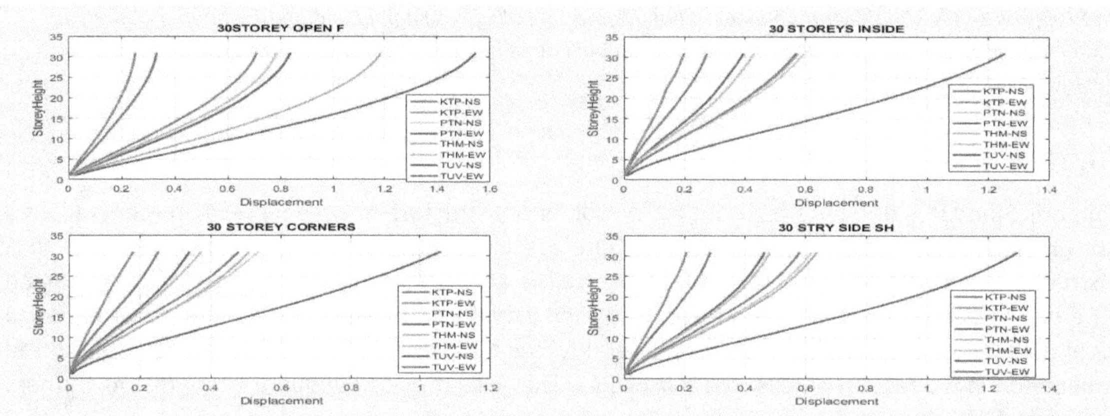

Figure 1: displacements after applying ground motion from Nepal Earthquake 2015

Figure one represents the displacement of structures models with respect to storey height when Time History Analysis is done from data obtained from Nepal Earthquake,2015 from four different stations. The curves on the extreme which show max displacements are obtained as such because of the frequency of the ground motion are in predominant mode of vibration w.r.t frequency of the structure.

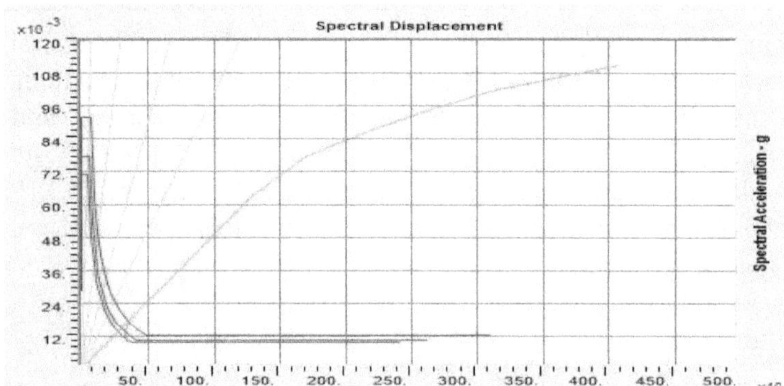

Figure 2: pushover curve by displacement control obtained from ground motion data.

Figure two showing the pushover curve obtained by displacement control from applying displacement obtained from ground motion data. The damage assessment is done after this after which we conclude the retrofitting techniques. With this the percentage of improvement can also be assessed.

References

Davaran A., Yousefzadeh Fard M., Nayeri Amiri S., Kashefi A. (2006) Comparison of concrete tall building behavior using an intermittent shear walls form in one frame with continuous shear walls

Surana M., Singh Y., Lang D.H. (2015) Seismic Performance of Shear-Wall and Shear-Wall Core Buildings Designed for Indian Codes

Fard D.M.Y., Amiri S.N., Kashefi A. (2006) An investigation on dynamic behaviour of shear walls on flexible foundation

Lee DG. (1988) An Efficient Element for Analysis of Frames with Shear Walls. In: Atluri S.N., Yagawa G. (eds) Computational Mechanics '88

Bungale S. Taranath Ph. D, (2010) Reinforced Concrete Design of Tall Buildings, Taylor and Francis group, Bacon Raton

Abstract ID: ITCSD-ES063

International Conference on
Innovative Trends in Civil Engineering for Sustainable Development (ITCSD - 2019)

Estimation of Seismic damage and retrofitting measures of ReinforcedConcrete framed buildings

A. Naveen[1], Chenna Rajaram[2]
[1]*Graduate student, School of Civil Eng., Rajeev Gandhi Memorial College of Eng. and Technology, Nandyal, India.*
[2]*Asst.professor, School of Civil Eng., Rajeev Gandhi Memorial College of Eng. and Technology, Nandyal, India.*
e-mail: drchenna78@gmail.com

INTRODUCTION

Earthquakes represent one of the most harmful and fateful natural disasters for humans. Over the last few decades, numerous earthquakes were responsible for the death and injury of lakhs of people as well as the damage or even collapse of many buildings. In particular, many reinforced concrete buildings, which compose a large part of worldwide construction portfolio, buildings undergo distress and very poor performance of buildings when subjected to seismic activities. The buildings which do not fulfil the requirements of seismic design, may suffer extensive damage or collapse against earthquake actions. The seismic evaluation reflects the seismic capacity of earthquake affected buildings for the study of future design practices against earthquakes.

Past researchers have demonstrated different strengthening methodologies and introduced some design practices. For example, strengthening of open ground storey RC structure by jacketing of columns, designed buttresses and addition of extra columns, a rational method was developed for the determination of strength requirement of open first storey (Hemant B. Kaushik, Durgesh C. Rai., at al 2009). The analysis compares seismic design practices of 16 different countries related to masonry infill RC frames and addresses the different issues (Hemant B. Kaushik, Durgesh C. Rai., et al 2005). Two existing damaged RC buildings due to 2011 Turkey earthquake were considered. A Non-linear time history performed using van earthquake ground motion record. The existing damages are compared with analytically computed damage and also evaluated the effect of shear walls of seismic behaviour of buildings (Tulay Aksu Ozkula, Ahmet Kurtbeyoglub., et al 2019).

In the present study, investigations were carried out for five storey (G+4) with five bays of RC building with bay width and height of building is 3m and designed as per IS 456:2000 Codal provisions. Push-over analysis is carried out for bare frame and open ground storeys to know the capacity of the structure. Later, the frame is strengthened with jacketing of columns, shear walls, shear walls with the variation of thickness and reinforcement ratios and bracings. At the end, the damage is calculated through energy methods (Rajaram et al., 2015; Vimala et al., 2014).

METHODOLOGY

Pushover analysis has its own advantageous for the seismic performance evaluation of structure because it is simple procedure to apply. Pushover analysis is a static Non-linear analysis method where the structure is subjected to monotonically increasing Lateral load under permanent vertical loads, the distribution is proportional to height raised to power 'k' (equivalent to 2). It helps in understanding the formation of flexural, shear cracks and plastic hinges in the structure. The non-linear static pushover analysis is performed using IDARC (Inelastic Damage Analysis of Reinforced Concrete), which requires knowledge of stress-strain model, plastic hinges and its location, moment-curvature relationship.

In this study, the analysis is carried out in displacement control with 4.0 % drift. These drifts are applied onto the structure in uniform pattern from ground floor to top floor. A step by step procedure of moment curvature relationship is implemented in the IDARC, it is carried out on the cross section ofconcrete by dividing fibres. The section is subjected to increments of curvature and the strain distribution is obtained from compatibility and equilibrium considerations.

RESULTS AND CONCLUDING REMARKS

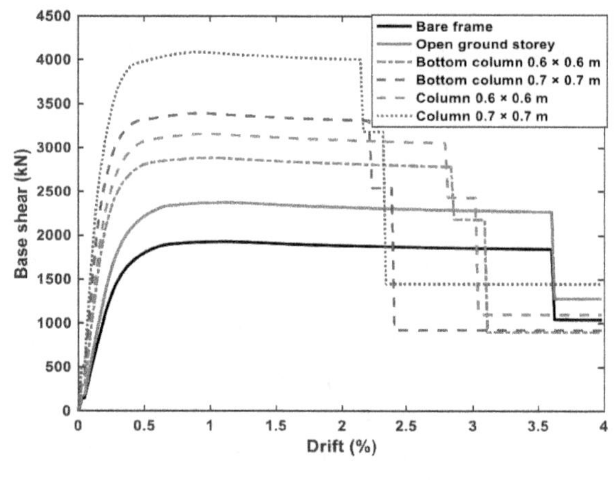

Fig 1: Pushover curves

Fig 2: Damage curves

The results are expressed in terms of pushover curves and damage. Figure 1 represents pushover curves of the frame before and after retrofitting. The Jacketing of columns is done by increasing the size of the columnsfrom 0.5 × 0.5 m to 0.6×0.6m, and 0.7×0.7m. It is observed that the base shear is increased to 30% and 50% for jacketing of column member at bottom from 0.5 × 0.5 m to 0.6×0.6m and 0.7×0.7m, respectively. Also, it is found that the base shear is increased to 45% and 80% for increasing the size of columns members from top to bottomfrom 0.5 × 0.5 m to 0.6×0.6m and 0.7×0.7m, respectively. At the end, the results are expressed in terms of damage through energy approach method (Rajaram et al., 2014; Vimala et.al, 2015).Figure 2 represents the damage curves developed for the bare frame before and after retrofitting. It can be concluded that,it is observed from the damage curves that the damage of the frame is reduced effectively after retrofitting of frame with jacketing of 0.7 × 0.7 m at bottom instead of retrofitting of all columns from top to bottom.

REFERENCES

Hemant B. Kaushik; Durgesh C. Rai; and Sudhir K. Jain, M.ASCE (2009)." Effectiveness of Some Strengthening Options for Masonry-Infilled RC Frames with Open First Story" Journal of structural engineering, 135(8), 925 – 937.

Kaushik, H. B., Rai, D. C., and Jain, S. K. (2006). "Code approaches to seismic design of masonry infilled reinforced concrete frames: A state of- the-art review." *Earthquake Spectra*, 22(4), 961–983.

Chenna Rajaram, Ramancharla Pradeep Kumar, Ajay PratapSingh, Kapil Mohan and Bal Krishna Rastogi," seismic damage estimation for port building: an energy-based approach" ISSE JOURNAL, 16(4), 13 – 24.

Tulay Aksu Ozkula, Ahmet Kurtbeyoglub, Muzaffer Borekcic, Basak ZengindAli Kocakc," Effect of shear wall on seismic performance of RC frame buildings" Engineering failure analysis 100, 2(32), 60 – 75.

Das, D., and Murty, C. V. R. (2004). "Brick masonry infills in seismic design of RC framed buildings. Part 2: Behaviour." *Indian Concrete Journal.*, 78(8), 31–38.

Crisafulli, F. J., Carr, A. J., and Park, R. (2000a). "Analytical modelling of infilled frame structures—A general review." *New Zealand Nat. Soc.Earthquake Eng. Bull*, 33(1), 30–47.

Title of the paper: Effect of silica modulus of sodium silicate solution on the Flyash and Ggbs based Geopolymer concrete

G.Vikas[1], T.D.Gunneswara Rao[2*]
[1]Research Scholar, Department of civil Engineering, National Institute of Technology, Warangal, India
[2]Professor, Department of civil Engineering, National Institute of Technology, Warangal, India
** e-mail: gugulothuvikas@gmail.com*

Introduction

The use of Geopolymer concrete to replace conventional concrete is becoming an environmentally friendly alternative construction technique. Activation of fly ash and Ground granulated blast furnace slag (GGBFS) with sodium hydroxide and sodium silicate solution is well described in the literature. But the alkaline activation of the GGBFS alone provides a concrete with a lower processing capacity and a quick setting character. Mallikarjuna Rao et al. 2015 reported the final setting time of GGBS paste prepared with alkaline solution having sodium hydroxide of 8M and 16M are 40 min and 60 min respectively. Therefore, there is a need of alternative activator in order to arrest the setting when GGBFS alone is used as binder. In this investigation, fly ash and GGBFS are used in different percentages for preparing the geopolymer mortars using Sodium silicate solution (waterglass) alone of two different silica modulus 1.99(Activator) and 2.92(Retarder) to assess the ability of this solution in arresting the quick setting aspect of GGBFS, consistency and setting time studies are conducted.

Previous studies have shown that higher silica modulus of sodium silicate can act as reactive medium as well as retarder by itself. Therefore in the present study, sodium silicate of silica modulus M_s 1.99 and 2.92 are selected as activators in the preparation of the Geopolymer mortars without using any additional chemical admixtures. Experimental program is also conducted to assess the strength development of mortars for different flyash-ggbs ratio(100-0, 50-50, 0-100) and different solutions(Ms 1.99 and Ms 2.92). The results indicated that these alkaline solutions increases the setting time and hence can be used in concreting.

Materials and Methods

Fly ash obtained from the NTPC Ramagundam, India and GGBS obtained from Toshali cements, Vizag, India was used in this work having specific gravity of 2.17 and 2.9 respectively. The two sodium silicate solutions used for alkaline activation in the present investigation has a silica modulus of SiO_2: Na_2O (M_s) = 1.99 with 28.98% SiO_2, 9.92% Na_2O and SiO_2: Na_2O (M_s) = 2.92 with 28.98% SiO_2, 9.92% Na_2O by weight which was procured from KIRAN GLOBAL Ltd, Chennai, India. River sand and crushed granite of 16mm nominal size were used as fine and coarse aggregates conforming specifications of IS 383: 2002.

As per IS: 4031 (Part IV & V), normal consistency and setting time tests of Geopolymer paste are determined using the Vicat's apparatus. Twelve cubes of each geopolymer mortar set with dimensions 100 mm×100 mm×100 mm are cast and tested in compression to determine 1, 3, 7 and 28-day compressive strength. Concrete samples are taken out of the moulds 24 h after casting and are left at normal room temperature curing conditions (27 ± 2°C, relative humidity 70%) till the day of the test.

Results and Concluding Remarks

Consistency and setting time- The normal consistency value is observed to be increasing with the GGBS percentage in the binder for both the cases. These values are observed to be relatively higher than that of OPC. The initial and final setting times are observed to be decreased with the increase in the GGBS, but the setting is slow for the paste Ms 2.92 compared to that of Ms 1.99. It is also observed that for 100% flyash paste the final setting time value for both the solutions is nearly the same but for 100% GGBS paste, the setting is more for Ms 2.92 i.e, 190 minutes when compared to Ms 1.99 i.e., 42 minutes.. This clearly shows that the solution Ms 2.92 gives a better workability for 100% GGBS mix.

Table 1. Consistency and setting times of Geopolymer pastes prepared with Activator and Retarder

Flyash-Ggbs(%)	Consistency(%)		Initial setting time(min)		Final setting time(min)	
	Ms 1.99	Ms 2.92	Ms 1.99	Ms 2.92	Ms 1.99	Ms 2.92
100-0	35	43	410	480	830	970
50-50	39	47	85	155	170	310
0-100	45	51	20	95	42	190

Compressive strength of mortars- The compressive strength is observed to be increasing with the increase in the GGBS percentage in the binder for both the solutions (Figure 1& 2). It can also be observed that the compressive strength is comparably more for the mortars prepared with the solution Ms 2.92.The use of sodium silicate activators leads to the formation of binders whose structure is mostly a calcium silicate hydrate(C–S–H) gel, which is the reaction product mainly responsible for increase inthe strength (Puertas F et al., 2011).

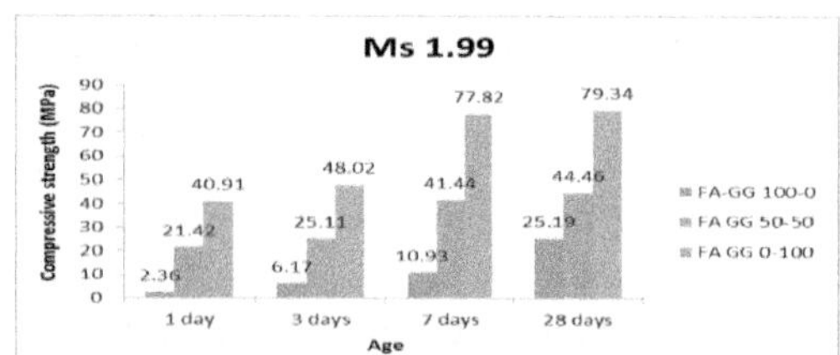

Figure 1. strength w.r.t Age for solution Ms 1.99

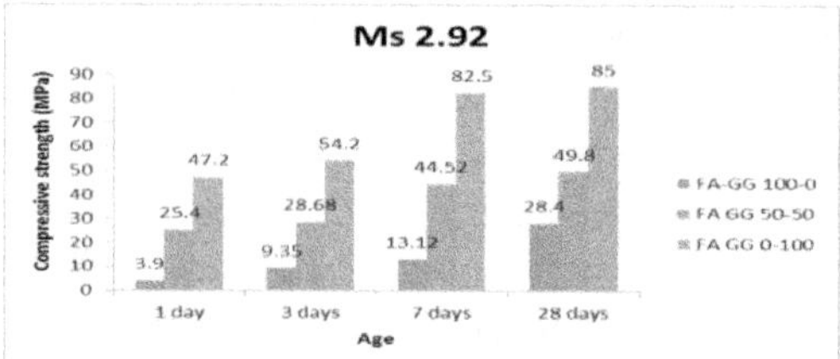

Figure 2 strength w.r.t Age for solution Ms 2.92

Results indicated that the range of compressive strength for these flyash-ggbs ratio for solution 1.99 varied from 25MPa to 80MPa and for 2.92, it varied from 28MPa to 85 MPa. 1 day strength for 100% flyash mix is observed to be 11.5% of 28 day strength and for 100% Ggbs, it is 53.54%. Half of the 28 day strength has been achieved in one day. The rate of gain of strength in flyash mortars is observed to be slow initially but it is more for later age whereas for ggbs, it is high initially and slow with the increase in the age. The results of present investigation have shown that Geopolymer mortars can be developed using sodium silicate solution of silica modulus 1.99 and 2.92 adopting ambient curing, which is more suitable for application in construction industry.

References

Douglas E, Bilodeau A (1991) Alkali Activated Ground Granulated blast-furnace slag concrete:Preliminary investigation. Cement and Concrete Research. V 21 P 101-108. https://doi.org/10.1016/0008-8846(91)90036-H

Glukhovsky V D (1980) High strength slag-alkaline cements. 7th Int congr Chem cem Paris, 3 , pp. V164-V168

IS: 4031–1988 (PART 4) (1988), Method of Physical Test for Hydraulic Cement—Determination of Standard Consistency of Cement Paste. Bureau of Indian Standards, New Delhi

IS: 4031–1988 (PART 5) (1988) Method of Physical Test for Hydraulic Cement: Determination of Initial and Final Setting Time. Bureau of Indian Standards, New Delhi

Mallikarjuna Rao G, Gunneswara Rao T D (2015) Final Setting Time and Compressive Strength of Fly Ash and GGBS-Based Geopolymer Paste and Mortar. Arab J Sci Eng. 40:3067–3074. https://doi.org/10.1007/s13369-015-1757-z

Malolepszy J, Petri M (1986) High strength slag-alkaline binders. In Proceedings of the 8th International Congress on the Chemistry of Cements, 4, 108-111.

Puertas F, Varga C (2014) Rheology of alkali-activated slag pastes. Effect of the nature and concentration of the activating solution. Cement and concrete composites. V 53 P 279-288. https://doi.org/10.1016/j.cemconcomp.2014.07.012

Serdar A (2014) Effect of activator type and content on properties of alkali activated slag mortars. Composites part B:Engineering v57 P 166-172. https://doi.org/10.1016/j.compositesb.2013.10.001

Talling B, Brandstetr J (1989) Present state and future of alkali-activated slag concretes. ACI Special Publication. V 114 P 1519-1546

Wang S D, Karen L (1994) Factors affecting the strength of alkali-activated Slag. Cement and Concrete Research. V 24 P 1033-1043. https://doi.org/10.1016/0008-8846(94)90026-4

Effects of Ground Granulated Blast Furnace Slag on Properties of Conventional Concrete

S Manissa[1*], K C Panda [2]
[1] *ITER, Siksha 'O' Anusandhan (Deemed to be University), Bhubaneswar, Odisha, India*
[2] *Professor, Department of Civil Engineering, Government College of Engineering, Bhawanipatna, Odisha, India*
* *s.manissa1993@gmail.com*

Introduction

In recent year, a universal trend has been occurred for advancement of durable, low-priced and pioneering construction materials with respect to conventional materials for building low cost, sustainable and energy proficient structures. The manufacturing process of cement emits considerable amount of carbon dioxide (CO_2). Therefore is an urgent need to reduce the usage of cement. Ground Granulated Blast furnace Slag (GGBS) is a by-product from steel industry. It has good structural and durable properties with less environmental effects. Some of the previous authors worked on GGBS based concrete such as Chidiac and Panesar (2008) carried out an investigation on concrete with GGBS as a replacement to cement. It was observed that the mechanical properties of concrete were influenced by the water content, quantity of GGBS and the curing period. The compressive strength improved as water binder ratio increased, due to better hydration. The increase in compressive strength was around 10% - 20% between 28 days to 120 days. Awasare and Nagendra (2014) experimental conducted on M20 Grade concrete by using partial replacement of cement with GGBS and found that at 30% replacement of cement with GGBS, achieves the maximum strength. Suresh and Nagaraju (2015) pointed out the usage of GGBS serves as replacement to already depleting conventional building materials and also used as an Eco Friendly way of utilising the product without dumping it on the ground. Exploitation of this material is not only decreasing the cost but also enhance the self compatibility. It reduced the labour cost, save time, increase design flexibility and durability of concrete. This paper presents the fresh and hardened properties of conventional concrete (CC) using GGBS as partial replacement of cement.

Materials and Methods

The materials used in this study are cement (OPC 53) Grade, natural coarse aggregate (NCA), natural fine aggregate (NFA) and GGBS. All the materials are tested as per the specification of Indian Standards (IS). Sand is used as natural fine aggregate which is passing through IS 4.75 mm sieve. Natural coarse aggregate (NCA) consists of rock fragments that are used in their natural state and is passing through IS 20 mm sieve. The specific gravity of NCA is 2.86 and fineness modulus 7.0. The concrete mix design is executed for the preparation of test samples (cubes, cylinders and prisms) as per guidelines IS: 10262-2009. The mix design is targeted for M40 grade of concrete and the water cement ratio is 0.40. The mix proportion for normal concrete mix is 1: 1.36: 2.87. In this study, total four mixes were prepared for conventional concrete mix with 0, 10, 20, 30% partial replacement of cement with GGBS.

Finally, tests have been organized to find out compressive strength, split tensile strength, flexural strength. Four different mixes of concrete mixtures were made by replacing 0, 10, 20, and 30% of cement with GGBS. Four mixes such as CM0G0, CM0G10, CM0G20 and CM0G30. CM0G10 indicate 90% cement, 10% of GGBS, CM0G20 indicate 80% Cement and 20% GGBS, CM0G30 indicate 70% of Cement, 30% GGBS. The individual materials were weighed and placed in the concrete mixer and it is thoroughly mixed in dry condition until the mixture become homogeneous. The mixing procedure is same for all the cement paste. The specified water of required amount for respective mix is then added during mixing. After mixing, the concrete were casted in steel mould and compressed using vibrator machine and remoulded after 24 hours then the specimen cured under water for 7, 28 and 90 days.

Results and Discussions

The experimental result of conventional Concrete (CC) for fresh concrete properties and hardened concrete properties is provided. The fresh concrete properties of CC i.e. slump test is carried out. The hardened concrete properties such as compressive strength, split tensile strength and flexural strength test is carried out. Figure 1 represents the compressive strength for different concrete mix for different proportion of GGBS.

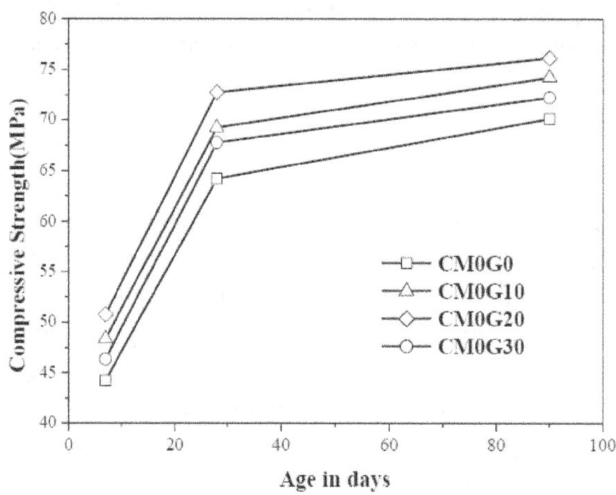

Figure 1: Compressive strength for different concrete mix for different proportion of GGBS

As the fresh concrete properties concerned; the slump value decreases as the partial replacement of cement with GGBS increases, so that the workability of concrete decreases. As the hardened concrete properties concerned; the compressive strength, split tensile strength and flexural strength increases as partial replacement of cement with GGBS increases up to 20%, thereafter the strength decreases. Whereas for 30% replacement of cement with GGBS, the compressive strength, split tensile strength and flexural strength is higher than the conventional concrete.

Concluding Remarks

The following conclusions may be drawn from the present study:

- The workability of normal conventional concrete mix is higher than the other concrete mixes. As the replacement of cement with GGBS increases, the workability decreases.
- The compressive strength, tensile strength and flexural strength of CMOG20 sample means 20% of GGBS is giving higher value than other three specimens.

Acknowledgments:

The author would like to thank ITER, Siksha 'O' Anusandhan (Deemed to be University), for the support of conducting the experimental work.

References

IS:10262:2009, Recommened guidelines for concrete mix design. BIS,India.

Chidiac S E, Panesar DK (2008) Evolution of mechanical properties of concrete containing ground granulated blast furnace slag and effects on the scaling resistance test at 28days. Cement and Concrete Composites 30 (2): 63-71.

Awasare V, Nagendra MV (2014) Analysis of strength characteristicsof GGBS concrete. International Journal of Advanced Engineering Technology. Vol. V, Issue IV: 82-84.

Suresh D, Nagaraju K. (2015) Ground Granulated Blast Slag (GGBS) in Concrete – A Review. IOSR Journal of Mechanical and Civil Engineering (IOSR-JMCE) 12(4): 76-82.

Numerical Study on Bolted Moment End Plate Connection Joining Hollow Tubular Sections

V Ramana Kollipara[1*], Prof. T.D. Gunneswara Rao[2]
[1 2] Department of Civil Engineering, National Institute of Technology,Warangal, India
** laxman.satya51@gmail.com*

Introduction

Steel framed structures consist of distinct members connected to each other for transfer of forces and moments from beam to column and beam to beam. Connections plays critical role in transfer of these loads without disturbing the effectiveness and durability of structures, structural elements. These are may be welded or bolted type. Especially bolted moment end plated connections, those are used mostly in all types of steel structures. As such that the connection designs impart the performance of the whole structure. Bolted end-plate type moment connections are used extensively in steel structures, e.g., buildings, bridges, water tanks and transmission towers. They have the advantages of requiring less supervision and a shorter assembly time than welded joints. The use of end-plate moment connections in multi-story, moment resistant frame construction is becoming more common because of advancements in design methods and fabrication techniques, both of which have resulted in decreased costs.

A numerical investigation is conducted to study the behaviour of extended end plate connections joining square hollow tubular sections as beam-beam splice connection. The deformation characteristics of extended end-plate connection is determined under pure bending condition. The numerical study was conducted using finite element package ANSYS. In this study a four bolted connection configuration is considered and also a parametric study is conducted by varying the parameters like thickness of end plate, diameter of bolt and dimensions of hollow tubular section. The behaviour of connection was studied based on its failure mechanism. Generally, the failure in the connection is caused due to the failure of bolts, failure of end plate, failure of specimen (hollow tubular section).Finally, the obtained numerical results are validated by comparing with the corresponding experimental results that are conducted at the structural engineering laboratory of NIT Warangal.

Materials and Methods

The numerical study consists of testing of hollow tubular beam-beam splice connection under the loading of pure bending by using ANSYS software. To attain the splice connection between beam-beam, the hollow tube is welded to one side of end plate and then end plate bolted to another end plate having same configuration. Finally, the fabricated member is subjected to four points loading as pure flexure loading. The materials used in numerical study is mild steel material with grade of 210MPa. This study involves the extended end plate type connection and the end plates are rectangular in shape. Two different diameters of bolts (M6 and M12) hexagonal bolts are used according to Indian standards. The position of bolts with respect to the flange and web of the section and the dimension details are given in Table 1 and connection configuration is shown in Figure1.

Table 1. Dimension Details of End Plate Connection

S.No.	Section	Dimensions of End Plate(mm)			Position of Bolt from flange or web of section(mm)			Diameter of Bolt (mm)
		W_p	D_p	t_p	a	c	s	
1	HTS1	75	110	6	10	0	20	6
2	HTS2	75	110	6	10	0	20	6
3	HTS2	100	140	5	20	0	20	12

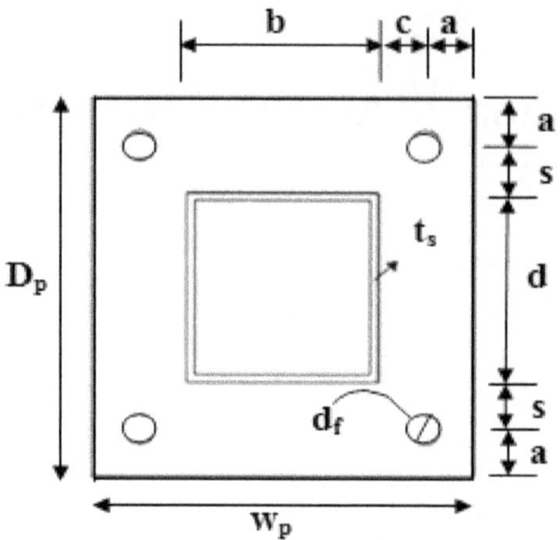

Figure 1. Configuration of End Plate.

Results and Concluding Remarks

Based on numerical study and comparison of numerical results with experimental data we observe the following conclusions:

- ✓ In all cases a good correlation is observed between both numerical and experimental results.
- ✓ In bolt failure specimen case we observed that more energy dissipation value in numerical study than experimental. This variation is may be due to experimental errors.
- ✓ The percentage variation between energy dissipation values calculated based on both experimental and numerical study have a less variation 10-15% only.

References

Ramana Kollipara, TDG Rao (2019) Experimental and Analytical Study for Plastic Moment Capacity of Beam–Beam Splice Connection. International Journal of Steel Structures, pp. 1-7.

Packer JA, Morris LJ (1977) A limit state design method for the tension region of bolted beam-column connections. *The Structural Engineer*, *55*(10):446-458.

Bayraktar D, Ay Z, Celik ID (2016) Experimental and numerical study on moment-rotation relations of welded end-plate tubular connections. *Journal of Constructional Steel Research*, *126*:63-73.

Wheeler AT, Clarke MJ, Hancock GJ, Murray TM (1998) Design model for bolted moment end plate connections joining rectangular hollow sections. *Journal of Structural Engineering*, *124*(2):164-173.

Wheeler AT, Clarke MJ, Hancock GJ (2000) FE modeling of four-bolt, tubular moment end-plate connections. *Journal of Structural Engineering*, *126*(7):816-822.

Vibration based structural damage detection using model updating techniques

Ashwin Ashok[1*], S. Arun[2]
[1] *M.Tech student, National Institute of Technology Karnataka, Surathkal, India*
[2] *Assistant Professor in Civil Engineering, Govt. Engineering College, Thrisssur, Kerala, India*
** e-mail: ashwinnashok@gmail.com*

Introduction

Structural Health Monitoring (SHM) aims to give a real time monitoring of the structure throughout its life, that helps in the diagnosis of the state of the structure. Traditional vibration-based damage detection studies employ natural frequencies and mode shapes as the damage indicators. Methods based on the measurement of natural frequencies are very attractive since this parameter can be determined by measuring at only one point of the structure. For localization of damage, techniques which makes use of higher derivatives of mode shapes have been employed. Such methods in fact requires a two-stage analysis, the first being the extraction of modal properties from structure dynamic response followed by the numerical computation of mode shape derivatives for estimation of mode shape curvatures. Hence the possibility of effect of damage on damage indicators getting masked are higher.

The present study explores the possibility of model updating techniques for identification and localization of damage. The stiffness properties of the numerical model of a base line structure (without damage) is updated based on time history of dynamic response obtained from a structure with simulated damage. The objective of the study includes development of suitable optimization algorithms for updating model for localising and detecting the extent of damage. The efficacy of these algorithms in detecting and localizing damages demonstrated for the case of simple portal frames.

Methodology

The model of the prototype structure was created and the parameters of the model were updated using measured vibration data by solving an optimization problem. The proposed methodology as shown in Figure 1, involves modelling the structure using Direct Stiffness Method, extraction of response using Newmark's Time integration method, and determining the location and extent of damage by solving optimization problem using Genetic Algorithm. A two-parameter optimization problem was adopted for damage detection. The extent of damage was simulated by a reduction in stiffness value of the element member. For damage localisation, the structure was discretized into a fixed number of elements and the element number was kept as the second variable in the optimization problem. The process of optimization involved minimization of an error norm defined as the difference between the measured response of the damaged structure and the simulated response from the numerical model of the structure.

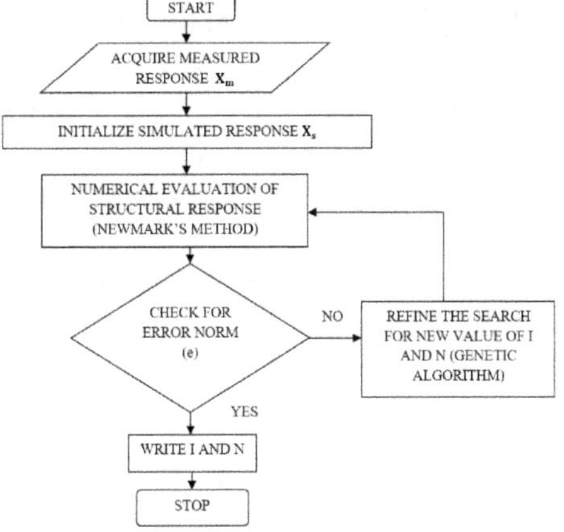

Figure 1. Flowchart of proposed methodology

The methodology was implemented in MATLAB. The efficiency of the detection algorithm was analysed by the determination of the minimum extent of damage that can be detected and the optimum number of sensors that are required for damage detection. Numerical experimentation was also conducted for the case in which the measured vibration response was corrupted by white Gaussian noise.

Results and Concluding Remarks

The damage was varied in a particular element and the relation between the extent of damage and the error in identified parameter representing the damage was studied. The algorithm was successful in detecting a damage extent of 5%. The error in damage identification increased as the extent of damage became smaller, as shown in Figure 2. In the case considered for simple portal frame, the optimum number of sensors to be placed at the nodes were found to be 3. For responses collected from 6 nodes, with noise in the measured vibration data, the minimum extent of damage detected by the algorithm was 10%.

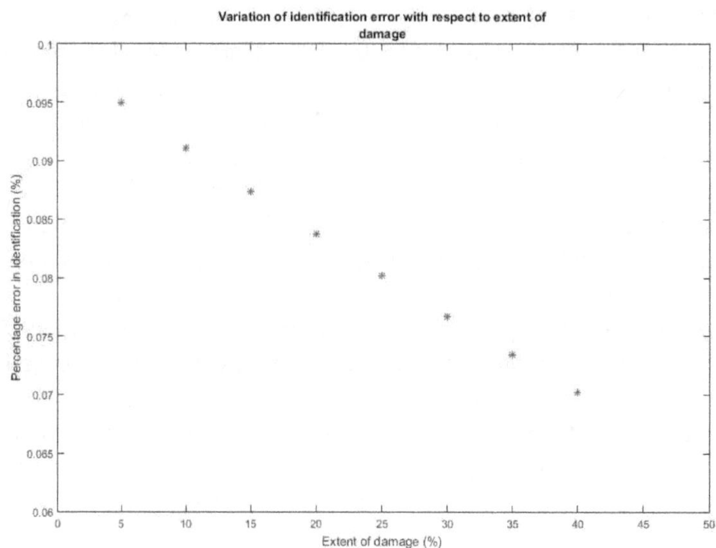

Figure 2. Variation of error in damage identification with extent of damage

Acknowledgments: This work was carried out as part of under-graduate project work of the first author. The contribution of other batch mates (Anju, Ajil, Aparna and Harshith) is acknowledged.

References

[1] Gere and Weaver (1990) Matrix Analysis of Framed Structures. Springer, US

[2] Finite element procedures (1996) Klaus-Jürgen Bathe. Prentice Hall International, US

[3] Brownjohn, J. M. W., and Xia, P. Q. (2000) Dynamic assessment of curved cable-stayed bridge by model updating. J. Struct. Eng., 126(2), 252–260

[4] Zhang, Q. W., Chang, C. C., and Chang, T. Y. P. (2000) Finite element model updating for structures with parametric constraints. Earthquake Eng. Struct. Dyn., 29, 927–944.

[5] Sohn, H., Farrar, C., Hunter, N. and Worden, K. (2003) A Review of Structural Health Monitoring Literature: 1996–2001, Los Alamos National Laboratory report, (LA-13976-MS)

[6] Rajasekaran S., VijayalakshmiPai G.A. (2003) Neural Networks, Fuzzy Logic and Genetic Algorithm. PHI Learning Pvt.Ltd , New Delhi

[7] Fan, Wei, and PizhongQiao (2011) Vibration-based damage identification methods: a review and comparative study. Structural Health Monitoring 10.1 (2011): 83-111

Evaluation of seismic damage parameters for reinforced concrete frame buildings.

Rachel Beulah.E[1], Chenna Rajaram[2]*

[1] *Graduate Student, School of Civil Engineering, Rajeev Gandhi Memorial College of Engineering and Technology, Nandyal, INDIA*
[2] *Assistant Professor, School of Civil Engineering, Rajeev Gandhi Memorial College of Engineering and Technology, Nandyal, INDIA*
* *e-mail: drchenna78@gmail.com*

Introduction

Schools are a gathering place for the new generation. They are anticipated to act as a refuge after an Earthquake. On April 25, 2015, a severe 7.8 Mw, shallow Earthquake with a focal depth of 15 km struck Nepal, Kathmandu, which is at a distance of 8 km from the epicentre. The field survey shows that 12 schools with various types of structures, including local characteristics reinforced concrete frames, mud-bonded or cement bonded timber and masonry frames were damaged due to an Earthquake. The Sankhu school building lies on the top of the Manichud Jatra Naghi hill near shivapuri National park, 25 kilometres north-east from Kathmandu. This school building is located 82 km from the epicentre of the 2015 Nepal earthquake. The building was severely damaged during the earthquake. Many researchers are investigating to find out the damage of the school building and few studies are discussed.

A detailed damage assessment has beendone using sensing technologies. In this study, 32 terrestrial laser scans are used to create a 3-dimensional point cloud of the building with the locations and geometry of the major cracks identified (Supratik Bose et.al., 2016).Various typologies of 12 school buildings have been selected to summarize the seismic damage of buildings. The effects of forward directivity, site configuration, earthquake sequence and a safe distance from a slope on to the buildings has been discussed. It is found that the major reason for the mega lost during these earthquake sequences could be soil foundation interaction, and few design problems (Hao Chen et al., 2017).During the field survey of the 2011 great east Japan earthquake in Miyagi, typical damage has been observed for RC school buildings. The reinforced concrete structures performed well, and the effect of the seismic retrofit has been found in the mitigation of damage, although severe damage to some seismically retrofitted buildings are noticed (Masaki et.al, 2012). Forthe current study, Sankhu school building is taken to assess the damage of the building due to near-field and far-field ground motions. The damage is expresseddue to forward directivity, fling effects, and near-field ground motions.

Methodology

A four storeyed two-dimensional structure of sankhu school building is initially designed as per IS 456:2000 and the building is modeled using Inelastic Dynamic Analysis of Reinforced Concrete Buildings (IDARC), a FORTRAN based tool, initially developed by Reinhorn in early 1990's. This tool is based on Finite Element Method (FEM). The building consists of 32 rectangular columns of size 450×450 mm and 28 beams of size 250×250 mm with each storey height of 3000 mm. The building is subjected to 38 ground motions from three earthquakes (1991 Uttarkashi, 2011 Sikkim and 2015 Nepal earthquakes). Out of which 10 records are near field ground motions and 28 records are far field ground motions. Also, the analysis is considered a fling effect, and directivity effects.

The general Dynamic equation for a building is given below.

$$[M]\{\ddot{U}\}+[C]\{\dot{U}\}+[K]\{U\}=-[M]\{\ddot{U}_g\} \qquad (1)$$

Where [M] is mass matrix; [C] is damping matrix; [K] is nonlinear stiffness matrix; ΔUand its derivatives are the incremental displacement, velocity and acceleration vectors respectively. The above equation is solved numerically using Newmark'sβ algorithm with average acceleration method (Chopra, 2001). Rayleigh proportional damping is used in this analysis and the coefficients αM and αK are calculated as follows:

$$[C]=\alpha_M[M]+\alpha_K[K] \qquad (2)$$

$$\alpha_M=\frac{2\xi_i\omega_i\omega_j^2-2\xi_j\omega_j\omega_i^2}{\omega_j^2-\omega_i^2};\quad \alpha_K=\frac{2\xi_j\omega_j-2\xi_i\omega_i}{\omega_j^2-\omega_i^2} \qquad (3)$$

Where $\xi_{i\ or\ j}$, and $\omega_{i\ or\ j}$ are the critical damping ratio and the circular frequency for mode 'i' and 'j'. In the IDARC program, the circular frequencies corresponding to the first and second modes are used for the Rayleigh damping type.

Results and Concluding

The below fig1 shows that the waves have displaced from the datum line(DL) indicating that the structure has entered into the nonlinear inelastic zone. Fig 2 indicates the damage of the structure due to nonlinearity. The figure 1, 2 are plotted from Uttarkashi ground motion, showing the significant damage to the structure.

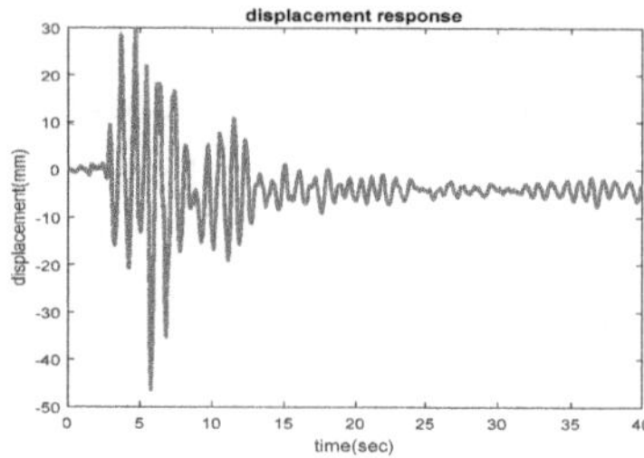

Figure 1: Displacement response

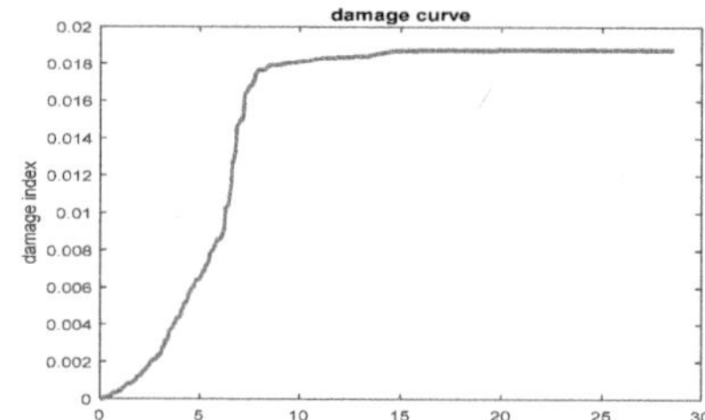

Figure 2: Damage curve

References

Supratik Bose et.al., (2016). "Structural assessment of a school building in sankhu, Nepal damaged due to torsional response during the 2015 Gorkha earthquake".University at Buffalo, Buffalo, NY, USA.

Hao Chen et al., (2017)."Seismic damage to schools subjected to Nepalearthquakes, 2015" Nat Hazards (2017) 88:247–284DOI:10.1007/s11069-017-2865-8.

Masaki et.al., (2012)."Damage to reinforced concrete school buildings in Miyagi afterthe 2011 great east japan earthquake".Proceedings of the International Symposium on March 1-4, 2012, Tokyo, Japan.

Energy Dissipation of RC Beam-column Joint Confined by different Types of Lateral Reinforcements

Ashish B. Ugale[1], Suraj N. Khante[2]*
[1] *Research Scholar, Applied Mechanics Department, GCOE-Amravati, Maharashtra, India*
[2] *Associate Professor, Applied Mechanics Department, GCOE-Amravati, Maharashtra, India*
* *e-mail: abugale@rediffmail.com*

Introduction

Beam column joints in a reinforced concrete moment resisting frame are essentially required for effective transfer of loads between the beams and columns. The failure of reinforced concrete moment resisting frame structures in recent earthquakes has highlighted the poor performance of beam column joints. Experimental and analytical studies have been carried out since last decades for understanding the behavior of joints under seismic conditions. The studies reveled that, the performance of joints is governed by the arrangement and type of reinforcement. Concrete confinement obtained by reinforcements ensures adequate development of bond stress for effective energy dissipation [3]. ACI code specify minimum area of joint transverse reinforcement and its maximum permissible spacing, however, this often leads to steel clog, difficulties in fabrication and concrete placement [1].

Headed bar having welded plate at one end is used as an alternative to L-shaped reinforcement to ease out fabrication and concrete placement issues. The headed bars not only provide the bond stress but also head bearing component in addition to development of diagonal compression mechanism. This leads to enhanced effective stress transmission and joint shear strength [4]. This finding suggests the possibility of reduction in joint transverse reinforcement when headed bars are used as longitudinal reinforcement [1]. However, the beam–column joint confined using transverse shear reinforcement and headed bars exhibits better seismic behavior. To investigate further, an experimental study has been carried out on joints to study energy dissipation capacity and shear strength by adapting types of hoop reinforcement.

Materials and Methods

The M25 grade concrete is prepared using 53 grade Portland Pozzolana Cement (PPC), Zone-I grade natural sand and course aggregates having maximum 20 mm size [2]. HYSD bars of Fe 450 grade 8 mm dia. are used as longitudinal reinforcement, whereas Fe 250 grade 6mm dia. bars are used as transverse reinforcement. Total six numbers of T-shape specimens are prepared in laboratory. Each of the specimens is subjected to cyclic and reverse cyclic loading using reaction frame as shown in Figure 1.

Figure 1. Experimental setup.

Results and Concluding Remarks

The tested specimens are investigated for the crack patterns developed during loading, the observed patterns are represented in Figure 2. The observed crack pattern indicates three types of failure, namely, beam flexure failure (BF), joint shear failure (JF) and combined beam-joint failure (BJF). Beam flexure failure (BF) was recognized as steady reduction in load-transfer after formation of plastic hinge. However, Joint shear failure (JF) is described by the continuous reduction in load-transfer before the formation of the plastic hinge. On the other hand the combined failure indicates the failure of beam and joint.

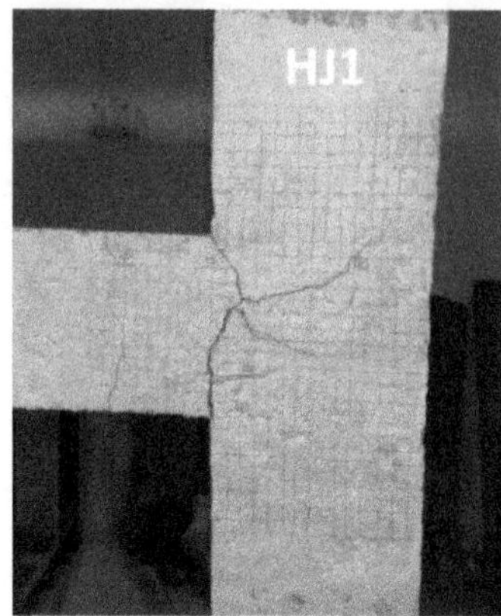

Figure 2. Crack pattern of specimen HJ1.

The hysteresis loops are obtained by plotting load against displacement. The shape of hysteresis loops indicates proper energy dissipation. The energy dissipated by specimens HJ_1SS, HJ_1DS, HJ_1TS, HJ_1VSH and HJ_1VDH is observed to be 4.08%, 10.60%, 0.73%, 1.54% and 2.64% more than that of HJ_1 respectively. The cumulative energy dissipated from all specimens is shown in Figure 3.

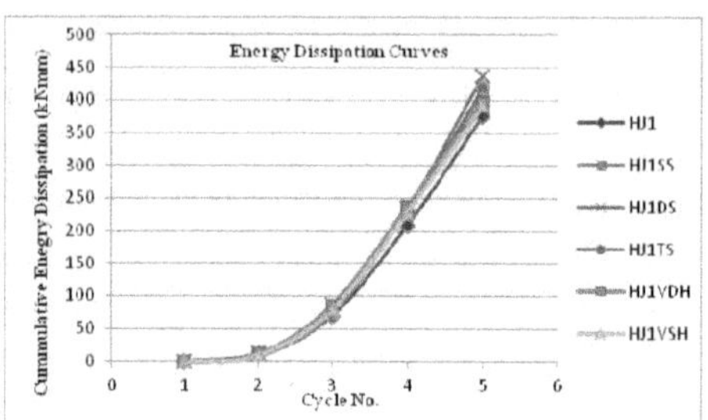

Figure 3. Cumulative energy dissipation.

The joint shear strength of specimen HJ_1 has the least value as compared to other specimens. Based on the capacity design concept, the estimated design shear force, $V_{jh,u}$ (Eq 4) is 46.43 kN for all the specimens. Specimen HJ_1, HJ_1SS, HJ_1DS, HJ_1TS, HJ_1VSH and HJ_1VDH exhibited 8.30%, 17.51%, 44.65%, 23.99%, 13.46% and 17.44% more shear force than the estimated value of shear force.

The experimental research is carried out with varying amount of joint hoops to investigate its effect on performance parameters of joints, such as shear strength and energy dissipation. The following conclusions are drawn from the experimental investigations.

- ✓ The amount of joint hoops influences the global reaction of the beam-column joint, particularly at high inelastic cyclic load reversals when shear demand at the intersection approaches the shear limit of the joint.
- ✓ The beam–column joint specimens without joint hoop reinforcement undergo large displacement with development of wider cracks as compared to other specimens.
- ✓ Increase in the joint hoop reinforcement enhances ultimate load-carrying capacity, energy dissipation, displacement ductility and joint shear strength.
- ✓ The headed bar facilitate better uniform distribution of compressive stress to the concrete, empower the advancement of a more extensive compressive strut in the joint and improve the joint shear capacity, reduction in joint transverse reinforcement.

References

[1] ACI (American Concrete Institute) (2008) Committee 318: Building Code Requirements for Structural Concrete and Commentary. ACI, Farmington Hills, MI, USA, ACI 318-08.

[2] BIS (Bureau of Indian Standards) (1970) IS 383-1970 (reaffirmed1997): Specification for coarse and fine aggregates from natural sources for concrete. Bureau of Indian Standards, New Delhi, India.

[3] Dhake PD, Patil HS and Patil YD (2015) Role of hoops on seismic performance of reinforced concrete joints. Structures and Buildings, Vol. 168, No. SB10, pp 708–717

[4] Chourasia Ajay and Gupta Shipali (2019) Influential parameters for headed bars in RC beam-column joint. Current Science, Vol. 116, No. 10

Light Weight Bricks or Blocks – A State of the Art Review

D.Nikhil Kumar[1*], Dr. P. Rathish Kumar[2]
[1] *Research Scholar, Department of Civil Engineering, NIT Warangal, Warangal, India*
[2] *Professor, Department of Civil Engineering, NIT Warangal, Warangal, India*
* *e-mail: nikhildegloorkar@gmail.com*

Introduction

The world is striving to develop materials that are sustainable, workable, have adequate strength and are economical. Light weight bricks or blocks come under the above said category that are revolutionizing the present construction industry. With the advent of framed structures, the role of walls is reduced to just an enclosure or a partition. For this reason, use of heavy clay bricks or stones are economically and structurally not feasible. There are many studies on the preparation of light weight bricks in the form of Autoclaved Aerated Blocks, Fly ash bricks etc. But, the study of light weight bricks or blocks has always been a fascinating subject due to the immense number of attributes it caters to. India is one of the leading producers of bricks, along with China, Pakistan, Bangladesh and Vietnam combined, produce 75% of the global demand of fired bricks. The replacement of traditional clay bricks with light weight bricks made of fly ash and other wastes is to reduce energy consumption involved in the manufacturing clay bricks. Also, it was reported that nearly 1000 million tons of solid waste is generated in India annually. This includes domestic waste, industrial waste, commercial waste etc. So, the usage of Fly ash, GGBS and other industrial wastes in producing light weight bricks would definitely reduce the environmental pollution and also decrease the energy consumption that would incur in the production of traditional fired clay bricks. The present study is a state of the art review on light weight bricks or blocks that would conglomerate various ideas and thereby give a possible recommendation for a new light weight brick or block generation.

Materials and Methods

The various alternative materials like flyash, GGBS, Paper waste etc can be used for the preparation of production of light weight bricks or blocks. These materials are not only eco-friendly, but, they also contribute to the GHG's by reducing the usage of cement.

Flyash is an industrial waste obtained from the combustion of coal in thermal power plants. Depending upon the CaO content it is classified as Class F and Class C. Class C for high amount of calcium oxide and Class F for low amount of calcium oxide. Also, in micro level it appears in the form of spheres that helps in the workability of the concrete or cement mix in which it is used.

GGBS is a by-product of the iron and steel industry. It is glassy in nature and due to its high reactivity it can replace cement to a great extent up to 30-60% to make Slag cement. Also, it has pozzolanic reactivity that makes it to be used as a mineral admixture in ready mix concrete as well.

Light weight aggregate can be natural like pumice and Scoria or they can also be artificial obtained from expanding shale, slate, perlite, Vermiculite etc. The usage of these aggregates in the making of concrete makes it less dense to as low as 300 Kg/m^3.

Foaming agent: The synthetic –enzyme based foaming agents, foam stability enhancing admixtures and specialised foam generating equipment made easy to incorporate stabilised air bubbles in mortar that making it light weight to a density as low as 300 Kg/m^3.

Method: The above materials depending upon their availability, different types of light weight bricks or blocks are manufactured. Some of the leading light weight bricks that are being manufactured currently are Autoclaved Aerated Concrete (AAC) blocks and Cellular Light Concrete Blocks (CLC). In AAC, the blocks are autoclaved for 10-12 hours and then cooled to get the desired blocks. In the case of cellular light weight concrete bricks, traditional concrete making methodology of mixing, casting, and then curing are considered and the desired blocks are cast. Also, technical skill is required in designing the mix proportions of light weight brick. Once the mix proportions are fixed, any unskilled labour under the supervision of semi-skilled labour can carry out the manufacturing process.

Cost: The cost analysis is performed as per Standard schedule of rates 2019-2020 of Telangana state. Table 1 shows the cost comparison between burnt clay bricks and other light weight bricks along with the number of bricks required for a 0.23 meter wall of 3.2 meter height.

Table 1. Cost to weight of bricks or blocks per running meter of wall for burnt clay brick and other light weight bricks.

Type of Brick	Density (Kg/m^3)	Weight (Kg)	Number of bricks per Running Meter (No's/Rm)	Cost per Rm	Cost/unit wt
Burnt clay brick (230mmx110mmx70mm)	1800	3.18	416	3325	1045.6
Fly Ash brick (290mmx225mmx140mm)	1400	12.79	81	2489	194.60
AAC Block (600mmx200mmx230mm)	600	17.11	27	3634	212.40
CLC Block (600mmx200mmx230mm)	775	21.40	27	6454	301.60

From the above (Table 1) it is clear that the cost/wt of AAC and CLC blocks is more compared to flyash bricks.

Advantages of alternate Light Weight Bricks: Along with the cost, the following some of the inherent advantages in using these light weight bricks irrespective of its type

- The low energy consumption and pollution free.
- The light weight bricks have a density as low as 800 kg/m^3 as compared to normally burnt clay bricks whose density is 1900 Kg/m^3. Due to this, a tremendous reduction in weight of walls leading to structural saving can be achieved in high rise buildings.
- The light weight bricks also show better thermal and sound insulation properties compared to burnt clay bricks. Also, the ease of working and casting to any shape is possible in case of light weight bricks which is a great challenge in case of burnt clay bricks.

Results and Concluding Remarks

1. The light weight bricks or blocks that are produced reduce the weight of wall to a great extent even up to 50% in terms of density and cost.
2. The blocks are light in weight, transportation and handling at site locations become easy. Thereby, cycle time of construction of walls is reduced in any framed type constructions.
3. Light weight bricks or blocks due to the presence of inherent voids provide the thermal and sound insulated environment.
4. Light weight blocks are helpful in constructing highly fire resistant structures that could withstand fire to about 5 hours and even more.
5. Compared to the traditional burnt clay bricks they cannot be used widely in the construction of load bearing walls. This can be mitigated by increasing the density of blocks reducing the entrained voids.

References

1. Tayfun Cicek, Yasin Cincin (2015) Use of fly ash in production of light-weight building bricks. Construction and Building Materials 94 (2015) 521-527.
2. Anant L, Murmu, A. Patel (2018) Towards sustainable bricks production: An overview. Construction and Building Materals 165 (2018) 112-125.
3. Balamurugan G, Chokalingam K, Chidambaram M, Aravindha Kumar M (2017) Experimental study on Light Weight Foam Concrete Bricks. International Research Journal of Engineering and Technology 4(4) 677-686.
4. Raut S, Ralegaonkar R, Mandavgane S (2013) Utilization of recycle paper mill residue and rice husk ash in production of light weight bricks. Archives of Civil and Mechanical Engineering 13(2013) 269-275.

MECHANICAL AND DURABILITY STUDIES ON HIGH PERFORMANCE SELF-COMPACTING CONCRETE CONTAINING WOLLASTONITE MICROFIBERS

J. Vishali[1], Durga Prasad R[2*], A. Jyothirmai[2]
[1]*M. Tech Scholar, Department of Civil Engineering, VNR VJIET, Hyderabad – 500090*
[2]*Assistant Professor, Department of Civil Engineering, VNR VJIET, Hyderabad – 500090*
** e-mail: vishalijalagam20@gmail.com*

Introduction

The effective utilization of mineral admixtures in cement concrete has proved its efficiency beyond doubt by improving the mechanical and durability characteristics, through reduced permeability characteristics. The effect of these mineral admixtures can be categorized though filler effect and pozzolanic effect. PawanKalla et al. (2015) had presented that the use of wollastonite microfiber, which is a naturally occurring alumino silicate material and had been quite significant in improving the performance characteristics. Moreover, studies on microstructure using SEM images revealed that there is densification of the cement matrix due to the addition of wollastonite microfibers. The current work is to study the mechanical, durability and corrosion characteristics of self-compacting cement concrete containing 5%, 10% and 15% wollastonite microfibers as a partial replacement of cement. However, the study is extended to evaluate the performance characteristics of wollastonite microfibers, when they are used as a partial replacement of cement in flyash based binary concretes containing 20% flyash. The results corresponding to compressive strength, resistivity, absorption-desorption, Rapid Chloride Penetration test (RCPT) and Accelerated corrosion cracking tests (ACC) conducted on self-compacting concrete have been evaluated and presented. An increase in the compressive strength is observed due to the addition of wollastonite microfibers in normal concrete in comparison to those containing flyash and wollastonite. Whereas, the durability characteristics of concrete containing flyash and wollastonite seems to be well above than the concrete containing wollastonite microfibers alone. This would establish a comprehensive understanding on the behaviour of wollastonite microfibers on self-compacting concrete due to the broad spectrum of experiments conducted.

Materials and Methods

Materials Used

A 53 grade ordinary portland cement procured from UltraTech Cement is used for the investigation. Fly ash is procured from RDC Ready Mix Concrete Pvt. Ltd.Wollastonite used in the study is obtained from Wolkem Industries from Rajasthan, India. Locally available river sand and 20mm crushed aggregates are used as filler materials. Potable water along with a poly-carboxylic ether based super plasticizer (SP) is used to improve the workability characteristics.The mixes used for the study are as shown in Table 1. The slump flow values indicate that all the mixes used for the study belong to class SF1 category of Self-compacting concretes with slump values ranging between 550-650mm.

Results and Concluding Remarks

Results corresponding to variation of strength and resistivity with respect to age are presented in figure 1(a) and (b). The results corresponding to Percentage absorption, RCPT and ACC test are presented in Figures 2(a) and 2(b). The results corresponding to binary concrete mixes containing wollastonite have shown an increase, whereas the results of ternary concrete mixes reveals a decrease in strength at 28 days, but has shown a substantial increase at the end of 90 days. However, the durability characteristics of concrete mixes containing flyash and wollastonite show a remarkable improvement in the durability characteristics in comparison to concrete mixes containing wollastonite alone.

Table 1: Mix proportions of experimental studies

Mix	Cement	Water	Fly ash	Wollastonite	Fine agg.	Coarse agg.	% SP	Slump flow
	kg/m^3	kg/m^3	kg/m^3	kg/m^3	kg/m^3	kg/m^3	%	(mm)
Cc	560	157	0	0	713	872	1.03	610
W5	532	184	0	28	713	871	1.03	580
W10	504	193	0	56	712	870	1.03	620
W15	476	203	0	84	711	869	1.55	650
F20W5	420	201	112	28	754	922	1.27	665
F20W10	392	192	112	56	753	921	1.27	610
F20W15	364	190	112	84	752	920	1.21	608

Fig 1(a) Fig 1(b)

Figure1. Compressive Strength and Resistivity of wollastonite based ternary concrete

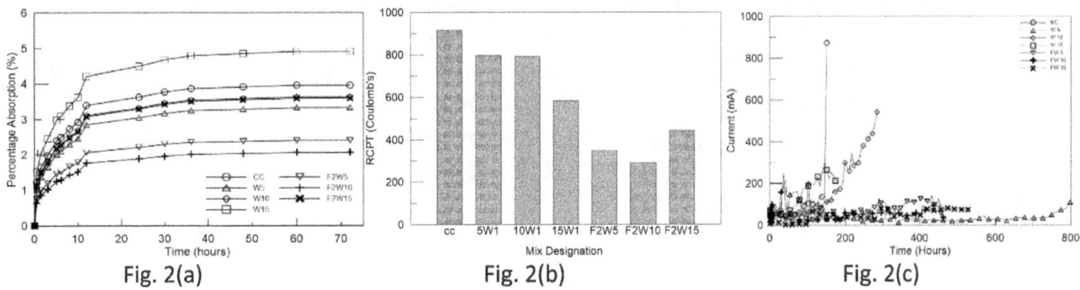

Fig. 2(a) Fig. 2(b) Fig. 2(c)

Figure 2. Results of Absorption studies, RCPT and ACC tests

The resistivity and ACC test values whichare measured exhibit a good correlation in terms of corrosion characteristics. The resistivity values of all concrete mixes indicate a very low rate of probable corrosion and the same could be inferred from ACC test, where the time taken for first crack to occur in concrete containing wollastonite and flyash was quite high in comparison to that of normal concrete. The report of entire results would be presented in the full-length paper.

References

1. Pawankalla[a], Aditya Rana[a], Yog Bahadur Cand b, Anurag Misra[c], Laszlo Csetenyi[d] (2015) Durability studies on the concrete containing wollastonite. Journal of cleaner production volume 87:PP 726-734. https://doi.org/10.1016/j.jclepro.2014.10.038
2. PawanKalla[a], AnuragMisra[b], Ramesh ChandraGupta[a], LaszloCsetenyi[c], VimalGahlot[d], AmarnathArora[e] (2013) Mechanical and durability studies on concrete containing wollastonite–fly ash combination. Construction and Building Materials volume 40: PP 1142-1150. https://doi.org/10.1016/j.conbuildmat.2012.09.102.
3. SeyedAlirez[a], ZareeiaFarshadAmeri[b], ParhamShoaei[c], NasrollahBahrami[a] (2019) Recycled ceramic waste high strength concrete containing wollastonite particles and micro-silica: A comprehensive experimental studyConstruction and Building Materials volume 201: PP 11-32. https://doi.org/10.1016/j.conbuildmat.2018.12.161.

A Comparative Study of the Compressive Strength Properties of Geopolymer Mortar made with Fly ash-GGBS and Fly ash-Red mud

M. Srinivasula Reddy[1]; Dinakar Pasla[2]; B. Hanumantha Rao[2]

[1] Department of Civil Engineering, GPREC, Kurnool, India -518007; E-*mail: sm26@iitbbs.ac.in*

[2]School of Infrastructure, IIT Bhubaneswar, India -752050.

1. Introduction

The rapid growth in urbanization and industrialization has lead to a huge rise in infrastructure development activities. In most of the infrastructure projects, concrete has become an essential component, which uses ordinary Portland cement as a binder component. The increased usage of cement in construction activities is resulting in large amount of CO_2 emissions. Cement industry is alone contributing to about 5 to 7% of global CO_2 emissions [1]. There is an immediate need to look for alternative binder material(s) to cement and reduce the global warming effect. In this regard, researchers have found a new technology called geopolymer technology to develop binders that can be used as a replacement to cement based binders [2].

Geopolymer binders can be developed by the alkaline activation of raw material(s) rich in aluminosilicates. By using geopolymer binders the usage of cement as a binder can be avoided completely in addition to an enhanced utilization of industrial wastes. But, several researchers have reported that the geopolymer binder's properties are influenced by many factors, some of them are alkaline activators type and concentration [3, 4], curing type, curing temperature [5], Na_2O to SiO_2 ratio [6], binder solids to liquid ratio [7], silicate modulus [8], curing time [9], Si to Al ratio in the mix [4], oxide composition of the raw material used [10]. Reddy et al., (2016) [10] specifically reported that, of all the influencing factors, oxide composition of the raw material(s) used plays a major role in determining the strength properties of the geopolymers. Therefore, keeping this in mind, efforts have been made in this present study to develop geopolymer binders using fly ash, slag, and red mud as binder solids in different combinations. For this purpose, a range of fly ash (FA) and red mud (RM) combinations and fly ash and GGBS combinations has been considered. The mix proportions includes, (i) FA-RM combination: 100% FA (M1), 90%FA10%RM (M2), 80%FA20%RM (M3), 70%FA30%RM (M4), 60%FA40%RM (M5), 50%FA50%RM (M6), (ii) FA-GGBS combination: 100% FA (M1), 90%FA10%GGBS (M7), 80%FA20%GGBS (M8), 70%FA30%GGBS (M9), 60%FA40%GGBS (M10), 50%FA50%GGBS (M11).

2. Materials and Methods

The materials used to develop the geopolymer binders are class F fly ash, ground granulated blast furnace slag (GGBS), red mud, sand, and alkaline activators as sodium hydroxide and sodium silicate. Class F fly ash and red mud is obtained from NALCO's Aluminium plant, Orissa, and GGBS is obtained from ACC plant, Goa. Natural river sand is used as a fine aggregate. Both NaOH pellets as well as Na_2SiO_3 gel used in this study are procured from the local supplier. The specific gravity of fly ash, GGBS, red mud and sand are found to be 2.1, 2.93, 3.05 and 2.63, respectively. The fineness modulus of fine aggregate is found to be 2.85 and falls under medium sand category. The chemical composition of sodium silicate gel used in this study is 10% Na_2O, 27.5% SiO_2 (Ms - SiO_2/Na_2O = 2.75) and 62.5% H_2O. The chemical oxide composition of the binder solids are shown in Table 1.

Table 1: Oxide composition of binder materials

Material	Oxide component (% by weight)							
	SiO_2	Al_2O_3	Fe_2O_3	CaO	Na_2O	K_2O	Other oxides	LOI
FA	61.92	28.1	4.15	0.89	0.36	0.8	3.3	0.48
GGBS	36.3	16.6	1.6	34.8	0.2	0.5	9.6	0.4
RM	14.5	25.26	45.31	1.19	4.5	-	2.74	6.5

In all the mortar mixes, the alkaline activator solution (AAS) to binder solids (BS) ratio was kept constant at 0.5. Similarly, the sodium silicate (Na_2SiO_3) to sodium hydroxide (NaOH) ratio and NaOH molarity were kept constant for all the mixes at 1.5 and 14M, respectively. All the specimens were cast as per the standard procedure. Different curing procedures were followed for FA alone, FA-RM based mixes, where they were cured at 70 C for 24 hours and then cured at open room temperature till the time of testing, and for FA-GGBS mixes, where they were cured at open room temperature immediately after casting to till the time of testing.

3. Results and Concluding Remarks

Tests are conducted at 7, 14, and 28 days of curing, and the results are plotted between the compressive strength and mix proportion, as shown in the Fig. 1. From the results, it can be observed that of all the three types of geopolymer binders i.e. FA alone, FA-RM, and FA-GGBS, FA-GGBS binders exhibits superior strength properties and also helps to produce geopolymer binders at ambient temperature curing.

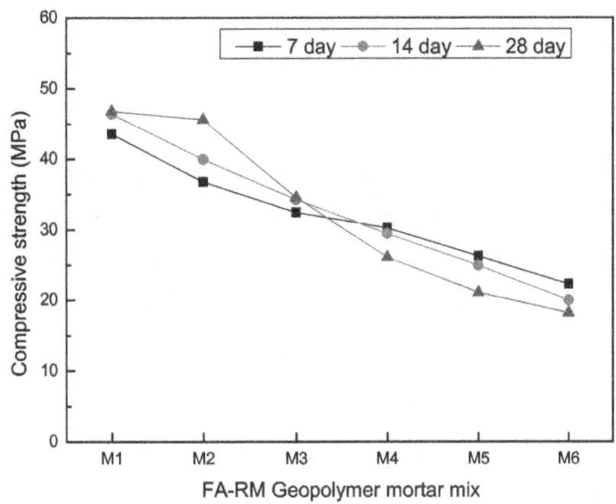

Figure 1: Compressive strength of FA-RM based geopolymer mortars with age

From this study it is found that the cement based binders can be completely avoided by replacing them with geopolymer binders. In addition, industrial waste such as fly ash, red mud and slag (GGBS) can be effectively utilized. Further, from the study it is found that using FA alone, geopolymer mortars having strength of about 50 MPa can be achieved, but it requires elevated temperature curing. Whereas, FA-RM based geopolymer mortars did not yield strength on par with FA alone geopolymers and in fact the strength decreases with an increase in the red mud content in the mix. Only advantage found with the red mud is the increase in the workability of the mix with the increased addition of it. While, FA-GGBS based geopolymer mortars yield very high strength properties than that of FA and FA-RM based geopolymer mortars. Moreover, GGBS helps to avoid elevated temperature curing. However, a maximum of 30% can be added in the mix, otherwise the setting time will become a major concern.

Acknowledgements: The authors are thankful to National Aluminium Company Limited (NALCO), Bhubaneswar, India, for funding the Research work. The financial support is greatly acknowledged.

References (few references cited here)

1. U.S. Geological Survey (USGS), Mineral Commodity Summaries, Feb. 2014. Report, http://minerals.usgs.gov/minerals/pubs/mcs/2014/mcs2014.

2. B.C. McLellan, R.P. Williams, J. Lay, A. van Riesse, G.D. Corder. Costs and carbon emissions for geopolymer pastes in comparison to ordinary Portland cement. J. Clean. Prod. 19 (2013) 1080-1090.

3. J. He, Y. Jie, J. Zhang, Y. Yu, G. Zhang. Synthesis and characterization of red mud and rice husk ash-based geopolymer composites", Cem. Concr. Compos. 37 (2013) 108-18.

4. 6. P. Duxson, S.W. Mallicot, G.C. Lukey, W.M. Kriven, J.S.J. Van Deventer, Geopolymer technology: the current state of the art. J. Mater. Sci. 42 (2007) 2917-2933.

Utilization of Industrial Wastes GGBS in Regular Concrete by Material Replacement Technique

Ramansh Bajpai[1], Deepesh Singh[2]

[1]Department of Civil Engineering, HBTU, Kanpur, India.

[2]Department of Civil Engineering, HBTU, Kanpur, India.

*email: ramanshbajpai786@gmail.com, dr.deepeshsingh@gmail.com

Introduction

Dumping of Industrial waste is the major problem that we are facing today, Cement industry is one of the largest carbon emission industries in the world which is responsible for the most concerned problem for our planet that is global warming. Concrete is the second most consumed material in the world after water and the waste produced from iron and steel industries is in tones, that can be partially utilized in the regular concrete in place of cement and thus it will reduce CO_2 emission at large scale. The experimental study has been done by partially replacing the OPC (Ordinary Portland Cement) by GGBS (Ground granulated blast furnace slag) in different percentages. This paper focuses on the technical specification and utilization of GGBS and small portion of RHA in concrete and producing a combined material of similar technical properties as the regular concrete. M. Dina et. al. (2017) had investigated the OPC replaced by High volume GGBS Pastes modifies with micro sizes metakaolin subjected to elevated temperature and found out that Compressive Strength before and after being exposed to elevated temperatures increased with micro metakaolin Content. *P.S Joanna et. al. (2014)* had conducted the experimental study on flexural behavior of partially replaced concrete by GGBS and concluded that the deflections under the service loads for the concrete beams with 40% GGBS were same as that of the controlled beams at 28 days testing. *Mathew et. al.* (2013) inferred that low strength of bottom ash GGBS based concrete is due to large particle size. *Arivalagan (2014)* had experimented on material replacement techniques. He replaced ordinary Portland cement with GGBS of 20% 30% and 40% in proportion with mix design of M35 at 7 and 28 days and came to conclusion that strength increases for 20% replacement. P.J. Wainwright et. al. (2000) had studied the influence of GGBS additions and time delay on the bleeding of concrete and concluded that the partial replacement of cement upto 55% GGBS increased the bleed capacity by 30%.

Experimental investigation on GGBS (ground Granulated Blast Furnace Slag) has carried out which is by product of iron industry and also can be used as a replacement for ordinary Portland cement in concrete. Use of GGBS as Cementitious binder reduces the cost of concrete and help to reduce the rate of cement consumption. The investigation is done for economical utilization of wastes from iron and steel industries and its applicability in concrete structures.

Materials and Methods

Ground Granulated Blast Furnace Slag (GGBS)is the byproduct of iron and steel industries. Iron ore, coke and lime stone are fed into the furnace and the resulting molten slag floats above the molten iron at a temperature of about 1500°C to 1600°C. The molten slag has composition close to chemical composition of Portland cement. It is the primary material that has been used for experimental purpose with small quantity of rice husk. Locally available OPC 43 grade cement and river Ganges sand of Kanpur is used for concrete mixtures. GGBS is delivered by Guru Corporation which deals with disposal of iron and steel industries wastes in Gujarat. Rice husk is locally available near kannauj city rice mill.

Eight different mixtures were prepared G0 to G8 that is 0 to 80 % GGBS replacement with 5 % of rice husk in every mixture. W/c is taken as 0.45 for M20 mix design and 0.45 for M40 (1:1.88:2.76) mix. Material is mixed in mechanical mixer at speed of 75 rpm and after mixing material is poured in Cylindrical and cube moulds. 6 samples for each mix is prepared for testing (7 and 28 days) with 3 and 5 layer compaction and vibrated for 40 seconds to remove air bubbles. The test is done after 7 and 28 days for compressive strength and split tensile strength also durability test is performed under different conditions to check the suitability of geopolymer (GGBS based) concrete use.

Results and Concluding Remarks

The prepared cubes and cylindrical samples are tested in compression testing machine after 7 and 28 days which gave the results as follows.

The compressive strength increases up to 50% replacement of cement with GGBS (45%) and rice husk ash (5%) after that it starts decreasing. The average compressive strength for 50% replacement of GGBS and RHA for M20 mix for 7 and 28 days respectively is 16.78 N/mm^2 and 25.13 N/mm^2 and for M40 mix it is 34.05 N/mm^2 and 45.87 N/mm^2. Geopolymer is tested for sulphate attacks and for its durability under severe conditions.

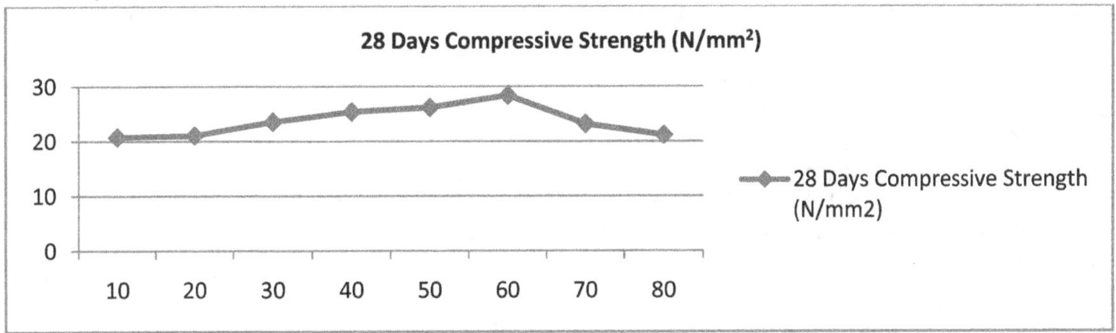

Fig, 1. Variation of Compressive strength for partially replaced cement by GGBS and RHA (M20 grade)

Durability test carried out by the investigation through acid attack test and chloride test with 1% Sulphuric acid and 3% hydrochloric revealed that when cement is replaced with 45% GGBS and 5% RHA in concrete is more durable in terms of durability factors than control mix as regular concrete losses nearly 7% compressive strength.

Average loss in weight of cube and compressive strength for acid resistant test with 1% Sulphuric acid by wt. of water is 2.21% and 5.70% respectively and average loss in weight of cube and compressive strength for Chloride attack test with 3% HCL by wt. of water is 2.97% and 6.53 % respectively.
The inference after testing is that nearly 50% of cement can be replaced by GGBS and rice husk ash which can be used for structural purpose and have good workability, strength and durability. Use of GGBS (45%) and RHA (5%) up to 50% is economical and can decrease the carbon emission in large scale and helpful in decreasing the global CO_2 emission.

Table 1. Use of GGBS replaced concrete for different types of structures

Type of Structures	**Usable GGBS Percentage**
For normal ground concrete structures	GGBS replacement is 30-35%
For Underground concrete structures with average strength	GGBS replacement is 35-45%
For marine structures and sewerage treatment plants	GGBS replacement is 50-65%

Table 1 describes the possible replacement of GGBS with cement for different types of structures. Concrete with 50% to 65 % GGBS is more durable for sulphate action and hence used for marine structures. For normal ground structures such as buildings, Offices etc. cement can be replaced up to 35% by GGBS, which has similar properties and more compressive strength than regular OPC based concrete.

Future scopes of GGBS are positive due to benefits in Durability, Sustainability, Appearance and strength obtains by partial replacement of GGBS with cement in concrete.

References

1. Alaa M. Rashad and Dina M. Sadek (2017),"An investigation on Portland cement replaced by high-volume GGBS pastes modified with micro-sized metakaolin subjected to elevated temperatures" , Elsevier-IJSBE, Vol. 6, pp. 91-101, 2017.
2. Arivalagan. S (2014), "Sustainable Studies on Concrete with GGBS as a Replacement Material in Cement", Jordan Journal of Civil Engineering, Vol 8, No. 3, pp 263-270.
3. W. C. Leung. Peter. and Wong H. D., "Final Report on Durability and Strength Development of Ground Granulated Blast furnace Slag Concrete ", Geo Report No. 258, no. SPR 1/2010, July 2010.
4. G. Singh, S.Das, A.A. Ahmed, S. Saha and S. Karmarkar (2015), "study of Granulated Blast Furnace Slag as Fine Aggregate in Concrete for Sustainable Infrastructure". Procedia- Social and Behavioral Sciences, Vol. 195, pp. 2272-2279, July 2015.
5. P.J. Wainwright and N. Rey (2000), "The Influence of ground granulated blast furnace slag (GGBS) additions and time delay on the bleeding of concrete" , Elsevier-Cement & Concrete Composites, Vol. 22, pp. 253-257, 2000.

Usage of Copper Slag & Nano-Silica: Enhancement of the Properties of High Strength Concrete

Mallikarjuna Reddy K[1*], M Naga Venkata Dayakar Reddy[2]
[1] *Assistant Professor, Department of Civil Engineering, GPCET(Autonomous), Kurnool, Andhra pradesh, India*
[2] *Assistant Professor, Department of Civil Engineering,BITS, Kurnool, Andhra pradesh, India*
** e-mail: kasireddymallikarjunareddy12@gmail.com*

Introduction

Concrete is an important part of society's infrastructure as it is one of the construction material and innovations are constantly being made in new types and applications for it. However, in recent years, the applications of High-Strength Concrete have increased, since it has now been used in many parts of the world. The growth has been possible as a result of recent developments in material technology and a demand for High-Strength Concrete. Artificially manufactured aggregates are more expensive to produce, and the available source of natural aggregates maybe at a considerable distance from the point of use, in which case, the cost of transporting is a disadvantage. Protecting the depleting natural sand resource and the shore line is a major concern of the day. It is essential today, to reduce excessive consumption of the natural river sand, thereby preventing sand mining. Thus, it is possible by utilization of industrial by-products as well as other waste materials in the production of normal concrete and High Strength Concrete. These products can be used as partial or full replacement of Aggregates. Copper slag is an industrial by-product material produced during the process of manufacturing of copper. About 2.2 tonnes of copper slag is generated, for every ton of copper production. Utilization of copper slag in applications such as aggregates has threefold advantage: eliminating the costs of dumping, reducing the cost of concrete, and minimizing air pollution problems.

Wei wu et al., (2010) observed that when copper slag was used to replace Fine Aggregate (up to 40% Copper Slag replacement), the strength of concrete increases when the surface water absorption decreases. Al-Jabri et al., (2011) stated that the water demand reduced by about 22% for 100% copper slag replacement. The strength and durability of High Strength Concrete improved with the increase in the content of copper slag of upto 50%. However, further additions of Copper Slag caused reduction in the strength due to increase in the free water content in the mix. Min Hong Zhang, et al. (2012) has used Nano-Silica(NS) to reduce setting times and increase early strength of concrete with high volumes of fly ash or slag, based on the experimental results produced by using Nano-Silica in pastes, mortars and concretes with about 50% of fly ash to cement. Pengkun Hou et al (2013) showed that the pozzolanic activity and hydration acceleration effect of colloidal NS (instead of NS powder) is higher when compare to than that of SF in the early age. Based on the available literature review, Copper Slag is chosen as a replacement material for Fine Aggregate. In the present work, Fine Aggregate is replaced with Copper Slag (i.e., from 0 % to 100%). Optimum mechanical properties were studied. By keeping those optimum percentage replacements constant, Cement is replaced with Nano Silica there after mechanical properties were found. Thus, significance of the present work is to prove that Copper Slag is the better replacement for Fine Aggregate.

Materials & Methods

The materials used in the present investigation are cement, flyash, coarse aggregate, fine aggregate, copper slag, and nano-silica. Experimental investigation has been carried out with a reference mix of M80 grade concrete which was designed as per ACI 211.4R-08. Fourteen concrete mixes were prepared. One is referred as reference mix. The other ten concrete mixes were casted, where fine aggregate was replaced with 10%,20%,30%,40%,50%,60%,70%,80%,90% and 100% copper slag by weight respectively. The optimum percentage of copper slag in concrete corresponding to maximum mechanical properties was identified. In the second phase, by keeping the optimum percentage replacement (40%) of copper slag constant, cement is replaced with nano silica 1%, 2%, 3%. Concrete mix specimens were evaluated for compressive strength, split tensile strength, flexural strength, workability, density and durability.

For analyzing fresh & mechanical properties of high strength concrete (M80) with Copper Slag, and Nano Silica different tests i.e. workability test, compressive strength test, split tensile strength test, modulus of rupture test are conducted. For Fresh Properties the Slump & Compaction Factor tests are conducted. To analyze the hardened properties, the compressive strength test is done on concrete cube of size 100mm x100mm x 100mm; Split tensile strength test is done on cylinder of size 100 dia x200mm height, Modulus of rupture test is done on beam of size 500mmx100mmx100mm. Finally for evaluating the durability properties, the chloride test (as per Indian Standard code 14959 (Part 2):2001) & the sulphate test (as per Australian Code 1012.20.1992) are done accordingly.

Results and Concluding remarks

In the present experimental study, Copper slag is used as a replacement material for fine aggregate and nano silica is used as a partial replacement material for cement. According to the analysis performed the following results & conclusions are drawn.

1. The maximum compressive strength of 101.66 N/mm^2 is achieved for 40% replacement of sand by Copper slag which is 8.57% more than that of reference mix for 28 days curing.
2. Maximum compressive strength is 105 N/ mm^2 is obtained for concrete mix with 40% Copper Slag and 2% replacement of cement by Nano-Silica which is 3.28% more than that of concrete mix without Nano-Silica for 28 days curing.
3. The maximum split tensile strength of 4.615 N/mm^2 is achieved for 40% replacement of sand by Copper Slag which is 20.8% more than that of concrete mix without Copper slag mix for 28 days curing.
4. Maximum split tensile strength is 4.774 N/mm^2 is obtained for concrete mix with 40% Copper Slag and 2 % replacement of cement by Nano-Silica which is 3.44% more than that of concrete mix without Nano-Silica for 28 days curing.
5. The maximum modulus of rupture strength of 6.04 N/mm^{2}is achieved for 40% replacement of sand by Copper Slag which is 17% more than that of reference mix for 28 days curing.
6. Maximum modulus of rupture strength is 6.36 N/mm^2 is obtained for concrete mix with 40% Copper Slag and 2% replacement of cement by Nano-Silica which is 5.29% more than that of concrete mix without Nano-Silica for 28 days curing.
7. The density of concrete increases with the increase in copper slag content in HSC, as Unit weight of Copper Slag more compare to Fine Aggregate.
8. Sulphate Content of High Strength Concrete (in which Fine Aggregate replaced with Copper Slag) is within the permissible limits (< 4% wt. of cement).
9. Chloride Content of High Strength Concrete (in which Fine Aggregate replaced with Copper Slag) is exceeding the permissible limits i.e., $\ngtr$ 0.6 kg/m^3, due to more chloride content observed in sand and Aggregate.

References

Wu.W, Zhang.W, and Ma.G (2010) Optimum content of copper slag as a fine aggregate in high strength concrete. Material design 31(6): 2878-2883. https://doi.org/10.1016/j.matdes.2009.12.037

Al-Jabri, K.S., Abdullah, H., Al-Saidy and Ramzi Taha (2011) Effect of copper slag as a fine aggregate on the properties of cement mortars and concrete. Construction and Building Materials 25: 933-938. https://doi.org/10.1016/j.conbuildmat.2010.06.090

Min-Hong Zhang, Jahidul Islam and Sulapha Peethemparan (2012) Use of nano silica to increase early strength and reduce setting time of concretes with high volumes of slag. Cem. and Con. Compo., 34, 650-662. https://doi.org/10.1016/j.cemconcomp.2012.02.005

Pengkun Hou, ShihoKawashima, DeyuKong, David J.Corr, JueshiQian, Surendra P.Shah (2013)Modification effects of colloidal nano SiO_2 on cement hydration and its gelproperty, Composites PartB: Engineering, https://doi.org/10.1016/j.compositesb.2012.05.056

Partial Replacement of Recycled plastic waste for Coarse and Fine aggregates in Concrete Paver Blocks

Dakshatha H P [1], Udhay M[1], Dr. G Kavitha[2]
[1] *PG Student, Department of Highway Technology, RASTA – Centre for Road Technology, Bangalore, India;*
[2] *Associate Professor, Department of Highway Technology, RASTA – Centre for Road Technology, Bangalore, India;*
** e-mail: dakshathahp@gmail.com, udhaymcv94@gmail.com, gkavitha@rastaindia.com*

Introduction

Plastics contribute to an enormous amount of solid waste and take centuries to breakdown in the land fill or the ocean. Disposal of these plastic wastes has become a serious problem globally. At the same time natural resources like aggregates which form an important part in Road construction are depleting day by day. Utilizing certain types of Plastic wastes as an alternative material in road construction can reduce the consumption of natural resources to some extent and at the same time solve their disposal problems. It also helps in maintaining a sustainable Environment.

In the present study attempts were made to replace natural aggregates by recycled plastic aggregates (RPA) in the manufacture of concrete paver blocks.

P. Chinnadurai et.al, (2017) conducted a study on Paver blocks using Waste plastic as a substitute for coarse aggregates. Waste plastics have been added incrementally to replace natural coarse aggregates by 0%, 2%, 4%, 6%, 8%, 10% and the blocks were tested for compressive strength. Results indicated that M20 Grade concrete had satisfactory strength value with 4% plastic replacement.

An attempt was made by Shubhankar Anant Bujone et.al, (2017) to use Fly ash and plastics in Paver blocks, the waste plastic was used as fibres in the concrete mix to improve the blend. Results showed that use of plastic in paver blocks could increase its strength by 30% - 40% and also proved to be cost effective.

Materials and Methods

a) **Cement** –OPC 53 grade *Birla super*
b) **Aggregates –** of different sizes, *procured from Chennigaraya stone crusher, Tumkur district*
c) **Plastic aggregates** - size passing 12mm and retained on 6mm sieve,
- Size less than 2mm,

Procured from Nayandahalli, Bengaluru.

METHODS

- Basic tests on the selected materials were done as per specifications.
- Mix design for M30 and M40 grade concrete was done as per IS: 10262-2009. Two sets of paver blocks were prepared.
    - First set of paver blocks - coarse aggregates of size passing 12mm and retained on 6mm were replaced with recycled plastic coarse aggregates in varying proportions of 5%, 10%, 20%, 30% and 40%.
    - Second set of paver blocks – Fine aggregates of size finer than 2 mm were replaced with recycled plastic fine aggregates in varying proportions of 5%, 10%, 20%, 30% and 40%.
- Tests such as I) Compressive Strength, ii) Flexural Strength, iii) Split Tensile strength, iv) Abrasion Resistance and v) Durability were conducted on M30 and M40 grade conventional and modified Paver blocks.
- Tests results for conventional Paver Blocks of M30 & M40 grade were compared with modified paver blocks and reported

Results and Concluding Remarks

Table 1.Results of the M30 and M40 grade conventional & modified paver blocks (for optimum replacement of 30% RPCA and 10% RPFA).

Experiment	Conv paver blocks		Modified paver blocks			
			M30		M40	
	M30	M40	30% RPCA	10%RPFA	30%RPCA	10%RPFA
Compressive strength (N/mm^2)	32	41.03	26.20	25.68	35.91	35.42
Flexural strength (N/mm^2)	7.10	6.72	5.10	4.55	5.90	6.60
Split tensile strength (N/mm^2)	4.50	4.62	2.68	2.92	3.01	3.28

*(*RPCA- Recycled plastic coarse aggregate, RPFA- Recycled plastic fine aggregate)*

- From Table 1, it can be observed that although the test results obtained for the modified paver blocks is less than that obtained for conventional paver blocks, yet it is within the desired tolerance limits. i.e. ± 5 N/mm^2. Hence it is concluded that replacement of coarse RPA up to 30% and Fine RPA up to 10% is effective in yielding optimum strength.

- Hence based on the present study, it is recommended to use recycled plastic aggregates as partial replacement to natural aggregates up to 30% for coarse and 10% for fine aggregates respectively in the construction of low volume roads. This paves a new way for disposal of waste plastic & helps in reducing pollution of our environment.

References

[1] Shubhankar Ananth Bujone, Sarang Shashikanth Pawar, (2017) use of Flyash and Plastic in Paver blocks. International Research Journal of Engineering and Technology, Vol 04 Issue: 11, ISSN: 2395-0056.

[2] B. Shanmughavalli, K. Goutham, P. Jeba Nalwin, B. Eshwaramurthy (2017) Reuse of Plastic wastes in Paver blocks. International Journal of Engineering Research and Technology, Vol 06 Issue: 02, ISSN: 2278-0181.

[3] IS 456: 2000 – Plain and reinforced concrete – code and practice," Bureau of Indian standards", 2000.

[4] IRC: SP-63-2004,Guidelines for the use of Interlocking Concrete Block Pavements"

[5] BIS: 15658, Indian Standards for Precast Concrete Blocks for Paving," Bureau of Indian Standards, New Delhi. 2006.

UTILIZATION OF LD SLAG AS A PARTIAL REPLACEMENT TO NATURAL COARSE AGGREGATES IN CONCRETE PAVER BLOCKS.

Harindra Kumar G [1], Dr. G Kavitha [2]
[1] *PG Student, Department of Highway technology, RASTA-Centre for Road technology, Bengaluru, India*
[2] *Associate professor, Department of Highway technology, RASTA-Centre for Road technology,Bengaluru*
** e-mail :* harindra1324@gmail.com, gkavitha@rastaindia.com

Introduction

The rising concern on depletion of natural resources associated with construction development, has led to the usage of unconventional materials for construction, in most of the countries. This results in low energy consumption and also helps in promoting sustainable environment. Slag is one such material obtained as a by-product in steel or iron manufacturing process. Wide utilization of slag in construction of Roads and buildings has paved a way for effective disposal of these by-products. Hence there is a need to understand the influence of slag when used in different types of road construction

In the present study an effort was made to determine the influence of LINZ DONAWITZ SLAG (LD slag) aggregates in concrete paver blocks by replacing natural coarse aggregates.

Jayakumar PT et.al, (2013) studied the use of steel slag as partial replacement to aggregate and sludge as a replacement to cement in concrete paver blocks. Results indicated that use of sludge up to 30% and slag up to 60% gave more strength and also proved to be economical for manufacturing paver blocks.

The study by Ahmed Abdelbary et.al, (2016) showed that paver blocks manufactured with EAF slag has higher resistance to abrasion and compressive strength compared to conventional paver block of M40 grade.

The main objective of the present study is to evaluate the strength properties and abrasion resistance of concrete paver blocks by replacing the natural coarse aggregate with LD slag aggregates in different proportions.

Materials and Methods

a) Materials

- **Cement** – OPC 53 grade, *Birla super*
- **Aggregates** – of different sizes , *procured from Chennigaraya stone crusher, Tumkur district*
- **LD slag aggregate** – passing 12mm sieve and retained on 6mm sieve, *procured from JSW Steel, Bellary*

b) Methods

- Basic tests on the selected materials conducted as per specification
- Mix design for M40 grade concrete as per IS: 10262-2009 (conventional mix using natural aggregates).
  Mix design for M40 grade concrete as per IS: 10262-2009 for replacement of LD slag in different proportions such as 5%, 10%, 20%, 30%, 40%, 70% and 100%
- Tests such as i) compressive strength ii) flexural strength iii) split tensile strength and iv) abrasion resistance were conducted on conventional and modified paver blocks and results were compared.

Results and Concluding remarks

Table 1. Results of conventional and LD slag paver blocks.

Experiment	Conventional Block	LD Slag @ 70% replacement
Compressive strength (N/mm^2)	41.03	48.87
Flexural Strength (N/mm^2)	5.29	7.7
Split Tensile Strength (N/mm^2)	4.62	5.6

- The compressive strength, flexural strength and split tensile strength increased for modified paver blocks up to 70% replacement of natural aggregates and there after it decreased. This could be due to rough surface texture and higher density of LD slag compared to natural aggregates which results in better binding between the slag particles and cement paste.
- Abrasion resistance of the paver blocks increased with the increase in slag proportion up to 70%.
- Hence it can be concluded that 70% replacement of natural aggregates with LD slag is optimum.
- Also, construction of concrete blocks with LD slag aggregates proved to be cost effective and can compensate to the lack of natural aggregates.

Reference

[1] Jayakumar P T, Ram Kumar.V.R, Suresh Babu.R, Mahesh Kumar. M (2013) "Experimental Investigation on Paver Blocks Using Steel Slag as Partial Replacement of Aggregate and Sludge as Partial Replacement of Cement". International journal of science and research (IJSR), India online ISSN: 2319-7064.

[2] R. Padmapriya, V.K. Bupesh Raja, V.Ganesh Kumar, J. Baalamurugan (2015) "Study On Replacement Of Coarse Aggregate By Steel Slag And Fine Aggregate By Manufacturing Sand In Concrete". International journal of ChemTech Research vol.8, No.4, pp. 1721-1729.

[3] RA.B.Depaa, Dr. T. Felix Kala (2017) "Experimental Study on Steel Slag as Coarse Aggregate in Concrete". International Journal on recent researches in science, engineering & technology, volume 5, issue 4, ISSN: 2348-3105.

[4] IS 456: 2000 – Plain and reinforced concrete – code and practice," Bureau of Indian standards", 2000.

[5] IRC: SP-63-2004,Guidelines for the use of Interlocking Concrete Block Pavements".

[6] BIS: 15658, Indian Standards for Precast Concrete Blocks for Paving," Bureau of Indian Standards, New Delhi. 2006.

Effect of Response Reduction Factor, R on Seismic Drift Response of Buckling Restrained Braced Frames

Muhamed Safeer Pandikkadavath[1*], Romanbabu Oinam[2]
[1] *Department of Civil Engineering, Government Engineering College Kozhikode, Kerala-673005, India*
[2] *Department of Civil Engineering, Indian Institute of Technology Tirupati, Andhra Pradesh-517506, India*
** e-mail: Corresponding Author (msafeerpk@geckkd.ac.in)*

Introduction

The Buckling Restrained Braced Frames (BRBFs) are special type of Concentrically Braced Frames (CBFs) with enhanced lateral load resisting capabilities. BRBFs are charcterized by Buckling Restrained Braces (BRBs) that can yield in both tension and compression under inelastic reversed cyclic loading conditions. BRBs have stable and balanced hysteretic response along with excellent low cylcle fatigue capacity and hence it can safeguard framed structural systems against unintended seimsic induced damages (Merritt *et al.,* 2003; Black *et al.,* 2004). Though BRBFs are effective to minimze the ground motion induced damages, it may cause larger residual drift responses to the building frames in the post earthquake regime (Sabelli *et al.,* 2003; Sahoo and Chao, 2014; Pandikkadavath and Sahoo 2017). The increased drift response is mainly due to the lower post yield stiffness of BRBs along with preferred moment free beam column joints in BRBFs (Fahnestock *et al.,* 2007). The larger Residual Drift Ratio (RDR) may affect the retrofitting and reusablity of structures negatively. Many researchers investiagted on this point and proposed methodologies to improve the drift reponses of BRBFs positively under seismic disturbances. It includes incorporation of dual frames with BRBFs along with back-up Momemnt Resisting Frames (MRFs), Stiffness based design approach with heavy clomuns (Sahoo and Chao, 2014) and use of BRBs with shorter yielding core lengths (Hoveidae *et al.*, 2015; Pandikkadavath and Sahoo, 2016). Though all these techniques showed effectiveness to minimize the RDR of BRBFs, each one has its on limitations interms of economy and feasibility. .

In the present investigation the drift response of BRBFs are evaluated by modifying the response modification factor, R suitably. Present United States design guidelines recommends a R value of 8 for BRBFs by relying on the higher ductility capacities of BRBs. The decrease in numercal value of R factor can contribute larger BRB cross sections having relatively better elastic and post yield stiffness to the BRBFs and hence this modifcation may improve the drift responses associted with the BRBFs.

Methodology

A 6-storey building frame with inverted-V bracing configuration designed as per current United States design procedure with applicable provisions (ASCE7-16, 2016) has been selected for the investigation (Sabelli *et al.,* 2003). These building frames are assumed to be located in Los Angeles, USA. The building has 6 bays (9.14 m each) along both directions in plan and has a storey height of 4 m for every floor except bottom one (5.49 m). Additionally the building consists of 6 numbers of braced frames along both directions, located at the periphery of the building plan. Figure 1 (b) and (c) shows the study building and frame details. The seismic weight of the building is found to be 68934 kN. For the present investigation, four different values for R namely, 8, 7 and 6 are used and the frame is designed for all the respective cases assuming the lateral loads are entirely resisted by BRBs only. Subsequently frames are named as 6VBRBFR8, 6VBRBFR7 and 6VBRBFR6 respectively.The non-linear models of BRBFs are created in SAP 2000 (2019). The P-Delta effect due to gravity load is accounted by providing a leaning column to the braced frame in the model. Sequentially Non-linear Static and Dynamic analyses are carried out. The approximate time period of the structure for the design is found to be in the range of 0.80s and the analytical result showed that it is in the range of 1.1s. Non-linear Time History (NTH) analysis is carried out under selected ground motion records after scaling them suitably.

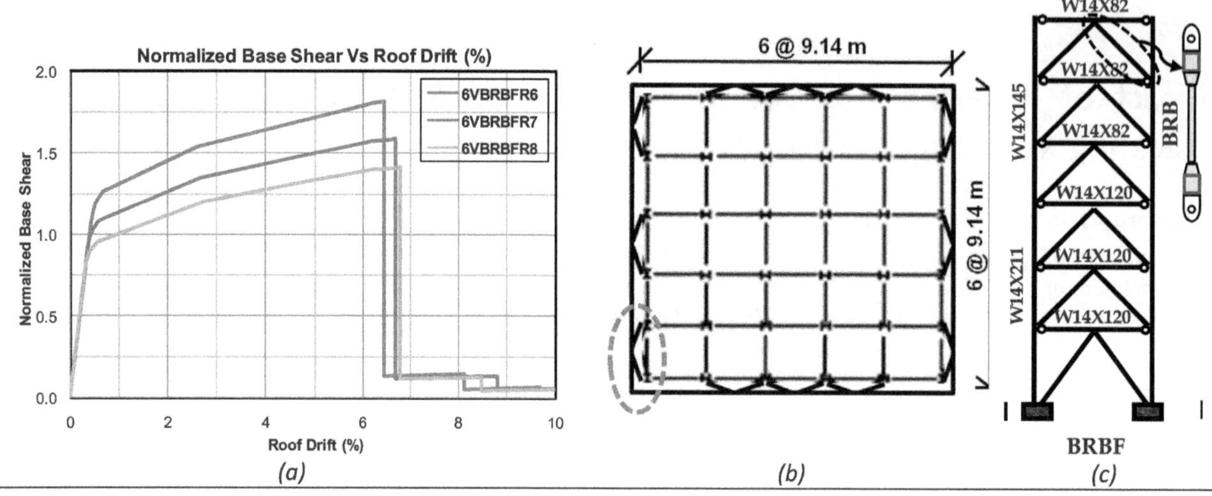

Figure 1. (a) Push-over curve for different frames, (b) Plan of the 6-storey building and (c) Details of study frame

Results and Concluding Remarks

The preliminary results showed that, elastic and post elastic stiffness of BRBFs are increased with respect to decrease of R value numerically (refer Figure 1(a)). In the same trend both Inter-storey Drift Ration (IDR) and Residual Drift Ratio (RDR) decreased under selected scaled ground motions. The result showed that this modification can enhance the reusability of the BRB frames after strong ground motions. It is also found that the ductility demand associated with BRBs also decreases in the same trend as that of drift responses. Hence this response can give reserve strength to BRBFs especially against secondary waves after main shocks. Though numerically decreasing the value of R improves the drift response of BRBFs under seismic disturbances, the cost of the frame system may increase slightly (owing to more use of steel). However this effect can be acceptable to the case where the framed structures damage vulnerability to severe seismic disturbances (that effectively increases the retrofitting cost).

References

ASCE/SEI 7-10. (2010), *Minimum design loads for buildings and other structures*, American Society of Civil Engineers;Reston, VA.

CSI. (2019), *SAP 2000 user guide*, Computers and Structures Inc.; Berkeley, CA.

Fahnestock, L. A., Ricles, J. M. and Sause, R. (2007). "Experimental evaluation of a large-scale buckling-restrained braced frame."*ASCE Journal of Structural Engineering,* 133(9), 1205-1214.

Hoveidae, N., Tremblay, R., Rafezy, B. and Davaran, A. (2015),"Numerical investigation of seismic behavior of short-core all steel buckling restrained braces", *Journal of Constructional Steel Research*, 113, 89-99.

Kiggins, S. and Uang, C.-M. (2006), "Reducing residual drift of buckling restrained braced frames as a dual system", *Engineering Structures,* 28(11), 1525–1532.

Merritt, S., Uang, C. M. and Benzoni, G. (2003). "Sub-assemblage testing of star seismic buckling-restrained braces." *Report No. TR-2003/04*, Department of Structural Engineering, University of California, La Jolla, CA.

Pandikkadavath, M. S. and Sahoo, D. R. (2016). "Analytical investigation on cyclic performance of buckling-restrained braces with short yielding core segments." *International Journal of Steel Structures*, 16(4), 1273-1285.

Pandikkadavath, M. S. and Sahoo, D. R. (2017). "Mitigation of drift response of concentrically braced frames using short yielding core BRBs", *Steel and Composite Structures*, 23(3), 285-302.

Sabelli, R., Mahin, S. A. and Chang, C. (2003). "Seismic demands on steel braced frame buildings with buckling-restrained braces." *Engineering Structures*, 25(5), 655-666.

Sahoo, D. R. and Chao, S. H. (2014). "Stiffness-based design for mitigation of residual displacements of buckling-restrained braced frames." *ASCE Journal of Structural Engineering*, 149(9), 04014229.

RESIDUAL STRESS-STRAIN CURVE AND MODULUS OF ELASTICITY OF CONCRETE SUBJECTED TO UNIAXIAL LOAD AND TEMPERATURE CONCURRENTLY

V.S.Vani[1]

[1] *Department of Civil Engineering/ Aditya Institute of Technology and Management (AITAM)/ Tekkali, India;* [2]
**drvsvani@gmail.com*

Introduction

Urbanization is increasing day by day due to migration of people from villages to towns and cities. These areas are overpopulated, demanding infrastructure for their basic needs like buildings, roads etc,. Due to scarcity of land multistory structures are taking their way, demanding lot of constructional resources. The naturally supplied constructional resources are in scarce compared to their demand. To meet the demand of present construction need, supplementary cementitious materials are required. Fly ash is one such industrial waste that may help sustainable development.

Multistory structures constructed with such materials must be durable as insisted by IS 456 - 2000. In these structures fire accidents are quite common. The structural concrete in such case exposed to temperature while it is sustained with load. In such stressed condition supplementary cementitious material should not fail. One of the most important parameter that governs design and performance of multi storey concrete structures is modulus of elasticity of concrete.

Most of the previous *studies* including Phan L T (1996), Petkovski (2010), Adam M. knaack et al. (2009) concentrated more on unstressed tests suggested by Phan L T (1996). Very few researchers studied the performance of concrete in stressed state considering Ordinary Portland Cement. Study on concrete containing PPC at elevated temperatures in stressed condition is very minimal with respect to residual stress-strain curve.

Hence the present research focus on the residual stress-strain curve and modulus of elasticity of concrete subjected to uniaxial load and temperature simultaneously. The temperatures are 27, 100, 300, 500 and 800 °C sustained for a period of half an hour, at stress levels 0.1 f_{ck} and 0.5 f_{ck} compared with unstressed concrete.

Materials and Methods

The experiment is carried out on fly ash based Portland pozzalana cement (PPC) in addition fine and coarse aggregate, to obtain M25 grade concrete as per the proportion given in Table 1. The concrete cylinders of size 15 x 30 cm are cast, demould and cured for 28 days. The specimens are removed from the curing tank and inserted in test setup shown in Figure 1. The set up contains Digital Universal testing machine of 100 ton capacity (UTES-100) to which a cylindrical furnace is attached.

The temperatures under study are 27, 100, 300, 500, 800 °C, stress levels are 0.1 f_{ck} , 0.5 f_{ck} sustained for a period of half an hour concurrently. Each time a particular temperature and stress level is chosen for study. After half an hour duration, the specimen is loaded for failure and residual stress-strain curves are obtained. The specimens in unstressed condition are exposed to temperature, later loaded to failure for obtaining residual stress-strain curves. Residual Modulus of elasticity is calculated based on the obtained residual stress- strain curves consider tangent modulus of elasticity.

Table 1. Concrete Mix proportions as per IS 10262-2009

Cement	Fine Aggregate	Coarse Aggregate	Water
1	1.6	3	0.48

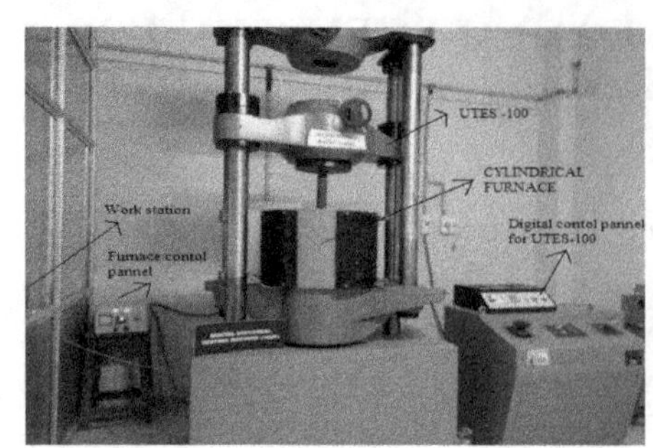

Figure 1. Specimen surrounded by Cylindrical Furnace in Digital UTES-100

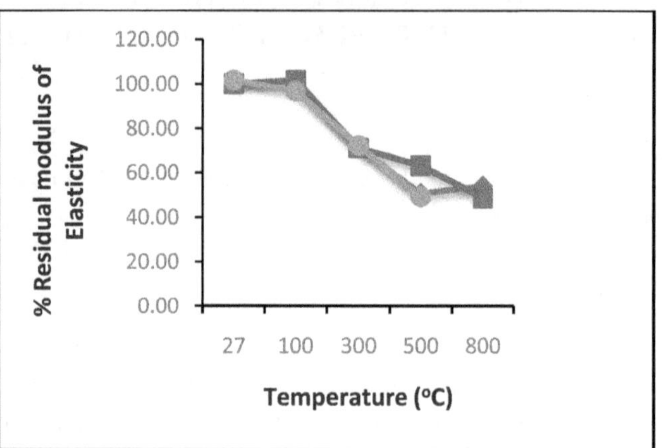

Results and Concluding Remarks

From the obtained test results, graphs are plot for stress ratio and temperature. Results indicated that the specimens stressed to 0.5 f_{ck} stress level are performing inferior to unstressed specimens, while 0.1 f_{ck} stressed specimens are showing superior results. At stress level 0.5 f_{ck} and temperature 800 °C, specimens could not sustain stressed condition. This may be due, at high temperature concrete is already weak due change in C-S-H gel to Cao/Sio_2, added to which it is stressed to 0.5 f_{ck} , for which the concrete fibers could not bear pressure and failed. At 0.1 f_{ck} concrete specimens exhibited creep due to slipping.

As the temperature increased peak stress ratio decreased (stress at any temperature to stress at room temperature of unstressed specimen) for stressed and unstressed specimens. As stress level increased to 0.1 f_{ck} the stiffness and toughness increased but both of them decreased at stress level 0.5 f_{ck} with respect to unstressed specimen. The study indicated that the strain correspond to peak stress does not change significantly for lower temperatures at all stress levels. The maximum and minimum peak stress is exhibited by concrete at 100 °C and 800 °C respectively.

Figure 2 Show variation of % Residual Modulus of Elasticity (RME) (Modulus of Elasticity at any temperature to Modulus of Elasticity at 27 °C of unstressed specimen multiplied by 100) in unstressed, 0.1 and 0.5 f_{ck} stressed concrete at all temperatures. The overall trend of % RME slightly disproved for 0.5 f_{ck} stressed concrete than unstressed and 0.1 f_{ck} stressed concrete all temperatures. At temperatures 800 °C the concrete failed. This may be due to high stress on fibres and reduction of strain at high temperatures lead to failure. But concrete stressed to 0.1 f_{ck} performed better may due to densification of paste. At 100 °C specimens exhibited high modulus of elasticity. The maximum and minimum residual modulus of elasticity is 101.74 % for 0.5 f_{ck} stressed specimen at 27 °C and 48.69 % for 0.5 f_{ck} stressed specimen at 800 °C respectively, of course at 800 °C 0.5 f_{ck} stressed concrete retained no modulus of elasticity.

Finally from the study it is concluded that 0.5 f_{ck} stressed specimen performed better compared to other specimens.

Acknowledgments: The author express her sincere thanks to Science and Engineering Research Board (SERB) with whose funding under Fast Track Young Scientist - Engineering and Science scheme, the present research work is completed.

References

Phan, L T (1996) Fire Performance of High-Strength Concrete: A Report of the State-of-the-Art, NISTIR 5934. Building and Fire Research Laboratory, National Institute of Standards and Technology, Gaithersburg, Maryland: 1-105.

M. Petkovski (2017) Effects of stress during heating on strength and stiffness of concrete at elevated temperature. Cement and Concrete Research 40(12): 1744–1755. https://doi.org/10.1016/j.cemconres.2010.08.016

Adam M. knaack, Yahya C. Kurama, David j. Kirkner (2009), Structural Engineering research report on Stress-Strain properties of concrete at elevated temperatures. Department of Civil Engineering and Geological Sciences, University of Notre Dame, Notre Dame, Indiana: 1-130.

IS 10262-2009, Guidelines for concrete mix design proportioning, Bureau of Indian Standards, New Delhi.

Durability of Self Compacting Concrete in Acidic Environment in cooperation with Mineral Admixtures

A. Deep Tripathi[1*], B. Rakesh Kumar[1], C. P. K. Mehta[1], D. Amrendra Singh[1]
[1] *Department of Civil Engineering, MNNIT Allahabad, Prayagraj, India*
** e-mail: drt.ald10@gmail.com*

Introduction

Concrete is most widely used building material in present era, so it is an important topic of interest for further research. SCC is a type of concrete which flows under its own weight and there is no need of vibration for compaction so SCC is developed to overcome workability problem which is a common problem faced today's in congested reinforcement. To improve the durability of Self Compacting Concrete (SCC), several mineral admixtures are used viz.; Fly Ash (FA), Metakaolin (MK), Rice Husk Ash (RHA), Ground Granulated Blast Furnace Slag (GGBS) etc. The Ordinary Portland Cement (OPC) was partially replaced by FA and MK in combination. Nitric acid (HNO_3) is one of the acids influencing the durability of concrete. Due to inclusion of mineral admixtures (pozzolanic materials) in the concrete, the liberation of calcium hydroxide stabilises, during the hydration process of the cement to form additional cementious material (C-S-H). The resultant binder matrix of concrete is chemically more resistant, by virtue of its dense microscopic pore structure (Monteny et al. 2001) [1].

This paper presents the results of an experimental programme to investigate and compare the durability of SCC without replacement as referral concrete (i.e. M-1) and SCC with combination of FA and MK (i.e. M-2). The M25 grade concrete cubes of size 100 mm were cast. The OPC was partially replaced by 25% by weight with mineral admixtures (i.e.15% FA + 10% MK). The workability of M-2 mix was found more in comparison to M-1. The cubes of both the M-1 and M-2 mixes were cured in Tap water as well as in 3% Nitric acid solution till 90 days. Several researchers concluded that there is a slower rate of acid attack on concrete that contains mineral admixtures (Harrison et al. 1987; Fattuhi et al. 1988; Mehta 1985; Kazuyuk et al. 1994) [2,3,4,5]. The compressive strength of cubes was determined after 7, 28, 56, and 90 days curing. For the visual assessments the photographs were taken to get some idea about how nitric acid affects the appearance of the concrete. The weight change characteristics were also considered.

Materials and Their Properties

Cement: The OPC of 43 grade (Brand- M P Birla) was used in this experimental study. The Physical properties of OPC i.e. normal consistency and specific gravity were found to be 27% and 3.14, respectively. The initial and final setting times were 45 and 480 minutes, respectively, which conforms to the requirements of IS 8112-1989. The compressive strength of 50 cm^2 cement mortar cube at 28 days was found to be 51.95 N/mm^2. **Fine Aggregate:** The natural river sand of rounded particle shape and smooth surface texture was used for making RC which conforms to zone II as per IS-383-1987. The specific gravity and bulk density were found to be 2.65 and 1680 kg/m^3, respectively. The fineness modulus was 2.7. **Coarse aggregate:** 10 mm and 20 mm sizes coarse aggregates, having specific gravity 2.66 and 2.7, respectively, and water absorption of 1.0% and 0.9%, respectively (conforming to IS 383-1987) were used. The fineness modulus for the coarse aggregates of sizes 10 mm and 20 mm was 6.7 and 7.2, respectively. **Fly ash:** Class F- FA of grey colour having specific gravity 2.13 and confirming to IS 3812-2000, procured from NTPC Unchahar, U.P., was used in this study. **Metakaolin:** MK (Off-white colour) was procured from Kaolin Techniques Pvt. Ltd, Bhuj, Kutch, Gujarat. The specific gravity and bulk density of MK was reported as 2.5 and 0.9 gm/mm^2 respectively. **Superplasticizer:** A polycarboxylic ether based Master Rheobuild 817RL superplasticizer having density approximately 1.08 and pH approximately 5.0 was used. Tap fresh potable water was used in the mix design and curing of specimens. **Acid:** The HNO_3 used for creating the acidic environment was of brand RENKEM and was procured from The Scientific Traders, Civil lines, Allahabad.

Expeimental Methadology

In this experimental study, SCC concrete mix of grade M25 was prepared using 43 grade OPC (RC) confirming to EFNARC specifications. Water/Binder ratio (W/B) of 0.43 and the total binder content was kept constant (450 Kg/m^3). The quantity of fine aggregate, coarse aggregate and dosage of superplastisizer were 890 Kg/m^3, 750 Kg/m^3 and 4.95 Kg/m^3, respectively. The final mix proportion was 1:1.98:1.66 (Binder: Fine aggregate: Coarse aggregate). Other mixes were also prepared in which OPC was replaced in part by FA and MK. Slump flow, T_{50} slump flow, V-funnel, L- box, U- box and J-ring experiments were performed to find the properties of fresh concrete mixes to act as SCC. 100 mm cubes of different compositions were cast. These samples were submerged in tap water as well as in 3% HNO_3 solution for curing periods of 7, 28, 56 and 90 days. The results of fresh properties of both the mixes are presented in Table. 1.

Table 1. Test results of fresh SCC

Sl. No.	Tests	Results		EFNARC limits	
		RC	FA+MK mixed SCC	Min	Max
1.	Slump flow	690 mm	710 mm	650 mm	800 mm
2.	T_{50} time	5.0 sec	4.0 sec	2 sec	5 sec
3.	V- funnel	11 sec	9.5 sec	6 sec	12 sec
4.	L-box (h_2/h_1)	0.9	0.8	0.8	1.0
5	U-box (h_2-h_1)	28 mm	27.5 mm	0	30 mm
6.	J-ring	10 mm	9 mm	0	10 mm

Results and Concluding Remarks

The followings were concluded from the experimental study;

- Above experimental study shows that use of FA and MK improves the durability of SCC.
- The workability of FA+ MK mixed SCC was found more in comparison to RC.
- The compressive strength of FA+ MK mixed SCC specimens cured in tap water was higher than the RC. The compressive strength of both the mixes was decreases in acidic solution for all the exposure periods; however, the decrease is lower in case of FA+ MK mixed SCC.
- It was found that in RC, the weight loss of 0.50, 0.85, 1.02, and 1.32% is found after 7, 28, 56 and 90 days curing in acidic environment respectively. For RHAC, the respective values are 0.43, 0.72, 0.92, and 1.01%.
- It was found that there is a gradual change in colour from grey to dark yellow, as the exposure period in acid solution increases.

References

1. J. Monteny, N. De Belie, E. Vincke, W. Verstraete and L. Taerwe. (2011) Chemical and biological test to simulate sulphuric acid corrosion of polymer modified concrete. Cement and Concrete Research, 31(9): 359-1365.
2. Harrison W.H. (1987) Durability of concrete in acidic soils and waters Concrete, 21(2): 18-24.
3. Freidin C. (1999) Behavoiur of Silica- Concrete based on quartz bond in sulpuric acid. Cement and Concrete Composites, 21(4): 317-323.
4. Mehta P.K. (1985) Studies of chemical resistance of low water / cement ratio concrete. Cement and Concrete Research, 15(6): 969-978.
5. Khayat, K.H., Assaad J., Daczko J. (2004) Comparison of field- oriented test methods to assess dynamic stability of Self – Consolidated Concrete. ACI Material Journal, 101(2): 168-176.

Durability of Rice Husk Ash Concrete in Acidic Environment

A. Amrendra Singh[1*], B. Rakesh Kumar[1], C. P. K. Mehta[1], D. Deep Tripathi[1]
[1] *Department of Civil Engineering, Motilal Nehru National Institute of Technology Allahabad, Prayagraj, U.P., India.*
* *e-mail: corresponding.amrendrasingh859@gmail.com*

Introduction

The main factor which affects the durability of the concrete are the constituent materials, micro and macro-structure of the concrete and the external environment. Rice Husk Ash (RHA) is one of the mineral admixtures which improves the durability of the concrete and Nitric acid (HNO_3) is the chemical which influences the durability of concrete. In this paper, the experimental results of both Normal Concrete (NC) and RHA Aided Concrete (RHAC) were compared from durability point of view. Grade of M25 of concrete was prepared using normal constituents for referral concrete and 15%, by weight of Ordinary Portland Cement (OPC) was replaced by RHA i.e. (RHAC).The cubes of both the NC and RHAC were cured in potable water as well as in 5% Nitric acid solution till 90 days. The compressive strength of cubes was determined after 7, 28, 56, and 90 days curing. The concrete incorporating 10% of the RHA as a OPC replacement had some what higher compressive strength and higher resistance to chloride-ion penetration compared with the RC of the same water-to-cementitious materials ratio [1]. The replacement with 30% of reburnt RHA leads to substantial improvement in the permeability properties of blended concrete when compared to that of unblended OPC concrete [2]. The effect of nitric acid on laterite concrete and reported that compressive strength significantly reduces with increase in acid concentration, immersion period and laterite content and the effect of richness of mix on resistance of laterized concrete to the acidic aggression becomes more pronounced at the highest (50%) laterite content [3].
To study the weight change characteristics, the weights of the samples were taken after 7, 28, 56, and 90 days. Visual assessments were also carried out and photographs were taken to get some idea about how Nitric acid affects the appearance of the concrete. The XRD analysis was carried out to estimate the effect of nitric acid on the constituents of concrete. The higher peaks of Calcium Silicate Hydrate(C-S-H) are detected in RHAC samples.

Materials and Methods

OPC of Grade 43 (Brand-Jaypee) was used in this experiment. The physical properties of OPC was determined as per IS 8112-1989. Normal consistency, initial setting time and final setting time of cement was 31%, 92 minutes and 240 minutes, respectively and 28 days compressive strength of mortar was 52.50 N/mm^2.The natural river sand was used for making referral concrete which conforms to zone II as per IS-383-1987 [8] and its fineness modulus, specific gravity and bulk density value were found to be 20492, 2.48 and 1680 kg/m^3 respectively. Coarse aggregate of size 10 mm and 20 mm conforming to IS 383-1987, were used. The bulk density values of 10 mm and 20 mm were 1590 kg/m^3 and 1560 kg/m^3, respectively. The specific gravity of 10 and 20 mm aggregates was found 2.67 and 2.7, respectively. The fineness modulie and water absorption of 10 mm and 20 mm aggregates was found 6.25, 7.27 and 1.2%, 1.1 %, respectively.
The concrete mix design was carried out as per the procedure given in IS: 10262 (2009). And mix proportion was Cement: FA: CA :: 1: 1.31: 2.46 with water content of 186 litre/m^3. The water/cement ratio was 0.40. OPC was replaced with RHA at different level of 10, 15, 20 and 25% on equal mass basis to get the optimum replacement level of OPC by RHA. Concrete cubes of size 150mm were cast to check the compressive strength after curing at 7 and 28 days. Concrete samples of size 100×100×100mm were cast as referral concrete and RHA mix concrete for further study. These samples were submerged in tap water as well as in 5% HNO_3 solution for curing periods of 7, 28, 56 and 90 days for each parameter.

Results and Concluding Remarks

From the experimental study following were concluded;

- The compressive strength of water cured RHAC is higher than the NC. The compressive strength of both the NC and RHAC decreases in acidic solution for all the exposure periods; however, the decrease is lower in case of RHAC.
- It was found that in NC, the weight loss of 0.54, 0.89, 1.23, and 1.35% is found after 7, 28, 56 and 90 days curing in acidic environment respectively. For RHAC, the respective values are 0.50, 0.74, 0.93, and 1.13%.
- There is a gradual change in colour from grey to dark yellow, as the exposure period in acid solution increases from 7 to 90 days. This might be due to the corrosion of the surface of concrete samples.
- The higher peaks of Calcium Silicate Hydrate(C-S-H) are detected in concrete samples containing RHA. There is a noise observed in the peaks of samples cured in acidic solution.

References

[1]. M.H. Zhang, R. Lastra and V.M. Malhotra. (1996) Rice-husk ash paste and concrete: some aspects of hydration and the microstructure of the interfacial zone between the aggregate and paste. Cement and Concrete Research, 26 (6): 963-977.

[2]. D.D. Bui, J. Hu & P. Stroeven. (2005) Particle size effect on the strength of rice husk ash blended gap-graded Portland cement concrete. Cement & Concrete Composites, 27: 357–366.

[3]. Olubisi Ige, Stephanie Barnett, John Chiverton, Ayman Nassif & John Williams. (2017) Effects of steel fibre-aggregate interaction on mechanical behaviour of steel fibre reinforced concrete. Cement and Concrete Science, 116 (4).

Eco-Friendly Utilization of Industrial Sludge as a Building Material: A Study of Steel Industries in Tarapur Region, Maharashtra

Tarang Jobanputra[1*], Vaidik Gajera[2], Gaurav Kapse[3]
[1, 2] Student, Manubhai Shivabhai Patel Department of Civil Engineering, Charotar University of Science and Technology, Changa, Gujarat, India – 388 421*
[3]Assitant Professor, Manubhai Shivabhai Patel Department of Civil Engineering, Charotar University of Science and Technology, Changa, Gujarat, India – 388 421
**e-mail: tarangjobanputra100@gmail.com*

Introduction

21st Century, a point, where we have reached a stage that clamours for protecting the planet have proliferated substantially. Increasing pressure on secured landfill sites and stringent environmental laws has made it necessary for companies to follow waste management practices. In India, there are a total of 41,523 industries generates about 7.90 million tonnes of hazardous waste annually, but about 63% of waste is being dumped in the landfills, and thus creating ecological problems*[1].

In the contemporary practices of reuse and recycling, many researches had been done to reuse the steel industrial waste in the construction, since the chemical components of many building materials are known to be compatible with iron (Khajuria et al. 2013, Yi et al. 2012). Also it is economically beneficial as well since material cost can be reduced.

In the present work, an intimate study had been done to reuse the steel industrial waste sludge (Generation rate = 500T+/month) namely, "Effluent Treatment Plant (ETP) Sludge" and "Phosphate Processed Bonder sludge", by taking into account the paradigm of Circular Economy. The components of study included the amount of waste generated, properties (both physical & chemical), research and its utilization in construction works, so as to reduce the load on the secured landfill sites of Taloja and subsequent cost associated.

Materials and Methods

For the purpose of making blocks, apart from two types of hazardous wastes, cement and lime for the binding purpose and grit dust and aggregates for providing bulk were used. Part replacement method was used by optimizing various materials and to arrive at a final engineered design. The practical study was divided into two parts:

1. Characterization of Sludge: As there was very little info available on the sludge, the properties of both the sludge were checked for its physical and chemical properties.

2. Experimental Trials for Mix Proportions: For the purpose of testing, the standard procedure of Batching, Mixing, Placing, Curing, and Testing was adopted (IS 516- 1959). The experimental trials adopted based on part replacement sequence, materials were optimized step-by-step, and the best proportion in terms of strength was selected. Trial 1-4 indicates optimization of ETP sludge with lime (Fig.1). Later, Trail 5-8 for Bonder sludge, Trail 9-12 for Aggregates, Trial 13-16 for Grit dust, and Trial 17-22 for Cement, were adopted as per the procedure (Figures not mentioned).

Results and Concluding Remarks

The Physical analysis revealed that moisture content of around 2/3rd and 1/4th of its weight was there in ETP and Bonder sludge respectively. The XRF and chemical analysis revealed a *high* concentration of Fe and Pb, and Fe and Zn were observed in ETP and Bonder sludge, respectively. A high concentration of Fe

*[1] The above information pertaining to year other than 2008, have not been scrutinized and verified by CPCB.

bolstered the claim to use steel industrial sludge, as it will have better bonding capacity with cement compounds. With this background, experimental trials were designed and tested for compressive strength.

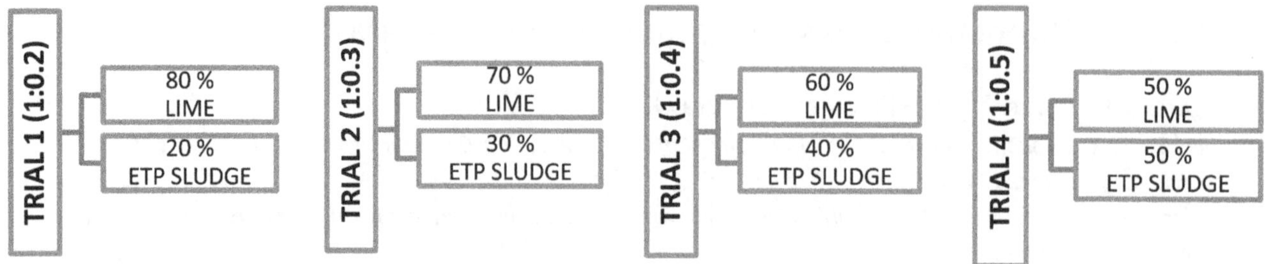

Figure 1.Optimization of ETP sludge with Lime

After the rigorous process of trial and testing, the final proportions, based upon their strength, were fixed from the results of Trial 17-22. The final results from the test showed a significant strength of 3.58 Mpa. Post studies revealed that the density, water absorption and toxicity of blocks were also found within limit.

Overall, these blocks made with steel industrial sludge were found lighter in weight (at least by 60% as compared to brickwork), which will lead to the more economical frame structure, and it can be used in non-structural elements. The experimental results indicated that there is an increase in strength with decreasing in lime content and increasing cement content, which indicated that the sludge material is more suitable for cement (Strength = 3.58Mpa) as a binder as compared to that of lime, but for a cost-effective solution, lime can be used (Strength = 3.29Mpa). The maximum strength was obtained with 44 % of Binding material (Cement + Lime), 22% Grit dust, and 35% of Hazardous Material (ETP + Bonder), and it was considered the best suitable proportion for the blocks.

References

1. Existing Waste Mangaement Scenario in India – Central Pollution Control Board, India. http://cpcbbrms.nic.in/writeread/HW%20Management.pdf
2. Khajuria R, and Siddique R (2014) Use of Iron Slag as Partial Replacement of Sand to Concrete. International Journal of Science, Engineering and Technology Research, Vol 3, Issue 6. http://ijsetr.org/wp-content/uploads/2014/06/IJSETR-VOL-3-ISSUE-6-1877-1880.pdf
3. Yi H, Xu G, Cheng H, Wang J, Wan Y and Chen H (2012) An overview of utilization of steel slag. Procedia Enviornmental Sciences 16 (2012) 791 – 801. https://doi.org/10.1016/j.proenv.2012.10.108

Durability Performance of high strength concrete with copper slag and carbon fiber admixture

kHima Bindu[1], G.Lalitha[2]

[1]Civil(structural),VNRVJIET,M-Tech,Hyderabad,India.
[2]CivilDepartment,VNRVJIET,,Assistantprofessor,Hyderabad,India
himabindu453@gmail.com[1] lalitha_g@vnrvjiet.in[2]

Introduction

Intent of present research work is to use copper slag as a partial substitute of fine aggregate as it is a contemporary imperishable building material. Along with copper slag, carbon fibers are used as reinforcing material in concrete..

In present scenario growth of introducing alternative materials in different forms for all kind of construction will limit the pressure on utilization of natural materials have been noticed, which will show impact on economic purpose of the project and also help in protection of environment. Now a days many industries are releasing by products and waste materials which is very difficult situation to dispose and dump which will show effect on health .Among the industrial waste copper slag is one such material is a by product produced during smelting of copper, though copper slag used in abrasive tool manufacturing and sand blasting industries ,left over waste is neither recycled nor recovered and can be used as a building material as affine or coarse aggregate replacement due to its sustainable, durable, superior ,soundness,toughness characteristics It is preferred as a partial substitute material of fine aggregate. Binaya patnayaik; Sheshadri shekhar(2015); stated by their experimental study the concrete containing copper slag has a good resistance against sulphate attack. The main objective of the study is to find the durability performance of concrete (M-60)With partial replacement of fine aggregate with copper slag in 0%,10%,20,30,%40%.50% various proportions along with constant 0.25% carbon fiber admixture for every mix. Acid attack test, Rapid chloride permeability test, Abrasion test were conducted to check the performance of concrete. The randomly oriented Carbon fibers as an admixture start arresting the inhibited cracks in concrete .It is used in repair of old buildings which has been effected by shear and flexure failure in the form of carbon fiber wrap.The most favourable characteristics such as High chemical resistance ,Low weight. High temperature tolerant made it to use as an admixture in concrete.

Materials

Cement : cement acts as a good binder in concrete formed by the heating of lime clay ,shell chalk forms hard rock and grounded into fine powder forms cementOPC-53 Grade cement was used .

Fine aggregate: Fine aggregate pass through 4.75mm was obtained from krishna river belong to category of zone-2

Coarse aggregate: The size of the aggregate used is 12.5mm which was purchased from dumping quarry yard in hyderabad.

Copper slag: The irregular,black,glassy and small granular slag have similar properties as river sand was obtained from Vivekananda chemicals,jagadgirigutta ,Hyderabad.

Carbonfiber: Chopped 6mm length carbon fiber with 7µm dia was obtained from vruksha composites, Tenali .

Water: potable tap water was used which is free from pathogens and bacteria.

Methodology

Sieve analysis : In order to determine the particle size distribution the sieves were placed in order and seived .The retained material was tested as per IS:2386(Part-3),1963.

Mix design

As Per ACI 211-1993Code-5 Mix design was prepared for high strength concrete. Quantities were calculated in kg/m^3.

Results and Concluding Remarks

Acid attack test :The cube samples of size 150mm was immersed in hydrochloric acid of 5M. After 28 days of normal curing .The pH was monitered regularly and maintained constant throughout the test the manuscript.The weight loss is measured and compressive strength is tested after drying that is after removal of cubes from acid solution. And average loss of strength is calculated. The test is carried as per ASTM-C452 guidelines.

Table- 1 The reduction of strength of specimen subjected to acid exposure

Sl. No.	Mixes	28daysf$_{ck}$ (N/mm^2)	Strength after immersion in 5NHCL(N/mm2)	% loss in strength after immersion in 5N HCL
			28Days	28Days
1	CS0	59	50	15.25
2	CS10	91.6	85.52	6.637
3	CS20	103.54	96.07	7.214
4	CS30	115.33	114.31	0.88
5	CS40	96.61	93.02	3.71
6	CS50	74.6	71.91	3.605

Rapid chloride permeability test: RCPT specimens of size 100mm dia and 50mm height was used to carry the test.The digital led display was placed which indicates the voltage available across concrete specimen.Nacl of 2.4M is filled in one chamber and 0.3 M NaoH is taken in another chamber. The chloride ion were forced to migrate through centrally placed vacuum saturated concrete specimen under an impressed DC voltage of 60volts.The resistance of concrete to chloride ion penetration has no bias because the value of this is defined by test method as per ASTM-C1202.After six hour exposure period the interpretation is larger the coulomb indicates more permeable the concrete specimen and low coulombs indicate concrete is low permeable. The passage of current through concrete specimen replaced with copper slag is very low and less permeable compared to nominal mix as per table-2.

Table-2 The amount of charge passed and rating of chloride permeability according to ASTM-C1202

Type of mix	Charge passed (coulombs)	Chloride ion penetration
Nominal mix	4051.8	High
CS10	921.6	Very low
CS20	522.9	Very low
CS30	345.6	Very low
CS40	391.5	Very low
CS50	1043.1	Low

Abrasion resistance of concrete (under water method): In my research study Abrasion specimen of 300mm diameter and 100mm height is casted for 6 different mixes. Abrasion resistance refers to the ability of materials and structures to withstand abrasion .It is a method of wearing down or rubbing away surface by means of friction.This procedure simulates the abrasive action of water borne particles(silt,sand,gravel,and other solids).The test results are evaluated according to ASTMC 1138-1997.

Table-3 The abrasion resistance of concrete.

Mix	% weight loss in Abrasion
CS0	0.280
CS10	0.0275
CS20	0.0331
CS30	0.0345
CS40	0.0345
CS50	0.0545

References

1)Binaya patnaik, sheshadri shekhar(2015) "Strength and durability properties of copper slag admixture concrete" in International research in engineering and technology, eSSN:2319-1163A .

2)Ayano T., and Sakata K., "Durability of concrete with copper slag fine aggregate", In: Proceedings of the 5th CANMET/ACI international conference on durability of concrete, SP-192; 2000. p. 141–58.

3)Caliskan S. and A. Behnood, "Recycling copper slag as coarse aggregate: hardened properties of concrete", Proceedings of seventh international conference on concrete technology in developing countries Malaysia (2004), pp. 91–98.

4)Reported by ACI committee 211 "Guide for selecting proportions of high strength concrete with Portland cement"ACI211.4R-93,pp.1-13,1998.

Development of fragility curve for existing soft storey RC building with varying soft storey heights using HAZUS method

A. Srikanth[1] and B. Narender[2*]
[1] *PG Student, Department of Civil Engineering, Anurag Group of Institutions, Hyderabad, India*
[2] *Associate Professor, Department of Civil Engineering, Anurag Group of Institutions, Hyderabad, India*
** e-mail: drnbodigece@cvsr.ac.in*

Introduction

Earthquakes are one of the most unpredictable and devastating natural hazards. They pose a hazard to community, property, and population. In 2001 Bhuj Earthquake, structures experience the damage due to multiple reasons one such reason is soft storey construction. In India, most of the Urban cities located in moderate to high seismic zone, where mid-rise and soft-storey buildings are predominantly constructed and designed to resist the gravity loads. These existing buildings are not safe for future events of earthquakes. Hence need to assess the vulnerability of existing buildings. The vulnerability of the buildings is measured using the fragility curves.

This paper presents a study on the seismic performance of existing soft storey buildings and to develop the fragility curves for 5 story building (mid-rise). The fragility curve is defined as the conditional probability which gives the probability of exceeding a predefined damage state for a given earthquake intensity measure. The soft-story effect of the building is defined by varying the height of the storey. In recent years some researches investigated the effect of soft storey on the response of the structures. Andrian and Andres 2015 investigates the seismic vulnerability of the soft storey irregular buildings. They employed the pushover analysis to study the vulnerability of the irregular buildings and investigate the performance of the building models with varying soft storey height with considering soft storey height of 3.38m, 3.5m, 4.0m, 4.5m, 5.0m, 5.5m, and 6.0m. Linda et al. 2017 assess the performance of the moment-resisting frame with fragility curves and compared it with the frame shear wall system. He defines the damage states based on HAZUS MH MR5 (Loss estimation methodology, developed by FEMA, 2010) as slight, moderate, extensive, and complete and based on ATC-40 (Seismic Evaluation and Retrofit of Concrete Buildings, Volume 1, 1996) as immediate occupancy, life safety, and structural stability. He uses the capacity spectrum method to develop the fragility curves.

Materials and methods

Five storeys reinforced concrete building is designed based on IS 456-2000 to resist the gravity loads. To see the effect of soft storey height, an infill frame with varying heights including 3.5m, 4.5m, and 5.5m is employed. To compare the infill effects, bare, infill and partial infill frames are selected. The gravity loads are considered as per IS 875 parts 1 and 2. A live load of 2.5kN/m^2 and 1.5kN/m^2 are assigned for stories and roof respectively as per IS 875 part 2. The concrete of M25 and reinforcement bars of Fe415 grades are employed in this study. A knowledge factor (material modification factor) of 0.8 proposed by IS 15988-2013 is employed to capture the minor deterioration in the structure. The unreinforced masonry infill walls are modelled as a diagonal strut based on provisions is given by IS 1893-2016 part 1. Compressive strength of 4.33 MPa of a brick prism is used in this study. Three-Dimensional Building models created and Analysis is carried out by using SAP2000.

Fragility curves were developed analytically for all the considered building models as per the procedure is given in HAZUS technical manual. The HAZUS technical manual develops the capacity spectrum methodology and uses the standard normal cumulative distribution function to derive the fragility curves. In this method, each fragility curve is characterized by a limit state (median) and uncertainty (standard deviation) values. The response of the building models is carried out in terms of pushover curves employing the pushover analysis and then the capacity curve is changed into the spectrum capacity in the format of ADRS (Acceleration Displacement Response Spectrum). The damage states for each capacity spectrum were defined based on HAZUS MH MR5 as slight, moderate, extensive and complete damage. Limit state (median) values for each damage state were obtained from the capacity spectrum in terms of Spectral displacement. These spectral displacement values were converted into Spectral acceleration from the spectrum capacity and then used to develop the fragility curves.

Results and concluding remarks

The results of this study present in terms of fragility curves. The fragility curve which describe an exceeding the predefined damage states for a given spectral acceleration (intensity measure) is derived in this study for building with varying soft-story height such as 3.5m, 4.5m, and 5.5m., for buildings with different frame type (bare, infill and partial infill) and for building with considering no deterioration and minor deterioration. The vulnerability of the building models is predicted in four seismic zones with the spectral acceleration of 0.1g, 0.16g, 0.24g, and 0.36g represent Zone II, III, IV, and V respectively.

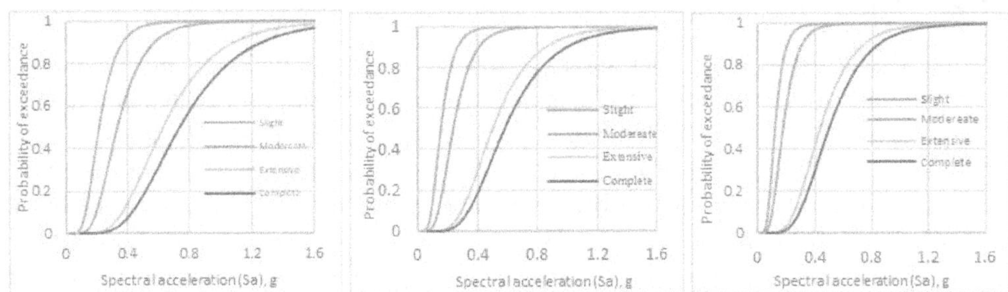

Figure 1: Fragility curve for an infill frame with a soft storey height of a) 3.5m, b) 4.5m, and c)5.5m.

From the fragility curves it is observed that, for the expected earthquake intensity in Zone V with an intensity measure of 0.36g of spectral acceleration, the probability of damage of an infill frame with soft storey height of 3.5m, 4.5m, and 5.5m for complete damage state is approximately 4.65%, 12.5%, and 24.6% respectively. Hence, from the results, it is concluded that for the same expected earthquake intensity, the probability of damage is increased as the height of the soft storey increases.

References

Andrian Fredrick C. Daya and Andres Winston C. Oretaa (2015). "Seismic vulnerability assessment of soft-story irregular buildings using pushover analysis." The 5th International Conference of Euro Asia Civil Engineering Forum (EACEF-5). Procedia Engineering 125 (2015) 925-932. https://doi.org/10.1016/j.proeng.2015.11.103

Linda astriana, Senot sangadji, Edy purwanto, S. A. christiawan. (2017). "Assessing the seismic performance of moment resisting frame and frame shear wall system using seismic fragility curves." Sustainable Civil Engineering Structures and Construction Materials, SCESCM 2016. Procedia Engineering 171 (2017) 1069-1076. https://doi.org/10.1016/j.proeng.2017.01.457

F. Hosseinpour, A.E. Abdelnaby (2017). "Fragility curves for RC frames under multiple earthquakes." Soil Dynamics and Earthquake Engineering 98 (2017) 222-234. https://doi.org/10.1016/j.soildyn.2017.04.013

F. cherify, M-N Farsi, S. Kaci, O. Belaidi, F. Taouche-Kheloui. (2015). "Seismic vulnerability or reinforced concrete structures in Tizi-Ouzou city (Algeria)." 1st International Conference on Structural Integrity. Procedia Engineering 114 (2015) 838-845. https://doi.org/10.1016/j.proeng.2015.08.037

Federal Emergency Management Agency (FEMA). HAZUS MH MR5, technical and user's manual, Federal Emergency Management Agency, Washington, DC. (2010).

Bureau of Indian Standards (BIS). Indian standard for plain and reinforced concrete code of practice. IS 456-2000. New Delhi, India. (2000).

Bureau of Indian Standards (BIS). Indian standard criteria for earthquake resistant design of structures. IS 1893-2016. New Delhi, India. (2016).

Keywords: soft storey Building, equivalent diagonal strut, knowledge factor, pushover analysis, fragility curves, seismic zones, and HAZUS.

Review on performance evaluation of Geopolymer concrete

Ishita Verma[1*] Shobha Ram[2] Nirendra Dev[3] Gaurav Chand[4]
[1]Graduate student, Civil Engineering, Gautam Buddha University, Greater Noida, U.P.
[2]Head of Department, Civil Engineering, Gautam Buddha University, Greater Noida, U.P.
[3]Head of Department, Civil Engineering, Delhi Technological University, Delhi
[4]Faculty, Civil Engineering, Gautam Buddha University, Greater Noida, U.P.
*shtverma22@gmail.com

Introduction

With the increasing growth of urbanisation and industries, the emission of CO_2 continues to rise. Cement, being the major constituent in building construction, contributes about 7-8% to the emission of green-house gases [1, 2015]. With sustainable development in view, emphasis has been provided in developing materials which not only provide similar mechanical properties as cement but are eco-friendly also. Geopolymer is one such material which was first introduced by the French scientist Joseph Davidovits in 1979. It is an alkali activated inorganic material composed of aluminium and silicon effluent containing base materials to serve as a binder. It is cost effective as a variety of low cost materials or industrial by-products such as Fly ash (FA), Bottom ash (BA), Volcanic ash (VA), High calcium fly ash (HCFA), Waste paper sludge ash (WPSA), Ground granulated blast furnace slag (GGBFS), Rice husk ash (RHA), Air cooled slag (ACS) Steel slag (SS), Metkaolin, Kaolin, Clay, Silica waste are used as base materials [2, 2018].

While Geopolymer shows zeolite and ceramic like chemical properties, it has an amorphous structure. For a Geopolymer to have appreciable properties, it should have base materials with enough spherical glass beads content and extremely amorphous structure. The alkaline solutions often used are sodium silicate (Na_2SiO_3), sodium hydroxide (NaOH) and potassium hydroxide (KOH). Jumrat et al. [3, 2011] reported that the flow values of the Geopolymer mortar decreases with an increase in the ratio of fly ash to alkaline solution and that of Na_2SiO_3 to NaOH. This occurs due to the high viscosity of Na_2SiO_3 hence more water has to be added to obtain a workable mix. Laskar and Talukdar [4, 2017] stated that type of alkali activator has a significant effect on the workability of Geopolymer mortar and working performance of the mix consisting of NaOH as alkali activator is superior to the mix containing both NaOH and Na_2SiO_3. Further, they reported that Geopolymer mortar having NaOH as alkali activator has longer setting time than mortar containing mixture of NaOH and Na_2SiO_3 as alkali activators. Geopolymer have high early strength due to its short hardening time of 2-4 hours. It attains about 70% of its ultimate strength with 4-hour curing showing similar strength characteristics of rapid hardening cement but having better physical properties [2, 2018]. The present paper reviews the development of compressive strength of Geopolymer concrete made with different base materials by three different researchers.

Materials and Methods

Okoye et al. [1, 2015] carried out the experiment by taking five mixes of Geopolymer concrete using Fly ash and Kaolin as base materials. Coarse aggregate of size 20 mm and 10 mm and river sand as fine aggregate were used. Mix 1 was prepared only using Fly ash, mix 2 was prepared using only kaolin, mix 3 had 90% by weight of Fly ash and 10% by weight of Kaolin and mix 4 & mix 5 had 50% each of Fly ash and Kaolin. They all were prepared using NaOH (14M) as alkali activator except mix 5, which had KOH (14M) as alkali activator. An OPC mix of grade M30 was also prepared for comparison with Geopolymer concrete mix. Naphthalene sulfonate (N.S) based superplasticizer for better workability was used. The Geopolymer concrete mix was prepared like conventional OPC mix. The compression tests were then carried out. Variations in the compressive strength of six mixes at 28 days were analysed.

Kumar et. al. [5, 2018] studied the mechanical properties of Geopolymer concrete by preparing three mixes with varied percentage in terms of weight of Ground granulated blast furnace slag (GGBFS) and Metakaolin. GPCG20M80, GPCG50M50, GPCG80M20 with ratio 80%-20%, 50%-50%, 20%-80% of GGBFS and Metakaolin respectively were prepared. Alkaline activator solution formed by NaOH (10M) and Na_2SiO_3 (10M) in the ratio 2.5 were taken. For better workability superplasticisers were also used. Then with the conventional method used for OPC, the Geopolymer concrete cubes were cast. The compressive strength was evaluated for a curing period of 7 days and then for 28 days of curing.

Rmaujee et. al. [6, 2017] investigated the mechanical properties of Fly ash based Geopolymer concrete. Low grade Geopolymer concrete (G20), Medium grade Geopolymer concrete (G40), High grade Geopolymer concrete (G60) were prepared and similarly M20, M40 and M60 grade cement concrete were prepared. Fly ash was used as the base material and NaOH in the flaked form was taken as the activator. The alkaline solution of 8-16 M was formed 24 hours prior to use by dissolving NaOH flakes in distilled water. The mass of coarse and fine aggregate used was about 75%-80% of the total mass. Polycarboxylic based superplasticizer were used for good workability. The compression testes were then performed.

Results and Concluding Remarks

From the works of Okoye et. al., it is observed that the strength of OPC mix was more than all the Geopolymer concrete mixes, except for the mix 4 having equal parts of Fly ash and kaolin. Mix 3, where only 10% of Fly ash was replaced by kaolin, showed lower compressive strength than OPC. Mix 1 and Mix 2, prepared from only Fly ash and kaolin respectively, showed favourable strength to OPC mix.

The results of Kumar et. al. [5, 2018] have shown encouraging results in terms of increase of compressive strength in all the mixes tested. GPCG80M20 had the maximum strength which underlines the importance of the contribution of GGBS if absolute values of compressive strength are considered. Increases in percentage of strength from 7 day to 28 day for the mix GPCG20M80, GPCG50M50 and GPCG80M20, were found to be 43.75% 33.74% and 31.55% respectively. The amount of Metakaoline is important if percentage increase in compressive strength from 7 days to 28 days is considered. It seems that the rate of increase is controlled by Metakaolin as it is a pozzolanic material.

The results of Ramujee et. al. [6, 2017] showed that Geopolymer concrete mix have similar compressive strength to that of cement concrete. It was reported that average compressive strength of G20, G40, G60 were 31.3, 50.6, 71.06 respectively and that of M20, M40, M60 were 27.5, 48.8, 68.6 respectively.

From this review study, following conclusions may be drawn.

(1) Geopolymer concrete exhibits similar mechanical properties to that of OPC.

(2) Results shows that materials which are rich in aluminium silicate such as Fly ash, metakaolin, GGBFS are suitable materials for Geopolymer concrete.

(3) Based on its mechanical properties and behaviour, Geopolymer exhibits some characteristics which enables it to be used as an engineered material.

References

[1] Okoye F.N, Durgaprasad J, Singh N.B. (2015) Mechanical properties of alkali activated flyash/Kaolin based geopolymer concrete. Construction and Building Materials 98: 685-691. www.elsevier.com/locate/conbuildmat.

[2] Zhanga P, Zhenga Y, Wangb K, Zhanga J. (2018) A review on properties of fresh and hardened geopolymer mortar. Composites Part B 152: 79–95. www.elsevier.com/locate/compositesb

[3] Jumrat S, Chatveera B, Rattanadecho P. Dielectric properties and temperature profile of fly ash-based geopolymer mortar. Int Commun Heat Mass Tran 2011;38(2):242–8.

[4] Laskar SM, Talukdar S. Development of ultra fine slag-based Geopolymer mortar for use as repairing mortar. J Mater Civ Eng 2017;29(5):1–11.

[5] Kumar P, Pankar C, Manish D, Santhi A.S (2018) Study of mechanical and microstructural properties of geopolymer concrete with GGBS and Metakaolin. Materials Today: Proceedings 5: 28127–28135.

[6] Ramujee K, PothaRaju M. (2017) Mechanical Properties of Geopolymer Concrete. Materials Today: Proceedings 4: 2937–2945.

Mix Proportioning of High Strength Concrete Using VSI Sand

R. R. Khartode[1*], D. A. Nikam[2], G. N. Narule[1]

[1]*Assistant Professor, Department of Civil Engineering, VPKBIET, Baramati, Pune.*

[2]*P.G.Student, Department of Civil Engineering, VPKBIET, Baramati, Pune.*

[*] *e-mail: rushikesh.khartode@vpkbiet.org*

Introduction:

High strength concrete plays an important role in present constructional activities such as high rise buildings, offshore structures, long-span bridges and structures at marine environment. These structures required high strength concrete for its stability and durability for a lifetime. Mineral admixtures such as silica fume are used to fill microscopic voids to get required strengths. To overcome the deficiency of river sand and to meet the requirement of the high strength concrete, it is required to formulate the high strength concrete using VSI sand.

The concrete mix was designed as per IS 10262:2019 to achieve a grade of M75. Nine different mix proportions were casted with 8%, 10% & 12% of silica fume with partial replacement of cement respectively whereas super plasticizer is used at 1%, 1.2% & 1.4%.

Materials used

1 Cement: Ordinary Portland cement of 53 grade confirming to IS 12269-1989, (Locally available brand Birla super) is used for this present study. The 28 days compressive strength of cement obtained is 63.50 MPa.

2 Coarse Aggregates: Crushed stone aggregates having a maximum size of 20mm size is used. Sieve analysis has been carried out by blending the aggregates of size 20mm to 12.50mm to get the good fineness modulus. The coarse aggregate having specific gravity 2.96 and fineness modulus 5.67 is used.

3 Fine Aggregates (VSI sand): VSI/Artificial sand is manufactured in vertical shaft impact crusher was collected from VSI crusher, Vasunde, dist. Pune. It is used as an alternative for river sand in the fully replacement of fine aggregates. The fine aggregate having specific gravity 2.94 and fineness modulus 4.54 is used.

4 Silica Fume: It is obtained from the Sudha enterprises, Pune. The specific gravity of Silica Fume is 2.2.

5 Superplasticizers: MasterEase 5801 as high range water reducing admixture was used provided by BASF Pvt. Ltd. Bhosari, Pune. Superplasticizer based on Polycarboxylate ether was used to impart additional desired properties to the high strength concrete.

6 Water: Potable tap water is used for mixing and curing which is free from deleterious materials.

Figure 1: VSI Sand.

Figure 2: Testing of Specimens.

Methodology

1) Collection of materials.
2) Testing of materials for their physical &chemical properties.
3) Proportioning of aggregates on the maximum density approach.
4) Mix design calculations for given cementitious content by blending of Silica fume in different percentages.
5) Finding water content for a given mix.
6) Carry out trial mixes to get required slump & homogenous mix without honeycombing & segregation.
7) The casting of samples for various cementitious contents with different percentages of silica fume and superplasticizer.
8) Testing of samples at 7 & 28 days age.

Results and discussions

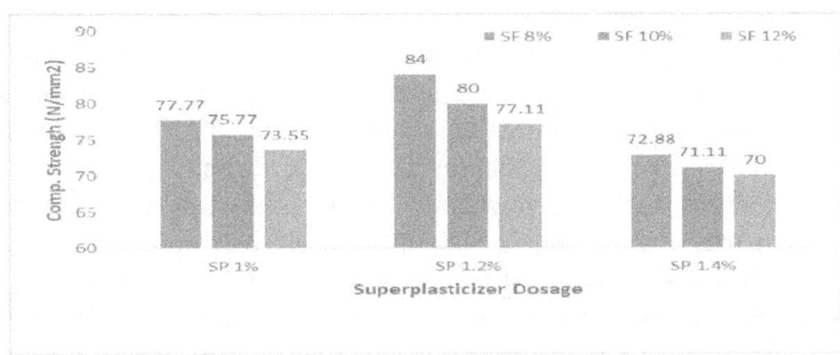

Figure 3: 28 Days Compressive strength results for w/b ratio 0.27

From the graph, it is observed that the compressive strength is maximum for a proportion of 8% silica fume and 1.2% superplasticizer with w/b ratio of 0.27. The compressive strength obtained is 77.77 N/mm^2 for an 8% slilica fume with a 1% superplasticizer. For a further increment of superplasticizer by 0.2%, the compressive strength is increased by 8.01%. But for the next increment of superplasticizer by 0.2%, the reduction in compressive strength observed is about 6.28% of its initial proportion.

Concluding Remarks

1. The addition of Silica fume up to 10%, leads to a significant increase in the characteristic strength of concrete.
2. The workability of concrete decreases with the addition of Silica fume as compared to the conventional mix.
3. Results shows that, the river sand can be fully replaced by VSI sand as per the experimental results.

References

[1] Adams Joe, A. Maria Rajesh, P. Brightson, M. Prem Anand (2013), "Experimental investigation on the effect of M-sand in high performance concrete" American Journal of Engineering Research, Volume-02, Issue-12, pp-46-51.

[2] Boopathi and J. Doraikkannan (2016) "Study on M-sand as a partial replacement of fine aggregate in concrete" International Journal for Research in Applied Science & Engineering Technology, Vol. 3, Special Issue 2, Page 746 – 749.

[3] Kulkarni D. B. and Khartode R. R. (2016) "Mix proportioning of HSC using manufactured sand" International Journal of Science and Research Volume 5 Issue 7, pp-423-429.

[4] Nimitha Vijayaraghavan, and A. S. Wayal (2013) "Effects of manufactured sand on compressive strength and workability of concrete" International Journal of Structural and Civil Engineering Research Vol. 2, No. 4.

[5] Prakash. P (2013) "Strength of concrete by partially replacing the fine aggregate using M-sand" Sch. J. Eng. Tech., 1(4):238-246.

[6] Priyanka A. and Dilip K. Kulkarni (2012) "An experimental investigation on the properties of concrete containing manufactured sand", International Journal of Engineering Research & Technology, Vol. III, Issue II, pp-101-104.

Damage Assessment of Reinforced Concrete multi-storeyed SMRF Buildings

Pritam Hait[1*], Arjun Sil[1], Satyabrata Choudhury[1]
[1] *Department of Civil Engineering, National Institute of Technology Silchar, Silchar, India*
* *e-mail: pritamhait15@gmail.com*

Introduction

According to earthquake history, North-East India and Himalayan region are the most seismic prone zone in India. In this seismic zone (zone-V) many seismic events occurred in the past and will happen in future. Numerous structures have been damaged and life has been deceased due to falling of structures which are vulnerable for a seismic event. Therefore a seismic investigation is essential in this region to determine storey wise damage index under the site-specific ground motion (GM). Structural damage has been quantified by several researchers in past decades such as for a reinforced concrete component, the damage was determined considering the combined effect of deformation and cyclic loading effect [1-2]. Further, DI was evaluated in terms of single or a combination of multiple demand parameters such as maximum deformation, modal parameters, dissipated hysteretic energy, inter-story drift, modal strain energy and initial to final resistance capacity of structures [3-5]. In this context, a comprehensive review has been done by the authors [6-7] to find out the limitation of work in this area. In that overview, it was suggested that a zonal basis [8-9] seismic damage evaluation is essential for a complete reliable damage assessment of structure that can be used in the future as design guidelines. Therefore in this study storey wise damage index and global damage index have been evaluated for 4, 8 and 12 storey SMRF buildings in seismic zone V in India.

Materials and Methods

There are several damage assessment methods are available among them Park-Ang [1-2] method is the preferred choice by the researchers as this method includes both deformation and energy effect and applicable on steel, RC and timber structure. This is why authors have adopted this method for storey wise damage assessment of the buildings. To perform the study a typical plan of 4, 8 and 12 storey reinforced concrete (RC) SMRF building have been considered. RCC design and seismic design were performed as per IS: 456-2000 and IS: 1893 2016 respectively. Capacity design principle i.e. all columns are stronger than beams by 1.4 times was followed in this design as per IS: 13920-2016. Non-linear time history (NLTHA) was performed in SAP2000 in both x and y directions to obtain the structural response. The buildings were analyzed under seven real earthquake ground motions converting into SCGM by Kumar software that matches the raw ground motion data with Indian response spectrum.

Results and Concluding Remarks

In this investigation, it was observed that ground storey is experiencing maximum damage and top storey is experiencing minimum damage for all buildings. Therefore ground storey should be strong enough to withstand the building in its position. It was also observed that ground storey DI (SDI) is higher than GDI therefore, ground storey DI should be considered as separate damage indicator than that of GDI. Storey wise damage index of all buildings in both directions have been plotted to observe the variation of damage with building height. Structural damage concentration has been observed in the lower storey due to the excessive moment in the beams. Moreover, among seven real earthquake GM, Berlongfer and Laisong causes maximum and minimum damage respectively due to highest and lowest strong ground motion duration (SGMD) [10]. The estimated results could be used in planning as design guidelines for the most seismic prone zone in India.

Acknowledgement

This research is funded by TEQIP III. Authors are grateful to TEQIP III for their support.

References

1. Park Young-Ji, Ang Alfredo H-S (1985) Mechanistic Seismic Damage Model For Reinforced Concrete. Journal of Structural Engineering 111(4): 722-739. https://doi.org/10.1061/ (ASCE) 0733-9445(1985)111:4(740)
2. Park Young-Ji, Ang Alfredo H-S, Wen Yi Kwei (1985) Seismic Damage Analysis of Reinforced concrete Buildings, Journal of Structural Engineering. 111(4): 740-757. https://doi.org/10.1061/(ASCE)0733-9445(1985)111:4(722)
3. J. V Amiri, QY Ahmadi, B Ganjavi (2008) Assessment of reinforced concrete buildings with shear wall based on Iranian seismic code (Third Edition). Journal of Applied Science 8(23): 4274-4283. 10.3923/jas.2008.4274.4283
4. Wong K. K. F. and Wang Y. (2001) Probabilistic Structural Damage Assessment and Control Based On Energy Approach. The Structural Design of Tall and Special Buildings 10(4): 283–308. https://doi.org/10.1002/tal.186
5. Guan H, Karbhari VM. (2008) Improved damage detection method based on Element Modal Strain damage index using sparse measurement. Journal of Sound Vibration 309(3-5): 465–494. https://doi.org/10.1016/j.jsv.2007.07.060
6. Hait Pritam, Sil Arjun, Choudhury Satyabrata (2018) Quantification of damage to RC structures: A comprehensive review. Disaster Advances 11(12): 41–59
7. Hait Pritam, Sil Arjun, Choudhury Satyabrata (2019) Overview of damage assessment of structures. Current Science, 117(1): 64-70. 10.18520/cs/v117/i1/64-70
8. Sil Arjun, Sherpa Dawa Zangmu, Hait Pritam (2019) Assessment on combined effects of multiple engineering demand parameters (MEDP) contributing on the shape of fragility curve. Journal of Building Pathology and Rehabilitation 4(5): 3-17. https://doi.org/10.1007/s41024-019-0047-7
9. Sil Arjun, Das Gourab, Hait Pritam (2019) Characteristics of FBD and DDBD techniques for SMRF buildings designed for seismic zone-V in India. Journal of Building Pathology and Rehabilitation 4(1): 1-18. https://doi.org/10.1007/s41024-018-0040-6
10. Özer, E., Soyöz, S., & Çelebi, M. (2012). Effect of Strong Ground Motion Duration on Structural Damage. 15th WCEE, 1-9.

Experimental Investigation on effect of Binder Index on Compressive & Flexural Strengths of GPC with different Alkaline to Binder ratios.

R. Shankaraiah[1], B.Sesha Sreenivas[2], D. Rama Seshu[3]

[1] *Research Scholar, Civil Engineering, Kakatiya University, Warangal,Telangana, India*

[2] *Prof & Principal, University College of Engineering, Kothagudem, Kakatiya University,Warangal, India.*

[3] *Professor of Civil Engineering, National Institute of Technology, Warangal, India.*

* *e-mail: shankarracharla@gmail.com, Cell No. 7093873666.(corresponding author)*

Introduction :

The quest for alternate sustainable technology in construction resulted in the development of alkaline activated binders with promising engineering properties and longer durability has emerged as an alternative to OPC. In this context the development of Geopolymer concrete (GPC) is being viewed as an emerging class of concrete material and could be the next generation concrete for applications in civil engineering infrastructure. The Geopolymer concrete is an environmental friendly material in sense that it uses the industrial by-products such as Ground Granulated Blast furnace Slag (GGBS) and Fly Ash (FA) along with alkaline activated solutions. The commonly used combination of alkaline activator solution is Sodium Hydroxide and Sodium Silicate. These rich in silica by - products form with alkaline solution a binder matrix to bound aggregate in the mixture and to produce the hardened concrete. Geopolymer concrete developed by Prof. Joseph Davidovits, is a new class of concrete that is attracting growing interest around the world due to its environmental and performance benefits compared to conventional Portland cement concrete. Davidovits (1970) reported the use of waste material like Fly Ash (FA) and Ground Granulated Blast furnace Slag (GGBS) and high alkaline solution as activators.

The experimental study consisted of determination of the Compressive Strength of Geopolymer Concrete (GPC) by casting and testing of cubes of size 150 X 150 X 150 mm and Flexure Strength of GPC by casting and testing of prisms of size 400 x 100 x 100 mm for different alkaline to binder ratios.

Materials and Methods:

Fly Ash and GGBS are used as Binders. The Robo Sand (RS) produced by stone crushing was used as fine aggregate and crushed granite of 20 mm nominal size was used as coarse aggregate. Sodium hydroxide solution and Sodium silicate solution are used as Alkaline Solution.

This paper presents an experimental investigation concerning the Compressive & Flexural strengths of GPC and its relation to a new parameter called the "Binder Index (BI)" for different Alkaline to Binder ratios. The Binder Index combines the effect of the GGBS to the Fly Ash ratio and the molarity of the Alkaline activator. The testing of specimens was carried out at the end of 28 days (28D) of outdoor curing. A total of 108 cubes and 108 prisms representing 3 different GGBS / FA ratios (0.25, 0.67, 1.5), 4 different molarities (6,8,10 & 12) of Alkaline solution, 3 different Alkaline to Binder ratios (0.64, 0.55 & 0.45), 3 identical specimens for each variation were cast and tested.

Results and Concluding Remarks:

The results have shown that the Compressive & Flexure Strength of the GPC is significantly influenced by varying the Binder Index and different Alkaline to Binder ratios. The results indicate that a non linear relation exists between the Binder Index and the Compressive strength of the GPC and the Binder Index and the Modulus of Rupture. There is a decrease in the strength of GPC as the Alkaline to Binder ratio decreases.

References:

1.Anuradha R. Sreevidya V. Venkatasubramani R. Rangan BV, "Modified guidelines for gepolymer concrete mix design using Indian Standard", Asian J Civil Eng (Build Hous), 13(3):353-364 (2012).

Potential Use of Waste-Glass Powder in Concrete: State of the Art

Deepa Paul[1*], Bindhu K.R.[2]
[1]Research Scholar, Department of Civil Engineering/College of Engineering Trivandrum, KTU, Kerala, India
[2]Professor, Department of Civil Engineering/College of Engineering Trivandrum, KTU, Kerala, India
[*]e-mail: deepapaul1@yahoo.com

Introduction

Effective substitution of building materials with cheaper and abundant substitute has been a favourite pursuit of decision makers and technologists to promote sustainability in construction field. Glass, being an abundant non-degradable waste material which requires low level technology is an ideal substitute for cement in construction industry. Similarity in chemical composition to cement makes the glass an alternative in the cementitious composites. Application of glass components in construction sector and commercial purpose has increased tremendously and in future it will be difficult to manage the non-degradable waste like glass from scrap generated from glass factories, from demolished building components and discarded bottles which cause considerable environmental pollution. Both environmental and economic advantages can be achieved by the utilisation of waste-glass in the construction industry.

Literature work pertaining towards the utilization of waste-glass as aggregate and as cement substitute in both mortar and concrete are reviewed in the present study. Related research revealed the successful transition of waste-glass usage from partial aggregate replacement to partial cement replacement in concrete. The synergistic effect of waste glass is observed in the creation of a green environment in the concrete construction sector with the proposed eco-cement. Partial substitution of cement by waste-glass powder (WGP) improves the mechanical and durability properties. The amorphous nature of glass powder causes it to fill the voids in concrete, thus making the material impermeable and more durable. The current scenario of applicability of waste-glass powder in cementitious composites and the gap area are identified in the study.

Materials and Methods

From the overview of the previous studies on the use of *waste-glass as fine aggregate replacement*, several drawbacks like ASR expansion, decreased workability, higher bleeding and segregation was observed (Rashad 2014, Saccani and Bignozzi 2010, Park et al.2004, Naik et al.2000, Polley et al.1998). Damaging expansion due to ASR is observed when *recycled glass is substituted as coarse aggregate* (Srivastava et al.2014, Topcu and Canbaz 2004, Johnston 1974, Kou and Poon 2009, Meyer et al.1996) as well. The study revealed that mechanical properties of cementitious composites were enhanced slightly than that of conventional concrete. It is observed that the challenges due to *ASR expansion* have been overcome in the review of literature pertaining to incorporating *waste-glass into concrete as supplementary cementing material* (Federico and Chidiac 2009, Kumarappan 2013, Ashutosh and Sangamnerkar 2015, Sombir 2017, Philips et al.1972). Decrease in particle size leads to a pessimum effect in ASR expansion as well as an increase in compressive strength. The presence of milled waste-glass powder (WGP) suppresses the ASR related expansion and results in enhanced durability characteristics (Nassar and Soroushian 2012, Ahmad 2002, Dyer and Dhir 2001, Shao et al. 2000, Meyer et al.1996). Increased compressive strength and improved durability is observed with cement replacement ratio upto 45% and 60% respectively (Du and Tan 2015, Kamali and Ghahremaninezhad 2015, Schwarz et al. 2008, Shi et al. 2004). Later studies (Du and Tan 2016) proved the presence of higher porosity in the mixes with high volume WGP and concluded that better mechanical properties can be ensured with an optimum replacement of 30% WGP.

Concerned with *mortar study*, the influence of glass powder with cement replacement of 10% enhances the mortar compressive strength and setting time and soundness observed is similar to unblended Portland cement (Ali et al., 2016, Lu et al. 2017, Matos and Sousa 2012, Rajabipour et al.2010). The amorphous WGP is considered as a safe and healthy material to replace cement and quartz powder in

UHPC and the optimum replacement of cement with glass powder was 20% in terms of concrete compressive-strength development (Soliman and Hamou 2016, Siad et al 2016). Better performance in compressive and flexural strengths is obtained with the inclusion of WGP as a supplementary cementitious material with recycled aggregates and shows pozzolanic properties that increase with decrease in particle size (Letelier et al., 2016).

Results and Concluding Remarks

A detailed review of prominent research on the use of WGP in concrete from the environmental point of view has been provided. Good agreement in results is obtained between the experimental values of waste-glass powder concrete (WGPC) with normal concrete.

Past studies revealed that, the consequences faced by glass wastes can be effectively managed by its application in concrete as an eco-cementitious material without affecting the structural integrity of concrete.The controlled production of cement and reduction in environmental contamination can be achieved by the application of WGPC, as a part of promoting green ecosystem. The social relevance of WGP lies in the mitigation of carbon emission by the partial substitution of cement in concrete and thus a measure to reduce environmental impact.

The present study concludes that crushed and reused concrete material can be effectively utilized for developing WGPC. Enhancement in mechanical and durability properties of WGPC was observed with an optimum of 30% substitution of WGP. Durability can be ensured from the past studies that WGPC has good resistance to acid/alkali attack.

According to the literature survey in the past decade, it can be concluded that the post- consumer glass in fine powdered form is a good pozzolan and exhibit competent mechanical and durability properties. Moreover, the generation of WGP neither produces any byproducts during the processing stage nor requires any sophisticated machineries and hence can be regarded as an eco-friendly material. For further consideration, an in-depth qualitative and quantitative analysis of WGPC is mandatory. The results drawn from the previous studies promote an effective foundation for future research.

References

Alaa M.Rashad, (2014) Recycled waste glass as fine aggregate replacement in cementitious materials based on Portland cement. Construction and Building Materials 72:340-357. https://doi.org/10.1016/j.conbuildmat.2014.08.092

Vikas Srivastava, S.P.Gautam, V.C.Agarwal and P.K.Mehta (2014) Glass wastes as coarse aggregate in concrete. Journal of Environmental Nanotechnology 3:67-71. https://doi.org//10.13074/jent.2013.12.132059

L.M.Federico and S.E.Chidiac, (2009) Waste glass as a supplementary cementitious material in concrete-Critical review of treatment methods. Cement and Concrete Composites 31:606-610. https://doi.org/10.1016/j.cemconcomp.2009.02.001

Roz-Ud-Din Nassar and Parviz Soroushian, (2012) Strength and durability of recycled aggregate concrete containing milled glass as partial replacement for cement. Construction and Building Materials 29:368-377. https://doi.org/10.1016/j.conbuildmat.2011.10.061

Hong Jian Du. and Kiang Hwee Tan., (2016) Properties of high volume glass powder concrete. Cement and Concrete Composites 75:22 -29. https://doi.org/10.1016/j.cemconcomp.2016.10.010

Ali A. Aliabdo., Abd Elmoaty M. Abd Elmoaty. and Ahmed Y. Aboshama., (2016) Utilization of waste glass powder in the production of cement and concrete. Construction and Building Materials 124:866 –877. https://doi.org/10.1016/j.conbuildmat.2016.08.016

N.A. Soliman and A.Tagnit –Hamou, (2016) Development of ultra-high-performance concrete using glass powder-Towards ecofriendly concrete. Construction and Building Materials 125:600-612. https://doi.org/10.1016/j.conbuildmat.2016.08.073

Viviana Letelier, Ester Tarela, Rodrigo Osses, Juan Pablo Cardenas and Giacomo Moriconi, (2016) Mechanical properties of concretes with recycled aggregates and waste glass. International Federation for Structural Concrete 18:40-53. https://doi.org/10.1002/suco.201500143

Andrea Saccani and Maria Chiara Bignozzi, (2010) ASR expansion behaviour of recycled glass fine aggregates in concrete. Cement and Concrete Research 40:531-536. https://doi.org/10.1016/j.cemconres.2009.09.003

Image-based analysis of concrete deterioration

V. Guru Prathap Reddy [1], B. Murali Krishna [1,2], L. Pranay Kumar [1], T. Tadepalli[1], P. Rathish Kumar[1], K. Gopi Krishna[1], M. Shashi[1],MVN Sivakumar[1]
[1] Department of Civil Engineering, NIT Warangal, India.
[2] Research Scholar
* *e-mail: vguruprathap@gmail.com*

Introduction

Concrete structures undergo deterioration naturally when affected by external environment. Concrete loses its strength and durability with the passage of time. In order to repair and rehabilitate these structures, the causes and extent of deterioration must be identified and quantified for solutions to be provided. The most common methods for condition assessment of structures are visual inspection and contact based non-destructive tests (NDTs) which can be conducted only for structures that are accessible. In cases when structure is inaccessible, image based condition analysis is feasible. This study identifies and quantifies concrete deterioration using digital image processing techniques. Digital image processing enables correlation of greyscale intensities with various deterioration measures such as mass loss, dimension loss and strength loss.

Materials and Methods

In the present experimental study, M30 grade concrete cubes (60 nos.) are cast using ordinary portland cement (OPC-53), fine aggregate, coarse aggregate, and water. After 28 days of curing, these 150x150x150mm concrete specimens are exposed to chemicals. For durability studies, 3 cubes each are immersed in plastic tubs containing HCl, H_2SO_4, NaCl and $MgSO_4$ at 5% concentration for periods of 3, 8, 15 and 30 days. After exposure to chemical attack, the photographs of all six faces of each cube are captured and greyscale intensity analysis is performed. The samples are weighed, measured and the compressive strength of concrete is obtained using rebound hammer test and destructive test to failure. This enables correlation between various image characteristics and deterioration measures like acid mass loss factor, acid attacking factor, acid strength loss factor, and acid durability loss factor.

Figure 1. Experimental set up

Results and Concluding Remarks

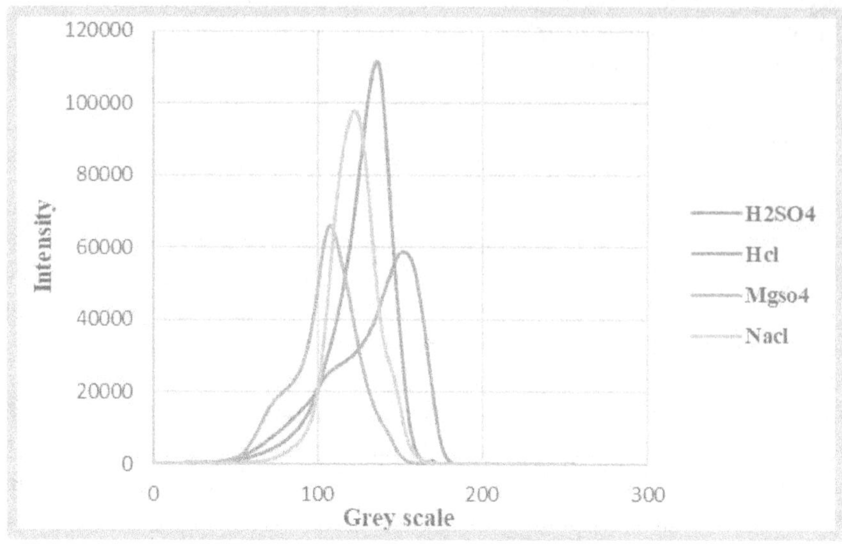

Figure 2: Grey scale Intensity for 30day duration

Table 1. *Age of immersion vs Acid Durability Loss Factor (ADLF).*

No of days exposure	HCl	H_2SO_4	NaCl	$MgSO_4$
0	0	0	0	0
3	0.04	0.05	0.00	0.00
8	0.34	0.78	0.04	0.05
15	2.78	4.47	0.44	0.33
30	14.05	23.93	1.79	1.08

Results indicate that the both *peak intensity of grey scale* as well as *mean grey scale value* are found to be good predictors of the extent of deterioration in concrete exposed to HCl and $MgSO_4$. Whereas, in the case of H_2SO_4 and NaCl, it is observed that both intensities of grey scale as well as mean grey scale are moderate predictors of extent of deterioration in concrete. The peak grey scale intensities enable identification of the individual chemicals. Relationships between the various deterioration parameters and grey scale spectrum obtained from digital images of concrete at various stages of deterioration, when exposed to four chemicals typically found in the environment, have been developed.

Acknowledgments: This study has been funded by the MHRD-IMPRINT project No. 7338.

References

1. Muthu, M., & Santhanam, M. (2018). Effect of reduced graphene oxide, alumina and silica nanoparticles on the deterioration characteristics of Portland cement paste exposed to acidic environment. *Cement and Concrete Composites, 91*, 118-137.https://doi.org/10.1016/j.cemconcomp.2018.05.005
2. Chand, M. S. R., Giri, P. S. N. R., Kumar, P. R., Kumar, G. R., & Raveena, C. (2016). Effect of self-curing chemicals in self-compacting mortars. *Construction and Building Materials, 107,* 356-364.https://doi.org/10.1016/j.conbuildmat.2016.01.018
3. Zhou, Y., Tian, H., Sui, L., Xing, F., & Han, N. (2015). Strength deterioration of concrete in sulfate environment: an experimental study and theoretical modelling. *Advances in Materials Science and Engineering, 2015*.http://dx.doi.org/10.1155/2015/951209
4. Annerel, E., & Taerwe, L. (2011). Methods to quantify the colour development of concrete exposed to fire. *Construction and building materials, 25*(10), 3989 3997. https://doi.org/10.1016/j.conbuildmat.2011.04.033

Decay of Blended Geopolymer Concrete in Acidic Environment

A.K.Mallinadh[1], T.Chandra Sekhar Rao[2*], N.V.Ramana Rao[3]

[1]*Ph.D Scholar, Department of Civil Engineering, Jawaharlal Nehru Technological University, Hyderabad*

[2] *Bapatla Engineering College, Bapatla*

[3]*National Institute of Technology, Warangal*

* e-mail: mallinadhkashyap@gmail.com

Introduction

Geopolymer concrete(GPC) is a most versatile futuristic concrete as it enables the researchers to utilize the industrial waste by product materials whose disposition is a major problem as an alternative to the cement concrete. Ordinary Portland Cement (OPC) is a primary binding material which is used in conventional construction practices. It was estimated that the manufacture of one ton of cement produces nearly one ton of carbon dioxide which emits into the atmosphere which causes the environmental pollution and leads to global warming. On par with the durability considerations of the ordinary Portland cement concrete, it was evident that in corrosive environments the OPC concrete starts to deteriorate. Henceforth, "Geopolymer Concrete" which was coined by Davidovits in 2005 [1] gained the interest of the researchers as an alternative material to the conventional cement concrete. Davidovits concluded that by the alkaline activation of the source materials like fly ash, ground granulated blast furnace slag, silica fume and metakaoline etc., with the polymeric reaction by sodium based or potassium based solutions. The inorganic alumino-silicate polymer made from the materials which are rich in silicon (Si) and aluminium (Al) materials by the polymerization process in a chemical reaction under alkaline condition gives the Geopolymer binding material.

Materials and Methods

Fly ash (Class F) was procured from Vijayawada Thermal Power Station (VTPS) and Ground Granulated Blast Furnace Slag was obtained from Vizag Steel Plant. Locally available River sand was used as fine aggregate. Coarse aggregate of sizes 20 mm,12 mm amd 6 mm were used in the present study so as to fill the voids. Distilled Water was used to prepare 12 Molarity concentration of Sodium Hydroxide Solution which was prepared prior to one day of casting. Sodium Silicate which is available in liquid form was used in the present study. The sodium Silicate to Sodium Hydroxide ratio adopted was 2.5.The Cube specimens after demolding were kept in different concentrations of Sulphuric acid, Hydrochloric acid and Sodium Chloride solutions for acidic exposure on high strength grade geopolymer concrete after 28 days,56 days and 90 days. The compressive strength of Geopolymer concrete has been evaluated on hydraulic compression testing machine of 2000kN capactiy. For the compressive strength test, cubes of size 150mm x 150mm x 150mm for H_2SO_4,HCl exposure and 100mmx100mmx100 mm for NaCl exposure are tested in compression.

Results and Concluding Remarks

Table: 1 Mix Design of M70 grade of Geopolymer Concrete

S.No	Constituents	Quantity(kg/m^3)
1.	Binder content(GGBS+Fly ash)	410.70
2.	Fine aggregate	770.00
3.	Coarse aggregate	1155.00
4.	Sodium Silicate solution	117.30
5.	Sodium Hydroxide solution	47.00
6.	Admixture (2%)	8.21
7.	Extra water (10%)	41.07

Table 2: Effect of various alkaline environments on compressive strength of M70 grade GPC

Mix Designation	Mix Specification	Weight of the Cube (Kgs)		Compressive Strength (MPa)		
		Initial	Final	28 days	56 days	90 days
M1	M70 CM	9.10	8.95	77.77	83.11	90.22
M2	M70-1%H_2SO_4	8.60	8.56	74.22	84.44	87.55
M3	M70-3% H_2SO_4	8.57	8.53	72.88	83.11	86.66
M4	M70-5% H_2SO_4	8.54	8.50	71.11	80.44	84.88
M5	M70-1% HCL	8.58	8.54	74.88	85.77	89.33
M6	M70-3% HCL	8.52	8.45	73.33	84.00	88.00

M7	M70-5% HCL	8.56	8.51	71.55	81.77	86.22
M8	M70-1% NaCl (100x100x100mm)	2.52	2.48	76.00	86.66	88.44
M9	M70-3% NaCl (100x100x100mm)	2.53	2.50	75.11	85.77	88.00
M10	M70-5% NaCl (100x100x100mm)	2.48	2.42	73.33	84.88	87.11

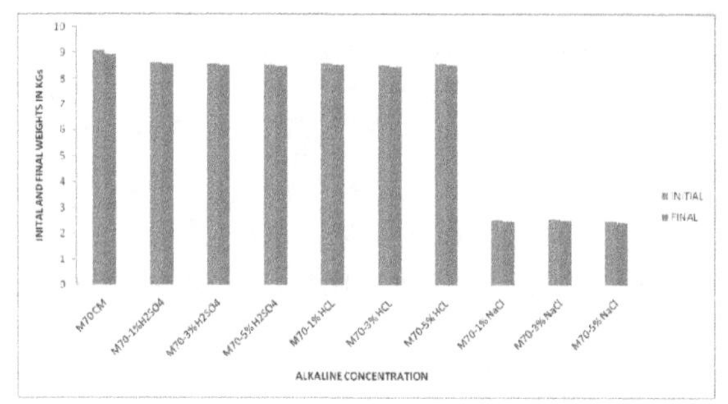

Fig.1 Effect of Various Alkaline Concentration on the Weights of M70 grade GPC

Fig2 Effect of various alkaline concentrations on the Compressive Strength of M70 grade GPC

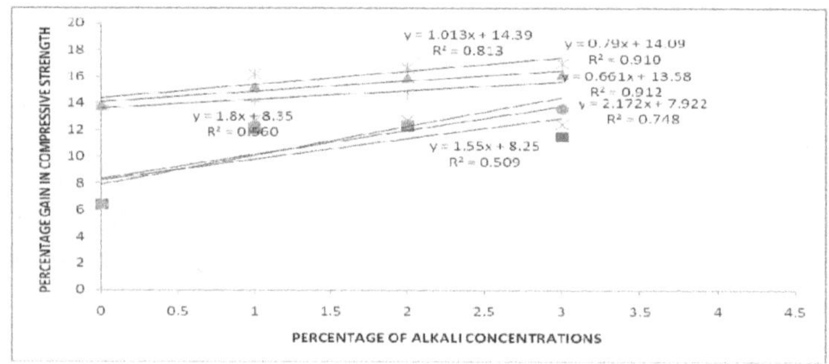

Fig.3 Regression Analysis for Percentage gain in compressive strength of concrete for the comparision of 28,56 and 90 days of ambient curing.

Conclusions

The experimental results stated that the GGBS and fly ash based geopolymer concrete showed an excellent resistance to sulphate and chloride attack on par with the controlled concrete. There are no compelling changes in the weight of the specimens and the compressive strength of the concrete upto 90 days of immersion in the acidic environment. The sulphuric acid exposure to the geopolymer concrete specimens damaged the surface of the specimens and the compressive strength was found to be decreased and the percentage decrease was found to be 4.6%,6.28%,8.5% for 28 days, 1.5% increase, 0%,3.21% for 56 days and 3%,3.94% and 5.91% for 90 days of immersion in the sulphuric acid of various concentrations when compared with the control mix.The Hydrochloric acid exposure to the geopolymer concrete specimens damaged the surface of the specimens and the compressive strength was found to be decreased and the percentage decrease was found to be 3.71%,5.70%,8% for 28 days, 3.1% increase, 1.05% increase,1.61% for 56 days and 1%,2.5% and 4.5% for 90 days of immersion in the Hydrochloric acid of various concentrations when compared with the control mix. The Sodium Chloride acid exposure to the geopolymer concrete specimens damaged the surface of the specimens and the compressive strength was found to be decreased and the percentage decrease was found to be 2.27%,3.45%,5.7% for 28 days, 4.1% increase, 3.2% increase,2.1% increase for 56 days and 2%,2.5% and 3.5% for 90 days of immersion in the Sodium Chloride acid of various concentrations when compared with the control mix.

References

1.Davidovits.J, "Geopolymer, Green Chemistry and Sustainable Development Solutions", France:Institute of Geopolymer, 2005,PP-9-15.

2.Rangan BV, Hardjto D. Development and properties of low calcium fly ash based geopolymer concrete. Research report GC-1. Faculty of Engineering, Curtin University of Technology, Australia. 2005: 35-50. 5.

3. Rangan BV, Wallah SE. Low-calcium fly ash based Geopolymer concrete Long term properties. Research report GC-2. Faculty of Engineering, Curtin University of Technology, Australia. 2006: 28-32.

4. Rajamane NP, Nataraja MC, Dattatreya JK, Lakshamanan N, Sabitha D. Sulphate resistance and eco-friendliness of geopolymer concrete. The Indian Concrete Journal. 2012: 13-22.

5. Bakharev T. Durability of geopolymer materials in sodium and magnesium sulphate solutions. Cement and Concrete Research. 2005(b): 1233-1246.

Use of Sugarcane Bagasse Ash in Concrete as Sand Replacement for Resistance to Sustained Elevated Temperatures

S. Jadon[1], S. Kumar[2*]

[1] *Department of Computer Aided Structural Engineering, IIIT Hyderabad, India*

[2] *Department of Civil Engineering, HBTU Kanpur, India*

* *e-mail: drsuniljadon@gmail.com*

Introduction

Depletion of natural resources in the present rapid urbanization and disposal problem of agricultural wastes have created opportunities for use of agro-wastes in construction industry as a replacement alternatives of cement, fine aggregates, coarse aggregates and reinforcing materials. The fibrous residue of sugarcane after crushing and extraction of its juice, known as bagasse, is reused as a fuel for heat generation that leaves behind ash which is known as Sugar Cane Bagasse Ash (SCBA). SCBA in concrete research has mostly focused on its use as a supplementary cementitious material due to high silica content. But, the availability of silica for pozzolanic reactions is doubtful as it is produced under uncontrolled high temperatures (>800^{0}C) and uncontrolled grinding of ash to suitable fineness.

Whilst there is considerable research into the use of SCBA in concrete as a replacement of cement (Gar et al. 2017), there has been limited research into the use of SCBA as a sand replacement in concrete (Arif et al. 2017). In the present investigation, a feasibility study is made to utilize the SCBA as a partial replacement of sand and examine its role in imparting resistance under elevated temperatures.

Materials and Methods

The experimental study aims to examine the suitability and effectiveness of SCBA as a partial replacement of sand in properties of concrete in fresh and hardened state. This also aims to study the effect of SCBA under sustained elevated temperatures when used as a partial replacement of sand.

Concrete ingredients include Portland pozzolana cement, Betwa river sand from Distt. Hamirpur (U.P.), 20 mm nominal size coarse aggregate from Distt. Mahoba (U.P.) and ash from sugar industry. The properties of concrete mix ingredients were tested as per the relevant Indian Standards. The grading of sand confirms to Zone II. The fineness modulus of SCBA (1.81) was less than the river sand (2.84). Concrete mix of M25 grade with slump 60$\pm$10 mm was prepared. Super plasticizer was used to maintain slump. SCBA replaced sand from 0-25% by mass fraction at 5% increments. 150 mm size concrete cubes were tested for compressive strength after 7, 28 and 56 days.

Most report confirms that concrete gains strength up to 300-400^{0}C. At higher sustained temperatures beyond 800^{0}C, surface cracks occur. Due to lack of research studies on the suitability and effectiveness of concrete under elevated temperatures when sand is partially replaced with SCBA, the effect of SCBA at sustained elevated temperatures was studied by subjecting concrete cubes of 28 days to elevated temperatures of 300^{0}C and 500^{0}C for 2 hours. The residual compressive strength was evaluated and compared with the controlled concrete samples at room temperature.

Results and Concluding Remarks

The use of SCBA as a partial replacement of sand provides additional pozzolanic property in concrete. While the presence of more finer particles in SCBA as compared to sand affects both fresh and hardened properties of concrete.

It was observed that the slump value of concrete using SCBA as partial replacement of sand reduces its workability with its increase and consequently super plasticizer is required to maintain the slump value.

Concrete with SCBA shows an increasing trend of compressive strength up to 10% replacement of sand than the controlled concrete, but further increase of SCBA decreased the compressive strength. It was further observed that with an increase in the sustained temperature, there was a gradual decrease in compressive strength of all mixes. The compressive strength of concrete was reduced slightly at 300^{0}C. But, the drop of compressive strength was significant at 500^{0}C ranging from 22% to 38% with respect to that at the room temperature. It was also observed that the use of SCBA imparts resistance to concrete against sustained elevated temperatures and an increase in the SCBA content appears to reduce the rate of decrease in compressive strength at high temperatures. The present study suggests that SCBA can be used as a replacement of sand in concrete. But, further work particularly in the pre-treatment by sieving and grinding may further enhance the filler and/or pozzolanic activity.

References

Gar PS, Suresh N, Bindigavile V (2017) Sugarcane bagasse ash as a pozzolanic admixture in concrete for resistance to sustained elevated temperatures. Journal of Construction and Building Materials 153: 926-936.

Arif E, Clark MW, Lake N (2017) Sugarcane bagasse ash from a high efficiency co-generation boiler as filler in concrete. Journal of Construction and Building Materials 151: 692-703.

Configuration of Lateral Force Resisting Elements for Minimization of the Torsion in Asymmetrical Buildings

Muhammed Masihuddin Siddiqui [1*], Prof. N. Murali Krishna [2]
[1] *Dept. of Civil Engg., Muffakhamjah College of Engg. & Technology, Hyderabad, T.S., India*
[2] *Dept. of Civil Engg, CVR College of Engineering, Ibrahimpatnam, T.S., India*
* *e-mail: masihuddin@mjcollege.ac.in*

Abstract: *- Structural analysis of asymmetric buildings due to seismic excitations is often very complex due to the presence of torsion as compared to symmetric buildings, which get merely translated. The past experience has clearly shown that the buildings with regular geometry as well as evenly distributed mass and stiffness in plan and elevation have undergone much less damage as compared to buildings with any or kinds of irregularities. The torsional moments acting on a building are influenced by the relative location of the centre of mass, the centre of strength and the centre of stiffness. Therefore, these centres of significance shall be located appropriately to minimise torsional forces, ensuring efficient functioning of the building. In this paper, the effects of strength eccentricity and stiffness eccentricity with respect to the mass centre of a building are evaluated and non-linear static analysis is carried-out to study the variation in base torsion. Thereafter, the structural configuration of the building is altered to find-out the best locations of mass, rigidity and strength centres to minimize the torsion. The study is carried-out on three different types of lateral load resisting building systems. The systems comprising of i) only columns, ii) only Shear walls and iii) columns with shear walls with varying values of stiffness and strength. The study is found to be very productive as for certain structural configurations, the base torsion is observed to be significantly reduced.*

Keywords: Strength eccentricity, Stiffness eccentricity, Configuration of centres, Asymmetric structure, Push over analysis, Genetic Algorithm

Introduction

Recent earthquakes have clearly shown that the irregular distribution of mass, stiffness and strength in a structural building cause serious damage to the structural systems. The damage reports of past earthquakes have amply indicated that one major cause of distress in asymmetrical building structures is torsion. To minimize the torsion, the distribution of mass, stiffness and strength shall be altered appropriately, yet meeting the functional / architectural requirements in full. Hence, the agenda of the structural engineer is to conceive an economical, safe, elegant and feasible structural system. The present practice of vulnerability assessment is by determining the degree of asymmetry in terms of the stiffness eccentricity. This parameter is found to be a very useful measure to correlate the seismic elastic response of asymmetrical buildings. However, when the system is excited into inelastic range, the yielding of the load resisting elements complicate the structural behaviour. In such a situation, stiffness eccentricity alone cannot be a good indicator of building's torsional response according to Aziminejad[1]. Sadek and Tso [3] have introduced the concept of strength eccentricity, the distance between the centre of yield strength to the centre of mass. Paulay[2] has contended that in the inelastic range, the strength eccentricity is more appropriate parameter for the torsional response of an asymmetrical building. The torsional response of a building is hence influenced by the relative location of the centre of mass, the centre of strength and the centre of stiffness. Therefore, the relative location of these centres of significance plays a very crucial role in minimising torsional forces on buildings. The centre of stiffness and the centre of strength depend on the sectional dimensions of the resisting elements. Hence to minimise the eccentricities, apt choice of sectional dimensions is a must.

Materials and Methods

To study the effects of variations of centres i.e., centre of strength & centre of stiffness, two-way asymmetric building with re-entrant corners having bays of un-equal lengths both in the X and Y- directions is analysed using Pushover Analysis in the latest version of SAP 2000. The total span length is 25m in X-direction and 16m in Y-direction. The plan is kept same for all buildings but with different lateral force resisting elements (LFRE) namely columns and/or shear walls. Three asymmetrical building systems with only columns (C), only shear walls (S) and columns along with shear walls (CS) as LFREs are used as basic models.

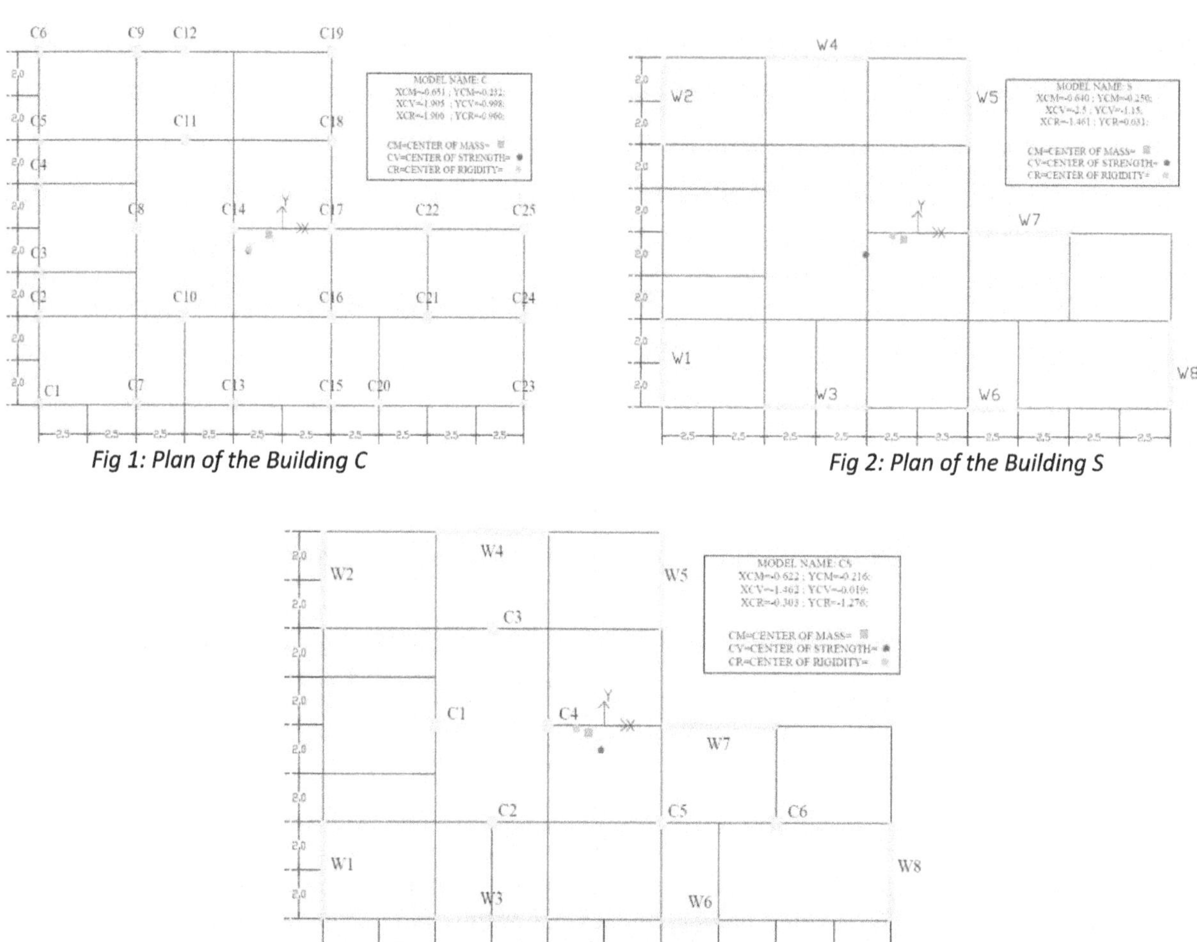

Fig 1: Plan of the Building C

Fig 2: Plan of the Building S

Fig 3: Plan of the Building CS

In order to configure the centres of significance, objective functions are defined in MATLAB package. The sectional dimensions of the lateral force resisting elements (LFRE) act as variables. The GA toolkit of MATLAB is used for optimising these objective functions. Making use of these optimal dimensions, the asymmetrical buildings are re-analysed and the torsion at the base is recorded and compared with the basic model.

For each category of building system, four models are generated including the basic model.

Model -1 (Basic) Sectional dimensions of the resisting elements i.e., columns and shear walls are of uniform size.

Model -2 (e_v = 0) Sectional dimensions of resisting elements are so chosen that the strength eccentricity is a minimum.

Model -3 (e_s = 0) Sectional dimensions of resisting elements are so chosen that the stiffness eccentricity is a minimum.

Model -4 (Proper Configuration) Sectional dimensions of resisting elements are so chosen that the centre of mass is in between centre of stiffness and centre of strength.

The comparisons are made of the three building systems and conclusions are drawn.

Results and Concluding Remarks

The present work evaluates the effectiveness of varying centres in the plan irregular buildings. The effectiveness is assessed with 'torsion at the base' as a parameter at the performance point of the building. Table 1 contains the results of the analysis carried out for the models as discussed earlier.

Table 1. Comparison of Base Torsion (in kN.m).

Models	Building C		Building S		Building CS	
	Px	Py	Px	Py	Px	Py
Basic	5517	10318	15463	21705	9259	15331
$e_v = 0$	4235	7900	13699	18900	7758	10848
$e_s = 0$	5043	9392	16196	20804	8016	11458
Proper Config.	4500	8500	12860	20250	7785	11063

The following conclusions can be the drawn from the above analysis

1. When the strength eccentricity is minimized, the base torsion decreased by 23.5% in the structure with only columns, by 13% in structure with only shear walls, and by 30%, in structure with both columns and shear walls as LRFEs.
2. When the centre of stiffness is coincided with centre of mass, the base torsion decreased by 9% in the structure with only columns, increased by 4% in structure with only shear walls, and decreased by 25% in structure with both columns and shear walls as LFREs.
3. Also when proper configuration of centres are made i.e., centre of strength and centre of stiffness made at equidistant and opposite with respect to centre of mass, the base torsion decreased by 18% in the structure with only columns, by 7% in structure with only shear walls, and by 28% in structure with both columns and shear walls as their LFREs.
4. Hence the maximum reduction of base torsion which was observed in the case of minimum strength eccentricity and the least reduction were observed in the case of models with minimum stiffness eccentricity.

References

1. Aziminejad,A., and Moghadam, S. (2009), "Performance of Asymmetric Multistorey Shear Buildings with Different Strength Distributions", Journal of Applied Sciences 9(6):1082-1089.
2. Paulay,T., (1996) "Seismic Design for Torsional Response of Ductile Buildings". Bulletin of the New Zealand National Society for Earthquake Engineering, Vol. 29, No. 3, pp. 178-198.
3. Sadek, A.W., Tso W.K., (1988), "Strength Eccentricity Concept for Inelastic Analysis of Asymmetrical Structures", Proceedings of 9th World conference on Earthquake Engg., August, Japan (Vol 5).

A Review on the Effects of Various Influence Factors in Formation of Geopolymer Binder

Suebha khatoon[1], A.K.Kaliluthin[2] ,Ashwin Raut[3]

[1]*Department of Civil Engineering, Lords Institute of Engineering and Technology, Hyderabad, India.*

[2]*Department of Civil Engineering, B.S. Abdur Rahman Crescent Institute of Sci. & Tech. Tamil Nadu, India.*

[3]*Department of Civil Engineering, Sree Dattha Institute of Engineering and Science, Hyderabad, India.*

E-mail: rehana.shaik91@gmail.com

Introduction

The tremendous rise in urbanization has led to the utilization of resources in the construction sector, resulting in unforeseen adverse environmental impact. Construction consume a lot of resources especially cement, hence its production has been a greater concern for environmental sustainability owing to the carbon emission. Thus, researchers are striving to develop the substitute having similar mechanical, physical, thermal and durability aspect leading to the development of geo-polymers.

Geo-polymers are amorphous three-dimensional alumina-silicate binder materials, named by Davidovits in 1970s [1]. They are inorganic polymers and a chain structure formed on a backbone of aluminium (Al) and silicon (Si) ions. Geopolymers can be synthesized by mixing alumina-silicate reactive materials with strong alkaline solutions. Any material that is rich in Si and Al in amorphous form can be a possible source material for geopolymer binder. Metakaolin is preferred due to its high rate of dissolution in the reactant solution and has easier control on the Si/Al ratio but its high cost prohibits its mass use in the geopolymer matrices. Low calcium fly ash (ASTM class F fly ash) is the preferred material than high calcium fly ash (ASTM class C fly ash). The presence of high amount of calcium in high calcium fly ash interfere the polymerisation process and alters the microstructure [2].

The paper aims to give in-depth critical analysis on the performance variables of research binders. The enhanced performance of geopolymer binder is due to the various parameters which should be addressed thoroughly. Also, the effects of these parameters have been assessed by various researchers, but still there is gap in understanding the specific design variables which can lead to better development of the material. The development of geopolymer based material is still complex procedure compared to the cement-based material. Thus, the understanding of the design variables can provide a roadmap for the industry to develop a better product having enhanced performance. Also, this research explores the variables alumino-silicate source materials contribution in development of geopolymer and it influence in performance.

Literature Review

The research work done over the years mostly points out that the nature of alumino-silicate source, liquid to solid ratio, sodium silicate to sodium hydroxide ratio, molarity of sodium Hydroxide and curing time and temperature are the key parameters which affects the development process and performance of geopolymer. The alumino-silicate sources such as fly ash, rice husk ash, metakaolin, silica fume, glass powder has been extensively studied by researchers. But, the consensuses on source material and its its proportion in the mix has been still under investigation. The optimal level for the alumio silicate source materials has been recommended by numerous researchers and it should be 20% to 50% of the mix.

Similarly, Molarity of NaOH (i.e., 8 to 14 M)[8.], Curing temperature (40°c to 70°c)[7], Liquid to solid ratio (0.52 to 0.58)[9], sodium silicate to sodium hydroxide ratio (2.2 to 2.8) [6,7] ,has been studied by various researchers and extensively discussed in this paper. All these parameters have been studied and standardized optimal composition has to be found out to get a standard mix design for geopolymer formation with various alumino-silicate sources material. The review also discusses the influence of these parameters on various properties which eventually will contribute to serviceability, durability, sustainability and applicability of the geopolymer based material in construction industry.

Results and Concluding Remarks

The paper reviews the literature extensively and identifies the gap in the standardization of the proportion to develop geopolymer material. The geopolymer material still not incorporated in the industry due to various factors such as lack of standardization in mix design and development, consistent availability of source materials. Various alumino-silicate materials such as Fly ash, GGBS, Silica fume, Glass powder, Metakaolin, Rice Husk Ash, Bagasse ash has been studied extensively by researchers, still there has been a gap in usage of these material on site.

The performance of the material is greatly influenced by various parameters which has been discussed in this paper. The setting time of the geopolymer is also a cause of concern during the design and development of geopolymer. The hurdles in the development of the geopolymer on the industrial scale if the issues regarding the availability of quality raw materials with consistent chemical composition, standardization of design and development of geopolymer is addressed with greater concern. There is also greater concern regarding the costing of the geopolymer material which has not been addressed in the reviewed literature, thus this issue also affects the commercialization the geopolymer based product.

References

1. *Davidovits J. Geopolymers and geopolymeric new materials. J Therm Anal 1989;35(2):429–41.*
2. *van Deventer JSJ, Provis JL, Duxson P, Lukey GC. Reaction mechanisms in the geopolymeric conversion of inorganic waste to useful products. J Hazard Mater 2007;139(3):506–1.*
3. *J. Davidovits, (1988b). Geopolymer Chemistry and Properties .*
4. *Palomo, M.W. Grutzeck, M.T. Blanco, Alkali-activated fly ashes: a cement for the future, Cem. Concr. Res. 29 (8) (1999) 1323–1329.*
5. *J. Davidovits, Geopolymers: inorganic polymeric new materials, J. Thermal Anal. Calorimetry 37 (8) (1991) 1633–1656*
6. *R. Prakash, Voraa ,V. Urmil, Dave, "Parametric Studies on Compressive Strength of Geopolymer Concrete", Procedia Engineering, Elsevier, 2013, pp-210-219*
7. *Kovalchuk G, Fernandez-Jimenez A, Palomo A. Alkali-activated fly ash: effect of thermal curing conditions on mechanical and micro structural development Part II. Fuel 2007;86:315–22.*
8. *Ren, X., & Zhang, L. Experimental Study of Geopolymer Concrete Produced from Waste Concrete. Journal of Materials in Civil Engineering, 31(7), (2019). 1943-5533.*
9. *Part Wei Ken, Mahyuddin Ramli, Cheah Chee Ban, An overview on the influence of various factors on the properties of geopolymer concrete derived from industrial by-products, Construction and Building Materials 77 (2015) 370 -395*
10. *Anant Lal Murmu, Anamika Jain, and Anjan Patel, Mechanical Properties of Alkali Activated Fly Ash Geopolymer Stabilized Expansive Clay, KSCE Journal of Civil Engineering (2019) 23(9):3875-3888.*
11. *Monita Olivia , Hamid Nikraz, Properties of fly ash geopolymer concrete designed by Taguchi method, Materials and Design ,Elsevier, 36 (2012) 191–198.*

Light Transmitting Concrete Blocks

Zeeshan Shah Khan[1], Mohammed Safiuddin[2*], Shaik Khaleel[1], Abdullah Sharieff[1]
[1]*Civil Engineering, Lords Institute of Engineering and Technology, JNTUH, Hyderabad, India*
[2]*Civil Engineering, Lords Institute of Engineering and Technology, JNTUH, Hyderabad, India*
** e-mail: msafiuddin@lords.ac.in*

Introduction

In the past, concrete was considered as a structural member element only, but the concept of concrete has changed today. Innovative and smart building materials like light transmitting concrete have come up. **Licrete Blocks** (also: **light-transmitting concrete**) is a concrete based building material with light-transmissive properties due to embedded light optical elements. Licrete blocks are used in fine architecture as a facade material and for cladding of interior walls. The proportion of the fibers is very small (up to 4%) compared to the total volume of the blocks. Moreover, these fibers mix in the concrete because of their small size. Therefore, the surface of the blocks remains as homogeneous concrete. The blocks can be customized and produced in various colours and sizes. According to IGBC (Indian Green Building Council), for building to be green building, 50% of day light is mandatory which accounts for 3 credits, these blocks will help in achieving this.

This project aims to produce a Light transmitting concrete meeting this criteria which would be stronger, aesthetically attractive, energy efficient and an eco-friendly green building material. These blocks would be in great demand in future.

Aron Losonczi, 2001 , a Hungarian Architect developed a special concrete. Either glass optical fibres or plastic optical fibers can be used with concrete in alternate layers to form light transmitting concrete. By increasing the percentage of plastic optical fiber, the light transmitted through the concrete is increased and optimum amount of strength and transmitted light can be achieved at 4 to 5% fibers in the concrete. **Shen Juan and Zhou Zhi, 2013** discusses the development of smart transparent concrete which reduces the power consumption the transparent concrete does not lose the strength parameter when compared to regular concrete. **Kashiyani Bhavin K., Raina Varsha, Pitroda Jayeshkumar, Shah Bhavnaben K., 2013**, studied light transmitting concrete, its various ingredients, manufacturing process construction, applications, advantages, disadvantages, etc. The thickness of optical fibers being 2 micrometres to 2mm. Alternate layers of POF and concrete are placed to form light transmitting concrete. **Bhushan Padma, Johnson D. et al. 2013,** constructed translucent concrete blocks using concrete and plastic optical fibers. They discussed the usage of these concrete blocks such as in the walls, ceilings to make it architecturally pleasing, illuminating speed bumps, use on sidewalks, on various interior and exterior surfaces of the buildings to make it aesthetically beautiful. Plastic optical fibers have various advantages such as they do not produce radiation, and are not affected by radio magnetic interference. Plastic optical fiber is by far the best replacement for glass. **Soumyajit and Avik 2013**, used six specimens of translucent concrete with varying P.O.F ratio as 1%, 2%, 3%, 4%, 5%, and 6% with the diameters of P.O.F as 1mm. Optical Power meter with a wavelength range of 400-1OO rmi. Incandescent lamp of 200W and Halogen lamp of 500W is taken to provide incident light. **Momin et al. 2014,** made light transmitting concrete samples with the help of Glass rods and optical fiber. Light transmittance was found to be 7.0 to 10.0% for optical fiber specimens and 0.2 to 1.5% with glass rods specimens. **Ahuja, Mosalam.M. Khalid et al,2014** investigated translucent concrete. They presented a geometrical ray-tracing algorithm to simulate light transmission properties of a panel of translucent concrete. It was concluded from the investigation that a tilt angle of 30 degree for the panel transmitted the maximum amount of light among all the tilt angles considered.

Materials and Methods

Initially two different fibres (nylon fibre and optical fibre) were used for checking the light transmission. Both these were used in casting licrete blocks, which were compared on the basis of compressive strength and light transmission. We have chosen optical fiber as it was transmitting more light as well as giving more compressive strength.

Special mould of size 100 mm x 100 mm x 100 mm were prepared for making LICRETE BLOCKS (Light Transmitting Concrete Blocks), and holes to the size of optical fiber were drilled as per Circle, Rectangle, and Triangle shape variants. For all variants, the distance was measured from the center of mould and in each variant, a total of 12 holes were drilled along the perimeter. Optical fibers of diameter 0.75 mm were inserted parallel to each other before filling the mould with concrete. Length of fibers must be longer than the size of mould so that it is comfortable to handle while casting concrete into the mould.

The concrete mix of M25 was designed using the IS codes ***IS 10262:2009 and IS 456:2000*** and materials were mixed in proper composition. OPC 53 Grade conforming to IS 8112 is used, the nominal maximum size of coarse aggregate was chosen as 10 mm so as not to damage the fibers and at the same time penetrate the space between fibers. All the raw materials used were tested individually for their properties, Specific Gravity of Cement, Fine Aggregate and Coarse Aggregate were determined. Water absorption test for Fine and Coarse Aggregate was conducted. Slump cone test (IS 1199), Compaction factor test (IS 5515) were conducted on fresh concrete before casting the moulds. Casted moulds were demoulded after 24 hours and cured in curing tank. The samples were tested for light transmission using Lux meter before testing them for compressive strength (IS 14858-2000) using calibrated compression testing machine of capacity 2000 kN at ages of 3, 7 and 28 Days.

Lux meter measures the intensity of light falling on its sensor. A box arrangement is made for this test. A 100 watt bulb is lit and placed at the wide end of the arrangement, while the sensor of the lux meter is placed at the converging end. Then, the licrete blocks of various variants are kept one by one between the bulb and the sensor. The new luminance value is noted and compared with the former luminance to calculate the percentage transmissibility of light through all the variations of samples.

Results and Concluding Remarks

Initially, a set of three cubes were casted using the nylon fiber as well as optical fiber, and these were tested for light transmissibility before testing for compressive strength. It was found that the samples with optical fiber gave a compressive strength of 77% more than the nylon fiber and better light transmissibility. Hence optical fiber was selected and used in different variants namely, (circular, rectangular and triangular).

The concrete mixes with optical fibers had an average slump of 75 mm indicating medium workability and a compaction factor ratio of 0.85, indicating acceptable compaction. Litcrete blocks having optical fibres in circular combination yielded the highest compressive strength increase of 25% when compared with the control (reference) sample of M25 due to availability of fibers in the centre of cube. The triangular variant transmitted 36% more light when compared with the circular and rectangular variants, due to the arrangement of optical fibers.

Reference

1. Shen Juan and Zhou Zhi, (2013), Preparation and Study of Resin Translucent Concrete Products, Advances in Civil Engineering: Volume 2019, Article ID 8196967, 12 pages, https://doi.org/10.1155/2019/8196967.
2. Kashiyani Bhavin K., Raina Varsha, Pitroda Jayeshkumar, Shah Bhavnaben K., (2013), A Study on Transparent Concrete: A Novel Architectural Material to Explore Construction Sector, International Journal of Engineering and Innovative Technology (IJEIT): Volume 2, Issue 8. pp 83-87.
3. Bhushan Padma, Johnson D. (2013) et al. Optical Fibers in Modeling of Translucent Concrete Blocks, International Journal of Engineering Research and Applications (IJERA): ISSN: 2248-9622, Vol. 3, Issue 3, pp.013-017.
4. Soumyajit and Avik (2013), Translucent Concrete, International Journal of Scientific and Research Publications: Volume 3, Issue 10, ISSN 2250-3153
5. Momin et al. (2014), Study on Light Transmittance of Concrete Using Optical Fibres and Glass Rods, International Conference on Advances in Engineering & Technology (ICAET-2014): pp.67-72
6. Ahuja, Mosalam.M. Khalid et al, (2015), Computational Modeling of Translucent Concrete Panels, Journal of Architectural Engineering: DOI: 10.1061/ (ASCE) AE.1943-5568.0000167.

An Appraisal on Shear Strength of Concrete for Different Codes

Ch. Manjula[1*], Sumanth Kumar B[2], D Rama Seshu[3]
[1, 2] Research Scholar, Department of Civil Engineering, National Institute of Technology, Warangal, India
[3] Professor, Department of Civil Engineering, National Institute of Technology, Warangal, India
*reddy.daughter319@gmail.com

Introduction

The shear in Reinforced concrete (RC) members has been recognized as most significant actions from the point of structural safety. Despite significant progress in the understanding and modeling of shear, it is regarded as one of the least understood but most important problems in reinforced concrete. The shear resistance of RC without any web reinforcement is generally attributed to three main mechanisms: the shear resistance of concrete in the un-cracked compression zone, the aggregate interlock at the cracked interface, and the dowel action of the longitudinal reinforcement. The magnitudes of these three mechanisms vary throughout the loading process and depend on cracking pattern and deformation. As the applied shear force is increased, the dowel action is the first to reach its capacity, after which a large shear is transferred to aggregate interlock. The failure of aggregate interlock necessitates a rapid transfer of shear to the concrete in the compression zone. The sudden transfer of shear to the concrete compression zone results in brittle failures. There is sometimes little warning before failure occurs and this makes shear failures in RC particularly objectionable.

Review of Codal Provisions

Numerous empirical and analytical models have been proposed to calculate the shear resistance of concrete in RC beams without shear reinforcement. In most of the codes of practice the shear strength of concrete is expressed through empirical equations resulting from experimental test outcomes. The factors considered to be influencing the shear resistance of concrete in most of the codes include: i) Strength of concrete ii) Longitudinal steel ratio iii) Shear span to effective depth ratio and iv) Size of aggregate. The codes of practice such as IS 456, BS 8110, ACI 318 and Euro code 2 consider into interpretation the effect of reinforcement ratio, effective depth and concrete compressive strength whereas Canadian code studies the shear strength to be a function of concrete compressive strength only. Model code 2010 reflects the shear strength of beams as a function of longitudinal strain in the web and size of aggregate. In most specific codes, shear strength of a reinforced concrete (RC) beam is the sum of the shear potential of the concrete component and the steel component is taken. This interpretation is for suitability only, even though the resistances offered by concrete and steel form parts of complex interactions. Though ample studies carried out on shear failure of concrete beam still it is debatable regarding the precise shear behaviour of reinforced cement concrete structure elements. In order to understand the effect of different parameters on the shear resistance of concrete it is necessary to carry out a comparative study of the models presented in different codes of practice regarding shear strength.

In view of the above an attempt is made in this paper to present an appraisal of shear strength of concrete beams with no shear reinforcement. The codes of practice considered in the study are IS 456-2000, BS 8110, Euro code 2 (EN 1992), Canadian code (CSA A23.3), ACI code 318 and Model code (FIB) 2010. Only the simplified methods specified in each of the codes have been used in this comparative study.

Review of different codes has been carried out and the shear stress of concrete has been calculated for beams without shear reinforcement. The corresponding variation of shear stress of concrete has been shown in Fig 1. The variation of concrete shear strength for different percentage of steel shown in Fig.2 is for a typical M25 grade of concrete and effective depth of 300mm.

Table 1 shows the comparison for IS 456 shear capacities, which indicates that the shear strength of concrete established by IS 456 is very much under estimated than other codes for constant % of reinforcement steel. Table 2 summarises that the FIB model 2010 is lower to other codes.

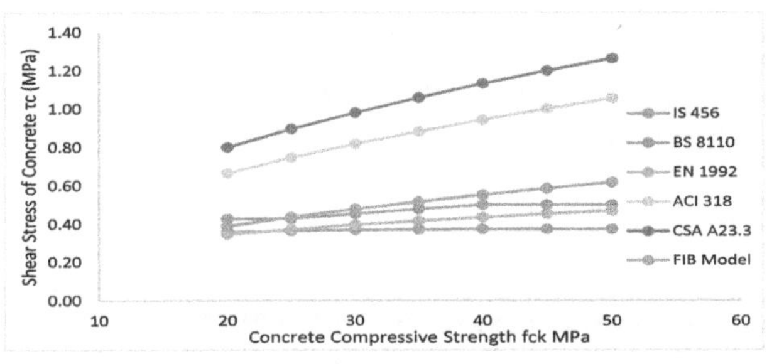

Figure 1. Variation of Concrete Shear Stress to Concrete Compressive Strength

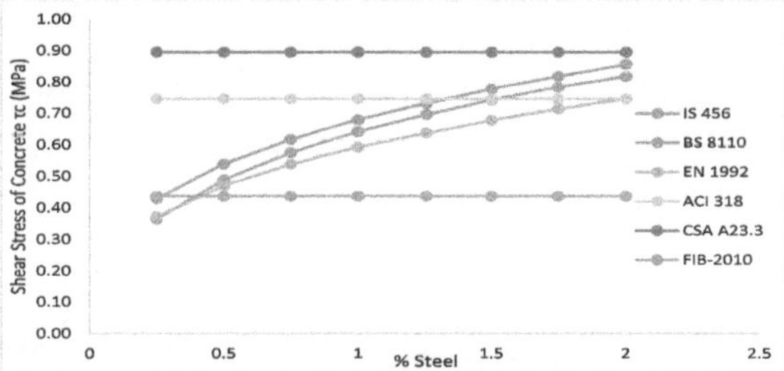

Figure 2. Variation of Shear Stress of concrete to % of Reinforcement Steel

Table 1. Comparison of Shear stress of IS 456 with different codes (Constant % of Reinforcement Steel)

S. No	Concrete Compressive Strength f_{ck} MPa	τc_{BS} / τc_{IS} BS 8110 - 1997	τc_{EN} / τc_{IS} EN 1992-1-1:2004	τc_{ACI} / τc_{IS} ACI 318 - 2014	τc_{CSA} / τc_{IS} CSA A23.3 - 2004	τc_{FIB} / τc_{IS} FIB Model - 2010
1	20	1.191	0.963	1.856	2.227	1.087
2	25	1.173	1.022	2.043	2.452	1.196
3	30	1.232	1.074	2.213	2.656	1.296
4	35	1.286	1.120	2.370	2.843	1.387
5	40	1.335	1.163	2.515	3.018	1.473
6	45	1.335	1.210	2.668	3.201	1.562
7	50	1.335	1.253	2.812	3.374	1.646
	Avg	**1.270**	**1.115**	**2.354**	**2.825**	**1.378**

Table 2. Comparison of Shear stress of IS 456 with different codes (Constant Grade of Concrete)

S. No	% of Reinforcement Steel	τc_{BS} / τc_{IS} BS 8110 - 1997	τc_{EN} / τc_{IS} EN 1992-1-1:2004	τc_{ACI} / τc_{IS} ACI 318 - 2014	τc_{CSA} / τc_{IS} CSA A23.3 - 2004	τc_{FIB} / τc_{IS} FIB Model - 2010
1	0.25	1.173	1.022	2.043	2.452	1.196
2	0.5	1.103	0.961	1.525	1.830	0.893
3	0.75	1.074	0.935	1.297	1.556	0.759
4	1	1.059	0.922	1.162	1.394	0.680
5	1.25	1.051	0.916	1.071	1.285	0.627
6	1.5	1.047	0.912	1.004	1.205	0.588
7	1.75	1.046	0.911	0.953	1.143	0.558
8	2	1.046	0.912	0.911	1.094	0.534
	Avg	**1.075**	**0.936**	**1.246**	**1.495**	**0.729**

Results and Concluding Remarks

An appraisal of shear strength of concrete without web reinforcement indicated that there is no consistent mechanical model for estimating the shear strength of concrete without shear reinforcement. The evaluation of the shear strength of concrete using different existing codes of practice showed that Indian standard code is highly underestimating the shear strength of concrete. Among 6 codes of practice, ACI 318,FIB 2010 and CSA A23.3 shear stresses are purely depends on concrete compressive strength and no other effect is considered. The study further indicates that the available approaches for shear design of concrete members without stirrups have still a great uncertainty and requires further research particularly in the context of increased use of new concretes.

References

ACI, A., 2014. 318–14. Building Code Requirements for Structural Concrete. American Concrete Institute, Farmington Hills, Michigan.

British Standard, B.S., 1997. 8110, ". The Structural Use of Concrete." British Standard Institution.

Canadian Standards Association, 2004. Design of concrete structures. Mississauga, Ont.: Canadian Standards Association.

Code, M., 2010. Fib model code for concrete structures 2010. Document Competence Center Siegmar Kästl eK, Germany.

Eurocode, C.E.N., 2004. 2: design of concrete structures–part 1-1: general rules and rules for buildings (EN 1992-1-1: 2004). Comité Européen de Normalisation, Brussels.

Standard, I., 2000. IS-456. 2000. Plain and Reinforced Concrete-Code of Practice", Bureau of Indian Standards.

Transportation Engineering

Comparison of Energy Consumption and Cost of Perpetual Concrete and Flexible Pavement

A. Abhijit Shinde[1*], B. Rohit Bodhgire[2], C. Vijay Kakade[3]
[1,2] *M.Tech Students, Department of Civil Engg., College of Engineering Pune, Pune, India*
[3] *Assistant Professor, Department of Civil Engg., College of Engineering Pune, Pune, India*
* *e-mail: shindeab17.civil@coep.ac.in*

Introduction

In India the flexible pavement and concrete pavement for the National highways and Expressways are designed for 15 years and 30 years respectively. Large amount of natural resources such as aggregate, bitumen, cement are utilized in the construction and maintenance of these roads. This will result in increase in emission of greenhouse gases. So construction of long lasting flexible and concrete pavement will be helpful in preservation of natural resources and consequently reduction in emission of greenhouse gases. Perpetual or long lasting pavements are generally designed for 50 years or more (Basu et al. 2013). The studies carried out in the past indicate that there is significant difference in energy consumption in construction of flexible and concrete pavement (Zapata and Gambatese 2005; Sreedhar et al. 2016). So that in a present study an attempt has been made to compare the energy consumption and cost of construction of perpetual flexible and concrete pavement. The study will help in identification of best type of pavement for construction of perpetual pavement.

Materials and Methods

The mix design of concrete used in design of PQC layer of perpetual concrete pavement is 1:1:2:0.4 (C: S: A: W). The flexural and compressive strength tests were performed on the concrete beams and cylinders cured for 28 days. The unconfined compressive strength test was performed on the cement treated base mix prepared with 5% cement and cured for 28 days for estimation of resilient modulus of cement treated base layer used in flexible pavement.

Pavement Design Details

Both flexible and concrete pavements were designed for 9863 commercial vehicles estimated for the design of Pune Ring Road in the state of Maharashtra. the design was done for subgrade having 3% CBR. The design life of perpetual flexible and concrete pavement was assumed to be 50 years.

Flexible Pavement

Flexible pavement design was done for pavement having granual base and/or cement treated base layers. The IRC 37:2018 was followed for design of flexible pavement.

Concrete Pavement

Concrete pavement design was done for unbounded concrete pavement having PQC, Dry lean concrete, Granular subbase and subgrade layer. The design was done by following IRC 58:2015.

Results and Concluding Remarks

The flexural and compressive strength of concrete was 6.21 MPa and 49.5 MPa respectively. The unconfined compressive strength of cement treated base mix was 3.11 MPa.

The result of energy consumption and thickness of different layers of concrete pavement and flexible pavements are given in Table 1.

Table 1. Design Thickness and Energy Consumption for Different Types of Pavement

Sr No	Pavement	Layer	Thickness (mm) (50 Years)	EE ($1*10^{13}$) J (50 Years)	Total EE ($1*10^{13}$) J	Thickness (mm) (30 Years)	EE ($1*10^{13}$) J (30 Years)	Total EE ($1*10^{13}$) J	Thickness (mm) (15 Years)	EE ($1*10^{13}$) J (15 Years)	Total EE ($1*10^{13}$) J
1	Concrete Pavement	PQC	245	2.55		239	2.49		232	2.42	
		DLC	150	1.48		150	1.48		150	1.48	
		GSB	225	1.46	5.49	225	1.46	5.43	225	1.46	5.36
2	Flexible Pavement with CTB Base	Bituminous Concrete	165	2.23		100	1.35		80	0.828	
		CTB	240	2.79		210	2.44		190	2.21	
		CT Sub-base	250	2.68	7.7	220	2.35	6.15	250	2.68	5.71
3	Flexible Pavement with GSB Base	Bituminous Concrete	408	4.22		340	3.52		280	2.9	
		GSB	750	5.24	9.46	700	4.89	8.41	700	4.89	7.79

From Table 1 it can be seen that perpetual pavement designed with base layer of granular material has highest energy consumption while concrete pavement has least energy consumption. Higher thickness of BC and GSB has resulted in significant increase in energy consumption in construction of flexible pavement with GSB as base layer. Increase in thickness of PQC layer with increase in design life is significantly less than increase in thickness of BC, CTB and GSB layers. This resulted in significant difference in energy consumption from concrete pavement and different types of flexible pavement.

Table 2. Construction Cost Comparison for Different Types of Pavement

Sr No	Pavement	Layer	Cost (Lakhs) (50 Years)	Total Cost (Lakhs)	Cost (Lakhs) (30 Years)	Total Cost (Lakhs)	Cost (Lakhs) (15 Years)	Total Cost (Lakhs)
1	Concrete Pavement	PQC	161		158		154	
		DLC	53	238	53	234	53	230
		GSB	23		23		23	
2	Flexible Pavement with CTB Base	Bituminous Concrete	128		77		62	
		CTB	34	199	30	140	27	125
		CT Sub-base	36		31		36	
3	Flexible Pavement with GSB Base	Bituminous Concrete	317	394	264	336	217	289
		GSB	76		71		71	

From Table 2 it can be seen that the construction cost of perpetual pavement designed with GSB as base layer is higher than cost of construction of pavement designed with CTB base and concrete pavement by 50% and 40% respectively. Cost of construction will increase by 1.68%, 37.2% and 26.7% for construction of perpetual concrete pavement, flexible pavement with CTB base and flexible pavement with GSB base respectively as compared to conventional design life of these pavements.

Conclusion

The increase in energy consumption and cost for construction of perpetual pavement as compared to pavement designed for normal design life is least for concrete pavement. So that concrete pavement is best option among different types of pavements considered for the construction of perpetual pavement.

References

Basu C, Das A, Thirumalasetty P, Das T (2013) Perpetual Pavement- A Boon for the Indian Roads. Procedia - Social and Behavioral Sciences 104: 139-248. doi: 10.1016/j.sbspro.2013.11.106

Zpata P, Gambatese A (2005) Energy Consumption of Asphalt and Reinforced Concrete Pavement Materials and Construction. Journal of Infrastructure Systems 11 (1): 9-20. DOI: 10.1061/(ASCE)1076-0342(2005)11:1(9)

Sreedhar S, Jichkar P, Biligiri K (2016) Investigation of Carbon Footprints of Highway Construction Materials in India, Transportation Research Procedia 17: 291 – 300. doi: 10.1016/j.trpro.2016.11.095.

IRC: 37, Guidelines for the design of flexible pavements, 4th Revision, Indian Roads Congress, New Delhi, India, 2018

IRC: 58, Guidelines for the design of plain jointed rigid pavements for highways, 4th Revision, 2015

Performance analysis of an uncontrolled and a controlled intersection in Silchar city, Assam

N. Chanda
[1] *Department of Civil Engineering, CMR Institute of Technology, Bangalore, India*
** nipachanda@gmail.com*

Introduction

The heterogeneous traffic is more diverse in nature due to lane changing and lack of lane discipline in India. Long queues on intersections often observed, causing huge fuel consumption as well as environmental pollution in the urban area beside considerable time loss. Even in countries where uniform traffic conditions exist, uncontrolled intersection analysis always presents a rather complex picture. This is mainly due to the unavoidable presence of critical gap in all capacity expressions.

One of the earliest studies on gap acceptance was done by Ashworth and Green (1966). Many studies (Sinha and Tomiak (1971), Radwan and Sinha (1980), Troutbeck (1992), Davis and Swenson (2004), Ashalatha et al. (2005), Ashalatha and Satish Chandra (2011)) have been conducted on capacity of the roadway intersections. Due to area restriction and type of vehicles (non-motorized class) many a times traffic studies at intersection suggest alternative methods to relieve traffic congestion. This article presents one such study to analyse the performance of one uncontrolled and one controlled (signal/traffic police) intersection in Silchar city of state Assam. The results are useful for improvement of those intersections for a congestion free traffic flow.

Materials and Methods

Uncontrolled intersection is studied using basic gap-acceptance concept and Critical-Gap through clearing behaviour to determine entry capacities of vehicle. The controlled (signal/traffic police controlled) intersection is studied to find out the delay time and queue length. The video graphic survey data are collected as following the flowchart (Fig. 1) and extracted (Fig. 2-3)

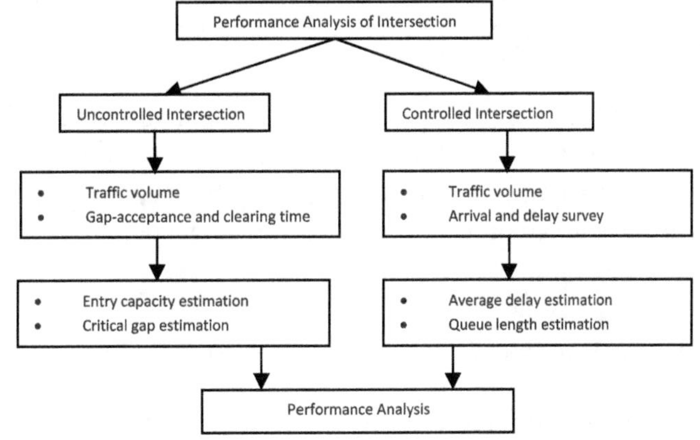

Figure 1. Flow chart of the study methodology

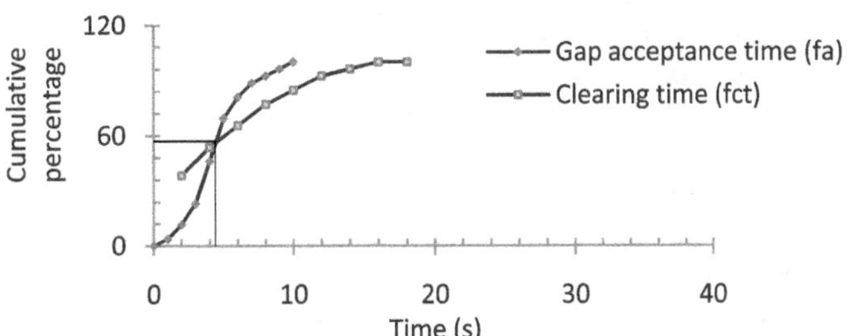

Figure 2. Cumulative frequency distribution curves for gap-acceptance time and clearing time

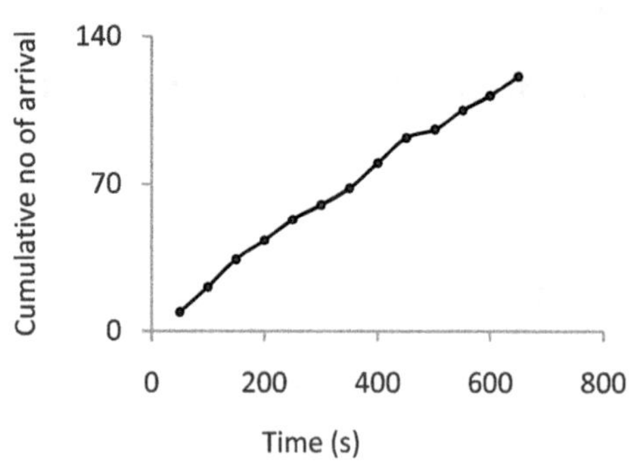

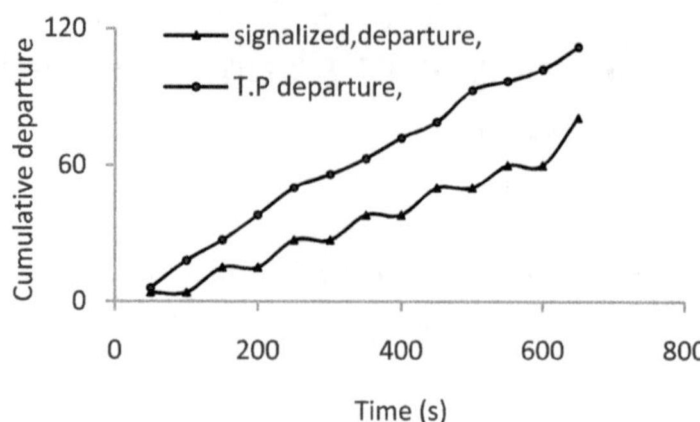

Figure 3. Arrival and departure of vehicle at controlled intersection (T.P stands for traffic police controlled)

Results and Concluding Remarks

Performance of uncontrolled intersection (Ambikapatti) is analyzed in terms of critical gap time (Table 1) and entry capacity for different conflicting traffic volume is determined whereas performance of controlled intersection (Premtola) is analyzed in terms of average delay and queue length (Table 2).

Table 1. Critical Gap-time and entry capacity of different vehicle for different conflicting volume

Right Turn From Minor Street						
Conflicting traffic volume (vph)	Critical Gap time in seconds			Entry Capacity		
	Car	Auto	Bike	Car	Auto	Bike
3922	4.72	4.39	2.98	27	37	209
3000	5.47	4.95	3.6	42	60	236
1938	6.24	5.49	3.74	112	153	544
676	8.37	6.08	4.41	514	689	1464
322	11	7.44	5	853	1010	2023

Table 2. Performance of controlled intersection in terms of average delay and queue length

Parameters	Saturated Condition		Un-saturated Condition	
	Signal control	Traffic police control	Signal control	Traffic police control
Average Delay (second)	166.322	41.322	129.775	37.921
Queue Length (vehicle no)	34	8	29	6

When the conflicting traffic volume is very high, vehicles from minor stream gets very less time to make a manoeuvre in uncontrolled intersection. The signalized intersection is also reflecting a long queue length and delay. Thus, the result obtained from the present study for both type of intersection leads the demand of revised lane width, larger turning radius, channelization for congestion free traffic operation.

References

Ashworth R, Green B D (1966) Gap acceptance at an uncontrolled intersection. Traffic Engineering and Control, Vol. 7, No. 11, pp. 676-678.

Sinha K C, Tomiak W W (1971) Section gap acceptance phenomenon at stop controlled intersections. Traffic Engineering, Vol. 41, No. 7, pp. 28-33.

Radwan A E, Sinha K C (1980) Gap acceptance and delay at stop controlled intersections on multilane divided highways. Journal of Institute of Transportation Engineers, Vol. 50, No. 3, pp. 38-44.

Troutbeck R J (1992) Estimating the critical acceptance gap from traffic movements. Physical Infrastructure Centre Research Report 92-5, Queensland University of Technology, Brisbane, Australia.

Davis G A, Swenson T (2004) Field study of gap acceptance by left-turning drivers, TRR 1899, Transportation Research Board, National Research Council, Washington, D.C., pp.71-75.

Ashalatha R, Chandra S, Prasanth K (2005) Critical gap at uncontrolled intersections using maximum likelihood technique. Indian Highways, Vol. 33, No. 10, Indian Roads Congress, New Delhi, pp. 67-74.

Ashalatha, R & Chandra S (2011) Critical Gap through Clearing Behavior of Drivers at Unsignalized Intersections. KSCE Journal of Civil Engineering (2011) 15(8):1427-1434. DOI 10.1007/s12205-011-1392-5

Effect of Signal Countdown Timer on Performance of Signalized Intersection

Nawal Kishor Singh[1], Arpan Mehar[2],

[1] *M-Tech Student, Transportation Division, Department of Civil Engineering, National Institute of Technology, Warangal, India*

[2] *Assistant Professor, Transportation Division, Department of Civil Engineering, National Institute of Technology, Warangal , India*

E-mail: om7659132@gmail.com

Introduction

Signalized intersections are introduced to increase efficiency and reduce traffic accidents. Various measures are being taken to increase the safety and efficiency of signalized intersections. Countdown timers are a technical measure and are becoming increasingly popular in many countries, including India. However, scientific studies in this field under Indian conditions are very limited and those available in the literature have compared different intersections with and without timers' studies in other countries in this area show mixed results. Furthermore, as driver behavior differs significantly from country to country, it is necessary to carry out separate studies from different geographical areas. Since Indian traffic is heterogeneous and undisciplined, special care has been taken to include these characteristics in the analysis. Based on previous results, it is clear that signal countdown timers (SCTs) are associated with a reduction in startup lost time, dilemma zone and red light violations (RLVs), according to some studies. There are other studies that contradict these findings. The main reason for these contradictory results could be the different traffic situation and the different driver behavior between the countries, even from one city to another. Studies in developing countries (China, Bangkok, Malaysia, India and Taiwan) have shown that the presence of green signal countdown timer (GSCT) and red signal countdown timer (RSCT) influences traffic patterns and driving behavior as well as efficiency and safety in different ways at signalized intersections. Caution should be taken when installing the signal countdown timers (SCTs) at urban intersections. It is interesting to note that in India, no technical study has been conducted to evaluate the effects on the efficiency and safety of intersections before installing this timer.

This study was conducted to investigate the impact of countdowns on various traffic characteristics and driver behavior, taking into account several measures of effectiveness (ME). The following goals are defined for this study.

- To analyze safety measure like approaching speed and red light violations (RLVs) at signalized intersections under Indian traffic conditions due to the presence of signal countdown timer.
- To analyze the changes in efficiency measures like start-up lost time and control delay at signalized intersections under Indian traffic conditions due to presence of signal countdown timer.

Study Methodology

To analyze the influence of different components of signal countdown timers on intersection efficiency and safety, two signalized intersections, Chengicherla X Road and Suchitra junction is chosen from Hyderabad city, India. At both signalized intersections traffic signals are installed with countdown timer already. To keep the geometric and traffic characteristics similar, data is collected with timer on and timer off condition at same intersection. Data is collected using RADAR gun and two video cameras on working day during same time period in presence and absence of timer.

The flow chart of methodology is as follows:

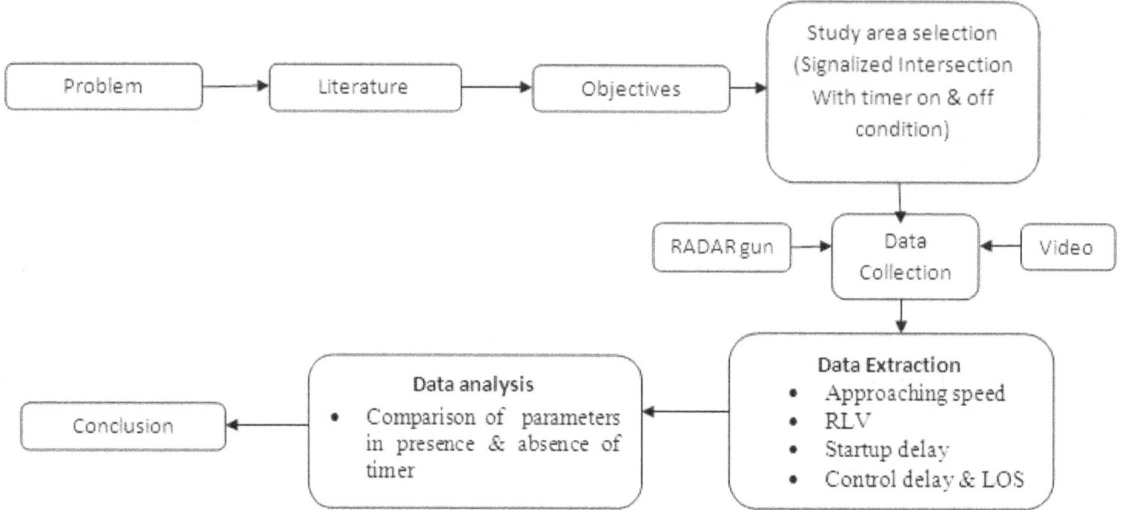

Results and Concluding Remarks

With respect to objectives of present study, following points may be concluded as-

1. There is significant increase in approaching speed during last 15 Sec. of green phase due to presence of signal countdown timer at both signalized intersections. In the presence of signal countdown timer, driver aware about the time of initiation of red phase, so they accelerate their vehicle in last few seconds of green phase in presence of timer so that they can cross the intersection within the remaining green time. Increase in approaching speed in presence of signal countdown timer cause speed limit violations which results reduction in traffic safety while crossing the intersection.
2. Average red light violations (RLVs) per cycle decreases at signalized intersections in presence of countdown timer. It suggest that there is decrease in red crossing of vehicle in timer on condition which can reduce the right angled crashes at intersections and helps to enhance the intersection safety.
3. After discharge headway and startup lost time (SULT) analysis it has been observed that there is significant decrease in SULT due to presence of countdown timer. In the absence of timer drivers waiting for green phase at front of queue takes more time to start and move as compared to timer on condition. Since driver in front of queue are aware about green phase initiation in timer on condition so they already prepare to cross the intersection which cause reduction in startup lost time.
4. Average control delay of signalized intersections decreases in presence of timer which results in increase of level of service (LOS) from F to E. So it can be concluded from above findings that signal countdown time helps to increase the efficiency of signalized intersections.
5. From the results of present study it can be concluded that there is effect of countdown timer on driver maneuvering behavior at signalized intersection.

References

Devalla, J., Biswas, S., and Ghosh, I. (2015)."The effect of countdown timer on the approach speed at signalised intersection". *Procedia Computer Science* 52, 920-925.

Fujita, M., Suzuki, K., and Yilmaz, C. (2007). "Behavior and Consciousness Analyzes on Effect of Traffic Signals Including Countdown Device". *Journal of the Eastern Asia Society for Transportation Studies*, 6, 2289–2304.

Harshitha, M. S., Agarwal, S., and Vanajakshi, L. (2012). "Headway Analysis at Signalized Intersections with and without Countdown Timer". *Highway Research Journal*, January, 33–40.

Effect of Pedestrian Characteristics on Traffic Performance at Uncontrolled Intersections using Micro Simulation

L. Govinda[1], D. Abhigna[2], S. Eswar[3], K.V.R Ravishankar[4*]
[1, 2, 3] Research Scholar, Department of Civil Engineering, National Institute of Technology Warangal, Warangal, India
[4] Assistant Professor, NIT Warangal, India
** e-mail: kvrrshankar@gmail.com*

Introduction

For shorter distances, walking is common across the world and it is most preferable option in the developing countries like India. The pedestrian risk taking behaviour and driver yielding behaviour varies from person to person. Width of the road, size of the city, climatic and temperature conditions, and time of the study play important role on pedestrian flow characteristics (Laxman et al. 2010). The age and gender both have significant effect on the observed safety-related behaviour. Male pedestrians are less likely to wait for the signal due to the factors such as issues of trust or propensity to take the risks of waiting (Bradbury et al. 2012). The spatial and temporal distribution of pedestrian-vehicle crashes vary for different pedestrian age groups and genders (Moridpour et al. 2018). Pedestrian crossing time was higher for elder and young pedestrians compared to other age pedestrians.

Even though the average waiting time can be measured in real life, the micro simulation software helps to analyse the situations for different volume combination and parameter settings. The travel time delay increases exponentially with the increase in vehicle volume (Istvan et al. 2014). The present study will discuss about the effect of pedestrian characteristics on both vehicular and pedestrian performance at uncontrolled intersections using micro simulation.

Study Area and Methodology

An uncontrolled intersection at 100 feet road from Warangal city, India was selected for the present study. A significant pedestrian crossing flow was observed at this location. Four hours (morning two hours and evening two hours) traffic data was collected from this location using videography method. The collected data was used to extract both traffic characteristics and pedestrian flow characteristics using Media Player Classic (MPC) video player and dimensional data intersection type, number of lanes, lane width, and crosswalk width were directly measured in the field. The extracted parameters from video were traffic volume with vehicle type, vehicle speed, travelling direction, pedestrian volume, age, gender, direction of travelling, and speed. Network was drawn in VISSIM for the selected location by giving the collected and extracted parameters as input values and calibration was done for both vehicle and pedestrian volumes. To calibrated traffic volume, driver behaviour (Wieddman 74 model) was used and for pedestrian's calibration, walking behaviour was used. Pedestrian characteristics used in this study were age, gender, crossing speed, and volume. Once the calibration was done, initially simulation was done by replacing male pedestrians (25%, 50%, 75%, and 100%) with female pedestrians, vice-versa. Similarly, VISSIM simulation was done for different pedestrian ages, crossing speeds, and total volumes (by replacing each field parameter with different parameter). From the evaluation-configurations in VISSIM, different traffic performance parameters were selected and the output files were imported into the MS Excel. From the results, vehicle and pedestrian performance were analysed with change in the pedestrian characteristics.

Results and Concluding Remarks

The observed field volume of vehicles and pedestrians were 10565 and 283 respectively and composition of each vehicle type and pedestrian type was given in the figure 1 below. Average travel time delay for vehicles and pedestrians for different pedestrian characteristics were shown in table 1 and table 2.

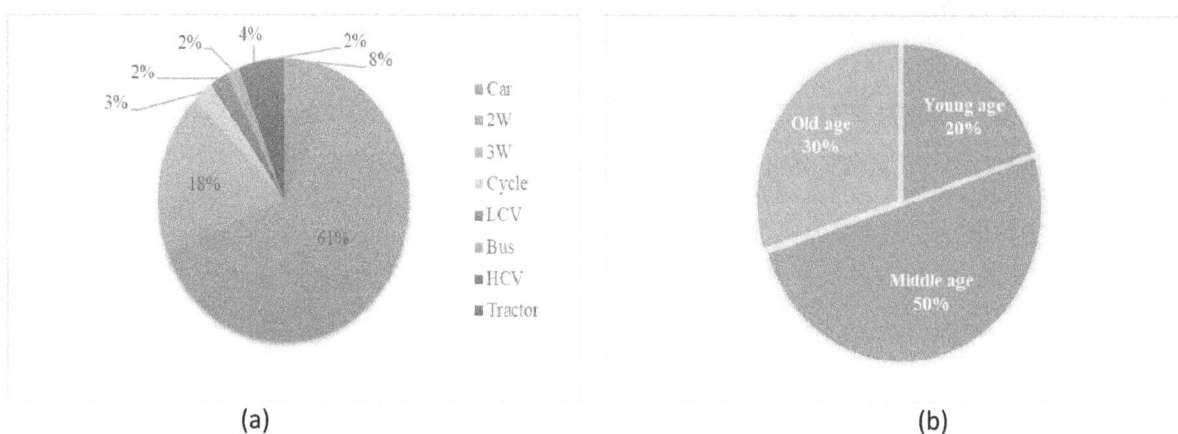

(a) (b)

Figure 1: Pie chart for (a) vehicular composition (b) pedestrian composition

Table 1: Pedestrian travel time delay variation with respect to age and gender

Pedestrian type	Pedestrian delay (seconds)	Vehicle delay (seconds)
Field	5.93	85.2
Calibrated	6.31	96.8
100% Male	5.31	92.49
100% Female	8.56	107.6
100% Young age	5.12	93.06
100% Middle age	6.84	97.36
100% Old age	12.54	117.91

Table 1: Pedestrian travel time delay variation with respect to pedestrian vehicle increase

Percentage increase	pedestrian increase		vehicle increase	
	Pedestrian delay (seconds)	Vehicle delay (seconds)	Pedestrian delay (seconds)	Vehicle delay (seconds)
25 percentage	10.93	103.36	12.22	109.12
50 percentage	13.68	110.27	19.52	124.18
75 percentage	17.94	119.25	29.44	138.25
100 percentage	23.16	132.02	48.31	162.74

The results showed that average travel time delay of both pedestrians and vehicles was higher for 100% female and old age pedestrians and lower for 100% male and young age pedestrians. Average travel time of both pedestrians and vehicles was decreases with the increase in the proportion of male pedestrians and young age pedestrians. The increase in either pedestrian or vehicular volume leads to increase in both pedestrian and vehicular delays. The performance of the intersection decreases with the increase in female, old age pedestrians and increase in the either pedestrian or vehicular volume.

References

1. Istvan. F. and Zsuzsanna. K.I., (2014) "Travel Time Delay at Pedestrian crossings based on microsimulation." periodica polytechnic 58(1): 47-53.
2. Ravishankar. K.V.R. and Nair. P.M., (2018) "Pedestrian risk analysis at uncontrolled midblock and un-signalised intersections." Journal of traffic and transportation engineering 5(2): 137-147.
3. Xuesong. W., Junguan. Y., Chris L., Zhuoran. J. and Shikai. Y. (2016) "Macro-level safety analysis of pedestrian crashes in Shanghai, China." Accident Analysis and Prevention: 12-21.
4. Rengarasu. T.M., Jayawansa. H.N. and Perera. G.P.W. (2012) "Estimation of Pedestrian walking speeds at controlled cross walks in Sri Lanka - a pilot study." Asian Conference on Electrochemical Power Sources: 1-6.

Calibration of Stairways connecting Foot over bridge in Intercity Railway Stations using Micro Simulation- VISWALK

S. Eswar[1*], L. Govinda[1], K.V.R. Ravishankar[2]

[1] *Research Scholar, Department of Civil Engineering, National Institute of Technology Warangal*

[2] *Assistant Professor, Department of Civil Engineering, National Institute of Technology Warangal*

* *e-mail: eswar.sala@gmail.com*

Introduction

Pedestrian flow characteristics are studied across the world on various pedestrian walking facilities- crosswalks, sidewalk, walkways, stairways etc. Results showed that the pedestrian speeds varies with respect to pedestrian characteristics, geometric characteristics, weather conditions, location and region. Pedestrian fundamental curves are similar for different facilities with varying coefficients signifying facility type affect (Hongfei et al 2009). Few studies (Shah et al. 2013 [a], Shah et al. 2013 [b], Jiten et al. 2016, Patra et al. 2017, Sala et al. 2017) are reported on pedestrian walking behavior on stairways in metro stations, sub-urban railway stations. Individual pedestrian characteristics and external conditions are observed to influence the pedestrian walking speeds. Some of the factors are age, culture, gender, shy away distance, temperature, travel purpose, type of infrastructure, walking direction (Vanumu et al., 2017). Under lower flow rates (<10 ped/m/min), walking speeds are governed by individual characteristics. Also walking speeds in afternoon are higher than the evening periods. Average walking speed of individual is almost same as average speed of pedestrian stream. (Shaha et al 2013, Shah et al 2013). Pedestrian walking speeds decreases with age and luggage. The walking speed of young and male pedestrians is greater than old and female pedestrians. Pedestrians without luggage move faster than pedestrians with luggage (Zhang et al 2009, Patra et al., 2016, Sala et al 2017). Walking speed is higher in the direction of major flow (Jiten et al 2014). Speed reduces gradually with increase in density and reduction in space (Jiten et al., 2016). Pedestrian walking speeds are higher on descending stairways than on ascending stairways (Lam et al 1995).

Pedestrian studies addressed the flow characteristics observed on metro station and sub-urban railway stations. Few Studies (Shah et al. 2013 [b], Jiten et al. 2016, Patra et al. 2017, Sala et al. 2017) are conducted to understand the flow characteristics of pedestrians on stairway in intercity railway stations. Unlike metro and sub-urban railway stations where passengers are daily commuters, passengers in intercity railway stations are rare visitors and makes long hauling trips. In metro and sub-urban railway stations, daily commuters are of higher proportions, Jiten et al (2016) reported 93% daily commuters carrying less or no luggage in sub-urban railway station. Also relatively higher proportions of pedestrians carrying luggage are observed in intercity railway stations. Thus the intercity railway station characteristics differ from metro and sub-urban railway stations. Hence pedestrian studies are to be conducted and the capacity is to be estimated corresponding to the pedestrian behavior. Due to pedestrian distribution to various foot over bridges spread along the platform, pedestrian flow characteristics at capacity ranges are difficult to attain under normal field conditions. In this respect, present study is framed to calibrate the stairways in intercity railway stations using micro simulation 'VISWALK' to replicate the field observations and compare flow characteristics on the two stairways to understand the effect of width.

Methodology

Video recording data of pedestrian flow on stairway was collected using a video camera mounted on suitable elevated position. Pedestrian traffic parameters flow, density and speed are extracted manually from playback videos. Pedestrian characteristics age: child (<15 years), young age (15-30 years), middle age (30-60 years) and elders/old age (>60 years), gender: male, female and luggage carrying (yes/no) are determined from visual interpretation. The extracted parameters on stairways are used as input to the VISWALK micro simulation.

Table 1: Default parameter value in VISWALK

Parameter	Default value
Tau	0.400
ReactToN	8.000
ASocIso	2.720
BSocIso	0.200
Lambda	0.176
ASocMean	0.400
BSocMean	2.800
VD	3.000
Noise	1.200
SidePref	None

The observed stairways were built as per the dimensions in VISWALK micro simulation software and the observed pedestrian volume, speeds, and relative flows were given in each direction as per the field relative volumes. Simulation was ran with the default calibration parameter values shown in **table 1** and MAPE (mean acceptable percentage error) was calculated for input and output pedestrian volumes. Calibration is continued with varying values of the parameters until the MAPE value for the pedestrian volume is less than 5%.

Results and Concluding Remarks

In this research work, pedestrian flow characteristics are collected from two stairways with different dimensions facilitating pedestrians to access FOB in Vijayawada intercity railway station. From the observed flow characteristics, the mean walking speeds on V_{ST1} and V_{ST2} are 34.26 m/min 40.25 m/min respectively. The mean walking speed is higher on wider stairway in both ascending and descending direction. Walking speeds are higher in descending direction than in ascending direction. The two stairways of different dimensions are calibrated using VISWALK micro simulation. With the calibrated parameters, the input volume is increased to 200% with an incremental interval of 25% and pedestrian flow characteristics are studied. Pedestrian flow characteristics variation on two stairways are compared to understand the effect of width.

Acknowledgment: This research study was funded by Science and Engineering Research Board, Department of Science and Technology, Government of India under "Fast Track Young Scientist-Engineering Science Scheme" with project designations SERB/F/1821/2014-2015 dated 18th June, 2014. Authors express their thanks and appreciation to the research project panel for their insightful comments and sanctioning of the project. We would like to extend our sincere gratitude to South Central Railway Officials for their cooperation in data collection.

References

1. Hongfei. J., Lili. Y. and Ming. T. (2009) "Pedestrian Flow Characteristics Analysis and Model Parameter Calibration in Comprehensive Transport Terminal." Journal of Transportation Systems Engineering and Information Technology 9(5): 117-123
2. Vanumu. L. D., Rao. K. R. and Tiwari. G. (2017) "Fundamental diagrams of Pedestrian Flow Characteristics: A Review." Eur. Transp. Res. Rev. (2017) 9: 49 DOI 10.1007/s12544-017-0264-6
3. Shah. J., Joshi. G. J. and parida. P. (2013a) "Behavioral Characteristics of Pedestrian Flow on Stairway at Railway Station." 2nd Conference on Transportation Research Group of India. Procedia-Social and behavioral Sciences 104(2013):688-697
4. Shah. J., Joshi. G. and Parida. P. (2013b) "Walking Speed of Pedestrian on Stairways at Intercity Railway Station in india." Proceedings of the Eastern Asia Society for Transportation Studies, Vol 9
5. Zhang. R., Li. Z., Hong. Ju., Han. D. and Zhao. L. (2009) "Research on Characteristics of pedestrian Traffic and Simulation in the Underground Transfer Hub in Beijing." Fourth International Conference on Computer Sciences and Convergence Information Technology pp: 1352- 1357. DOI 10.1109/ICCIT.2009.222
6. Jiten. S., Gaurang. J., Purnima. P. and Shriniwas. A. (2014) "Impact of train Schedule on Pedestrian Movement on Stairways at Suburban Rail Transit Station in Mumbai, India." Advances in Civil Engineering. Volume 2015. http://dx.doi.org/10.1155/2015/297807
7. Lam. W. H. K., Morrall. J. F., Ho. H. (1995) "Pedestrian Flow Characteristics in Hong Kong" Transportation Research Record 1487, 56- 62
8. Patra. M., Sala. E., Ravishankar. K. V. R. (2017) "Evaluation of Pedestrian Flow Characteristics across Different Facilities Inside a Railway Station", Transportation Research Procedia 25(2017), 4763- 4770.
9. Sala. E., Patra. M., Ravishankar. K. V. R. (2017) "Analysis and Comparision of Pedestrian Flow Characteristics variation on Passageway, Stairway and Escalator in Intercity Railway Station", International Journal of Transportation Engineering and Traffic System, Vol 3(2); 8-19.
10. Jiten, S., Gaurang, J., Purnima, P. and Arkatkar, S., (2016). "Effect of stairway width on pedestrian flow characteristics at railway stations." Transportation Letters, 8(2), pp.98-112.

VISSIM Based Analysis of Road Network in Bengaluru.

Vinutha K S[1*], Mr.Mruthyunjay S K[2], Dr. Vivek R Das[3]
[1] *M Tech Student, Department of CTM, Dayananda Sagar College of Engineering, Bengaluru, India*
e-mail: vinuthaskandgal@gmail.com
[2] *Assistant Professor, Department of CTM, Dayananda Sagar College of Engineering, Bengaluru, India*
e-mail: mruthyunjaysk@gmail.com
[3]*Professor,Department of CTM, Dayananda Sagar College of Engineering, Bengaluru, India*
e-mail: vivekdurgadath@gmail.com

Introduction

The complexity in the traffic flow behavior and the complications in performing experimentations in real-world traffic makes computer simulation software a significant analysis tool in traffic engineering. The study aims to analyze the traffic of a critical road stretch from Banashankari to silk board, which is currently facing traffic issue due to the ongoing metro project, under adverse traffic demand and to provide the solution for the same. Objective includes use of simulation technique to analyze the traffic flow operational state of a road network in a microscopic level under the peak traffic demand. The study efforts to use simulation method to determine the influence of various traffic organization plans, to confirm the capacity of the roadway, to find the bottlenecks and congestion, and thus, to provide technical assistance for the proposal of the alternative solutions. Many studies have been conducted to make an attempt in efficiently modelling the Indian heterogeneous traffic. Zhengyang Lu et al. 2016 has showed the method of calibrating the model by a video-based approach. Mathew and Radhakrishnan 2010, has conducted a case study at signalized intersection and found the microsimulation parameters that are more likely to affect the Indian non lane based mixed traffic condition at signalized intersection with high volume by sensitivity analysis. Siddharth and Gitakrishnan 2013 has tried to model the Indian heterogeneous traffic by finding the more significant parameters of VISSIM calibration model using ANOVA technique and automatically calibrating the model using COM interface.

Methodology

The study stretch was selected from Banashankari to silk board junction which comprises of 5.2 km located in Bengaluru city. The present travel time is found to be 40 minutes by bus. The stretch comprises of 12 major intersections with 6 signalized and 6 unsignalized intersections. The data collection for one hour was done using videography survey at all the intersection. The calibration and validation of the network model is done considering vehicle travel time as a performance measure. To find the operational performance of the present route, the data was fed into VISSIM software followed by the calibration and validation. Based on the evaluation results, three alternative routes are proposed.

Analysis

The following analysis was done in VISSIM with present scenario and proposed scenario. This is explained below.

1. Network performance result analysis: This mainly analyses the delay, travel time and speed. It was found that average delay of all the vehicles was decreased by 184 seconds, speed was increased from 5.0 kmph to 9.2 kmph and stops per vehicle was decreased from 20 to 12.
2. Data collection result analysis: The data collection results analysis was performed at 15 selected stretches and it was found that the acceleration at these stretches was increased after the alternative route proposal with substantial increase in the speed of the vehicles and decrease in queue delay.
3. Vehicle travel time results analysis: Vehicle travel time measurements which were predefined in VISSIM give the vehicle travel time. It was found that the average travel time between Banashankari to silk board was 40 minutes initially which later got reduced to 18 minutes after the alternative

route proposal. The vehicle travel time from Udupi garden to Vishnu park before proposal was 23 minutes which was decreased to 7 minutes via outer ring road and also for the vehicles travelling through alternative route, the travel time was found to be around 8 minutes. The travelling time of buses travelling route 1 was found to 22 minutes which was 46 minutes in the existing route from Banashankari to silk board. For alternative 3, the average travel was found to be 28 minutes. Similarly, the travel time between the intermediate stretches was also reduced to an extent.

4. Delay results analysis: For the predefined travel time measurement, the stop delay, vehicle delay and number of stops per vehicle were found out in VISSIM. The vehicle from Banashankari to silk board were facing delay of 28 minutes which later got reduce to 7 minutes. From silk board to Banashankari, the delay has been reduced from 24 minutes to 19 minutes.
5. Queue results analysis: Queue results analysis give the average and maximum queue length and queue stops at the upstream of each arm of the intersection. The results obtained before and after the proposal shows that the there is a considerable reduction in the queue length and stops at some arms.
6. Node results analysis: The operational performance of the intersection is found out by the node results analysis. Based on the results, the LOS at intersections namely Sangam circle, metro deviation, 9th main road, Marenahalli, Central, Ragigudda, East end circle and Jayadeva junction is increased. The LOS of intersections namely Raghavendra Swamy mutt, 9th block Jayanagar, Udupi garden and Vishnu park remains the same with improved results in vehicles delay and queue length.

Conclusions

- From the traffic analysis, it was found that the Banashankari to Silk board stretch which is one of the most critical stretches in Bengaluru having high traffic demand is facing vulnerable congestion and bottlenecks because of the the ongoing metro construction.
- The attempt was made to minimize the traffic demand in the Marenahalli and outer ring road by suggesting a scenario with three routes

Figure 1. The Overall Network with Proposed(blue) and Existing Route(maroon)

- The operational performance of the alternative routes was found to be satisfactory with the diverted traffic alone. Hence, the routes can deal with their existing traffic along with the diverted traffic by adopting some management policies like initialization of signals, shifting of bus stops from the intersections, one-way traffic and restriction of commercial vehicles at some road stretches etc.

References

1) Manraj Singh Bains, Balaji Ponnu, Shriniwas S Arkatkar (2012) Modeling of Traffic Flow on Indian Expressways Using Simulation Technique, 8th International Conference on Traffic and Transportation Studies, Changsha, China.
2) Siddharth S M P, Gitakrishnan Ramadurai (2013) Calibration of VISSIM For Indian Heterogeneous Traffic Conditions, 2nd Conference of Transportation Research Group of India.
3) Tom V. Mathew and Padmakumar Radhakrishnan (2010), Calibration of Microsimulation Models for Non-Lane Based Heterogeneous Traffic at Signalized Intersections, Journal of Urban Planning and Development, ASCE
4) Zhengyang Lu, Ting Fu, Liping Fu, Sajad Shiravi, Chazhe Jiang (2016). A Video Based Approach to Calibrating Car Following Parameters in VISSIM In Urban Traffic, International Journal of Transportation Science and Technology.
5) Dong LIN, Xiaokuan YANG, Chao GAO (2013) VISSIM – Based Simulation Analysis on Road Network of CBD In Beijing, China, 13th COTA International Conference of Transportation Professionals.

Effect of Aggregate Shape Factors on the Performance of Bituminous Mixtures

B. Giri Babu[1], Kakara Srikanth[2*], Venkaiah Chowdary[3]
[1] *Transportation Division/Department of Civil Engineering /NIT Warangal/ Warangal, India*
[2] *Transportation Division/Department of Civil Engineering /NIT Warangal/ Warangal, India*
[3] *Transportation Division/Department of Civil Engineering /NIT Warangal/ Warangal, India*

* *e-mail: sri717004@student.nitw.ac.in*

Introduction

In bituminous mixtures, aggregates are combined with bitumen to form a compound material. By weight of total bituminous mix, aggregate generally account for approximately 95 percent. Even though the volume occupied by aggregates within the bituminous mix is relatively higher than the bitumen, the relative cost of aggregates is less and contributes significantly towards the performance of the pavements. Thus, a detail characterization of properties of aggregates is important in the design of bituminous pavements (Masad et al. 2007). Based on the dimensions of aggregate particles, three properties are defined and they are shape, angularity, and surface texture (Barrett 1980). Aggregates are classified into different shapes or form based on the ratio of its dimensions. Flat or elongated aggregates (where the ratio of largest dimension to least dimension is high) tend to break when load is applied, i.e., during compaction or during trafficking. Based on the round or angular edges of aggregate, angularity property is described. Surface texture property is used to classify the aggregates based on its surface roughness. Surface texture is an important property which defines the bitumen retention on aggregate surface. These three properties are independent to each other (Rousan, 2004) and they are important, as they influence the performance and serviceability of bituminous mix (Brown et al. 1989; Kandhal et al. 1992). Aggregate morphological characteristics are very complex and cannot be characterized adequately by any single test. Masad et al. (2011) proposed an image analysis based methodology to quantify the sensitivity of bituminous mix performance to aggregate shape. Thus, an objective characterization of aggregate is required to quantify the effect of aggregate geometry on the performance of bituminous mix. In the present study, the aggregates are classified based on various indexes proposed in the literature and natural aggregates are replaced by aggregates of different shapes in different proportions in the mix. The bituminous mix properties are then determined to find better performing aggregate shapes, and to find best aggregate characterization method.

Materials and Methods

Strong, durable, proper shape and size aggregates are desirable for a monolithically acting bituminous layer in flexible pavement. The physical properties of aggregates which defines strength, toughness, hardness, shape, and water absorption are determined by laboratory investigation. For the present experimental investigation, aggregates are considered from one source, and Dense Graded Bituminous Macadam (DBM) grading – II has been adopted in this study. Only Viscosity Grade (VG) - 30 bitumen is used in the study and DBM grading – II mix proportion is fixed by Marshal mix design according ASTM D1559 guidelines.

To determine the physical properties of aggregates, basic laboratory tests including crushing strength, hardness, toughness, durability, and specific gravity were performed. Further, the aggregates were quantified based on the shape tests. The shape tests include determination of flakiness index and elongation index. Aggregates are classified in to four different shapes i.e., disk, blade, rod, and cube base on flatness ratio and elongation ratio. In this shape analysis particle longest diameter (dL), the intermediate diameter (dI), and the shortest diameter (dS) are determined. Flatness ratio is the ratio of dS to dI and elongation ratio is the ratio of dI to dL. The combined effects of particle shape and surface texture of an aggregate is determined using Aggregate Particle Shape Index which is defined in ASTM D 3398. Aggregate Particle Shape Index is a quantitative measure of the aggregate shape and texture characteristics that may affect the performance of bituminous mix.

The Optimum Bitumen Content (OBC) for a bituminous mix is determined were by Marshall mix design method. In this method the bituminous samples were prepared with bitumen content varied in increments of 0.5% in the range of 4.5% to 6 % of bitumen. OBC is the bitumen content that induce 4% air voids in the compacted bituminous mix. For the same compacted samples, the void in mineral aggregate (VMA), and the void space in coarse aggregate (VCA) are also determined. The performance of DBM mixes was studied with different shape aggregates and different proportions (0%, 10%, 20%, 30%, 40%, and 50%) replacing the original aggregates with four shape aggregates. Further, Indirect Tensile Strength (ITS) of bituminous mixtures was determined.

Results and Concluding Remarks

In all the DBM mix samples prepared with different shape aggregates, the Marshall stability values were observed to highest for the mixes prepared with cubical aggregates. Further, the stability found to increase with increase in proportion of cubical aggregates up to 30%. This phenomenon could be due to the cubical particles exhibit higher interlocking and internal friction, compared to flat, thin, and/or elongated particles. The other Marshall parameters such flow, voids filled with bitumen also found to increase with increase in proportion of cubical aggregates. In contrast to this, the parameters such as air voids and VMA found to increase with increase in proportion of blade shape aggregates. This could be due to the gaps between the aggregates within the bituminous mix is less in the case of blade shape aggregates.

The peak ITS is observed at the 50% replacement of different aggregate shapes for the DBM mix. Cubical shape aggregate attained the maximum value and lowest is observed for blade shape aggregates. This could be due to the blade aggregates tend to break down excessively during compaction. Similarly, particle index value is observed to be highest for cubical aggregates and the same higher ITS values are observed for cubical aggregates comparative to the other shapes. Thus, it can be concluded that particle index value quantifies the effect of shape of aggregate on the performance of bituminous mix.

References

ASTM. (1989). Test method for resistance of plastic flow of bituminous mixtures using Marshall apparatus. ASTM D1559, West Conshohocken, PA.

ASTM. (2006). Standard Test Method for Index of Aggregate Particle Shape and Texture. ASTM D3398, West Conshohocken, PA.

Brown, E.R., McRae, J.L, and Crawley, A.B. (1989) Effect of aggregate on performance of bituminous concrete. ASTM STP 1016, Philadelphia, 34-63.

Barrett, P. J. (1980). The shape of rock particles, a critical review. Sedimentology, 27, 291-03.

Kandhal, P.S., Khatri, M.A., and Motter, J.B. (1992) Evaluation of particle shape and texture of mineral aggregates and their blends. Journal of Association of Asphalt Paving Technologists, 61, 217-240.

Masad, E., Al-Rousan, T., Bathina, M., McGahan, J., & Spiegelman, C. (2007). Analysis of aggregate shape characteristics and its relationship to hot mix asphalt performance. Road Materials and Pavement Design, 8(2), 317-350.

Masad, E., Little, D., & Sukhwani, R. (2011). Sensitivity of HMA performance to aggregate shape measured using conventional and image analysis methods. Road materials and pavement design, 5(4), 477-498.

Rousan, T. M. (2004). Characterization of aggregate shape properties using a computer automated system. Ph. D. Dissertation, Dept. of Civil Engineering, Texas A&M University, College Station, TX.

Evaluating the Lateral and Longitudinal Deviation of the Vehicles Approaching at the signalized intersection.

Kishore. S.R[1], Vivek R Das [2*]
[1] *M Tech Student, Department of CTM, Dayananda Sagar College of Engineering, Bengaluru, India*
e-mail: kishoreramesh202@gmail.com
[2]*Professor,Department of CTM, Dayananda Sagar College of Engineering, Bengaluru, India*
e-mail: vivekdurgadath@gmail.com

Introduction

The vehicles approaching the signalized intersection stop line tends to manoeuvre between the vehicles so as to reach the stop line and to utilize the signal green time at the soonest. This phenomenon increases the distance travelled by the vehicles at the node, abiding the rules of following the queue and lane discipline, increased conflict points, causing deceleration to other large sized vehicles in the queue, increased queue length and also affects the discharge rate of the vehicles at the nodes and thus affecting the efficiency of the intersection. The main objective of the study is to determine the lane changing behaviour of the vehicles at the intersection. In this study, an attempt was made to determine the position and the path followed by the vehicles while approaching the intersection stop line by means of vehicles trajectory data. The main analysis involves, the determination of the position of the vehicles during its movement, and distance travelled by the vehicles while approaching the intersection stop line, which showcases the information of vehicles Longitudinal movement (X) along the length of the road and Lateral movement (Y) along the width of the road. The following studies have been conducted to analyse the irregular movement of the vehicles at the intersection comprising of heterogenous traffic conditions. Amit Agarwal, Gregor Lammel (2015), has proposed the model to simulate the behaviour of small sized vehicles in congested regime called as seepage action where the small sized vehicles seeps through gaps between the large size vehicles and approach to the intersection stop line. Gurcan Comert (2015) estimated queue length for probe vehicles at isolated intersection. This study develops the penetration level at the node and vehicles rate of arrival in finding queue length at remote traffic intersection. A closed form logical expressions and variance of these parameters were originated. The various derived estimators were compared based on square error losses. The number of cycles, estimation of penetration rate, impact of combination of these parameters on queue length were estimated. Amit Agarwal and Gregor Lammel (2016), extended the work of modelling the seepage behaviour of smaller vehicles in mixed traffic conditions using agent-based simulation. In order to allow the vehicles to seep, the traditional first in first out queue model is modified such that in free flow regime, faster vehicles can overtake slow vehicles, and in congested regime slower vehicles can overtake faster vehicles and the model is validated with the help of fundamental diagrams.

Methodology

The Hope farm junction is one of the highly congested signalized intersection in the city due to its increasing vehicular demand where high queue lengths are observed during the peak hours. The videography survey was conducted by means of aerial drone to capture the top view of the intersection and thus ensuring every vehicle utilizing the facility is captured during the survey. In this study an attempt has been made to analyse the movement of vehicles at the intersection by means of the software traffic data extractor developed by IIT Bombay. Initially the inventory survey was conducted at the intersection to measure all the static dimensions of the intersection and to mark all the permanent objects present at the sidewalks like light poles, tress etc. The distance between the objects and the distance between the stop line and the objects were measured which would also assist in determining the queue length formed by the vehicles during the red time. The video was uploaded in the software and vehicle trajectory information was extracted for each vehicle by fixing the grid size according to the dimensions of the carriage way. In the grid size of fixed dimension, the movement of different class of vehicles, there positioning and distance travelled longitudinally and laterally during the particular time interval within the grid was traced.

Analysis

The following analysis was made by using the traffic data extractor

1. The positioning of the vehicles and the time duration of the movement of vehicles within the grid was extracted and the vehicular movement and its deviation in longitudinal and lateral direction within the grid was traced by means of vehicle trajectories obtained by the traffic data extractor.
2. The vehicular movements longitudinally and laterally were noted within the grid in terms of meters, and the values were compared with the defined grid size

Conclusions

The longitudinal (X) and lateral deviation (Y) of vehicles from selected standard length is given in the table below. This is a measure of lane discipline. Lateral deviation is found maximum for two wheeler and least for car. Except for Arm 4 longitudinal deviation is more for two wheelers. The study shows that even in junction were vehicle speed is less lane indiscipline occurs.

Representation of longitudinal and Lateral deviation of vehicle in each phase of the intersection

Vehicle Type	ARM 1 (X,Y)		% deviation in (X,Y)		ARM2 (X,Y)		% deviation in (X,Y)		ARM 3 (X,Y)		% deviation in (X,Y)		ARM 4 (X,Y)		% deviation in (X,Y)	
	X= 60m	Y= 3.64m	X	Y	X =60m	Y= 3.4m	X	Y	X= 60m	Y= 3.6m	X	Y	X= 60m	Y= 3.55m	X	Y
2-W	70.61	6.45	18	77	69.9	5.84	17	72	72.4	6.1	21	69	67.5	5.98	13	68
Auto	69.65	4.1	17	13	68.6	3.4	14	0	71.2	3.1	19	0	68.7	3.9	15	10
LCV	71.21	4.3	19	12	64.5	3.9	7	15	62.0	4.3	3	19	67.2	4.7	12	32
Car	67.7	3.7	13	2	63.0	2.0	5	0	67.6	2.3	13	0	72.6	3.1	21	0
Bus	63.45	3.9	6	7	67.44	4.68	12	38	71.2	4.58	19	27	74.2	5.3	24	49
HCV	62.8	3.96	5	9	63.5	3.95	6	16	61.8	4.05	3	12	63.2	3.98	5	12

References

1) Amit Agarwal, Gregor Lammel. Seepage of Smaller Vehicles under Heterogeneous Traffic Conditions. The 4th International Workshop on Agent-based Mobility, Traffic and Transportation Models, which includes methodologies and applications, ABMTRANS 2015, Salzufer 17-19, 10587 Berlin, Germany.
2) Wang, J., Yang, J., Li, Q., and Wang, Z, The effect of nonstandard vehicle's special behavior on mixed-traffic modelling. The applications of advanced technologies in Transportation Engineering (2004), 589–594.
3) Amit. Agarwal, Gregor. Lammel. Modelling seepage behavior of small vehicles in mixed traffic conditions using an agent-based simulation. Transport in Dev. Econ. 2.2, pp. 1-12. DOI: 10.1007/s40890-016-0014-9.
4) S. Aupetit. S. Espie, and S. Bouaziz, "Naturalistic study of riders' behavior in lane–splitting situations," Cognition, Technology & Work, pp. 1–13, 2014.
5) M. Sperley and A. J. Pietz. Motorcycle lane-sharing: Literature review, "Tech. Rep, OR-RP-10-20, Oregon Department of Transportation Research Section,2010.

Travel Time Prediction in Mixed Traffic Conditions Using Non Linear Modelling

Jaya Krishna Jammla[1*], Ravi Shankar KVR[2]
[1]*Research Scholar, Transportation Division, Civil Engineering Department, NIT Warangal, Telangana, India*
[2]*Assisstant Professor, Transportation Division, Civil Engineering Department, NIT Warangal, Telangana, India*
* *e-mail: jayakrishna5800@gmail..com*

Introduction

Travel time is considered to be qualitative check for any road network. Based on the travel time of the corridor, any improvement plans of the corridor can be implemented. Hence, travel time prediction plays crucial role in traffic management. Many studies have been done on travel time prediction. Model has been developed using log normal distributions from the data collected from automatic vehicle locators (Kazagli and Koutsopoulos, 2013). Artificial neural network model has been developed using mean speed and maximum continuous acceleration from the car (Li and McDonald, 2002). From link detector data, a time varying arterial travel time model has been developed (Lu and Chang, 2012). Dynamic travel time model has been developed for buses using the GPS data (Fan and Gurmu, 2015). These studies have been done in homogeneous traffic conditions. In countries like India, with mixed traffic condition, travel time prediction modelling is mostly done on buses (Vanajakshi et al. 2008; Padmanaban et al. 2010; Kumar and Vanajakshi, 2014) and using car probe vehicle (Kumar et al. 2014).

In this study, travel time prediction model has been developed from the data obtained from probe vehicles using mobile GPS application.In India with mixed traffic conditions, two wheeler and three wheeler occupy significant traffic share than car for private mode of transportation. So, probe vehicles used are car, 2 wheeler and 3 wheeler. Mobile application used is Open street map (OSM) tracker which is available in android operating system of mobile phones and is a free application from google play store.

Materials and Methods

Study area selected for this study is from Kazipet station to Warangal station via Hanmakonda. The total road stretch is 14.7 km and has 14 intersections. Out of these intersections, 7 are signalized and remaining are un-signalized. Persons carrying mobile phone with the application will be travelled in the above mentioned modes. Three modes will start simultaneously at the starting point and move individually. Data is collected for 5 weekdays during different traffic flows like peak, intermediate and free flow conditions. Geometric data and signal data are collected manually.

After data collection, data is extracted from the applications. Data is stored in '.gpx' format which can be opened in Microsoft excel. Data obtained will be in terms of coordinates (latitude and longitude), speed and time.Distance between points is calculated using haversine formula given below

$$d = 2 \times r \times \sin^{-1}\sqrt{sin^2\frac{a2-a1}{2} + \cos a1 \times \cos a2 \times sin^2\frac{b2-b1}{2}} \quad (1)$$

Where d is the distance between coordinates, r is the radius of earth, a1,a2 are latitudes and b1, b2 are longitudes in radians. From the distance, speed and delay, travel time prediction model has been developed.

Results and Concluding Remarks

A multiple linear regression model has been developed in statistical package for the social sciences (SPSS) software from thedata extracted to calculate travel time (TT) in seconds from the individual parameters distance (D in m), speed (V in m/s) and stop time (S in seconds) and is given below.

$$TT = 41.69 + 0.12*D - 5.17*V + 1.04*S \quad (2)$$

The root mean square error (RMSE) value calculated for the above model is 25.68. As the value is higher, the non-linearity between the dependent and independent variables are checked. Then the model has been modified in SPSS based on the general equation for travel time given in equation (3).

Travel time = distance/speed + delay (3)

Based on the non-linearity between time to the distance and speed, equation 3 has been modified as below.The RMSE value obtained for equation 4 is 16.48.

$$TT = 0.36* (D^{0.91}/V^{0.26}) + 1.15*S \quad (4)$$

Equation 2 and 4 are obtained from 70% of the data collected and remaining 30% is used for validation.

The non-linear model predicts travel time better than the linear regression model. Since most of the studies uses car or bus as probe vehicle, this study suggests that 2 wheeler and 3 wheeler can also be used as probe vehicles for travel time data collection. If the travel time of private transport modes are available with the policy makers, then public transportation can be improved accordingly thereby encouraging public transport.

Acknowledgments: Authors would like to acknowledge the research assistance provided by the Department of Science and Technology, Government of India.

References

Kazagli E, Koutsopoulos HN. (2013) Estimation of arterial travel time from automatic number plate recognition data. Journal of the Transportation Research Board, 2391: 22–31.

Li Y, McDonald M (2002) Link travel time estimation using single GPS equipped probe vehicle. The 5th International Conference on Intelligent Transportation Systems. 3–6 September 2002, Singapore.

Lu Y, Chang G. (2012) Stochastic model for estimation of time-varying arterial travel time and its variability with only link detector data. Transportation Research Record: Journal of the Transportation Research Board. 2283: 44–56.

Fan W, Gurmu Z. (2015) Dynamic travel time prediction models for buses using only GPS data. International Journal of Transportation Science and Technology. 4(4): 353–366.

Vanajakshi L, Subramanian SC, Sivanandan R. (2008) Travel time prediction under heterogeneous traffic conditions using global positioning system data from buses. IET Intelligent Transport Systems. 3(1): 1–9.

Padmanaban RPS, Divakar K, Vanajakshi L, Subramanian SC (2010) Development of a real-time bus arrival prediction system for Indian traffic conditions. IET Intelligent Transport Systems. 4(3): 189–200.

Kumar SV, Vanajakshi L (2014) Urban arterial travel time estimation using buses as probes. Arabian Journal for Science and Engineering. 39: 7555–7567.

Satyakumar M, Anil R, Sivakumar B (2014) Travel time estimation and prediction using mobile phones: a cost effective method for developing countries. Civil Engineering Dimension: Journal of Civil Science and Application, 16(1): 33–39.

Effect of roadside friction elements on traffic characteristics of urban arterials - a systematic review and a pilot study

Shishodiya Ghanshyam Singh[1*], S. Vasantha Kumar[2]
[1] *Ph.D Scholar, School of Civil Engineering, VIT University, Vellore – 632014, Tamilnadu, India.*
[2] *Associate Professor, School of Civil Engineering, VIT University, Vellore – 632014, Tamilnadu, India.*
* *e-mail: shishodiyaghanshyam@gmail.com*

Introduction

In developing countries like India, the growth of population has led to the increase in the vehicular population which is showing adverse effects on the existing urban arterials. Various steps are being carried out by the Government to improve the conditions of the existing facilities and increase the vehicle carrying capacity with the improvement in the level of service. One of the important factors for not reaching the objective of free flow of traffic on the urban arterials is the presence of side friction. Side friction is the interference caused to the moving vehicles on the carriage way due to various factors like on-street parking, roadside encroachment, presence of garbage bins, stopping of buses, pedestrian movement on carriageway, entries and exists from approach roads, etc. These factors on the carriage way affect the normal traffic flow and thereby causing speed reduction and travel time increase. These side friction parameters leads to capacity and level of service (LOS) reduction which finally results in poor performance of facilities. The situation is even worsen in a heterogeneous type of traffic condition prevailing in countries like India. This paper reviews what has been studied in the area of impact of roadside friction elements on traffic characteristics of urban arterials in India and abroad. A pilot study was also carried out to understand the reduction in road width due to side friction elements along the two important highways in the city of Vellore, India.

Materials and Methods

As the present study is primarily a systematic review of studies in the area of road side friction elements and its impact on traffic characteristics, the literature review was carried out in two aspects: one is the study of side friction elements/factors considered and another is the methodology used to study its impact on traffic characteristics. The various friction types considered in existing studies were listed out and the friction elements that are very peculiar in Indian roadways but were not considered also were identified. The studies on impact of side friction on traffic characteristics were analyzed with respect to the traffic parameters considered, i.e., whether only speed reduction was focussed and/or capacity/LOS reduction also was addressed. The mathematical techniques used to quantify the impact were also investigated.

For the pilot study, two important highways that passes through Vellore, namely, NH-234 (Mangalore Villupuram national highway) and SH-59 (Katpadi-Tiruvalam state highway) were considered. Even though they are national and state highways, they serve as major arterial roads for the city of Vellore, as they carry heavy local traffic during peak hours in addition to bypassable traffic. A stretch of about 3 kms with road widths ranging from 4.8 to 6.8 m (one side) was taken into account for side friction measurement. Using measuring tape, the road widths were measured at the locations where the side frictions were present and the type of friction was also noted down along with the GPS coordinates. The obtained measurements were then used to analyze the reduction in road width due to side friction elements.

Results and Concluding Remarks

Based on the detailed study of literature, 12 side friction elements were found to adversely affect the facilities and they were pedestrian on the carriageway, vehicles stopping and parking manoeuvre, bus stops, wrong movements of the vehicles, entry and exit manoeuvre, different types of stalls, electric pole on the road, waste and dustbins, police barricades, lack of lateral clearance and lane discipline, roadside cutting activities and slow moving vehicles like bicycle, rickshaw van, etc. Among them the pedestrian movement, bus stops, on street parking of the vehicles, wrong movements of the vehicles, entry and exit manoeuvre and slow moving vehicles were considered in larger extent (Ashish 2017; Pallavi and Mehar 2018; Salini and Ashalatha 2018). Besides it, there are few more peculiar activities of side friction happening in India like a strip of sand particles found on carriageway edges, pot holes, rain water stagnating on the roads, hoardings which not only encroach the carriageway and footpath but also obstruct the sight of the vehicle drivers, drainage water spilling out on the carriage way due to overflowing, vendors sitting on the edge strips of carriage way should also be taken into consideration. However not much studies were found addressing the above peculiar side friction factors found in India. More research needs to be carried out to address how all these side friction elements affect the smooth flow of traffic. It is important to investigate how these side friction elements affect the speeds of various types of vehicles in different types of urban roads such as arterial roads, collector streets and local roads.

The methods used to study the impact of side friction factors can be broadly divided into two groups. One is relating friction parameters with the speed observed at stretches where side friction elements are present. Regression models were developed to quantify the relationship and with the developed models it would be easy to predict the speed for varying levels of side friction. The second group falls in development of speed-density and speed-flow curves for with and without side friction cases. Traffic flow models like Greenshields, Greenberg, etc. were fitted and using which capacity and LOS reduction due to side friction were calculated. Few studies have used microsimulation using VISSIM to study the impact of side friction on traffic flow. Some researchers have used indices like road side friction index to quantify the effect of side friction on heterogeneous traffic flow. Though there are many studies reported but still the research on study of side friction factors and its impact on traffic flow is in nascent stage especially under Indian traffic conditions and hence this necessitates need for further exhaustive research. The results of pilot study revealed that from 5% to almost 50% of the roadway was reduced due to various side friction elements such as sand particles, electric poles, on-street parking, dwell time of buses, road side shops, banners, barricades and potholes. A photograph showing the presence of sand particles and banner can be seen in Fig.1. These side friction elements not only reduce the effective road width but also raise questions on motorists and pedestrians safety.

Figure 1. Presence of sand particles (left) and unsafe banners in the study stretch

References

Ashish D (2017) Influence of crossing pedestrians at undesignated locations on capacity of 4-lane urban midblock sections. Journal of Traffic and Logistics Engineering 5(1): 10-14. https://doi.org/10.18178/jtle.5.1.10-14.

Pallavi G, Mehar A (2018) Analysis of side friction on urban arterials. Transport and Telecommunication 19(1): 21–30. https://doi.org/10.2478/ttj-2018-0003.

Salini S, Ashalatha R (2018) Analysis of traffic characteristics of urban roads under the influence of roadside frictions. Case Studies on Transport Policy, https://doi.org/10.1016/j.cstp.2018.06.008.

Role of Transit Oriented Growth Centres in Transformation of the Urbanization Process for Hyderabad Metropolitan Region

A. Chandrakumar Yadav Jala[1], B.Kumar M[2], C. Prashanth Shekar Lokku[3*], D. Prasad CSRK[4]

[1] *Resear Scholar, Department of Civil Engineering, Osmania University, Hyderabad, India*
[2] *Professor, Department of Civil Engineering, Osmania University, Hyderabad, India*
[3] *Research Scholar, Department of Civil Engineering, National Intitute of Technology Warangal, India*
[4] *Professor,Department of Civil Engineering, National Intitute of Technology Warangal, India*

* *e-mail: prashanth.lokku@student.nitw.ac.in*

Introduction

Transit Oriented Growth Centre (TOGC) is the similar concept to that of Transit Oriented Development (TOD). TOD is becoming popular in recent years for developing countries as lessons from the developed countries. In this line, TOGCs are captivating little bit bigger scale to convert city as a global city. Whereas TOD concept is deals with the development within the city and make it more liveable. Hyderabad Metropolitan Region (HMR) is the capital and one of the largest cities in country is considered for this study. Located along the Musi River, it is one of the mega cities and globally known for IT and quality human resources in knowledge sector. HMR has emerged as a futuristic city amidst social, political and economic transition of the region. As a part of its development initiative, Hyderabad Metropolitan Development Authority (HMDA) has taken up the development of the 8-laned 158 km access controlled expressway called Outer Ring Road (ORR) with a Right of Way (ROW) of 150m. To promote planned development and to curb haphazard and ribbon development along the ORR, with hierarchal road network and framing special development regulations for the areas falling within 1 km belt on either side of ORR.

Given the large scale interaction of various modes of transport at the interchange locations and significant transhipment of goods and passengers, the interchange locations have been envisaged as the nodes of growth. Hence, it is found imperative to develop these nodes by encouraging the transit oriented activities as also the resident activities that have high reliance on the transport connectivity. As an initial step, towards developing TOGCs along ORR interchanges in HMR are considered to identify the role of TOGCs for their constructive and planned development.

Objectives

The main objective is to identify and suggest suitable theme based development for each TOGC considering their role in the overall development of Hyderabad as an emerging global city.

Methodology

With this view, careful attempt has been made to conceptualize the approach of this exercise. The sequential approach adopted is presented in Figure 1. Firstly for any planning project it is necessary to understand the overall development plans and policies of study area are to be understood thoroughly. Growth perspective of the study area is also has given equal importance in planning era, which will have clear cut idea about proposed development plans. In parallel, study about issues and lacking with respect to growth are to be read. So that location of TOGC's are going to be much easier while proposing. After finalising the locations, it is necessary to understand the sustainable measures via Physical, Socio-economic and Environmental characteristics of each TOGC. Based on these judgements, finally role of each TOGC is to be addressed in the overall future development.

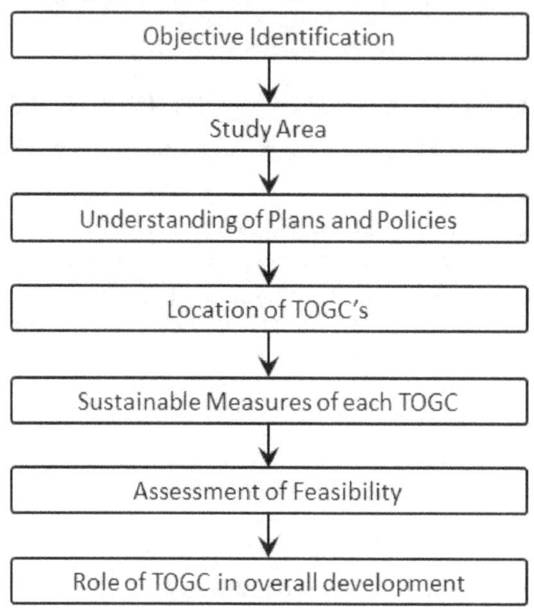

Figure 1. Flow chat for Methodology adopted

Results and Concluding Remarks

Total 13 TOGCs are proposed. The role in the development of the city primarily due to their location and proposed transport/transit connectivity. As envisaged in the master plan these TOGCs will "have the potential to grow into self – sustainable transit oriented developments connecting the core city with adjoining areas". Each of these proposed TOGCs are envisioned to develop based on either one or multiple-core economic activity, attract population and generate employment to develop as self-sustained communities. It is also to be noted that these TOGCs will have a certain level interdependence which is aided by their linkage through the outer ring road. The TOGCs are also to be connected with city through the transit systems as proposed in the CTS (CTSHMA, 2011).

The identification of these core activities is based on factors including existing population, location, connectivity, physical and social infrastructure, economic activity, and most important of all - regional linkages. Hyderabad city, which includes the erstwhile MCH area and the rest of GHMC area will continue to act as the core of the region and grow while the TOGCs will act as centres with specialised activities and renewed impetus from the government and private sector investments. Development of these TOGCs will result in a reorganisation/restructuring of economic activities in HMR and also in easing off the development pressure that is currently being exerted over the core area. For example TOGC Bonguluru is proposed to be a science and electronic city. Envisioned to be a Media city, a Logistics/ auto hub, textile park, Pedda Amberpet prominently marks its position towards the eastern side of the ORR.

Acknowledgments: We thank HMDA for sharing the data in smooth execution of my research work.

References

A Report on, Comprehensive Transportation Study for Hyderabad Metropolitan Area (CTSHMA), Hyderabad Metropolitan Development Authority, 2011.

A Report on "Transit Oriented Growth Centres at ORR interchanges in Hyderabad Metropolitan Region", Hyderabad Metropolitan Development Authority, 2014.

R. Cervero, C. Ferrell, and S. Murphy, TCRP Research result digest: Transit oriented development and Joint development in the United States: A literature Review. Transportation Research Board of the National Academics, Number 52, 2002, pp. 1–144.

Transit Oriented Development Best Practices Handbook, Calgary Land Use Planning and Policy, Calgary, 2004.

Analysing the Travel behaviour for compact and sprawl Area

A. Prashanth Shekar Lokku [1*], B. Yezzu Chandrasekhar[2], C. Prasad CSRK[3], D. Abhigna D[1]
[1] *Resear Scholar, Department of Civil Engineering, National Intitute of Technology Warangal, India*
[2] *Student, Department of Civil Engineering, , Osmania University, Hyderabad, India*
[3] *Professor, National Intitute of Technology Warangal, India*

** e-mail: prashanth.lokku@student.nitw.ac.in*

Introduction

Urban growth in developing countries is very high compared to developed countries. There has been an increase in vehicular growth rate year by year and the compact cities become more compacted where the sprawl areas mostly inclined to use private vehicles. To understand the travel behaviour characteristics of these two extreme areas in a city is become essential. In this line, Warangal city has been considered and focused on two extreme areas called compact and sprawl areas. Bifurcating the area into compact and sprawl is also a challenge and it is been included in this study along with the specifications followed. By considering the effects, it would be helpful to create and maintain sustainable and efficient transport systems.

Need and objectives

It is necessary to understand the effect of Compact and Sprawl areas on travel behaviour characteristics. Because in compact cities trip frequency will be higher due to the trip distance are low. This leads to congestion in compact cities. When it comes to the sprawl areas, most of the commuters rely on private vehicles. A model should be developed for the compact and sprawl areas for mode choice.

The following are the objectives to be achieved during the work:

- Analyse the travel behaviour characteristics in the study area in terms of trip generation, mode choice, trip frequency and trip length etc.
- Analyse the influence of non-urban form factors like Age, Gender, Income and Vehicular ownership on mode choice
- Develop the mode choice model for both compact and sprawl areas

Methodology

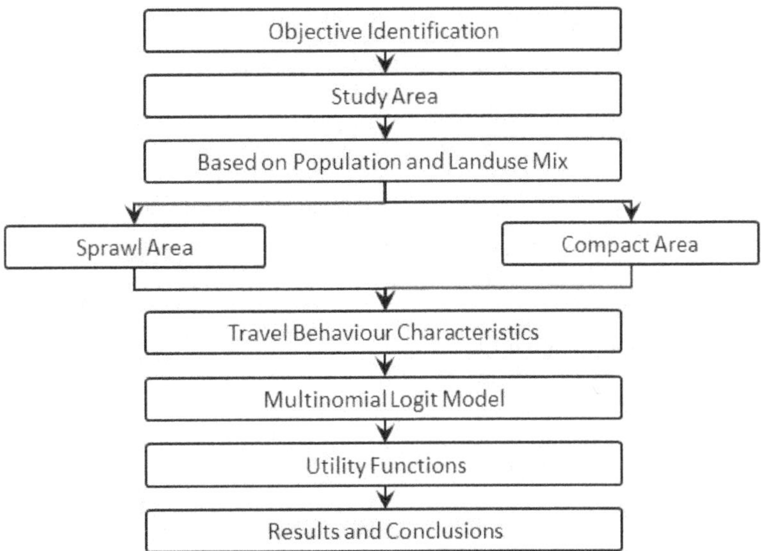

Figure 1. Methodology Flow chart

In the proposed research four compact areas and six sprawl areas are considered for the study. Travel behaviour characteristics considered for the research includes mode choice, trip length, travel cost and trip frequency. In Indian cities auto rickshaws are easily accessible as mode of travel than any other mode. So, it is also included in mode choice variable. A total six mode choices are considered. A Multinomial mode choice model is developed for compact and sprawl areas taking Auto-rickshaw as a reference category by using IBM-SPSS.

The sprawl has four elements: a population which is distributed in low density development, independent houses, shops, and workplaces. Poor connectivity of roads and poor access. Absence of well-defined activity centers, such as downtowns and town centers. Multinomial logit models are developed using SPSS (Version 16.0).

Results and Concluding Remarks

Several households characteristics like number of people in a house, vehicles owned by individual households, Income Age, Gender, Travel length, Trip frequency were selected which affects the mode choice in both compact and sprawl areas. Auto-rickshaw mode is dominated in both compact and sprawl areas. In compact areas the income levels higher than the sprawl areas due to most of them are employees. Mode choice depends on independent variables travel length and travel cost were most significant than other variables. Multinomial logit model is developed with Auto-rickshaw as a reference category. Travel length and travel cost were more significant in choosing auto-rickshaw as a mode choice in sprawl areas. Travel cost is more significant in choosing auto-rickshaw as a mode choice in Compact areas. 74% of the people in compact area are more likely to use two wheeler compared to Auto-rickshaw.

Utility equations for compact area
U_{TW} = 4.729 + 1.181 (TRIP LENGTH) – 0.937 (TRIP COST)
U_{C} = 3.43 + 1.22 (TRIP LENGTH) – 0.871 (TRIP COST)
U_{cycle} = 17.845 + 8.916 (TRIP LENGTH) – 43.388 (TRIP COST)
U_{BUS} = 1.507 + 0.994 (TRIP LENGTH) – 0.551 (TRIP COST)
U_{WALK} = 21.868 + 6.301 (TRIP LENGTH) – 46.45 (TRIP COST)
Utility equations for Sprawl area
U_{TW} = 0.652 -0.081 (TRIP LENGTH) – 0.009 (TRIP COST)
U_{C} = 6.26 - 0.388 (TRIP LENGTH) – 4.413 (TRIP COST)
U_{cycle} = 10.804 – 7.009 (TRIP LENGTH) – 4.435 (TRIP COST)
U_{BUS} = -0.554 - 0.054 (TRIP LENGTH) – 0.918 (TRIP COST)
U_{WALK} = 4.187 -27.25 (TRIP LENGTH) – 0.164 (TRIP COST)

In sprawl areas people are most likely to use Car than other modes. Due to more the travel length people more likely to use private vehicles. For long distances car is main mode for sprawl areas. 64% of the people in sprawl area are more likely to choose car over Auto.

These results can be used, while planning for the long term transportation related planning strategies.

References

Boarnet, M. G., & Sarmiento, S. (1998). "Can land-use policy really affect travel behaviour? A study of the link between non-work travel and land-use characteristics". Urban Studies, 35(7), 1155-1169.

Ben-Edigbe, J., & Rahman, R. (2010). "Multivariate School Travel Demand Regression Based on Trip Attraction". World Acad. Sci. Engg. Technol, 66, 1695-1699.

Dieleman, F. M., Dijst, M., & Burghouwt, G. (2002). "Urban form and travel behaviour: micro-level household attributes and residential context". Urban studies, 39(3), 507-527.

Ewing, R. (1995). "Beyond density, mode choice, and single-purpose trips". Transportation Quarterly, 49(4).

Influence of Aggregate Shape Factors on the Performance of Bituminous Mixtures

B. Giri Babu[1], Kakara Srikanth[2*], Venkaiah Chowdary[3]
[1] *Transportation Division/Department of Civil Engineering /NIT Warangal/ Warangal, India*
[2] *Transportation Division/Department of Civil Engineering /NIT Warangal/ Warangal, India*
[3] *Transportation Division/Department of Civil Engineering /NIT Warangal/ Warangal, India*
* e-mail: sri717004@student.nitw.ac.in

Introduction

In bituminous mixtures, aggregates are combined with bitumen to form a compound material. By weight of total bituminous mix, aggregate generally account for approximately 95 percent. Even though the volume occupied by aggregates within the bituminous mix is relatively higher than the bitumen, the relative cost of aggregates is less and contributes significantly towards the performance of the pavements. Thus, a detailed characterization of properties of aggregates is important in the design of bituminous pavements (Masad et al. 2007). Based on the dimensions of aggregate particles, three properties are defined including shape, angularity, and surface texture (Barrett 1980). Aggregates are classified into different shapes or form based on the ratio of its dimensions. Flat or elongated aggregates (where the ratio of largest or shortest dimension of an aggregate particle deviates significantly from its mean dimension) tend to break when load is applied, i.e., during compaction process or due to movement of vehicles after the road construction. Based on the round or angular edges of aggregate, angularity property is described. Surface texture property is used to classify the aggregates based on its surface roughness. Surface texture is an important property which defines the bitumen retention on the aggregate surface. These three properties are independent to each other (Rousan, 2004) and they are important, as they influence the performance and serviceability of bituminous mixtures (Brown et al. 1989; Kandhal et al. 1992). Aggregate morphological characteristics are very complex and cannot be characterized by any single test. Masad et al. (2011) proposed an image analysis based methodology to quantify the sensitivity of bituminous mixture performance to aggregate shape. Thus, an objective characterization of aggregate is required to quantify the effect of aggregate geometry on the performance of bituminous mixtures. In the present study, the aggregates are classified based on various indices proposed in the literature and natural aggregates are replaced by aggregates of different shapes in different proportions within the bituminous mix. The bituminous mix properties are then used to determine better performing aggregate shapes and proportion in order to find the best aggregate characterization method.

Materials and Methods

Strong, durable, proper shape and size aggregates are desirable for a monolithic bituminous layer in a flexible pavement. The physical properties of aggregates which define strength, shape, and water absorption are determined by laboratory investigation. For the present experimental investigation, aggregates are considered from one source, and Dense Graded Bituminous Macadam (DBM) grading – II has been adopted in this study. Only Viscosity Grade (VG) - 30 bitumen is used in the study and DBM grading – II mix proportion is fixed by Marshal mix design according ASTM D1559 guidelines.

To determine the physical properties of aggregates, basic laboratory tests including crushing strength, hardness, toughness, and specific gravity were performed. Further, the aggregates were quantified based on the shape tests. The shape tests include determination of flakiness index and elongation index. Aggregates are further classified in to four different shapes i.e., disk, blade, rod, and cube base on flatness ratio and elongation ratio. In this shape analysis particle longest diameter (dL), the intermediate diameter (dI), and the shortest diameter (dS) are determined. Flatness ratio is the ratio of dS to dI and elongation ratio is the ratio of dI to dL. The combined effects of particle shape and surface texture of an aggregate is determined using Aggregate Particle Shape Index which is defined in ASTM D 3398. Aggregate Particle Shape Index is a quantitative measure of the aggregate shape and texture characteristics that may affect the performance of bituminous mix.

The Optimum Bitumen Content (OBC) for a bituminous mix is determined by Marshall mix design method. In this method the bituminous samples were prepared with bitumen content varying in increments of 0.5% in the range of 4.5% to 6 % of bitumen. OBC is the bitumen content that induce 4% air voids in the compacted bituminous mix. For the same compacted samples, the voids in mineral aggregate (VMA), and the voids in coarse aggregate alone fraction (VCA) are also determined. The performance of DBM mixes was studied with different shapes of aggregates and different proportions (0%, 10%, 20%, 30%, 40%, and 50%) replacing the original aggregates with four shapes of aggregates. Further, Indirect Tensile Strength (ITS) of bituminous mixtures was determined.

Results and Concluding Remarks

In all the DBM mix samples prepared with different shapes of aggregates, the Marshall stability values were observed to be highest for the mixes prepared with cubical aggregates. Further, the stability found to increase with increase in proportion of cubical aggregates up to 30%. This phenomenon could be due to the cubical particles exhibiting higher interlocking and internal friction, compared to flat, thin, and/or elongated particles. The other Marshall parameters such as flow, voids filled with bitumen were also found to increase with increase in proportion of cubical aggregates. In contrast to this, the parameters such as air voids and VMA found to increase with increase in proportion of blade shape aggregates. This could be due to the gaps between the aggregates within the bituminous mix being less in the case of blade shape aggregates.

The peak ITS is observed at the 50% replacement of different aggregate shapes for the DBM mix. Cubical shape aggregate attained the maximum ITS value and lowest ITS value is observed for blade shape aggregates. This could be due to the fact that blade shaped aggregates tend to break down excessively during compaction. Similarly, particle index value is observed to be highest for cubical aggregates and the same higher ITS values are observed for cubical aggregates compared to the other shapes. Thus, it can be concluded that particle index value quantifies the effect of shape of aggregate on the performance of bituminous mix.

References

ASTM. (1989). Test method for resistance of plastic flow of bituminous mixtures using Marshall apparatus. ASTM D1559, West Conshohocken, PA.

ASTM. (2006). Standard Test Method for Index of Aggregate Particle Shape and Texture. ASTM D3398, West Conshohocken, PA.

Brown, E.R., McRae, J.L, and Crawley, A.B. (1989) Effect of aggregate on performance of bituminous concrete. ASTM STP 1016, Philadelphia, 34-63.

Barrett, P. J. (1980). The shape of rock particles, a critical review. Sedimentology, 27, 291-03.

Kandhal, P.S., Khatri, M.A., and Motter, J.B. (1992) Evaluation of particle shape and texture of mineral aggregates and their blends. Journal of Association of Asphalt Paving Technologists, 61, 217-240.

Masad, E., Al-Rousan, T., Bathina, M., McGahan, J., & Spiegelman, C. (2007). Analysis of aggregate shape characteristics and its relationship to hot mix asphalt performance. Road Materials and Pavement Design, 8(2), 317-350.

Masad, E., Little, D., & Sukhwani, R. (2011). Sensitivity of HMA performance to aggregate shape measured using conventional and image analysis methods. Road materials and pavement design, 5(4), 477-498.

Rousan, T. M. (2004). Characterization of aggregate shape properties using a computer automated system. Ph. D. Dissertation, Dept. of Civil Engineering, Texas A&M University, College Station, TX.

Laboratory Evaluation of Construction and Demolition Waste as an Alternate Material in Various Flexible Pavement Layers

R. Sudhakar[1], Kakara Srikanth[2*], Venkaiah Chowdary[3]

[1] *Transportation Division/Department of Civil Engineering /NIT Warangal/ Warangal, India*

[2] *Transportation Division/Department of Civil Engineering /NIT Warangal/ Warangal, India*

[3] *Transportation Division/Department of Civil Engineering /NIT Warangal/ Warangal, India*

** e-mail: sri717004@student.nitw.ac.in*

Introduction

Exploitation of natural resources by the construction industry over decades has led to the depletion of the natural aggregates. Thus, recycling of used materials has become an essential part of the construction industry. Utilization of recycled materials can benefit the environment by natural resource conservation, reduced landfill, and reduced energy consumption. Taking into account these benefits, many studies have reported usage of demolished building waste in road construction. Studies have reported the usage of Recycled Concrete Aggregate (RCA) as coarse aggregate for concrete (Ho et al. 2013), in pavement unbound layers (McCulloch et al. 2017), and in bituminous mixtures (Giri et al. 2018). The other major demolition waste is bricks. Studies have reported the usage of crushed brick aggregates in pavement base and sub-base layers (Hou et al. 2016). However, the physical, chemical, and mechanical properties of these materials are different from the properties of the original aggregates (Mills-Beale and You 2010). Thus, a detailed study on the performance of these materials as aggregates in various layers of the pavement is required.

In this study, two types of construction and demolition waste materials i.e., Recycled Concrete Aggregate (RCA) and Crushed Brick Aggregate (CBA) are used with the combinations of natural aggregates in bituminous layers and granular layers of flexible pavement, respectively. Bituminous layers with RCA were found to have less strength (Gallego 2012), higher water absorption (Shen and Du 2004), lower density due to presence of cement motor (Rafi et al. 2010) and, high optimum bitumen content (Aljassar et al. 2005) compared to layers with natural aggregate alone. Similarly, CBA has low density, high porosity, and lower strength compared to natural gravel (Hou et al. 2016). Thus, the effectiveness of these materials in bituminous layer and granular layers was investigated by carrying out through extensive laboratory tests.

Materials and Methods

This study seeks the possibilities of the usage of RCA and CBA as highway construction materials. RCA and CBA were used with the combinations of natural aggregates at different replacements (0%, 25%, 50%, 75%, and 100%) as aggregates. The effectiveness of these materials was investigated by carrying out extensive laboratory tests. This study includes sieve analysis, basic test properties of aggregates, Marshal stability test and indirect tensile strength test for RCA and grain size analysis, Atterberg's limits, compaction test and California Bearing Ratio (CBR) test for CBA. RCA and CBA were collected from a demolition waste source. RCA was investigated for use in binder course material in Dense Bituminous Macadam (DBM) grading - II and CBA as sub-base course material.

Important characteristics of aggregate particles include shape, gradation and Los Angeles abrasion value, base permeability, density and moisture content. The shape and grading of aggregates affect the shear strength, which is an important performance related property. Thus, properties of RCA are compared with the properties of natural aggregate. The desirable bituminous mix with different proportions of RCA was determined using Marshal mix design. For each laboratory design mix gradation, two specimens were prepared at each bitumen content starting from 4.5 % in increments of 0.5 percent in accordance with ASTM D1559 using 75 blows/face compaction and the Optimum Binder

Content (OBC) was determined. Six specimens were prepared at OBC at five replacements of RCA, and the specimens were tested for the Indirect Tensile Strength (ITS).

Clay brick was collected from the same demolition waste source and were crushed manually using a hammer to produce both coarse and fine aggregates ranging from 26.5 mm to 0.075 mm and are referred to as CBA in this study. The gradation of CBA used in sub-base course was selected from the MORTH specifications. Modified Proctor's compaction test was carried for all the replacement ratios of the soil including 0%, 20%, 40%, 60%, 80% and 100% of CBA to determine the Optimum Moisture Content (OMC) and the corresponding maximum dry density. The CBR test was performed on the specimens at the OMC for different replacement ratios of CBA.

Results and Concluding Remarks

Based on the tests on physical properties of aggregates it was observed that CBA exhibited the highest water absorption value, followed by RCA and natural aggregate. Similarly, natural aggregate exhibited the highest density, followed by RCA and CBA. The physical properties of CBA replacing natural aggregates in unbound layers is found to be satisfactory according to MORTH specifications. Similarly, RCA also found to satisfy the physical requirements of aggregates as specified by MORTH.

The observed physical properties of blended RCA with original aggregates up to 75% replacement of RCA were found to be suitable for binder course construction. The Marshal stability and flow values of the bituminous mix up to 75% of RCA is found to be satisfactory. ITS remained more or less constant up to 50% of RCA and later decreased with increase in RCA proportion.

With increase in CBA content, the maximum dry density decreased and the OMC increased. The increasing trend was observed with the increase in replacement of CBA. The CBR gradually increased with increase in CBA. The CBR of CBA is higher than natural gravels, and this could be due to lower coarse to fine aggregates volume ratio in CBA compared to gravel material. Thus, replacement rate up to 20% of CBA is suitable for roads with low volume of traffic and from 40% to 100% of CBA replacement are suitable for roads with relatively higher volume of traffic. Replacement rates from 80% to 100% of CBA can satisfy the requirements of road traffic exceeding 2 MSA, i.e., for high volume roads.

References

ASTM. (1989). Test method for resistance of plastic flow of bituminous mixtures using Marshall apparatus." ASTM D1559, West Conshohocken, PA

Aljassar, A.H., Al-Fadala, K.B., and Ali, M.A., (2005) Recycling building demolition waste in hot-mix asphalt concrete: a case study in Kuwait. Journal of Material Cycles and Waste Management, 7 (2), 112–115.

Gallego, J., Perez, I. and., Pasandin, A.R., 2012. Stripping in hot mix asphalt produced by aggregates from construction and demolition waste. Waste Management & Research, 30 (1), 3–11.

Giri. P.J., Panda. M., and Sahoo, U.C., (2018) Performance of Bituminous Mixes Containing Emulsion-Treated Recycled Concrete Aggregates. Journal of Materials in Civil Engineering, 30 (4), 04018052.

Ho, N. Y., Lee, Y. P. K., Lim, W. F., Zayed, T., Chew, K. C., Low, G. L., and, Ting, S. K. (2013). Efficient utilization of recycled concrete aggregate in structural concrete. Journal of Materials in Civil Engineering, 25(3), 318-327.

Hou, Y., Ji, X., Zou, L., Liu, S., and Su, X. (2015). Performance of cement-stabilised crushed brick aggregates in asphalt pavement base and subbase applications. Road Materials and Pavement Design, 17(1), 120–135.

Mills-Beale, J. and You, Z., (2010). The mechanical properties of asphalt mixtures with recycled concrete aggregates. Construction and Building Materials, 24 (3), 230–235.

McCulloch, T., Kang, D., Shamet, R., Lee, S. J., & Nam, B. H. (2017). Long-term performance of recycled concrete aggregate for subsurface drainage. Journal of Performance of Constructed Facilities, 31(4), 04017015.

Rafi, M.M., Qadir, A., and Siddiqui, S.H., (2010). Experimental testing of hot mix asphalt mixture made of recycled aggregates. Waste Management & Research, 29 (12), 1316–1326.

Shen, D.H. and Du, J.C., (2004). Evaluation of building materials recycling on HMA permanent deformation. Construction and Building Materials, 18 (6), 391–397.

Development of passenger car units for estimating saturation flow at signalized intersections in mixed traffic conditions

Sushmitha Ramireddy[1*], Eswar Sala[1], Ravi Shankar K V R[2]
[1] *Research scholar,Transportation Division, Civil Engineering Department, National Institute of Technology, Warangal*
[2]*Assistant Professor, Transportation Division, Civil Engineering Department,National Institute of Technology, Warangal*
** e-mail: susmitharamireddy@gmail.com*

Introduction

Intersection is the general area on the road network where vehicles moving in different directions with different speeds want to occupy the same space at same point of time. Proper design of signals is very important in order to minimize delays and congestion to vehicles. Passenger Car Units, saturation flow and delay are the important parameters that must be considered in the design of signalized intersections. The non-uniformity in different categories is generally unified with the help of passenger car units. Saturation flow is defined as the flow which occurs during stable moving platoon of vehicles during green interval. The saturation flow is estimated as PCU/hr or PCU/hr/lane. This paper aims in developing PCU values for different categories of vehicles and estimating saturation flow at signalized intersections in mixed traffic conditions. INDO HCM 2017 is been developed for mixed traffic conditions considering data from metropolitan cities. But, it is necessary to study the the suitability of those pcu values for medium sized cities. The developed pcu values in the present study are compared with those given in both IRC SP 41 as well as INDO HCM 2017.

Branston David and Zuylen Henk van 1997, estimated site specific PCU values using headway ratio method. Hadiuzzaman et al. 2008, developed the saturation flow model at signalized intersection for non-lane based traffic. PCU values were found using a synchronous regression method and they used the ROAD NOTE 34 method for calculation of saturation flow. Shrestha Sambridhhi and Anil Marshini 2014,developed PCU values for different categories of vehicles using multiple linear regression method. Patel et al. 2015, derived PCU values using speed to area ratio and multiple regression approach. Chand et al. 2016, conducted a study on saturation flow rate at selected intersections in Delhi. Dynamic PCU method was adopted to convert the heterogeneous traffic into homogeneous one.

Study area and data

Three signalized intersections are selected for the present study from Hyderabad city. All are four legged signalized intersections. Traffic data is collected using video graphic method. Geometric data is collected using tape and signal control details are measured using stop watch. The names of signalized intersections are Panty Circle, RTC Cross Road, Chaderghat Intersection. All Intersections have all type of vehicle. Satellite view of all study area has shown below in figure 1.

Figure 1. Satellite view of the three intersections from Hyderabad

The headway method is used in the present study for estimating passenger car equivalents. "In this method, the time headways of different categories of vehicles (only when the front vehicle and rear vehicle are exactly in the same lane and the one following the other) ie the time difference between passage of front bumper of the lead vehicle and the front bumper of the following vehicle" are considered for the purpose of estimating PCU values. The vehicles are categorized into two wheeler, three wheeler, car, light commercial vehicle and heavy vehicles. The categories such as jeep and van are included in the category of car and bus is included in the category of heavy vehicles. The PCU of vehicle category i is given in equation 1.

$$\text{PCU of the vehicle category i} = \frac{\text{mean headway of vehicle category } i \text{ to } i \text{ combination}}{\text{mean headway of car to car combination}} \quad (1)$$

The necessary condition for obtaining PCU values is that the sum of mean headways of car to car and two wheeler to two wheeler must be equal to sum of mean headways of car to two wheeler and two wheeler to car (this condition is also applicable for other categories of vehicles like three wheeler, LCV and HCV). If this condition is not satisfied the following correction given in equation 2 (Empirical relation) is need to be applied.

$$\text{Correction factor } C = \frac{abcd(w-x-y+z)}{abc+abd+acd+bcd} \quad (2)$$

Where, w,a=mean headway and number of headways when car following car, x,b=mean headway and number of headways of type i vehicle following car, y,c=mean headway and number of headways car following type i vehicle, z,d= mean headway and number of headways of type i vehicle following type i vehicle.

Classified vehicle volume count is taken for every 5 second count interval during saturated green intervals. The average values of all saturation flow values during saturated green intervals is considered as final saturation flow for that particular approach.

A multiple linear regression model is developed for estimating saturation flow based on correlation analysis and is given in equation 3.

$$S = (76.1 \times G) + (13.3 \times P\text{-}tw) - (37.5 \times P\text{-}rt) + 46.2 \quad (3)$$

Where, S= saturation flow in PCU/hr, G= green time in seconds, P-tw= proportion of two wheelers and P-rt= proportion of right turning vehicles. The following table shows the significant variables.

Results and Conclusions

In the present study, Passenger car equivalents for various categories of vehicles are been developed at signalized intersections under mixed traffic conditions. The PCU values obtained using headway ratio method in the present study are close to those suggested by INDO HCM when compare to those suggested by IRC SP 41. The reason may be that in the present study the intersections are selected from metropolitan city i.e Hyderabad and the INDO HCM is also been developed by considering 18 metropolitan cities. So there may be a similarity in various aspects like traffic calming measures, land use characteristics, traffic composition, turning proportions etc., A multiple linear regression is developed in the present study using field observed data. The saturation flow values obtained using the developed model is close to those obtained from the field observations. So, it is concluded that the developed PCU values and saturation flow model can be used for similar kind of intersections for better estimation of saturation flow.

References

1. David Branston and Henk van zuylen.,1997,"The estimation of saturation flow, effective green time and passenger car equivalents at traffic signals by multiple linear regression", Transport research volume 12.
2. Indian Road Congress (IRC)., 1994."Guidelines for design of at grade intersections in rural and urban areas. IRC special publication No.41, Indian Road Congress, New Delhi, India.
3. Indian Highway Capacity Manual (Indo HCM)., 2017. " Council of Scientific and Industrail Research (CSIR)", New Delhi.
4. Md Hadiuzzaman et al 2008"Saturation flow model at signalized intersections for non lane based traffic", Canadian Journal of Transportation.
5. Pinakin Patel et al.,2015" Effect of mixed traffic characteristics on saturation flow and passenger car units at signalized intersections. European Transport.
6. Shrestha and Anil., 2014. "Development of saturation flow and delay model at signalized intersection of Kathmandu", proceedings of IOE Graduate conference, 2014.

Estimation of Passenger Car Unit and Capacity Using Infrared Sensor Data on Multilane Divided Intercity Highways

Sandeep Singh[1*], Rajesh Kumar Panda[1], Dinesh Kumar Saw[2], & S. Moses Santhakumar[1]
[1] *Department of Civil Engineering, National Institute of Technology Tiruchirappalli, Tiruchirappalli, India*
[2] *Department of Civil Engineering, Sardar Vallabhbhai National Institute of Technology, Surat, India*
* *e-mail: sandeepsingh.nitt@gmail.com*

Introduction

Estimation of passenger car unit (PCU) value is very much important for traffic capacity analysis and other relevant applications such as measurement of the level of service (LOS) and the development of microscopic and macroscopic traffic flow models (Raj et al. 2019). In developed countries, various methods were developed for estimating PCU values for different types of facilities which carry a lesser degree of heterogeneity in traffic. In developing countries like India, a higher degree of heterogeneity in traffic can be seen because of the plying of various categories of vehicles, so these methods are not analogous for mixed traffic. On the other hand, traffic flow characteristics in developing countries are quite different from those in developed countries.

Due to this heterogeneity in the Indian traffic condition, the conventional method of data collection like videography becomes tedious and time-consuming both during data collection and data extraction process. This is overcome by the new technological device like the transportable infrared traffic logger (TIRTL), which is an automated Infra-Red (IR) sensor-based device which is capable of recording traffic parameters like lane-based classified volume count, speed, headway, spacing, gap, clearance, etc. The main aim of this study is to estimate the passenger car unit and capacity using the data recorded by TIRTL and to develop macroscopic models. The data were collected using the TIRTL device setup across one direction of the multilane divided intercity highway (NH-45) connecting the Chennai city with Tiruchirappalli city. Using the collected data, the PCU model was developed and capacity was estimated based on the Indian Highway Capacity Manual (Indo-HCM-2017) method. Later the acquired PCU values and capacity values are compared with recommended values of the Indian Road Congress Manual (IRC) -64 (1990) and the Indian Highway Capacity Manual (2017).

Materials and Methods

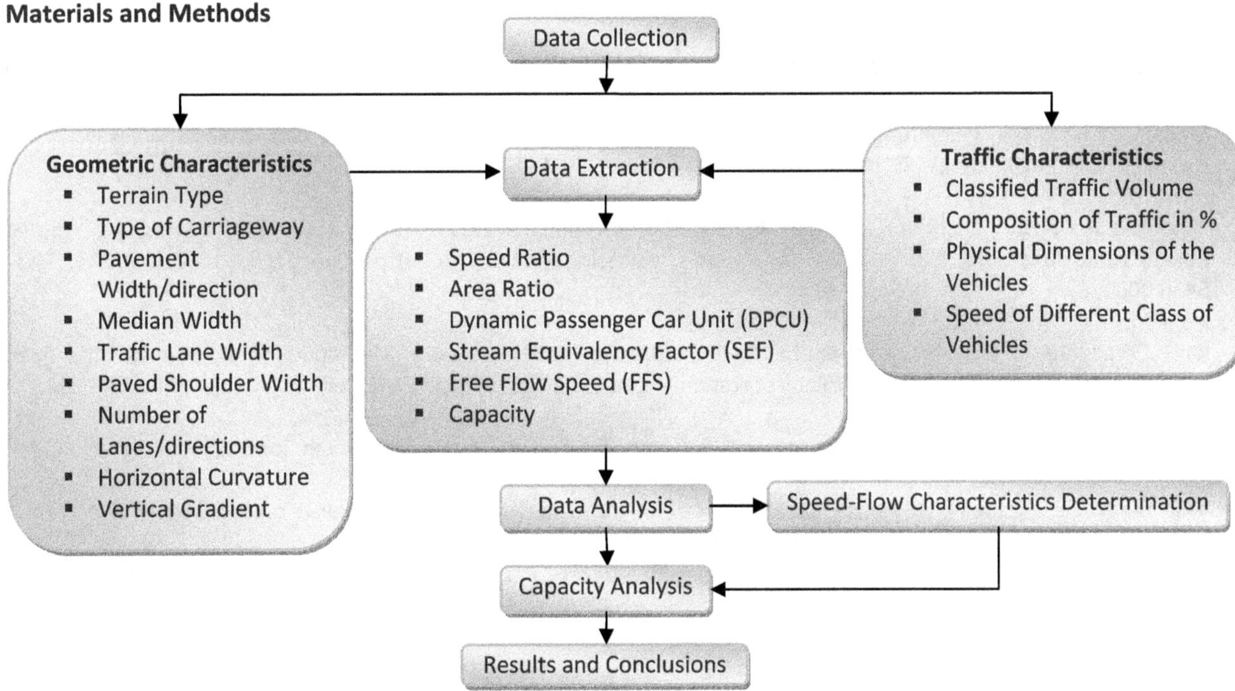

Figure 1. Flow Chart Showing Study methodology

DPCU is estimated based on the speed-area ratio method (Chandra and Sikdar 2000). Using this concept, dynamic PCU values are estimated at different flow levels and the speed flow diagram (based on Linear Greenshields theory) is plotted to find the capacity in each direction of the multilane divided intercity highway as shown in figure2.

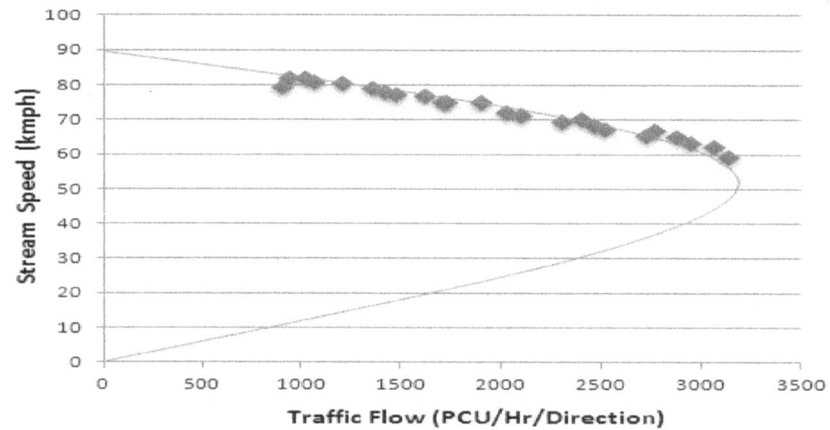

Figure 2. Speed Flow Relationship for Chennai-Tiruchirappalli multilane divided intercity road

Results and Concluding Remarks

The major contribution of this research paper is the development of a relationship between operating speed and flow to determine the capacity. This relationship will be extremely useful to field engineers, researchers and planners to get the capacity of the mid-block section of multilane divided intercity roads using the transportable infrared traffic logger (TIRTL). It will save a lot of time and money spent on data collection and data extraction processes.

References

Arasan VT, Koshy RB (2005) Methodology for modeling highly heterogeneous traffic flow. Journal of Transportation Engineering ASCE 131(7): 544-551. DOI: 10.1061/ASCE 0733-947X (2005) 131:7 (544).

Chandra, S, Sikdar P. K (2000) Factors affecting PCU in mixed traffic situation in urban roads. Road Transportation. Research 9 (3): 40–50.

Indian Highway Capacity Manual (Indo-HCM-2017), CSRI CRRI, New Delhi India.

Mehar A, Chandra S, Velmurugan S (2015) Effect of Traffic Composition on Capacity of Multilane Highways. KSCE Journal of Civil Engineering, 42(2): 1-8. DOI: 10.5604/01.3001.0010.0528.

Penmetsa P, Ghosh I, Chandra S (2015) Evaluation of Performance Measures for Two-Lane Intercity Highways under Mixed Traffic Conditions. Journal of Transportation Engineering ASCE 141(10).https://doi.org/10.1061/(ASCE)TE.1943-5436.0000787.

Raj P, Sivagnanasundaram K, Asaithambi K, Ravi Shankar AU (2019) Review of Methods for Estimation of Passenger Car Unit Values of Vehicles Journal of Transportation Engineering 145 (6). https://doi.org/10.1061/JTEPBS.0000234.

Transportation Research Board Highway capacity manual, Fifth Edition, National Research Council, Washington, DC.

Webster N, Elefteriadou I (1999) A simulation study of truck passenger car equivalents on basic freeway sections. Transportation Research 33B:323–336. https://doi.org/10.1016/S0965-8564(98)00036-6.

A Novel Approach to Use Waste Plastic in Bituminous Mixtures

Arjun Kumar Tirumali[1], Bhanuchander Konda[2*], Venkaiah Chowdary[3]
[1] *Transportation Division/Department of Civil Engineering /NIT Warangal/ Warangal, India*
[2] *Transportation Division/Department of Civil Engineering /NIT Warangal/ Warangal, India*
[3] *Transportation Division/Department of Civil Engineering /NIT Warangal/ Warangal, India*

** e-mail: arjunkt@student.nitw.ac.in*

Introduction

India produces approximately 8 million tonnes of plastic products every year out of which 4000 to 5000 tonnes of waste plastic is generated per day (Government of India 2000). Plastics are mostly used for wrapping packages, shopping bags, garbage bags, fluid container, clothing, toys, household and industrial products, building materials, etc. Most of the plastics are classified into thermoplastic and thermoset (IRC:SP:98-2013). The most widely adopted method of waste plastic disposal is through landfilling or incineration. Both these methods are not safe for disposal of plastic because dumping of plastic in landfills create problems including waterline clogging, clogging the drainage, reducing soil microbial activity, and affecting water recharge. Such dumping may produce harmful gas like methane which causes greenhouse effect (Sangita et al. 2011). Incineration is normally carried out above 700 °C and at such high temperatures, harmful gases such as carbon monoxide and carbon dioxide are produced. These gases cause air pollution and global warming. To deal with and mitigate such environmental pollution, plastics are widely being used in the construction of the flexible pavements especially in bituminous layers.

Blending of waste plastic with bituminous mixtures can be done in two processes such as the dry process and the wet process. In the case of a dry process, shredded plastic is added to preheated aggregate to form a thin plastic layer over the aggregate and bitumen is poured over plastic coated aggregates to prepare the bituminous mix. In the case of wet process, shredded plastic is added to bitumen at high temperature and mixed at high shear rates to form plastic modified bitumen and this plastic modified bitumen is used for the preparation bituminous mix (IRC:SP:98-2013). Even though obtaining a homogeneous blend of bitumen with waste plastic is a difficult task, some of the researchers reported changes in the properties of modified bitumen in terms of penetration, softening point, ductility and complex modulus (Al-Hadidy and Yi-qiu 2009; Naskar et al. 2010). Several researchers demonstrated that the use of low-density polyethylene (LDPE) in bituminous mixtures using wet process improved Marshall Stability, indirect tensile strength and moisture susceptibility compare to the conventional bituminous mixtures (Al-Hadidy and Yi-qiu 2009; Panda and Mazumdar 2002; Punith and Veeraragavan 2007). Panda and Mazumdar (2002) demonstrated that usage of LDPE in the bituminous mixtures using wet process improved resilient modulus and fatigue life compared to the control mixtures. Punith and Veeraragavan (2007) reported that usage of LDPE in bituminous mixtures improves resilient modulus and resistance to permanent deformation compared to control mixture.

However, the major disadvantage of using waste plastic in the preparation of bituminous mixtures through wet process is the separation problem. That is, shredded plastic will not blend uniformly in bitumen and is likely to get separated which can be evaluated through the separation test specified for modified bitumen. Hence, in this study dry process is adopted in the preparation of bituminous mixtures. IRC:SP:98 (2013) provides guidelines for blending of waste plastic with bituminous mixtures in the dry process. When this process is adopted in this study to prepare the bituminous mixtures by adding the shredded waste plastic to all the sizes of heated aggregate, it was observed that there is little effect in terms of improvement in the volumetric properties of the bituminous mix. Thus, a novel approach is adopted in this study where the coarse aggregates alone is coated with shredded waste plastic and the fine aggregates are used as such without coating with waste plastic to determine the optimum bitumen content apart from evaluating the pertinent volumetric properties.

Materials and Methods

Materials used in this study included locally available aggregates, VG20 grade bitumen and shredded LDPE plastic. All the basic properties of aggregate and VG20 grade bitumen were determined as per MoRTH (2013) and IS:73 (2013) specifications, respectively. Bituminous concrete grading 2 (BC II) with mid aggregate gradation was selected as per the MoRTH (2013) specifications. The Marshall mix design was carried out by adding LDPE plastic at blending proportions of 0%, 2%, 4%, 6% and 8% by weight of bitumen where the 0% blending proportion represents the control mix. For the preparation of bituminous concrete mixture with LDPE plastic (LBC II), coarse aggregate and fine aggregate were heated separately at the mixing temperature of bitumen i.e. at 150 °C. Then the shredded waste plastic was added to coarse aggregate alone fraction and mixed thoroughly to coat the aggregates uniformly with LDPE plastic. The waste plastic coated aggregates were added to the fine aggregates by maintaining the same mixing temperature. Subsequently, heated bitumen was poured over the aggregate mixture and thoroughly mixed to get a uniform coating of bitumen. Marshall specimens were prepared by applying 75 blows on each face of the specimen and tested as per ASTM D6927-15.

Results and Concluding Remarks

The optimum dosage of LDPE to bituminous concrete mid gradation was found to be 6% by weight of bitumen. Further, this dosage of LDPE satisfied the specification requirements of MoRTH (2013) for bituminous concrete. Higher and lower dosages of LDPE failed to satisfy the requirements of MoRTH (2013). Even though several studies reported the usage of waste plastic with bituminous mixtures in dry process, the novel approach adopted in this study overcomes several difficulties faced by the user intending to blending waste plastic in dry process. Thus, waste LDPE can be successfully used to minimize the environmental problems created by waste plastics to a greater extent.

References

Al-Hadidy, A. I., and Yi-qiu, T. (2009). "Effect of polyethylene on life of flexible pavements." Construction and Building Materials, Elsevier Ltd, 23(3), 1456–1464.

ASTM D6927. (2015). Standard Test Method for Marshall Stability and Flow of Asphalt Mixtures. American Society for Testing Materials, American Society for Testing and Materials.

Government of India. (2000). Solid Waste Management Handling Rules - Government of India.

IRC SP 98. (2013). Guidelines for the use waste plastic in hot Bituminous mixes (Dry Process) in wearing courses. indian road congress, Indian Road congress, New Delhi, India.

IS 73. (2013). Indian Standard PAVING BITUMEN — SPECIFICATION (Fourth Revision). Bureau of Indian Standards, New Delhi, India.

MoRTH. (2013). Specifications for road and bridge works. Indian Roads Congress, New Delhi, India.

Naskar, M., Chaki, T. K., and Reddy, K. S. (2010). "Effect of waste plastic as modifier on thermal stability and degradation kinetics of bitumen/waste plastics blend." Thermochimica Acta, Elsevier B.V., 509(1–2), 128–134.

Panda, M., and Mazumdar, M. (2002). "Utilization of Reclaimed Polyethylene in Bituminous Paving Mixes." Journal of Materials in Civil Engineering, 14(6), 527–530.

Punith, V., and Veeraragavan, A. (2007). "Behavior of Asphalt Concrete Mixtures with Reclaimed Polyethylene as Additive." Journal of Materials in Civil Engineering, 19(June), 500–507.

Sangita, Reena, G., and Verinder, K. (2011). "A Novel Approach to Improve Road Quality by Utilizing Plastic Waste in Road Construction." Journal of Environmental Research and Development, 5(4), 1036–1042.

Speed-Flow-Density Relationship – A Case Study on Shivamogga City

Syed Yaseen Afshad[1*], Naveen Navale[2], Arun V[3]
[1] *Traffic Engineer/ Transport Planner, Muhel Consulting Ltd. Doha, Qatar*
[2] *Graduate Engineer, ARCADIS – INDIA*
[3] *Assistant Professor, Dept of Civil Engineering, JNNCE – Shimoga, Karnataka, India*
** e-mail: afshad09@gmail.com*

Introduction

Traffic Flow depends upon the driver's movement and the interactions done by the vehicles in between two points. Traffic Flow behaviour cannot be predicted only by the driver's movement which is more difficult to analyse. The basic parameters of traffic flow are speed, density and flow which are most essential to design, plan and operate the roadway facility [Parameswaran and G. Asaithambi] and these parameters would also help to determine the capacity of a roadway facility [C. Achyuta, R. Swamy, and C. R. Munigety]. Some models to determine the speed –density and flow relationship are given below:

- Green shield's macroscopic stream model: v = vf – (vf / kj) k
- Greenberg's logarithmic model: v = vo ln(kj/ k)
- Underwood's exponential model: v = vf* e(-k/ko)

The present study was carried out on a selected four-lane divided urban road in Shivamogga city. The speed- density- flow data was collected on the study stretch by manual counts covering both peak and non-peak hours, Based on the R^2 value, the best fitting model was selected for the determination of capacity from speed - flow relationship.

Materials and Methods

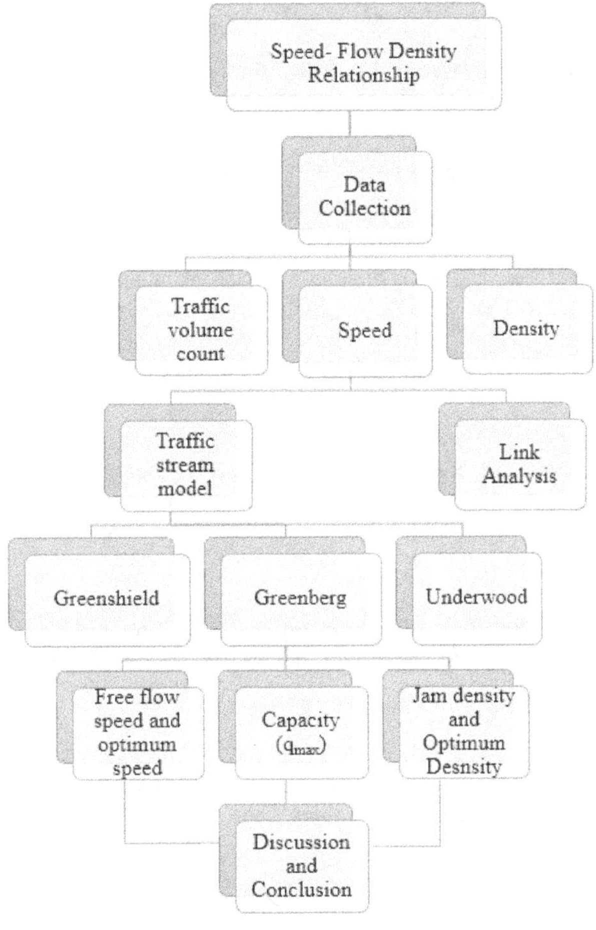

Results and Concluding Remarks

Table 1: Speed, density and Capacity model

Road No.	Models	R2	Capacity	V_f	V_o	K_j	K_o
	Greenshield	0.34	940	48.86	24.09	76.94	39
1	Greenberg	0.28	3613	NA	8.86	1107.52	405
	Underwood	0.32	1127	49.82	18.18	NA	61.5
	Greenshield	0.38	1165	49.61	24.72	93.92	47
2	Greenberg	0.34	10462	NA	7.33	3875.8	1430
	Underwood	0.37	1463	50.2	18.51	NA	79.2
	Greenshield	0.03	3728	38.62	19.11	386.1	195
3	Greenberg	0.03	NA	NA	NA	1.58e^10	NA
	Underwood	0.03	5220	38.67	14.3	NA	366.93
	Greenshield	0.14	3161	39.58	19.75	319.39	160
4	Greenberg	0.11	NA	NA	NA	1.58e^10	NA
	Underwood	0.14	4417	39.61	14.72	NA	303.13

R^2 of 0.38 was found to be the highest for Greenshield's model indicating a capacity of around 1165pcu/hr/lane and free flow speed of about 50 kmph. From the present study it was found that the single regime models used to explain the basic relationship between speed and density cannot be applied as R^2 values are found to be low to define a acceptable relationship in turn depicting a need to go for higher statistical model or simulation with possible impacts of changing the standard capacity values that are typically assumed in calculations.

References

[1]**Chandra S and Upendra Kumar** (2003), "Effect of Lane Width on Capacity under Mixed Traffic Conditions in India", Journal of Transportation Engineering, DOI: 10.1061/ (ASCE) 0733-947X (2003)129:2(155)

[2] **Castillo, J. M. De., and F. G. Benítez**.(1995) "On the Functional Form of the Speed-Density Relationship-I: General Theory." Transportation Research Part B, vol. 29, no. 5, 1995, pp. 373–89, doi:10.1016/0191-2615(95)00008-2.

[3] **Lum, K. M., et al.** (1998)"Speed-Flow Modeling of Arterial Roads in Singapore." Journal of Transportation Engineering, vol. 124, no. 3, 2002, pp. 213–22, doi:10.1061/(asce)0733-947x(1998)124:3(213).

[4] **Gerlough D. L. and Huber M. J.** (1975) "Traffic Flow Theory - A Monograph". Transpn. Res. Board, Special Report 165, Washington D. C

[5] **Harold Greenberg,** (1959) "An Analysis of Traffic Flow". Operations Research 7(1):79-85

[6] **Drake J. S., Schafer J. L. and May A. D.** (1967) "A Statistical Analysis of Speed-Density Hypotheses". Highw. Res. Rec. 156, 53-87

[7] **Swamy, C. A. R., C. R. Munigety, and M. V. L. R. Anjaneyulu**. (2016). "Passenger car unit based on influence area." In Proc., 12th Transportation Planning and Implementation Methodologies for Developing Countries. Maharashtra, India: IIT Bombay.

[8] **A. Parameswaran and G. Asaithambi,**(2016) "Capacity estimation of uncontrolled intersections in mixed traffic: comparison of gap acceptance procedure and additive conflict flow technique," Proceedings of the 12th International Conference on Transportation Planning, and Implementation Methodologies for Developing Countries, IIT Bombay, India.

[9]**Ashish Dhamaniya and Satish Chandra** (2014) "Midblock Capacity of Urban Arterial Roads in India" IRC, Indian Roads-A Review of Road and Road Transport Development, ISSN 0376-7256, Page-39

Modeling of Time Headway Distribution on Multilane Divided Intercity Highways under Mixed Traffic Condition

Sandeep Singh[1], Nidhi Kathait[2*], M. Kishore[3] S. Satheesh[3], K. Gunasekaran[3] & S. Moses Santhakumar[1]

[1]*Department of Civil Engineering, National Institute of Technology Tiruchirappalli, Tiruchirappalli, India*
[2]*Department of Civil Engineering, Indian Institute of Technology Roorkee, Roorkee, India*
[3]*Department of Civil Engineering, Anna University, Chennai, India*
** e-mail: nidhikathaitar@gmail.com*

Introduction

One of the most fundamental microscopic parameters in traffic engineering is the vehicular time headway. Time headway is defined as the time elapsed between bumper to bumper of successive arrivals of vehicles at a reference point of measurement on a road segment in a lane. Time headway is influenced by prevailing traffic flow characteristics of the freeway segment. To address any of the traffic problems effectively, accurate vehicle metrics like time headway is essential. Hence, the method for time headway measurement should be accurate. This paper attempts to model the theoretical time headway distributions for different flow levels for multilane divided intercity highways under mixed traffic condition.

The data related to vehicle class, speed, flow and headway are collected using the infra-red sensor-based traffic detector instrument, named as the Transportable Infra-Red Traffic Logger (TIRTL). This device is easy to use and collect time headway data under the non-lane-based heterogeneous traffic conditions prevailing in developing countries such as India. The speed distribution profile is evaluated for various classes of vehicles and also the time headway distribution profiles are determined for both the direction of traffic flow.

Materials and Methods

Headway is one of the important parameters to be used in modeling and analysis of road traffic, especially in traffic simulation studies (Lelitha Vanajakshi et.al., 2016). The present methodology used by the India Highway Capacity Manual (Indo-HCM-2017) for developing Dynamic Passenger Car Unit (DPCU) for multilane divided highways uses traffic data observed in the real field using video cameras. The data collection, extraction and processing of this collected data for its analysis becomes time-consuming, which is overcome by the using TIRTL instrument.

The proposed methodology seeks to estimate separate DPCU values for various classes of vehicles based on the method recommended by Indo-HCM-2017. The present study analysed time headway distribution data acquired using the TIRTL instrument, under Indian traffic conditions for different traffic flow levels.

To find a best-fitted model for different flow rate levels, many probability density functions are then tried. Accordingly, five statistical distribution models which fitted the frequency histograms in order of best fit are considered in this study. This is based on the goodness-of-fit tests like Chi square test and Kolmogorov–Smirnov (K–S) test. The best fit distributions are Log-logistic distribution, Log-Pearson-III distribution, Burr distribution, Weibull distribution and lognormal Distribution.

The distribution profile of speed of all vehicle classes against the frequency of vehicles and speed vs time headway analysis on multi-lane divided intercity road is shown in figure 1 and figure 2 respectively.

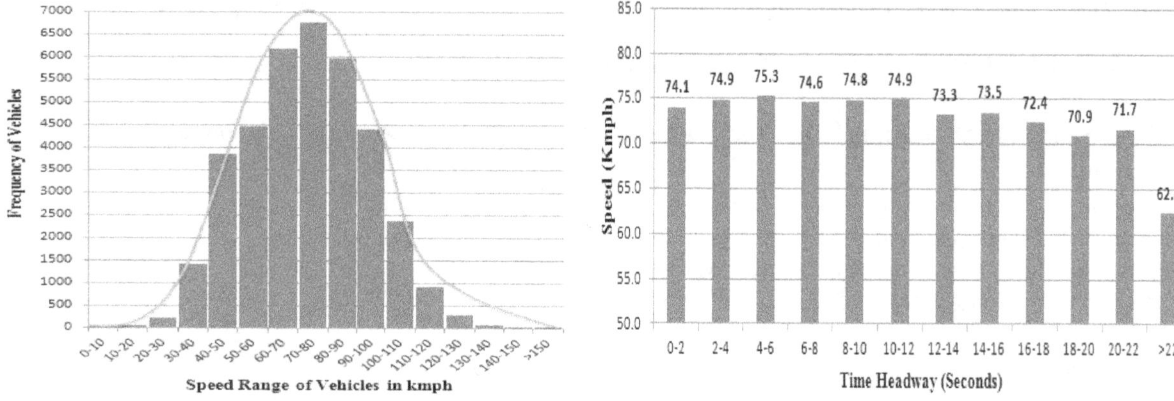

Figure 1. Distribution profile of speed of all vehicle classes

Figure 2. Speed Vs Time Headway analysis

Results and Concluding Remarks

The DPCU values for various classes of vehicles by Indo-HCM-2017 method are found and are as follows: Two-Wheeler-0.25, Three Wheeler-0.93, Car-1.0, Small Commercial Vehicle-1.56, Light Commercial Vehicle-1.76, Medium Commercial Vehicle-3.85, Heavy Commercial Vehicle-5.65, Multi Axle Vehicle-7.26, Tractor with/without one Trailer-5.63 and Bicycle-0.36.

The time headway analysis results show that time headway follow different distribution patterns under different traffic flow conditions. This study also shows that there exists a strong correlation between the microscopic parameter headway and the speed of vehicles, which is more dynamic in nature. This paper can provide a better understanding of the headway distribution for uninterrupted traffic facilities. The findings can be used for developing micro-simulation models for two-way traffic with four-lane divided roads under mixed traffic conditions and also can be used in capacity estimation and traffic safety analysis.

References

Maurya A K, Das S, Dey S, Nama S (2016) Study on Speed and Time-headway Distributions on Two-lane Bidirectional Road in Heterogeneous Traffic Condition. 11th Transportation Planning and Implementation Methodologies for Developing Countries, Mumbai, India.

Al-Ghamdi A S (2001) Analysis of time headways on urban roads: case study from Riyadh. Journal of Transportation Engineering 127(4): 289-294. http://dx.doi.org/10.1061/(ASCE)0733-947X(2001)127:4(289).

Badhrudeen M, Ramesh V, Vanajakshi L (2016) Headway analysis using automated sensor data under Indian traffic conditions. Procedia Social and Behavioral Science (17) 331-339. https://doi.org/10.1016/j.trpro.2016.11.103.

Pueboobpaphan R, Park D, Kim Y, Choo S (2012) Time headway distribution of probe vehicles on single and multiple lane highways. KSCE Journal of Civil Engineering (2013) 17(4):824-836. https://doi.org/10.1007/s12205-013-0212-5.

Chandra S, Kumar R (2001) Headway modelling under mixed traffic on urban roads. Road and Transport Research 10(1):61-71. TPMDC 2014, 10-12 December 2014, Mumbai, India. doi: 10.1016/j.trpro.2016.11.084.

Effect of exclusive bus lane on travel time for urban mid-blocks

Punith B. Kotagi[1*], Rohineesha A S[2], Likhitha M[2], Varsha H M[2]
[1] *Assistant Professor, Dept. of Civil Engg., The national Institute of Engineering, Mysuru, Karnataka, India.*
[2] *Under Graduate Student, Dept. of Civil Engg., The national Institute of Engineering, Mysuru, Karnataka, India.*
* *e-mail: punithbkotagi@nie.ac.in*

Introduction

In most of the developing nations such as India, traffic flow is heterogeneous in nature with mix up of different vehicles having varying static and dynamic characteristics and vehicle in such stream follow no-lane discipline. According to Annual Report by Ministry of Road Transport and Highways, number of registered vehicles in India is increasing at a rate of 12 % per year. This is mainly due to rapid urbanisation which in turn leads to various traffic problems such as peak hour congestion, delay, pollution, and accidents further reducing the capacity of existing road system. In most of the cases existing road capacity cannot be improved by constructing additional lanes due to lack of resource and right of way. In such cases traffic control and management measures play crucial role to improve the capacity of existing system. One such traffic management measure is provision of exclusive bus lane by allocating a reserved lane on urban arterials to exempt buses from other private vehicles. The effect of these Exclusive bus lanes (XBL) on travel time under mixed traffic flow can be analysed through microscopic traffic simulation packages such as VISSIM. Previous studies show that, provision of XBL cuts down the journey time and travel time gets reduced which motivate the people to become more reliable on mass transportation system (Yang and Wang, 2009; Vedagiri and Jain, 2012; Raj et al., 2013; Syed et al., 2016; Abdilfatah and Wahid, 2017).

Methodology

Initially the study parameters (composition, volume, speed, travel time) which are essential for the analysis are identified. Then reconnaissance survey is conducted in several parts of Bengaluru city, Karnataka. Based on this, two mid-block sections are selected for the present study which satisfies the ideal conditions such as no on-street parking, no cross roads, no side interferences. Video-graphic traffic data is collected from these study sections during peak hours of a typical weekday. To identify the peak hour, traffic survey is carried for entire day. Study parameters are extracted from the video with the help of Adobe Premium Pro. By using microscopic traffic simulation package VISSIM, effect of XBL on travel time is studied. Two study locations are considered for this project of which both the study locations are at the heart of Bengaluru city: (a) C.V. Raman road, connecting Mekhri circle to C.P.R.I bus stop. Length of the stretch is 77m and carriageway width is 17.40 m. Data was collected in the peak hour during 5.00 PM to 6:00PM on typical weekday. (b) Hebbal road (Bellary road), connecting Mekhri circle bus stop to Hebbal. Length of the stretch is 55m and carriageway width is 20m. Different vehicle types considered for the study are two-wheelers, three-wheelers, four-wheelers and bus. For C.V. Raman road, the volume count of vehicles during the peak hours comprised of 2684 motorized vehicles per hour. It is observed two-wheelers dominated the total flow (49.76 %) followed by four-wheelers, three-wheelers and then bus. The average speed of 48.05 kmph is observed for the two-wheelers and the least average speed was 43.48 kmph of the buses. In case of Hebbal road, volume count of vehicles during the peak hours is of 8421 motorized vehicles per hour. It is observed four-wheelers dominated the total flow (43.95 %) followed by two-wheelers, three-wheelers and then bus. The average speed of 25.96 kmph is observed for the two-wheelers and the least average speed was 23.10 kmph of the three-wheelers.

Implementation of Exclusive Bus Lane in VISSIM

To study the effect of provision of exclusive bus lane under the assumed roadway condition, for the purpose of simulation, an exclusive bus lane is introduced by the side of the median on the considered stretch of road, and this roadway is given as input to the model by holding the traffic volume and composition to be same as the field data. The Bus lane width is fixed as 3.5m. The model is validated for the field scenario by calibrating. The parameters considered for the validations are the composition and speeds. Here validation for volume is not considered because, for midblock sections volume validation is not necessary. The mean absolute percentage error for simulated and observed speed and composition is less the 10 % for both the study location, implying the developed model represents field conditions satisfactorily. In VISSIM, during the input of vehicles, all other type of vehicles except buses was blocked in the median side lane making that lane exclusively for the movement of buses. The composition and speeds were kept constant for the further simulation of bus lane system. Observed Speed variations are as shown in table 1 and table 2. From table 1, the negative improvement indicates the decrease in the speed of other vehicles. The percentage of decrease in speed is within the 5% which is which is within the limits. There is significant improvement of speed of buses, from 43.42 kmph to 55.05kmph which is 26.78%. From table 2, in the second study section the traffic volume was 8421 which represents the high traffic flow which results in low speed. The speed other category of vehicles excluding buses are observed to decrease but within allowable limit. Whereas the speed of bus was increased drastically from 23.11kmph to 26.97kmph which is 14.14% improvement.

Table 1 Improvement in speed after introduction of bus lane of study location 1

Type	Observed speed (kmph)	Simulated speed (kmph)		% Improvement
		Before	After	
Car	22.85	23.77	23.80	0.14
Bus	23.11	23.63	26.97	14.14
Auto	23.10	23.33	22.97	-1.56
Bike	25.96	25.75	25.44	-1.19

Table 2 Improvement in speed after introduction of bus lane of study location 2

Type	Observed speed (kmph)	Simulated speed (kmph)		% Improvement
		Before	After	
Car	47.70	47.39	45.53	-3.92
Bus	43.48	43.42	55.05	26.78
Auto	43.72	44.46	42.92	-3.46
Bike	48.05	48.45	48.13	-0.66

Summary and Conclusion

After the implementation of XBL in VISSIM, in study location 1 the speed of buses (26.78%) improved compared to four-wheelers (-3.92%), three-wheelers (-3.46%) and two-wheelers (-0.66%). Whereas, in study location 2 the speed of buses (14.14%) improved compared to four-wheelers (0.14%), three-wheelers (-1.56%) and two-wheelers (-1.19%). After the implementation of XBL in VISSIM, in study location 1 travel time of buses (1.34secs) improved compared to four-wheelers (-0.24secs), three-wheelers (-0.23secs) and two-wheelers (-0.04secs). Whereas, in study location 2 travel time of buses (0.95secs) improved compared to four-wheelers (-0.01secs), three-wheelers (-0.12secs) and two-wheelers (-.08%). The travel time of other vehicles except bus not effected by introduction of XBL.

References

1. Arasan, V. T., &Vedagiri, P. (2009). *Planning for dedicated bus lanes on roads carrying highly heterogeneous traffic* (No. 1429-2016-118615).
2. Sekhar, C. R., &Velmurugan, S. (2013). Micro simulation based performance evaluation of Delhi bus rapid transit corridor. *Procedia-Social and Behavioral Sciences, 104*, 825-834.
3. Umar Syed, Sujesh D Ghodmare, Bhalchandra V Khode. (2016). Performance impact of various bus priority strategies using VISSIM
4. Vedagiri, P., &Arasan, V. T. (2009). Estimating modal shift of car travelers to bus on introduction of bus priority system. *Journal of transportation systems engineering and information technology, 9*(6), 120-129.
5. Vedagiri, P., & Jain, S. (2012). Simulating Median Side Bus Lanes in Indian Traffic Conditions. In *Proceedings of the International simulation conference of India, IIT Bombay* (pp. 02-04).

Capacity Analysis of Roundabout by Indo-HCM and SIDRA Intersection

Revansidda[1*], Mruthyunjay S K[2]
[1*] *M Tech Student, Department of CTM, Dayananda Sagar College of Engineering, Bengaluru, India*
e-mail: ssiddusajjan@gmail.com
[2] *Assistant Professor, Department of CTM, Dayananda Sagar College of Engineering, Bengaluru, India*
e-mail: mruthyunjaysk@gmail.com

Introduction

In 1950, 30% where as in 2000, 47% and presently 54.4% of whole world population lives in urban area. This will increase up to 60% by the year 2030. The urbanization is caused by rapid growth of industries and leads to increase in population density of cities. Bengaluru one of the metropolitan cities of India, which has a population of about One crore, making it as a megacity and the third-most populous city and fifth-most populous urban agglomeration in India. Presently there are 13.01 lakh cars, 1.35 lakh taxis, 46.54 lakh two-wheelers and 1.71 lakhs of autos in the Bengaluru city, all this with a population of about one crore (*Published in The Economic Times-14th March 2017). The effective operation of urban intersections are very much important, which will affect capacity, delays to vehicles, operational efficiency and the safety of whole road network. With growing traffic volume , intersection jam happens constantly, an effective way to solve problems is to search for appropriate control mechanism with help of traffic analytical and simulation softwares.

Anna-Karin Ekman 2013 has shown that Sidra Intersection can be calibrated in three different methods, *i*)Manual Calibration of gap parameters, *ii*)Manual calibration of Environment factors and *iii*)Automatic calibration of environment factor-based optimization. And author compared results of all three calibration methods and concluded that the automatic calibration method performs better but requires more details and more time processing. Ying Liu et.al 2013 Have shown that Under the influence of conflicting flows, the traffic operation at roundabouts was complicated, based the results, they concluded that vehicle velocity of outer circulating lane is lesser than inner circulating lane. Xuanwu Chen et.al 2016 Used various software packages including RODEL, SIDRA and VISSIM to estimate several performance measurements, such as capacity, queue length, and delay, compared with the collected field data. With the comparison, they have found that all the three software packages overestimate multi-lane roundabout capacity before calibration. So, it is very much necessary to Calibrate the software used for analysis. D. Muley et.al 2014 Analysed multi-lane Four-legged modern roundabout operating in Muscat using SIDRA model. And he used Critical gap and follow-up headways for calibration. And based on analysis results he proposed to convert the present unsignalized roundabout into a Metered one.

Methodology

After study of available literatures and publications, a location and methodologies are selected for analysis. *Ashoka Pillar monument* - a roundabout is selected, methodologies selected were *Indo-HCM* and *Sidra Intersection*. The geometric data like width of approaches, radius of central island, entry angle etc., are collected. Classified traffic volume count is also made at the study location. Collected traffic volume is then converted into PCU by multiplying with equivalent PCU conversion factors given by Indo-HCM. By the methodology of Indo-HCM, roundabout is analysed, Capacity and LOSs are determined. And obtained *gap parameters* from *standard formula* given by *Indo-HCM* method are used for the calibration of Sidra by *method of Manual Calibration of gap parameters*. Since the thesis work is carried out in trail version of Sidra, Other two methods of calibration can not be developed in Trail versions. Sidra is used for Present Indian Scenario by modifying model parameters(passenger car equivalents-given by Indo-HCM) in Parameter settings tab while analysing.

SIDRA: *"Signalized and Unsignalized Intersection Design and Research Aid"*. It is an Intersection analysis program developed by ARRB Transport Research. Ltd, in Australia for capacity, timing and performance analysis of intersections

Analysis and results

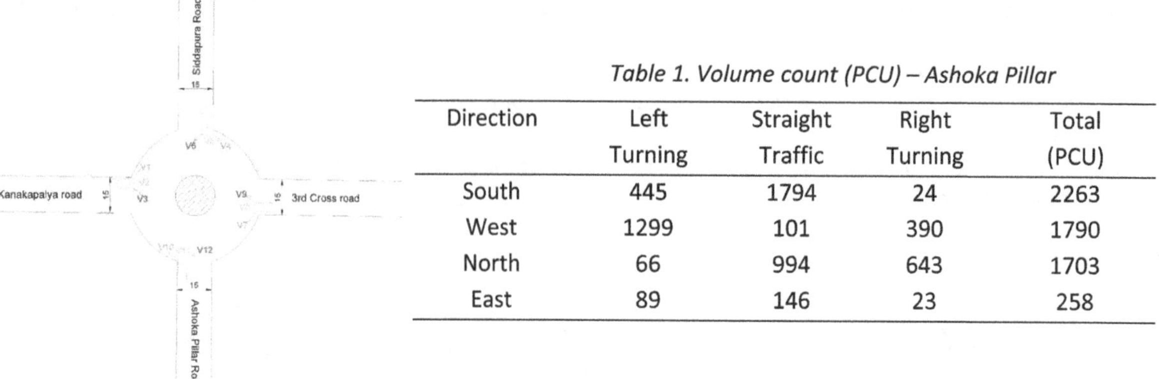

Table 1. Volume count (PCU) – Ashoka Pillar

Direction	Left Turning	Straight Traffic	Right Turning	Total (PCU)
South	445	1794	24	2263
West	1299	101	390	1790
North	66	994	643	1703
East	89	146	23	258

Figure 1. Ashoka Pillar Monument

Level of service obtained from both methodologies are tabulated as below in table 2.
LOS obtained from Indo-HCM method is based on delay, and Sidra method is based on volume-capacity ratio(Degree of saturation).

Obtained capacity of Ashoka pillar is 6508 PCU/h and 2851veh/h by Indo-HCM and Sidra Intersection respectively (Sidra gives results in units of Veh/h whereas Indo-HCM gives results in PCU/h).

Table 2. Analysis Results

Parameters	Indo-HCM	Sidra Intersection
LOS	C	F

Conclusions

- LOS of Ashoka pillar roundabout is C and F by Indo-HCM and Sidra Intersection respectively.
- Sidra Method of analysis was easier when compared to Indo-HCM.
- Sidra Analysis shows, the selected roundabout is serving with LOS F, but its not true when compared to Indo-HCM. So, further calibration of software is necessary to attain more accurate results.

References

[1] Anna-Karin Wkman (2013) Calibration of Traffic models in SIDRA.

[2] Ying Liua, Xiucheng Guoa, Dewen Konga and Hao Lianga (2013) Analysis of Traffic Operation Performances at Roundabouts: 13th COTA - International Conference of Transportation Professionals.

[3] D. Muley, H. S. Al-Mandhari (2014) Performance Evaluation of Al Jame' Roundabout Using SIDRA: World Academy of Science, Engineering and Technology International Journal of Architectural and Environmental Engineering Vol:8, No:12, 2014.

[4] Xuanwu Chen, Ming S. Lee (2016) A case study on multi-lane roundabouts under congestion: Comparing software capacity and delay estimates with field data: Journal of traffic and transportation engineering (English edition) 2016; 3(2): 154 -165.

[5] Traffic Modelling Guidelines - Sidra Intersection 7 (2017) Published by Government of South Australia (Department of Planning, Transport and Infrastructure.

[6] Indo-HCM: CSIR – Central Road Research Institute, New Delhi.

STUDY ON PEDESTRIAN BEHAVIOUR AT INTERSECTIONS

Kavya S[1*], Mr.Mruthyunjay S K[2],
[1] *M Tech Student, Department of CTM, Dayananda Sagar College of Engineering, Bengaluru, India*
e-mail: kavya.2011pes@gmail.com
[2] *Assistant Professor, Department of CTM, Dayananda Sagar College of Engineering, Bengaluru, India*
e-mail: mruthyunjaysk@gmail.com

Introduction

Pedestrian crashes is becoming a main problem in India. This may be because of increase of population in metropolitan cities, rapid urbanization, and people may have lack of knowledge on traffic rules and regulations. In India, census data(referred from article published in times of India newspaper) have shown that the fatalities rate increasing from 12330 in 2014 to 20457 in 2017, jumping nearly to 66%. Pedestrians stands as the vulnerable road users and they are not protected from accidents. The pedestrian safety has become least priority while designing the roadways. In highly populated cities where pedestrian travel are high at intersections with minimum amount of safety. The current study attempts to evaluate the pedestrian crossing behaviour like crossing speed, gap, pedestrian comply with signals and pedestrian vehicular interactions under the mixed traffic condition and to find the influencing factor that affects the pedestrian crossing behaviour, based on the statistical tests. Previously many studies has attempted to find the pedestrian behaviour at intersections. The literature work referred for the present work are as followed: Marisamynathan, Vedagiri Perumal (2011): In this paper the crossing behavior of pedestrians like their crossing speed, the pedestrian behaviour of compliance with signal and pedestrian vehicular interaction are studied and factors influencing the behaviour are studied based on the statistical tests. Akash Jain, Ankit Gupta , Rajat Rastogi (2014): This study examines about characteristics of pedestrians like gender. Age, carrying baggage and their effect on pedestrian crossing behaviour. Amin Mirza Boroujerdian Morteza Nemati (2016): The aim of the study was to assess the factors influences pedestrian and vehicle conflict behavior, by development of logistic regression model.

Methodology

Six intersections at different locations are selected for the study in Bengaluru city, out of which three were signalized junctions and other three are un-signalised junctions. The peak hour was found by conducting 12 hour pedestrian count survey at the study area. One hour videography survey was conducted at each site during the peak hour. Two video cameras are used to collect data from selected intersections. Collected data were analysed using ANOVA test, student-t test, Pearson's correlation test. The significant factors affecting the pedestrian crossing behaviour was found from the statistical tests. In SPSS 24 software regression model is developed.

Data Analysis

Total 647 and 518 pedestrian samples were clearly observed from video recording at Signalised and un-signalised intersections respectively. The video recording offer information about pedestrian crossing volume, crossing time , pedestrian appearance (like age , gender, and group) , crossing behaviour such as walking , running, alone or group crossing and crossing speed variations) crossing location (whether using cross walk or not), pedestrian crossing (whether pedestrians cross during green phase or non-green phase) and pedestrian vehicular interaction while crossing the road.

1. **Pedestrian Crossing Speed:** The crossing speed variation is defined as the difference between the 85th and 15th percentile speed. A new factor termed as Crossing speed deviation factor (CSDF) is established and defined as the ratio of the crossing speed variation and average crossing speed.
2. **Pedestrian crossing compliance:** Generally in signalised intersections, pedestrians should not attempt to cross the road during pedestrian non green phases. This is done to avoid interaction or

accidents between pedestrian and the moving automobiles. At the time of survey it was observed that few pedestrians were having a noncompliance behaviour with the signal. At unsignalised intersections the compliance is considered with the utilization of cross-walk by the pedestrians while crossing the junction.

3. **Pedestrian Vehicular Interaction:** In signalised intersections during pedestrian non green phases, pedestrian and vehicular interactions occur due to pedestrian non-compliance behaviour and during pedestrian green phase the ped-Veh interaction occurs due to drivers' non-compliance behaviour. At unsignalised intersections the Ped-Veh interaction will occur either by pedestrian or by the automobile driver. The pedestrians who are interacting with the vehicles while crossing road are considered.
4. **Logistic regression Model:** To indicate the significant factors analysis has been made by developing a logistic regression model and odds ratio is used to describe the effect between two groups.The probability of parameter 'i' with choosing a behaviour 'c' is based on the independent variable (Xi) and their relationship is expressed as

$$p(Yi = c) = \frac{\exp(\beta cXi)}{\sum l = 0 \exp(\beta lXi)}, c = 0,1$$

Conclusions

- Pedestrian gender, age and group are factors which has more effect on crossing speed variations at intersections.
- At Signalised junction's Gender, Group (i.e more than one or two pedestrians) and no of pedestrians are the significant factors which affects the pedestrian compliance behaviour.
- At un-signalised intersections, Gender was the most significant factor affecting pedestrian compliance. Female pedestrians were not likely to use the crosswalk compared to male pedestrians.
- At signalised intersections approaching vehicle is the main factor affecting the pedestrian-vehicular interaction
- Pedestrian traveling in group and approaching vehicle direction was found to be significant factors affecting the pedestrian vehicular interaction in un-signalised intersections.

References

1) Marisamynathan, Vedagiri Perumal (2014) Study on pedestrian crossing behavior at signalized intersections Journal of Traffic and Transportation Engineering(English Edition),1 (2) :103-110).
2) Akash Jain, Ankit Gupta, Rajat Rastogi, (2014) Pedestrian Crossing Behaviour Analysis at Intersections, International Journal for Traffic and Transport Engineering, , 4(1): 103 – 116 .
3) Jitendra Singh Yadav , Anuj Jaiswal, Raman Nateriya , (June-2015) Modelling Pedestrian Overall Satisfaction Level at Signalised Intersection Crosswalks International Research Journal of Engineering and Technology (IRJET) e-ISSN: 2395-0056 Volume: 02 Issue: 03.
4) Amin Mirza Boroujerdian , Morteza Nemati (2016) Pedestrian Gap Acceptance Logit Model in Unsignalized Crosswalks Conflict Zone International Journal of Transportation Engineering, Vol.4/ No.2/.
5) www.satisticshowto.datasciencecentral.com last accessed on 2019/05/23.
6) https://statisticsbyjim.com/anova/ last accessed on 2019/05/20.
7) www.ibm.com last accessed on 2019/04/28.

Optimization of signal cycle length using VISSIM - A Case Study

Ch.Vasavadatta[1], J.Venkateswara Rao[2*], R.Gokulan[3]
[1] P.G. Student, Department of Civil Engineering, GMR Institute of Technology, Rajam, India;
[2] Professor, Department of Civil Engineering, GMR Institute of Technology, Rajam, India;
[3] Assistant Professor, Department of Civil Engineering, GMR Institute of Technology, Rajam, India;
** e-mail: vasavadatta9@gmail.com*

Introduction

Due to the heavy rate of migration from rural areas to nearby cities, there is a rapid population growth there by leading to overcrowding of routes especially at junctions of road. The congestion of roads are occurring due to the increase in motor vehicles usage day by day. Traffic in developing countries such as India, Taiwan, and Vietnam is heterogeneous in nature. Heterogeneous traffic is characterized by a wide mix of vehicles having diverse static and dynamic characteristics (Siddharth et al. 2013). Especially in India the heterogeneous traffic condition is seen where the motor vehicle move with less lane discipline (Tom V. Mathew et al. 2010). So, this traffic congestion can be reduced in a systematic way by means of current technology available. One such alternative of existing technology is use of microsimulation softwares such as VISSIM, AIMSUN, and SIDRA, etc. In this paper, a method and results on finding sensitive parameters, automatic calibration and optimizing the signal cycle length using of VISSIM model have been proposed using data from Hanumanthawaka intersection in Visakhapatnam. With the help of Latin hypercube sampling and Python programming language, the random sets of parameters are generated. With the VISSIM COM interface, the external program has been written in Python programming language and automatic simulation was done in VISSIM. The optimum combination of parameters was obtained from the automatic simulation during the calibration process. With the variation of cycle length and comparing with the Field delays, the optimum cycle length has been observed as 210 sec.

Methodology

- In the present study, Hanumanthawaka Junction in Visakhapatnam is chosen for the case study. The required traffic data is acquired from the site such as geometric data through the total station, volume counts were done manually for every five minutes interval from morning 8:00 AM to 12:00 PM. Two weeks of data was collected and found peak hour as 9:05 AM to 10:05 AM and with the video data the Speed of every individual type of vehicle is extracted through kinovea software.
- Micro-simulation software "PTV VISSIM (64 bit) 11.00-09 Thesis (Academic License)" is used in this project because of its ability to model diverse vehicle categories and model non-lane based traffic. Any model created in VISSIM needs to be calibrated so as to sufficiently represent field conditions. In calibration, the default parameter values are changed until the error between the actual and simulated measure like flow or travel time is less than the required threshold value.
- A model has been created in the software that represents the same field geometry. VISSIM has a large number of parameters which affects its driving behaviour. So, a Latin hypercube sampling program has been written in Python programming language and 150 random sets of parameters have generated that need to simulate in VISSIM.
- Manually calibrating VISSIM by changing all the sensitive parameters and simulating the model to get the errors between the actual and simulated measure is time-consuming. VISSIM has a COM interface which can be used to call and simulate VISSIM externally through a code. An external python programming language was written to run these sets of parameters automatically.
- For every set of results, the error has been calculated and the minimum error set of parameters are chosen such that the results have replicated with the field. At this point, the calibration is completed with minimum error. The after with varying the volume the validation is done. By this, the calibration and validation of the model are completed.
- The main objective of this paper is to optimize the signal cycle length to reduce the overall delay at

the intersection because of its heavy traffic in peak hours. In the present field condition, 185 cycle length has been adopted where the overall delay for 185 cycle length is simulated in the software. With increasing in the cycle length and observing the delay results of all the sides and increasing the signal timing for the side which has got more delay likewise new signal timing was given and simulation was done for one hour for every new signal cycle length.

- The delay was measured with the simulation run results obtained for every signal cycle length on VISSIM model. The delay results are compared with the 185 sec cycle length result. Where optimum cycle length is proposed such that the overall junction delay will be reduced. The detailed results are shown in the full paper.

Results and Concluding Remarks

The simulation results obtained from VISSIM were shown in the below figures. Figure 1 shows the results when simulates with default parameters with an error of 46.90%. With the optimum values of parameters, the calibration results have shown 7.02% in Figure 2 and validation has shown 1% error in Figure 3. The optimum cycle length has shown in Figure 4 and it is observed as 210 sec.

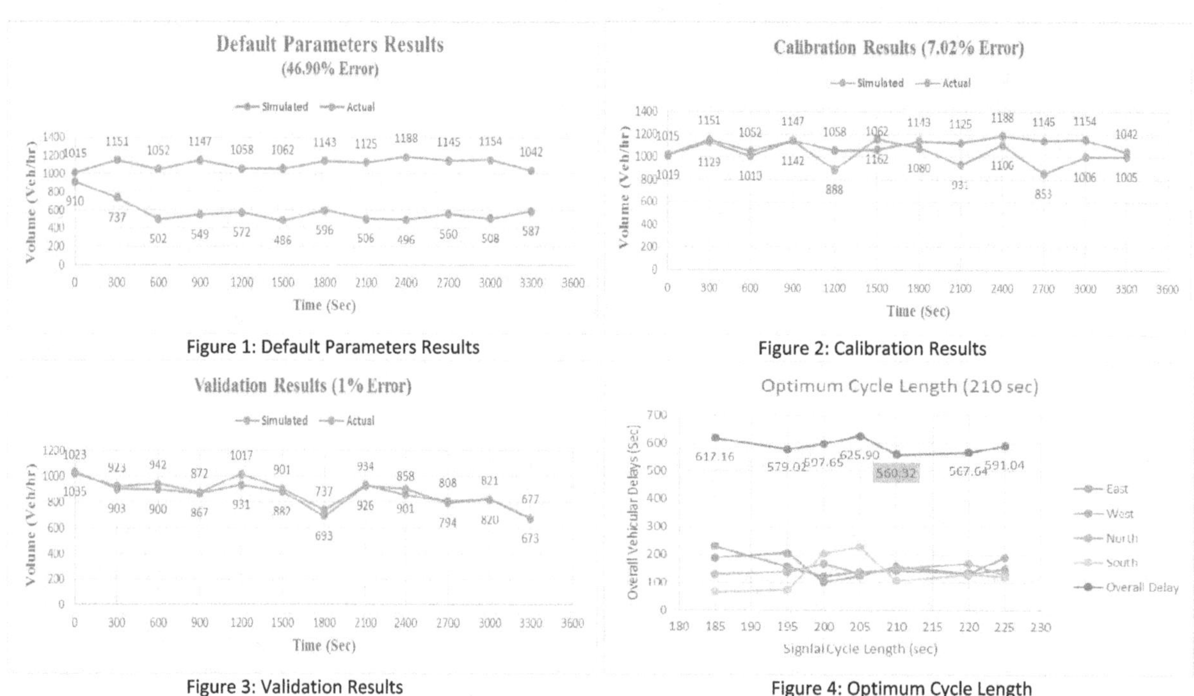

Figure 1: Default Parameters Results

Figure 2: Calibration Results

Figure 3: Validation Results

Figure 4: Optimum Cycle Length

From the above-mentioned results, it is concluded that the error has been reduced from 46.90% to 7.02% with the method adopted by using Latin hypercube sampling and Python programming language. In the microsimulation software, the calibration and validation is the major segment and it has been achieved. The validation results show a 1% error. After considering different cycle lengths and comparing the delay results with the field cycle length the signal cycle length of 210 sec has given a reduction in overall delay of the intersection.

References

Park, B. and Schneeberger, J. (2002). Microscopic Simulation Model Calibration and Validation: Case Study of VISSIM Simulation Model for a Coordinated Actuated Signal System. Transportation Research Record: Journal of the Transportation Research Board, 1856(1), pp.185-192.

Mathew, T. and Radhakrishnan, P. (2010). Calibration of Microsimulation Models for Nonlane-Based Heterogeneous Traffic at Signalized Intersections. Journal of Urban Planning and Development, 136(1), pp.59-66.

Siddharth, S. and Ramadurai, G. (2013). Calibration of VISSIM for Indian Heterogeneous Traffic Conditions. Procedia - Social and Behavioral Sciences, 104, pp.380-389.

Use of shredded plastic in DBM mix and evaluation of its laboratory performance

P R Vikas[1*], Venkatesh B R[2*], Dr. G Kavitha[3*]
[1] *M.Tech student, RASTA-Center for Road Technology*
[2] *M.Tech student, RASTA-Center for Road Technology*
[3]*Associate Professor, RASTA-Center for Road Technology*

* *e-mail: vikaspr2017@gmail.com, venkateshbrvenki@gmail.com, gkavitha@rastaindia.com*

Introduction

Plastic has its application in almost all the sectors including construction Industry. Recycling of plastic is a big issue, which has resulted in serious environmental effects. One way of handling this issue is by using plastic as a modifier in the bituminous mix. It can be used either in the form of recycled plastic aggregate or shredded plastic. Plastic is generally used as a modifier by wet process where it is mixed with binder and then the modified binder is added to the heated aggregates while preparing the bituminous mix. PMB is a modified binder which have proved to be performing better than the neat bitumen. However, the performance of a plastic modified bituminous mix by dry process is still not known. Hence in the present study, an attempt is made to study the influence of shredded plastic (LDPE) when it is mixed with heated aggregates by dry process and then the plastic-coated aggregates are mixed with hot bitumen.

A study by Punith V S and Dr. A Veeraragavan shows that by adding different ratios of plastic (2.5, 5.0, 7.5 and 10% by weight of bitumen) to 80/100 paving grade bitumen, the properties can be enhanced. Indirect tensile strength, dynamic creep and resilient modulus tests were conducted in which polyethylene added bitumen showed better performance compared to conventional mix.

The main objective of the present study is to evaluate the laboratory performance of plastic modified DBM mix prepared by dry process and to determine its life cycle cost.

Materials

1) **Aggregates:** Coarse and fine aggregates of different sizes to meet the gradation requirement of DBM grade II were procured from Chennigaraya stone crushers, Tumkur. The aggregates complied with the specifications mentioned in MORTH 5th revision.
2) **Bitumen:** It is a by-product of petroleum which should be a paving grade complying with IS 73-2013, for present study VG-30 grade is used, which was procured from Mangalore refineries.
3) **Plastic:** It should be of shredded type and conform to the size passing 2.36mm and retaining on 600-micron sieve, the type of plastic used is LDPE. It was procured from K K plastics, Bangalore added in the dosage of 8% and 10%.

Methods

- Basic tests for aggregate and bitumen were conducted.
- Proportioning of aggregates to meet the desired gradation of DBM, grade II was obtained by Rothfutch's method – A: B: C: D= 35:15:20:30.
- The plastic waste was collected, cleaned and dried, which was later shredded into small pieces of 2-3mm size, which was then added to the heated aggregates.
- The plastic-coated aggregates were then mixed with hot bitumen, the resultant modified plastic mix was tested.

- The following tests were conducted for conventional and modified plastic mix viz, (i) Marshall stability – (ASTM D6927-15) (ii) Indirect tensile strength – (ASTM D6931-17) (iii) Rutting test by Wheel Rut Tester (AIM 579) and (iv) Fatigue test by Repeated Loading Testing Machine of capacity 5T was used and the parameters were compared.

Results and Concluding Remarks

Table 1: Experimental results of conventional and modified plastic mix

Parameters	Conventional	Modified plastic mix	Remarks
Marshall Stability	1461kg	1799kg	For 10% dosage
OBC by Marshall Test	5.3%	4.98%	For 10% dosage
Indirect tensile strength	12.14kPa	15.79kPa	For 10% dosage
Tensile strength ratio (TSR)	82%	89%	For 10% dosage
Rut depth by Wheel rut tester for 10000 passes	4.3mm	3.2mm	For 10% dosage
No. of repeated load cycles until failure (Fatigue test)	2240	4053	For 10% dosage
Initial cost of construction for a 2 lane road of 7.5m width and 1km length	Rs.1,66,40,942	Rs.1,66,76,798	Additional cost of Rs.35,856
Maintenance cost for 12 years	Rs.1,10,06,400	Rs.55,03,200	Savings of Rs.54,67,344

- OBC obtained for conventional mix is 5.3%, whereas for 8% and 10% dosage it is 5.1% and 4.98% respectively. Hence there is saving of 0.32% binder at 10% plastic dosage.
- Marshall stability was high by 23% for plastic modified DBM mix compared to conventional DBM mix
- The Tensile Strength Ratio (TSR) was higher by 7% for plastic modified DBM mix compared to conventional DBM mix.
- Rut depth for 10% plastic modified DBM mix was only 3.2 mm after 10,000 cycles compared to 4.3 mm rut depth in conventional DBM mix, indicating that plastic modified mix have better rut resistance.
- Fatigue life improved by 81% for 10% plastic modified mix compared to conventional DBM mix, hence has better resistance to fatigue under repeated load applications.
- The initial cost of construction of pavement with modified plastic was higher by just 0.22% compared to initial cost of conventional pavement, considering 2 lane road of 7.5 m width and 1 km length. However, the life cycle cost assuming various maintenance interventions (in the absence of field performance data) works out to be lesser by 20% for plastic modified flexible pavement compared to conventional flexible pavement.
- From the present study as the laboratory performance of the DBM mix has improved at 10% dosage of LDPE plastic and also since savings in the overall maintenance cost is envisaged, it is recommended to adopt 10% as optimum dosage of plastic for the DBM mix. However, there is a need to evaluate field performance of the mix by constructing trial stretches before using it on large scale.

References

[1] Amit Gawande, Bharsakale G R, Range V C, Saurabh Tayde, Zamre G S (2012) Utilisation of waste plastic in asphalting of roads SRCC: 2(2), 147-157 Akola, India

[2] Dr. Malik Shoeb Ahmad (2014) Low density polyethylene modified dense graded bituminous macadam. International Journal of Engineering Trends and Technology (IJETT) - Vol 16 No.8

[3] Punith V S, Veeraragavan A (2003) Laboratory fatigue studies on Bituminous concrete mixed utilising waste shredded plastic modifier. Proceedings of 21st ARRB Transport research and 11th Road Engineering Association and Australia (REAAA) conference, Cairns, Australia

[4] IRC: SP: 98-2013, 'Guidelines for the use of waste plastic in hot bituminous mixes (Dry process) in wearing courses' Indian Road Congress, Karma Koti Marg, Sector-6, R.K. Puram, New Delhi

Investigation of rheological properties of waste plastic modified bitumen

N. Sai Rahul Reddy[1*], V. Sunitha[2]
[1] *Post Graduate Student, Department of Civil Engineering, NIT Tiruchirapalli, Trichy, India;*
[2] *Assistant Professor, Department of Civil Engineering, NIT Tiruchirapalli, Trichy, India;*
[*] *e-mail: nagurusairahul@gmail.com*

Introduction

The transportation demand is escalating rapidly due to the growth in population. Road network takes a significant share among different modes of transportation. So, studies on the performance of pavements have become predominant. Binder plays an essential role in the adequate performance of the pavements. Even though conventional binders perform well under ordinary conditions, due to increase in rainfall and temperature variations, freight movement and axle weight, unmodified binders were failing to meet the needs of the mounted traffic demand [3].In the present work, an attempt is made to improve the properties of the conventional binder by adding waste plastic. The usage of plastic has been rising every year, and its disposal has become a major problem. Plastic usage is likely to exceed 400 million tonnes by 2020. To overcome this waste plastic threat, reusing of waste plastics concentrated with bitumen has become a great advantage for decreasing environment hazards as well as to construct sustainable pavements. Studies were conducted since the late 90's on the usage of waste plastic in road construction. Based on the literature review, it was concluded that both the physical and rheological properties of binders were improved when mixed with waste plastic.

Feng Zhang and Changbin Hu (2015) compared the effect of different types of waste plastic as the modifiers to the crumb rubber. In their study, they used LDPE, PP, and LLDPE as the modifiers and found that the addition of linear low-density polyethylene (LLDPE) predominantly improves the high-temperature performance of the binder [1]. Praveen Kumar & Rashi Garg (2009) investigated the rheology of bitumen modified with waste plastic fiber. Praveen Kumar & Rashi Garg concluded that a substantial rise in properties of bitumen could be observed when bitumen was modified with plastic fibers [2].

Materials and Methods

The effect of waste plastic on the rheological properties of bitumen at various frequency and amplitude is investigated in this paper. The waste plastic includes all types of plastic such as Low-density polyethylene (LDPE), Polypropylene (PP), Polystyrene (PS), etc. IRC- SP: 98 – 2013 has specified that addition of waste plastic in small doses, about 5-10%, by weight of bitumen helps in substantially improving the Marshall stability, strength, fatigue life and other desirable properties of the bituminous mix. So, the control binder (VG30) modified with 6%,7% and 8% waste plastic content by weight of bitumen were tested and compared their performance. The test performed in this study includes the conventional tests and Dynamic Shear Rheometer (DSR).

The waste plastic modified binder was obtained from the "**CMR Bitplast**", a company which involves in the production of plastic modified binders through the wet process. The homogeneously blended waste plastic bitumen concentrated tablets packed in a drum, where the plastic percentage will be 36.36%. The mix proportions to obtain 6%, 7%, and 8% plastic modified bitumen has been done in the laboratory from the following calculations.

$$\frac{\mathrm{X}\,(\%)}{36.36(\%)} \times \mathrm{Y}\ (\mathrm{gm}) = \mathrm{Z}\ (\mathrm{gm})$$

Where, X = required percentage of plastic in bitumen (6%, 7% and 8%)
Y = total weight of plastic modified bitumen (300 gm)
Z = weight of CMR Bitplast sample to be taken
300 – Z = weight of VG30 to be taken

Results and Concluding Remarks

Conventional tests demonstrated that the addition of plastic increased the stiffness of the binders. Complex shear modulus (G*) and phase angle (⍰) of the binders were analyzed at a wide range of 40°C to 100°C. From the DSR test results, lesser phase angle (⍰) at higher temperatures indicates that the binder can uphold its elastic characteristic when modified with waste plastic. The SHRP rutting parameter G*/sin⍰ was increased with an increase in plastic content, which demonstrates that the binders can withstand rutting even at high temperatures. A frequency sweep between 0.1Hz – 30Hz was performed at 25°C on the binders, and the results showed that loss modulus (G'') was decreased for plastic modified binders which indicate the improved resistance to fatigue damage

Even though the initial cost of plastic modified bitumen is high due to the additional process of shredding the waste plastic and installing a special mixing plant, it would be highly economical in the long run as it improves the durability of the pavement and thereby reduce the maintenance needs.

References

1. Feng Zhang and Changbin Hu (2015) The research for crumb rubber/waste plastic compound modified asphalt. Journal of Thermal Analysis and Calorimetry 124(2): 729-741. http://doi.org/10.1007/s10973-015-5198-4

2. Praveen Kumar & Rashi Garg (2009) Rheology of waste plastic fiber-modified bitumen. International Journal of Pavement Engineering. http://doi.org/10.1080/10298430903255296

3. Robert N. Hunter, Andy Self and John Read (2015) "The Shell Bitumen Handbook." ICE Publishing, London.

Impact of Land use, Socio-Economic Profile & Metro Rail on BRTS Ridership: A Case of Surat

Dr. Pankaj Prajapati[1*], Disha J. Pachchigar[2]

[1]*Associate Professor, Civil Engineering Department, Faculty of Technology and Engineering, The Maharaja Sayajirao University of Baroda, Vadodara, India*

[2] *Planning Assistant, Surat Urban Development Authority, Surat, India*

* *E-mail: pankaj-ced@msubaroda.ac.in*

Introduction

In urban transportation, Bus Rapid Transit System (BRTS) is the key public transport to reduce the conundrum of traffic congestion in cities like Surat. It is observed in most of the Indian cities that modal share of public transportation is being reduced and it is mostly replaced by private mode of travel. Though BRTS is operational since 2013 in Surat city, its existing ridership is 86,122 riders/day (Surat Sitilink, 2017), which is less than the predicted ridership of 2,89,224 riders/day for 2016. (Comprehensive mobility plan Surat, 2008). This study was carried out with the aim to find out ways to increase BRTS ridership. The impact of different factors like land use, socio-economic profile of commuters and proposed metro rail was analyzed to make conclusion and recommendations. The study was carried out along two BRTS corridors; one of the most successful route of BRTS (Railway station to Kamrej) and another of the less successful route of BRTS (SVNIT to ONGC) with land use mapping in GIS and regression analysis based on primary and secondary data. The impact of land use was analyzed through GIS mapping and regression analysis, at 400 m buffer on both the sides of two mentioned routes of BRTS. To establish the relationship between socio-economic profile and BRTS ridership, primary data was collected for trip purpose, housing ownership, income group and population density along two routes of BRTS. Lastly, to find impact of upcoming metro rail on BRTS ridership, questionnaire survey asking willingness to shift to proposed Metro was carried out. The analysis of the study reveals that (i) industrial and commercial land uses were main reason to generate higher ridership of BRTS. (ii) Along the most successful route of BRTS 21% of Low Income Group, 46% Middle Income Group people are living and 57% of High Income Group people are living along less successful route of BRTS. (iii) from existing 74% BRTS users have shown their willingness to use Metro rail which may reduce BRTS ridership. This study can be helpful to transportation planner and urban planner during finalizing new routes of BRTS.

Data and Methods

Railway station to Kamrej route contributes higher ridership of 1901 passenger/day/km among all routes and SVNIT to ONGC contributes less ridership of 217 passenger/day/km. (Surat Sitilink, 2017). One of the objective of the study is to determine existing ridership of BRTS. To fulfil this objective, BRTS ridership data of each route was collected from Surat Sitilink office, Surat Municipal Corporation.

Second objective was to examine relationship between land use and BRTS ridership along existing BRTS routes. With the help of, ArcGIS software, land use mapping was done along 1 km from both BRTS routes. Also, regression analysis was carried out to identify relationship of land use and BRTS ridership for the said routes. The BRTS ridership has following relationship with considered parameters:

$$R = 1115.81 + 17.1\ LRESI + 69.7\ LC + 70.61 LI + 11.82\ LREC - 31.92\ LA - 46.46\ LV + 10.08 LE + 3.15\ LP + 5.81\ LM$$

Where, R - BRTS ridership, β_0 – Constant, LRESI – Residential Land use LC – Commercial Land use LI –Industrial Land use, LREC – Recreational Land use, LA – Agriculture Land use, LV – Vacant Land LE

- Educational LP – Public Purpose LM – Mixed use

Third objective was to examine relationship of socio-economic profile of commuters and BRTS ridership. Primary data like trip purpose, income group and housing typology were collected from BRTS users, city bus users and non BRTS users.

Forth objective is to determine impact of upcoming Metro Rail on existing BRTS ridership. To fulfill this objective, secondary data of metro rail routes was collected. Primary data of willingness survey of using metro rail was carried out.

Results and Concluding Remarks

- As per the analysis, Industrial land use, commercial land use, recreational land use and residential land use (LIG, MIG) are responsible to increase BRTS Ridership and vacant land use is one of the factors for less BRTS Ridership.
- As per primary survey, 74% BRTS users said yes to use Metro rail. 72 % city bus users deny for using metro and the reason behind this is, Metro rail will not pick up and drop near home like city bus. The 38% of persons who travelled longer distance of 13-16 km said yes for metro. 45% workers said yes for Metro Rail.
- As per the analysis of this study, following provisions can be provided in BRTS Planning Guide, which is published by ITDP (Institute for Transportation and Development Policy). While providing station location of BRTS, priority should be given to following type of land use.
    - Industrial land use
    - Recreational Land use
    - Commercial land use
    - Residential Land use where LIG and MIG are living.
- When more than one public transportation is proposed in any city like Surat, the routes of public transportation should be provided such a way that, it will not overlap each other. If two public transportation runs on same route, the ridership of former public transport may be affected by other public transport on the same route.

References

Comprehensive Mobility Plan and Bus Rapid Transit System Plan (2008), Surat.

Surat Sitilink Ltd, Surat Municipal Corporation, (2017), Surat.

ITDP Institute for Transportation and Development Policy

Understanding Crowding in Public Transport

Prajapati Pankaj[1*], Shah Malay[2]
[1]*Associate Professor, Civil Engineering Department, Faculty of Technology & Engineering, The Maharaja Sayajirao University of Baroda, Vadodara, India*
[2]*M.E. student, Civil Engineering Department, Faculty of Technology & Engineering, The Maharaja Sayajirao University of Baroda, Vadodara, India*
* *e-mail: pankaj-ced@msubaroda.ac.in*

Introduction

Public transport plays an important role in a city's economy and its social equity. Crowding in public transport is a major issue for commuters around the world. Crowding in public transport is becoming a growing concern as demand grows at a rate which is outstripping available capacity. To tackle the problem of level of service, it is important to understand the utility associated with a public transport services. Traditionally, this has been measured using factors such as time and monetary cost of travel. However, the level of comfort of users is not given enough importance. In the case of public transport, this includes the number of passengers that have to share a bus or train, the quality of seats and the smoothness of the ride, among many others. The relevance of these qualitative aspects for public transport policy is expected to increase over time in both developing and developed economies. As the income of commuters increase, public transport users are likely to give more value to quality and comfort features, along with reduction in travel time. Crowding must be taken far more seriously than at present. The current chronic crowding in all the major cities that gave evidence, which is unacceptable and must be addressed. This study is based on an analysis of Ahmedabad BRTS crowding scenario that have derived estimates of the value of crowding in standing passenger per square meter. This study also measures the generalised cost of travel considering comfort as one of the important factors moderating the cost. This research examines public perception of crowding and its correlation with passenger's willingness to pay for comfortable ride.

Methodology and data collection

This study is based on primary data, which has been collected for two different BRTS routes. The routes selected are based on their spread in whole city, types of land-uses and available ticketing data. The secondary data like bus schedule, routes and ticketing data were provided by Ahmedabad Janmarg Ltd. (Special Purpose Vehicle). Two routes were selected for the study which were Regional Transport Office (RTO) Circle route (201) and Iskon-Naroda (8). A total number of 500 passengers were surveyed on these routes; 250 on each route. They were surveyed on the bus stands while waiting for the bus. On each route, half of the passengers were surveyed during morning peak (8.30 AM to 10.30 AM) and half of the passengers surveyed during evening peak (5.30 PM to 8.00 PM). The commuters were asked to choose perceived crowding scenario from a set of different crowding levels (in pax/m^2) from pictures (Varghese V. & Adhvaryu B, 2016). The crowding levels shown to commuters were presented in four different levels of perceived density. The survey form also involved questions on commuters' perception of crowding inside the bus, extra waiting time they willing to spend for more comfortable ride, extra fare they willing to spend for seat availability inside the bus, trip data and related socio-economic data. The response of perception of commuters about the crowding gives the perceived density (objective measure of standing passengers/m^2) on the routes. Observed density inside the bus (from station to station) was also observed by travelling inside the bus on both the routes for one week in both the morning and evening peak. The survey forms were designed to offer commuters two time bids (Haywood L. & Koning M. (2013) for their willingness to spend additional time at the bus-stand for a more comfortable ride. Considering the responses from both the bids, time multiplier values were calculated. Then the linear equation is developed to verify relation between total travel time and the time multiplier. The same data set will use to find the best fitting model using other logit and exponential functions.

Results and Concluding Remarks

The data collected from the two BRTS route helped in finding the levels of overcrowding existing in the selected routes. The average value of crowding (by combining both the route, both the peak) was observed to be 3.8 standing passengers per square metre. The data also helped in capturing the public perception of crowding, the perceived density (by combining both the route, both the peak) of the commuters was found to be 4.9 standing passengers per square metre. The reason behind high perceived density may be possible that what commuters perceive is not the actual density. The study presented will be helpful for planners and policy makers to improve the quality of the existing BRTS system and improvement plan can be done accordingly. This much higher density is not acceptable for the efficiency of public transport. So there are some policy we can derive from this work are as follows: route diverting buses into the some particular station to other station (in peak hours), where travel demand is very high (Example: Shivranjani to Anjali). Increasing fares 15 to 25% (as people willing to spend extra for comfortable ride) during the peak hours is another option that will help the urban transport bodies recover the extra investment BRTS body will make to increase the number of bus on the busy routes. As we have derived from our survey that total 47.4% of user from the respondents willing to pay average 5.6 rupees extra for seat inside the bus. Time multiplier value found 1.43 by combining both the routes, which conclude that passengers are willing to spend 43% extra time for the more comfortable ride. We can also arrange one extra bus per half hour (in all busy routes in peak hours) which stop only at high demanding station observed during survey. This cost of extra buses can be used to justify subsidies and investments made on the part of Government. We should also consider the impact of crowding on demand and supply from the early stages of the appraisal of public transport projects, as the design of the system and the estimation of demand and social benefits rely on the multiple dimensions of the crowding phenomenon are accounted for in the formal assessment of projects. Most important policy recommendation is that designer and planners of public transport system should consider comfort as the one of the most important factor (along with travel time and cost) for the utility of the public transport service.

References

Varun Varghese & Bhargav Adhvaryu. (2016) measuring overcrowding in Ahmedabad buses: costs and policy implications. Transportation Research Procedia (17): 145 – 154. https://doi.org/10.1016/j.trpro.2016.11.070

Li, Zheng, & Hensher, David A. (2013) Crowding in Public Transport: A Review of Objective and Subjective Measures. Journal of Public Transportation, 16(2)

Luke Haywood & Martin Koning, 2013 Estimating Crowding Costs in Public Transport, Discussion Papers of DIW Berlin 1293, DIW Berlin, German Institute for Economic Research.

Laboratory Evaluation of Recycled Concrete Aggregates for Base and Sub-base Course of Low Volume Roads

[1] B. Sudharshan Reddy[1], G.Sreenivasa Reddy*, C. Sashidhar
[1] *Department of Civil Engineering, Narasimha Reddy Engg College, Hyderabad, Telangana-India*
[2] *Department of Civil Engineering, KSRM College of Engg, Kadapa, Andhra Pradesh-India*
[3] *Department of Civil Engineering, JNTUA College of Engg, Anantapur, Andhra Pradesh-India*
* *E-mail: sudharshan2055@gmail.com*

Introduction

India has undergone much industrialisation, followed by urbanisation. This process needed a considerable road network spread all across the country. Our nation has been quite prosperous in achieving this, but a lot of natural resources have been used to serve the purpose. As a result, there is an urgent need to shift the focus from conventional resources to non-conventional resources. They can be even wastes generated from the demolition of buildings which possess the required strength and other engineering properties and also should be cost-effective and a sustainable resource. The proposed research work would aim to evaluate the recycled concrete aggregates for base and sub base course of low volume roads. As 100% Recycled Concrete Aggregate (RCA) has high porosity, water absorption and low strength, Natural Aggregate(NA) and Recycled Concrete Aggregate(RCA) are mixed in proportions of 0%, 25%, 50%,75% and 100% and performance tests like Modified Proctor's Test and California Bearing Ratio (CBR) tests are conducted and the best replacement ratio is evaluated. Firstly, the necessary tests like Gradation, Aggregate Impact Value (AIV), Water absorption, Specific Gravity, Los Angeles Abrasion are conducted for NA, and RCA and the results are compared with the Ministry of Roads Transport and Highways MoRTH, 2013 specifications. It is concluded that considering all the factors, 50 % NA and 50% RCA mix is suitable for use in the base and sub-base course applications for rural roads.

Materials and Methods

In the current study, RCA and NA are used. NA comprise of crushed stone, sand and gravel. RCA is a term that describes crushed concrete or asphalt from construction debris which is reused in other building construction, road base, cement concrete and other infrastructure projects. The primary source for recycled aggregates is construction and demolition waste. For the preparation of samples, demolished building waste was identified and collected from the Bahadurpalli, Medchal district, Hyderabad. Later it was pulverised for further utilisation. So replacing NA with RCA is considered to be more advantageous, RCA shows up similar performance as of NA given the proper modification or treatment. The necessary tests were performed on both NA and RCA and compared the result. The necessary tests, namely Sieve analysis, Blending, Shape tests, Specific Gravity, Water Absorption, Aggregate Impact value, and Los Angeles, were conducted on RCA and NA, respectively. Further, the performance tests like Modified Proctor test determine the MDD and OMC, and CBR was done to check the mechanical strength conducted on the various replacement of NA with RCA. The RCA and NA are blended with all sizes with a nominal size of 19mm.

Results and Concluding Remarks

Table 1. Basic test result for natural aggregate and recycled concrete aggregate

Parameter	Natural Aggregate	(MoRTH, 2014) Specifications for Natural Aggregate	Recycled Concrete Aggregate	(MoRTH, 2013) Specifications for RCA
Specific gravity	2.87	2.5-3.2	2.38	2.5-3.2
Aggregate Impact value (%)	20.77	30 (max.)	37.16	30 (max.)
Water Absorption (%)	0.48	2.0 (max.)	3.76	2.0 (max.)
Los Angeles Abrasion (%)	31.12	40 (max.)	40.694	40 (max.)
FI and EI Index **(%)**	27.68	35 (max.)	34.55	35 (max.)

Table 1 shows the results of fundamental tests conducted on natural aggregate and RCA. The natural aggregate satisfies the limits specified by the MoRTH 2014, whereas the RCA does not satisfy the limits for Grading II of granular subbase materials specified by MoRTH 2014 as presented in Figure1.

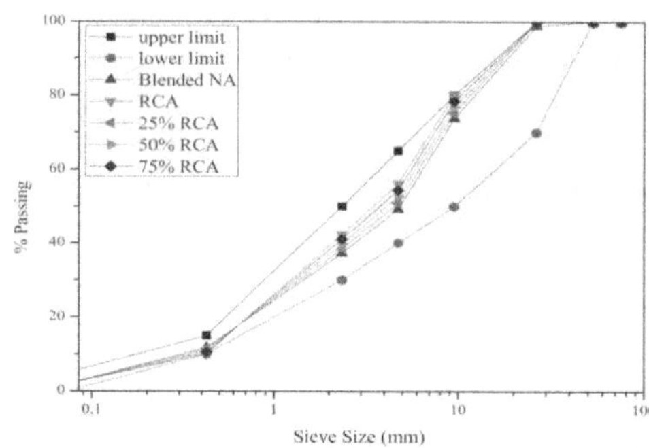

Figure 1. Grain size distribution curve

From the basic tests conducted, it is observed that 100% of RCA has higher values of water absorption, AIV, abrasion value than 100% NA and lower specific gravity than 100% NA. With the different percentage of replacement of NA with RCA OMC, MDD and CBR value were determined. Table 2 summarises the performance tests results as presented below.

Table 2. Summary of performance tests results.

Blend Proportions	(OMC) (%)	MDD (g/cc)	Soaked, CBR (%)
100% NA	7.68	2.225	28.22
75% NA and 25% RCA	8.42	2.148	44.0
50% NA and 50% RCA	10.08	2.136	92.06
25% NA and 75% RCA	10.89	2.113	91.8
100% RCA	11.58	2.09	46.46

From the results of Modified Proctor's Compaction test, it is observed that as the percentage replacement of RCA with NA increases, OMC increases and MDD decreases. Therefore, the CBR test samples are cured for two days and then tested for unsoaked and soaked conditions. From the CBR results, it is observed that 50% NA and 50% RCA and 75% RCA and 25% NA has higher soaked CBR values than unsoaked condition. Considering all the factors, 50% NA and 50% RCA mix are suitable for use in the sub-base course application. It is concluded that 100% of RCA has higher values of water absorption, aggregate impact value, abrasion value than 100% NA, and lower specific gravity than 100% NA. From the results of Modified Proctor's Compaction test, it is observed that as the percentage replacement of RCA with NA increases, OMC increases and MDD decreases and it is observed that, 50% NA and 50% RCA and 75% RCA and 25% NA has higher soaked CBR values than unsoaked condition. Finally, 50% NA and 50% RCA mix are suitable for use in the sub-base course applications. The reason for RCA has higher CBR value than the natural aggregates CBR value is due to the self cementing properties of RCA. The current research is in progress, and all the performance tests like resilient modulus, rutting and fatigue analysis is under progress for the various proposed blends.

References

Sonawane T. R and Pimplikar S. S (2013) Use of recycled aggregate in concrete. International Journal of Engineering Research and Technology (IJERT), 2(1), 1-9.

Arulrajah A, Disfani M. M, Horpibulsuk S, Suksiripattanapong C, and Prongmanee N (2014) Physical properties and shear strength responses of recycled construction and demolition materials in unbound pavement base/subbase applications. Construction and Building Materials, 58, 245-257.

Ayan V, Limbachiya M. C, Omer J. R, and Azadani S. M. N (2014) Compaction assessment of recycled aggregates for use in unbound subbase application. Journal of Civil Engineering and Management, 20(2), 169-174.

Ardalan N, Wilson D, and Larkin T (2017) Laboratory Performance of Recycled Concrete Aggregate as Base-course Material. In Australasian Transport Research Forum (ATRF), 39th, 2017, Auckland, New Zealand.

OSD performance of vehicles manoeuvre on un-channelized roadway: Case study in Nellore, India

Uma Sai Krishna.P[1], Purushottham.D[2*], Sai Pradeep.P[3]
[1] *Civil Engineering/KL University/Vaddeswaram, Vijayawada, India;*
[2] *Geethanjali Institute of science & Technology, Nellore,A.P*
** e-mail: umasai.paluru@gmail.com*

Introduction

Traffic studies on Indian roads grow into more complex due to increase in vehicle usage. In India traffic is highly mixed in nature with wide variations in the static and dynamic characteristics of vehicles. Traffic movements mainly overtaking becoming complex operation especially in non lane based traffic where the vehicles use the opposing lane to overtake the slower vehicles with the presence of oncoming vehicles from opposite direction. The main objective of the paper is to present the overtaking characteristics of vehicles on undivided road in Nellore city which is one of the smart cities in India.

Materials and Methods

For this purpose, details of overtaking data were collected on a two-lane two-way undivided road by means of moving car observer method and registration plate method. Two types of overtaking were observed from the analysis which is flying and accelerative.The overtaking characteristics of all types of vehicles under mixed traffic conditions were observed and mathematically modelled.
The data extracted and analyzed were the acceleration characteristics, speeds of the overtaking vehicles, overtaking time, overtaking distances, safe opposing gap required for overtaking, flow rates, overtaking frequencies, types of overtaking strategy, and types of overtaking and overtaken vehicles.

Results and Concluding Remarks

The results obtained from this study will be useful to understand the overtaking behaviour of vehicles in mixed and non-lane discipline traffic conditions. It was found that the mean values of overtaking distances for accelerative overtaking is higher compared to flying overtaking. The time taken for the flying overtaking is lesser when compared to the accelerative type of overtaking.

References

GeetimuktaMahapatra[1], Akhilesh kumar Maurya[2] **Study of Vehicles Lateral Movement in Non-lane Discipline Traffic Stream on a Straight Road, Procedia - Social and Behavioral Sciences**

Volume 104, 2 December 2013, Pages 352-359

Abdul-Mawjoud and Sofia, 2014,A.A. Abdul-Mawjoud, G.G. Sofia,Passing behaviour on rural two-lane highways Al-rafidain Engineering, 22 (2) (2014), p. 123

Evaluation of Existing Pavement for Hot In-Place Recycling and Design of Recycling Mixes

Dr. Amarendra Kumar. Sandra
Associate Professor, Civil Engineering Department, IIIT RK Valley, RGUKT-Andhra Pradesh.
E-mail: *amarendra123@gmail.com*

Introduction

In the recent years, Hot In-Place Recycling (HIPR) technique is gaining popularity due to rapid increase in construction rate and maximum percentage utilization of existing pavement. However, the suitability of HIPR for an existing pavement depends on its structural strength, surface condition, variation in asphalt content and change in aggregate gradation w.r.t to the originally adopted gradation (Button, J.W. et. al; 1999 and Russell, M. et. al.; 2010). Usually in the existing pavements, due to movement of traffic over a period of time, the adopted aggregate gradation becomes finer and the crucial aggregate fraction passing 75 micron sieve normally high and can make the HIPR option unsuitable. Also, in some cases, the variation in asphalt content over length of the road and the physical properties of the existing asphalt make HIPR unsuitable. Hence, deciding particular pavement for HIPR option is critical and it affects the long term performance of the pavement (Hafeez, I. et.al. 2014). After deciding the suitability of pavement for HIPR, estimating the percentage of new mix to achieve the required gradation and selection of appropriate recycling agent is crucial in recycling mix design.

Methodology

As a part of the periodic maintenance, it is proposed to strengthen the existing 4 - laning Tambaram – Tindivanam section of NH-45 for a length of 10 km (20 lane km) comprises both LHS and RHS by HIPR. In the process, initially, pavement condition mapping was done and based on it three Homogenous Sections (HS) having similar distresses were been identified. Then, the structural strength of the pavement was evaluated by conducting Falling Weight Deflectometer (FWD) studies. Based on the deflection values arrived from FWD, the remaining life of the pavement was found to be 25 MSA, 28 MSA and 36 MSA respectively on the identified three sections. With the available life, the pavement can sustain more than 5 years, however, the functional condition needs to be improved. In such a case HIPR found to be more appropriate.

To adopt HIPR, the physical condition of the existing bituminous concrete in terms of its gradation, bitumen content and volumetric properties were evaluated. On the identified sections, 10 asphalt concrete cores per km length were collected i.e. one for every 100m of length in a zig zag manner. The depth of the core is equal to the thickness of the bituminous surface layer. While coring, it was noticed that the bigger size fractions were broken and hence, correct representation of asphalt content and grading was not possible to achieve. Therefore, in addition to the cores, ripped Reclaimed Asphalt Pavement (RAP) was also collected at random intervals covering entire length of the sections.

The ripped RAP was softened and extracted for bitumen content and aggregate gradation. The average bitumen content on HS-1 is 4.26%, on HS-2 is 4.68% and on HS-3 is 5.27% and its standard deviation on three homogenous sections is 0.93, 0.38 and 0.43 respectively. The aggregate gradations obtained on a particular homogeneous section were averaged and checked its deviation as compared to the specification limits of MoRTH. Figure 1 show the aggregate gradation obtained on HS-2. As per the Figure 1, the existing aggregate gradation is towards upper side (finer side) of the specified band and in particularly, the aggregate fractions from 2.36mm to 19mm sieve are beyond the upper band. This is due to the reason that aggregates have undergone abrasion under the movement of traffic. From the extracted solvent, bitumen was recovered using Rotary Evaporator and its physical properties such as penetration, softening point and Performance Grade (PG) was determined.

Form the cores lot, 6 cores from each Homogenous section were identified and tested to find its density, volumetric properties and elastic modulus at 35^{0}C. The air voids in the HS-1 and HS-2 are in the range of 3 to 4.5% and in HS-3 they are in the range of 2 to 4%. The resilient modulus values of cores in all homogeneous sections are in the range of 1050 Mpa to 1600 Mpa. Based on the variability (statistical) of the asphalt content, aggregate size fractions and volumetric properties, it is proposed to adopt HIPR option for the HS-2 and HS-3 and rejected for the HS-1.

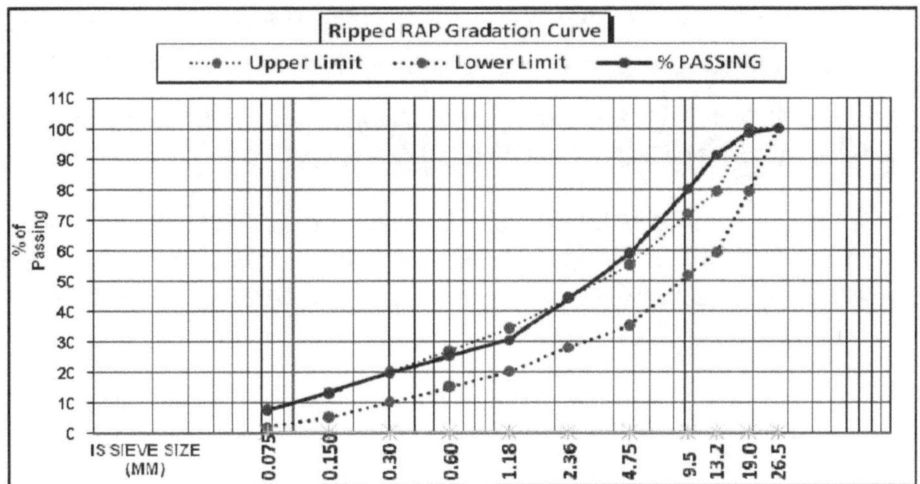

Figure 1: Ripped RAP Gradation

Having taken a decision on acceptance of HIPR, systematic procedure has been followed in designing the recycled mix. This involves in fixing of Performance Grade (PG) of bitumen by adding rejuvenator to the recovered asphalt, deciding percentage of asphalt and virgin aggregate fractions to be blended. The existing BC layer was constructed with VG30 bitumen and PG grade of VG30 bitumen was found to be PG64. In the present case, the recovered bitumen PG grade is found to be PG76 in all homogeneous sections and hence, the rejuvenator (VG10 bitumen) of various percentages was added to the recovered bitumen to achieve PG64. In case of HIPR, existing pavement need to be heated to soften the surface layer up to the required depth. In this process, bitumen oxidation takes place at higher rate and due to this bitumen becomes stiff as compared to the recovered bitumen form ripped RAP, hence, premature failures are possible after the completion of HIPR. Therefore, while adding the rejuvenator, care has been taken so that the PG grade of the bitumen is lower than PG64.

As a measure of verification of the design mix, the rejuvenator and required virgin aggregate was added to the heated RAP and Marshall Specimens were prepared. The volumetric properties of the mix were determined and checked for its compliance. To check the homogeneity and success of the HIPR after its completion, again the cores were taken at random locations from the pavement surface and mix properties such as air voids and elastic modulus were determined.

Conclusions

- HIPR is suitable for the pavements sections on which no much variation in asphalt content and aggregate gradation over a length.
- Based on the variation in asphalt content and aggregate gradation, on the selected three road sections, HIPR was proposed for two sections and for the third section HIPR was not suitable.
- The success of HIPR depends on how meticulously the rejuvenator and different aggregate proportions were added. To check this, the properties of mixes were evaluated by taking the cores at random interval after the construction and found that the properties were within the specified limits.

Key Words: Pavement Condition, Hot In-Place Recycling (HIPR), Variation in Asphalt Content, Recycling Mix Design

References

ASTM D7369-09 (2009); "Standard Test Method for Determining the Resilient Modulus of Bituminous Mixtures by Indirect Tension Test".

Button, J.W., Estakhri, C.K., Little, D.N. (1999) "Overview of Hot In-Place Recycling of Bituminous Pavements" Transportation Research Record 1684, TRB, National Research Council, Washington D.C. pp. 178–185.

Hafeez, I., Ozer, H. and Al-Qadi, I. (2014) "Performance characterization of hot in-place recycled asphalt mixtures", Journal of Transportation Engineering, Vol. 140, No. 8, pp. 04-14.

Russell, M.; Jeff S. Uhlmeyer; Joe DeVol; Chris Johnson; Jim Weston (2010) "Evaluation of Hot In-Place Recycle" Report No. WA-RD 738.1, Washington State Department of Transportation Materials Laboratory, Olympia, Washington. https://www.wsdot.wa.gov/research/reports/fullreports/738.1.pdf

Asset Management in Highway Infrastructure Decision making

Pattabhiram Yegulla[1], Dr. Amarendra Kumar Sandra[2*], Dr. P. Sravana[3]
[1] Highway Design Engineer, Druta Designs, Hyderabad
[2] Associate Professor, Civil Engineering Department, IIIT RK Valley, RGUKT-AP
[3]Professor and HoD, Civil Engineering Department, JNTU, Hyderabad
** Corresponding Authror e-mail: amarendra123@gmail.com*

Introduction

Road asset management offers a discipline for integrating asset data in a way that facilitates its use for decision-making. It provides a framework for managing road networks using a long-term perspective, rather than the relatively short-term view currently adopted by many highway authorities. The main objective of asset management is to improve decision-making processes for allocating funds among an agency's assets so that the best return on investment is obtained (Jana, S. et.al., 2008; Frangopol, D. M. and Liu, M., 2007). To achieve this objective, asset management embraces all of the processes, tools, and data required to manage assets effectively. For this reason, the asset management is also defined as "a process of resource allocation and utilization" (AASHTO, 1999).

Key benefits to organisations that adopt formal asset management systems and practices include: Improved understanding of service level options and requirements, Minimum life cycle (long term) costs for an agreed level of service are identified, Better understanding and forecasting of asset-related management options and costs, Managed risk of asset failure, Improved decision making based on costs and benefits of alternatives, Clear justification for forward works programmes and funding requirements, Improved accountability over the use of public resources, Improved customer satisfaction and organisation image.

Asset management planning enables asset owners to demonstrate to their customers and other stakeholders that services are being delivered in the most effective manner. It also provides a basis for evaluating complex service price/quality relationships in consultation with customers.

Case Study

The paper discusses the case study of Gujarat Road Management System (GRMS) developed for Gujarat Government Roads & Buildings Department and Planning. The GRMS has the eight key modules *viz.* Road Information System (RIS), Traffic Information System (TIS), Pavement Management System (Practical and HDM -4 based), Routine Maintenance Management System (RMMS), Bridge Management System (BMS), Environment and Social Information System (EIS), Accident Information System (AIS), Monitoring and Evaluation System (MES).

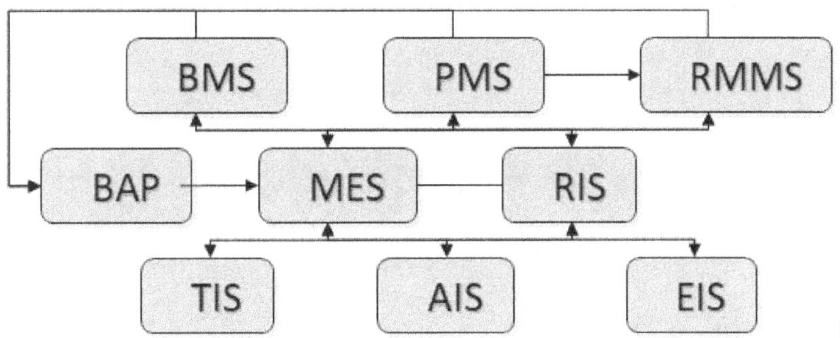

The main features of the various models, the data required for developing various modules and methodology adopted for collecting the data has been discussed. Based on the input data, the desired output for each of the module is generated.

Conclusions

When the funds available for maintenance of infrastructure are limited, it is necessary to take the decision judiciously. In the present study, information such as pavement network data, traffic information, existing pavement distress data, road roughness, structural condition data in terms of deflection and pavement composition, routine and periodic maintenance information, inventory and condition of the existing structures and accident data was collected and inputted in the HDM-4 framework. With the input data, detailed schedule has been developed to carry out various activities such as maintenance of road to improve functional / structural condition, maintenance of structures, increasing the capacity of road, rehabilitation/replacement road furniture and precautions to be taken to improve the safety.

References

AASHTO (1999); "The Maintenance and Management of Roadways and Bridges" American Association of State Highway and Transportation Officials, Washington, DC.

Frangopol, D. M. and Liu, M. (2007); "Maintenance and management of civil infrastructure based on condition, safety, optimization and life-cycle cost",. *Structure and Infrastructure Engineering*, 3(1): 29–41.

Jana Šelih, Anžej Kne, Aleksander Srdić and Marjan Žura (2008) "Multiple-criteria decision support system in highway infrastructure management", Transport, 23:4, 299-305, DOI: 10.3846/1648-4142.2008.23.299-305

Redesign of Palace Road Bengaluru as per TENDER SURE Guidelines.

Krutika M Huragi[1*], Mr.Srinivasa V[2], Mr. Madhav M P[3]

[1] *M Tech Student, Department of CTM, Jawaharlal Nehru National College of Engineering, Shimoga, India*
e-mail: huragikrutika@gmail.com

[2] *Assistant Professor, Department of CTM, Jawaharlal Nehru National College of Engineering, Shimoga, India*
e-mail: srinivasajetty.v@gmail.com

[3] *VP Projects,Alcon Consulting Engineers Pvt. Ltd., Bengaluru, India*
e-mail: madhavp@alconsurvey.com

Introduction

The urban roads nowadays have set an example for intense congestion of traffic due to the rapidly increasing population which in turn increase the number of motorized vehicle users. Bangalore being one among the fastest developing cities of our country is also experiencing the same problems and is yet to get worsened in the upcoming years. On the other hand the safety of pedestrians and cyclists is least neglected and as a result of this they are most oftenly prone to be the part of accidents. The Tender SURE Project which has been initialized in the present decade is an attempt to bring necessary changes in the urban roads execution by addressing various major problems related to traffic, pavement condition, travel lanes, footpaths, sub-terrain utilities, drainage, etc. The project is also concerned with the redesign of existing roads by providing wider and uniform travel lanes, providing enough space for bus bays and improving the geometric conditions of the existing roads and providing suitable street furniture. Mr. Vidyadar Patil B, Mr. Kilabnur Pramod, Mr. Madhav M P (2017) conducted a case study on Subedhar Chatram Road, Bangalore for the redesign of the road under Tender SURE guidelines by providing an overlay according to IRC:SP 76-2015 and have given details regarding the existing traffic conditions, pedestrian facilities and parking facilities. Similarly Basavraj Kabade, K.T.Nagaraja, Swathi Ramanathan, A.Veeraraghavan and P.S.Reashma (2018) have conducted studies on 3 stretches namely Vittal Malya Road, Residency Road, St. Mark's Road and Cunningham Road to find out the adequacy of pedestrian facilities on the Tender SURE Roads. For the redesign of the present road AutoCAD Civil 3D software is made use as it provides an easy and effective design approach.

The project overcomes the issue of repeatedly digging the roads for repair and maintenance and thus arrives at an once for all solution. The motto of the project is to expend additional and right relatively than to expend much extra by investing for wrong numerous times. Thus, in return this project struggles for improving the condition of present urban streets and finally to improve the quality of lives of people, as good road networks provide a greater accessibility to all the basic and important needs of people improving their growth and life condition. White topping is one among the prominent methods of pavement rehabilitation which is in practice in these years. It can be taken as a permanent solution for the entire design life of the pavement as the once properly rehabilitated white topping pavement can serve as long as for 30 years. Keeping in view the construction and maintenance cost of laying a cement concrete overlay, thin white topping overlay can be effectively laid on the existing asphalt road. A detailed design procedure of thin white topping overlay is provided for the estimated amount of traffic and its future growth.

Methodology

The study stretch was selected from SBI Circle to Chalukya Circle which comprises of 1.5 km located in Bengaluru city. A detailed reconnaissance and pavement condition survey was conducted to know about the existing pavement condition. It was then followed by the pavement strength analysis or BBD survey, traffic analysis for understanding the existing traffic flow and peak hour pedestrian volume count to determine the adequacy of present footpaths. Finally detailed geometric design and the cross sections were obtained by AutoCAD Civil 3D and then a detailed pavement design was carried out to determine the thickness of the rigid overlay.

Analysis

The following studies and analysis were carried out:

1. Reconnaissance and condition survey: This mainly analyses the existing pavement condition and the surface features along the alignment. It was found that the pavement condition was fair.
2. Pavement Strength Evaluation: This was done to evaluate the strength of the pavement to provide an overlay and a characteristic deflection value of 1.015 was obtained as an average of both sides.
3. Traffic analysis: Classified traffic volume count was carried out in both the directions for 3 days 24 hours and the percentage composition of vehicles and the peak hour volume was found. The results showed that the maximum amount of traffic in the stretch composed of two wheelers and the peak hour was 5 pm to 6 pm.
4. Pedestrian volume count: Pedestrian count was made for the stretch during the peak hours and the density, LoS was determined. The results showed that the present stretch showed the LoS A.
5. Geometric Design: A detailed geometric design was done using AutoCAD Civil 3D software. It involves the importing of points, creating surfaces, alignment, cross sections, profiles and assembly.
6. Pavement Design: The pavement design involves the design of thin white topping overlay and is carried out with the help of IRC: SP 76-2015 Guidelines for the design of thin and ultra thin whitetopping.

Conclusions

- The redesigning of the Palace Road according to the Tender SURE guidelines is to bring about several advantages with regard to the increase in carriageway width which in turn increases the capacity, widened footpaths for safe and efficient pedestrian movement, laying of sub terrain utilities underground which prevents the further digging of roads for repairs.
- The pavement design by providing thin white topping overlay has been proved beneficial and hence the same technique is adopted for the present project stretch. The detailed calculations have shown that a thickness of 170mm is required to cater the existing and upcoming increase in traffic.

References

1) Basavaraj Kabade, K T Nagaraja, Swami Ramanathan, A Veeraraghavan, P S Reashma (2018) Improvement to Pedestrian Walkway Facilities to Enhance Pedestrian Safety- Initiatives in India, World Academy of Science, Engineering and Technology, International Journal of Transport and Vehicle Engineering, Vol: 12, No: 3.
2) Mitesh D Patel, Prof. P.S. Ramanuj, Bhavin Parmar and Akash Parmar (2012) Whitetopping as a rehabilitation method: A Case Study of Budhel-Ghogha Road, International Journal of Advanced Engineering Research and Studies, Volume 1, Issue IV, E-ISSN: 2249-8974.
3) Mr. Vidyadhar Patil B, Mr. Kilabanur Pramod and Mr. Madhav M P, (2017) Preparation of Detailed Project Report for Redesigning SC Road as per Tender SURE Guidelines, IJRTI, Volume 2, Issue 11, ISSN: 2456-3315.
4) S.A.Raji, A. Zava, K. Jirgba and A.B.Osunkunle (2017) Geometric Design of a Highway using AutoCAD Civil 3D, JMEST, Volume 4, Issue 6, ISSN: 2458-9403.
5) T. Subramani, Study on Existing Pedestrian Traffic and Facility Study on Major Roads in Salem and Formulate Suggestions for its Improvements, IJMER, Volume 2, Issue 3, May-June 2012, ISSN: 2269-6645.

DRIVER BEHAVIOR ANALYSIS USING TRAJECTORY DATA AT URBAN MID-BLOCK SECTION FOR MIXED TRAFFIC CONDITIONS

Sowjanya[1*], Dr.S. Moses Santhakumar[2]

[1] *Civil Engineering Department, G.Pulla Reddy Engineering College, Kurnool, India*

[2] *Civil Engineering Department, National Institute of Technology, Tiruchirappalli, Inda*

* *e-mail: sowjinitt@gmail.com*

Introduction

For any study related to traffic, a good understanding of the real traffic observed is essential on Indian roads, it is challenging to measure microscopic characteristics as the flow patterns are involved with not much lane discipline. Driver behavior on road sections is one of the most complex phenomenon, to study, particularly under mixed traffic environment. This is also interconnected with the competence of the road section and the traffic flow characteristics of the road section at macroscopic as well as microscopic level. With the involvement of multiple vehicle classes, high-resolution trajectory data is necessary for exploring vehicle following, lateral movement, and seeping behavior under varying traffic flow states.

While dealing with homogeneous traffic, it is easy to collect and extract microscopic and macroscopic traffic characteristics. In heterogeneous traffic conditions due to lack of proper data collecting techniques and the level of difficulty involved, very few researchers have attempted studying these characteristics.

Narayanaraju et al. (2018) have studied vehicle behavior in a mixed traffic environment. They analyzed the macroscopic and microscopic flow characteristics, including the relative velocity, relative spacing between the vehicles both in the longitudinal and lateral direction. Venkatesan Kanagaraj and Gowri Asaithambi (2016), evaluated vehicle following behaviour and concluded that vehicles in the mixed stream-in particular motorcycles move unsubstantial in the lateral direction. Bangarraju and Ravi Shankar (2016) worked on the lateral distance, keeping driver behavior in mixed traffic conditions with little lane discipline. They have concluded that Lane changing frequency is influenced by the density, flow and mean speed of the traffic stream and the lane by lane vehicular arrivals are not very independent.

With this background, an access controlled mid block road section was selected for video data collection. The videos were collected from the urban roadway in the Kurnool district of Andhra Pradesh. The length of the stretch is straight with 120m, and the width of roadway is 10.5m. Data was collected by dividing the stretch into four equal parts each of length 30m. The data was extracted to know the variations in terms of longitudinal and lateral speeds, accelerations, decelerations, vehicle following and lane changing behavior of the drivers.

Objectives

The main objectives of this study include

- To extract trajectories of vehicles at the mid-block study location under heterogeneous traffic conditions
- To analyse the lane changing and vehicle following behaviour of the driver on the mid block section in heterogeneous traffic conditions

Gaps in literature review

The studies were carried out to analyse the safety and lane changing behavior of the driver and to extract the vehicle trajectories, were done on national highways, multi lane roads and on bridges, i.e., on control points. Only a few studies were found in urban areas to describe the acceleration and deceleration, longitudinal and lateral behavior of vehicles. Most of the research works were carried out without considering the lateral speeds and lateral accelerations of the vehicles on selected road sections. There is a need to study the behavior of drivers on urban mid blocks, considering the above mentioned characteristics

Methodology

Kurnool, one of the most populous districts of the state of Andhra Pradesh, has seen a considerable increase in traffic over the last few years.

Government General Hospital is located in the central part of Kurnool city having heavy traffic. Road (six lanes divided Highway) dividers have been provided on the road. Traffic flow videos are taken by providing CCTV camera on foot over bridge located at the height of 8.0 m from the surface of the road. Traffic is recorded continuously for a period of 5 hours from 8:00 AM to 1:00 PM to covering upto 120m from the Foot Over Bridge located to towards the Kurnool-Cadapa canal, popularly known as K-C canal to assess the volume, speeds, trajectories, and lane change behaviour of drivers of different types of vehicles.

For analysing the traffic characteristics, the peak 15 minutes data was considered. The vehicle trajectories, flow, speed were extracted using Traffic Data Extractor(TDE), for every 1.0 second resolution. To minimise the error of parallax, the video is cropped and enlarged for every 30m, and the lateral movements corresponding to the longitudinal distance of the vehicle were reported. The data extracted were smoothened by moving average method to minimise human errors. The lateral amplitude of the vehicles of various types was analysed. Relative distances, vehicle following and lane changing analysis was done using Kinovea software. For this, the median side was taken as the origin with lane -1 followed by lane-2 and lane-3. Statistical analysis was also done to know the significant differences between the various vehicle types in travel speeds and accelerations using SPSS software.

Conclusions

In this paper, an attempt has been made to study the traffic characteristics of mixed traffic. One significant component in the driver behavior modeling is the data collection, which is especially important in the lane-changing research. A detailed set of vehicle trajectory data was collected in an urban midblock road section in the Kurnool City of Andhra Pradesh, to estimate, longitudinal and lateral speed and acceleration values. Statistical analysis was also done to know the significance level for 99%. The resulting data were studied for combined traffic flow characteristics and variables related to the lateral movement of the vehicles. The results siggest directions for developing a driving behaviour model for mixed traffic streams. Based on the study and analysis of data observed, the following broad conclusions have been arrived at.

1. The main characteristic of mixed traffic is the presence of significant numbers of vehicles of various types in the stream. There are substantial differences in the flow characteristics between these vehicle types. The results bring out the differences in travel speeds, accelerations, choice of lateral position on the roadway, and almost all other measures that were studied.

2. It was found that smaller vehicles show more lateralmovement freedom compared with heavy vehicles and their presence may result in deviation of lanewise movement, particularly whenever they experience delay.

3. Lateral characteristics of the traffic are one of the important characteristic in case of no lane discipline road. Due to the high lateral movement there exists frequent accidents and congestion in Indian roads. Hence to avoid these, lane discipline is an important criteria to be considered.

References

1. Munigety, C. R., V. Vivek, and T. V. Mathew, 2014 "Semi-automated Tool for Extraction of Microlevel Traffic Data from Video graphic Survey", Transportation Research Record, Journal of the Transportation Research Board, No. 2443, Transportation Research Board of the National Academies, Washington, D.C., pp. 88–95.

2. Narayanaraju, Pallavkumar, Aayushjain, 2018 "Application of Trajectory data for investigating the vehicle behaviour in mixed traffic environment", Transportation Research Record, Journal of the Transportation Research Board.

3. Venkatesan Kanagaraj, Gowri Asaithambi, Tomer Toledo, and Tzu-Chang Lee, 2015, "Trajectory Data and Flow Characteristic of Mixed Traffic", Transportation Research Record: Journal of the Transportation Research Board, No. 2491, Transportation Research Board, Washington, D.C., pp. 1–11.

Effect of waste materials as alternative of natural resources for sustainable bituminous paving mixes

Jagadeesh Gopalam[1], Jyoti Prakash Giri[2*], Mahabir Panda[3]
[1] *P.G. Student, Department of Civil Engineering ,GMR Institute of Technology, Civil Department, Rajam, India;*
[2] *Assistant Professor, Department of Civil Engineering ,GMR Institute of Technology, Civil Department, Rajam, India;*
[3] *Professor, Department of Civil Engineering,National Institute of Technology, Civil Department, Rourkela, India;*
** e-mail: jagadeesh1329@gmail.com*

Introduction

Sustainable development offers various scopes of civil constructions for gaining importance to the noticeable and appropriate reasons. In different transportation construction industries, engineers from industry, and researchers are all craving this for bituminous paving as well as concrete mixes for highway projects which transfer to a sustainable future. In the flexible pavement, a reflective number of innovative materials and technologies are being developed to ascertain their suitability for the design, construction and maintenance of these pavements. Warm mix asphalt (WMA), recycled asphalt pavement (RAP), fly ash, bottom ash and shingles are some of the materials that transportation researchers believe to have hold the future to sustainability in the bituminous highway construction industry in conventional way (Mills-Beale and You, 2010). Utilization of these sustainable materials or techniques have resulted tremendous benefits for the conservation of natural resources, optimizing the landfill use and saving the project cost corresponding to the waste dumping charge (Giri et al., 2018). Another potential material that few researches have attempted for its re-use in bituminous paving mixes, is recycled concrete aggregate (RCA). (RCA) which is basically a demolition waste originated from the concrete structures due to either attaining their expected life or due to other reasons. The large scale generation of RCA is a common feature mostly in urban areas and large project sites. On the contrary, RCA has been found to have offered promising results when utilized as base aggregate material for sub base layer in bituminous pavement and in Portland Cement Concrete (PCC) pavement worldwide (Motter et al., 2015). However, studies on the use of RCA in bituminous paving mixes are limited. This fact has motivated the authors to investigate and explore the practicality implication of RCA as replacement of coarse aggregates in bituminous paving mixes and take up a comparative study with respect to the conventional bituminous mix. In this experimental research, the RC aggregates have replaced fully the coarse aggregate portion in the bituminous mix with two different types of fillers namely fly ash and stone dust addition to two types of bitumen such as VG 30 and VG 40. The impact of aggregate, filler and bitumen type on bituminous mix performances has been evaluated in terms of conventional Marshall properties, moisture sensitivity and resilient modulus values.

Materials and Methods

In the present experimental investigation, to study the effects of aggregate, bitumen and filler types on the performances of dense bituminous macadam (DBM) mixes, eight combinations with two different types of aggregates, i.e., natural aggregate (NA) and recycled concrete aggregate (RCA), two types of bitumen, i.e., VG 30, and VG 40 and two types of fillers, i.e. stone dust (SD) and fly ash (FA) are tried. The properties of all the materials used in this study have been determined in the laboratory. The test results satisfy all the parameters according to Ministry of Road Transport and Highways (MoRTH, 2013). From the test results, it is observed that RCA has a bit lower specific gravity as well as higher water absorption value compared to conventional aggregate. For evaluation purposes, different performance tests on each type of mix such as Marshall properties, indirect tensile strength, moisture susceptibility [in terms of tensile strength ratio (TSR) and retained stability (RS)] and resilient modulus test, have been taken up. For these performance tests, all the eight bituminous mixes were prepared with 5% bitumen content as per Marshall procedure. The mixes prepared with NA, two types of filler (SD and FA) and two types of bitumen (VG 30 and VG 40) have been given nomenclature as NSD 30, NFA30, NSD 40 and NFA 40. Similarly, the mixes made with RCA as coarse aggregate have been named as RSD 30, RFA30, RSD 40 and RFA 40.

Results and Concluding Remarks

The test results of DBM mixes prepared with different combinations are presented below through Figure 1 through Figure 4. The Marshall parameters such as Marshall stability, flow value and air voids for the eight mixes compacted with 5% bitumen are presented in the Table 1.

Table 1. Marshall parameters of different DBM mixes

Mix type	Marshall Stability (kN)	Flow value (mm)	Air voids (%)
NSD 30	13.66	3.35	4.18
NFA 30	12.89	3.45	4.29
RSD 30	15.89	3.1	4.10
RFA 30	16.10	3.15	4.18
NSD 40	15.28	3.3	3.63
NFA 40	14.29	3.35	4.21
RSD 40	17.61	2.68	3.94
RFA 40	17	2.82	4.15

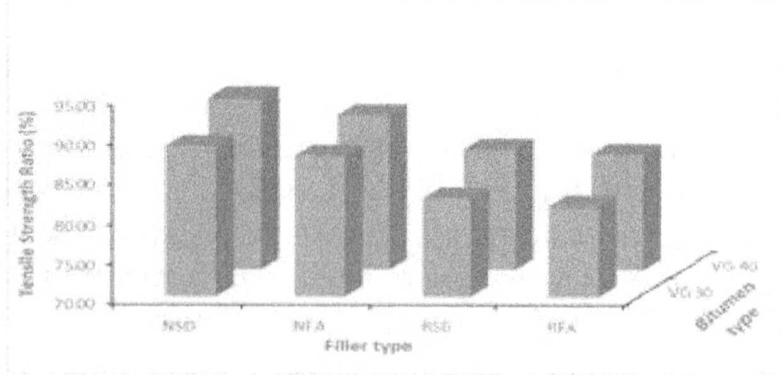

Figure.1 TSR value for different DBM mixes

Figure.2 RS value for different DBM mixes

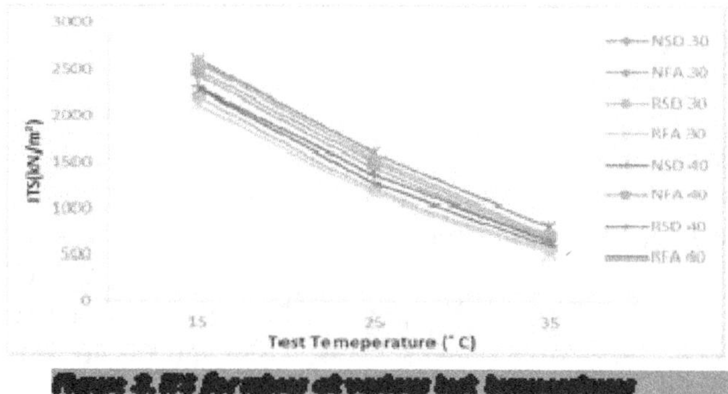

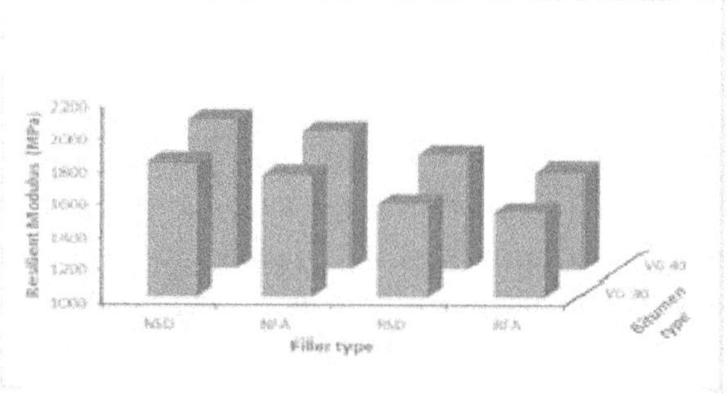

Figure 4. Resilient modulus value for DBM mixes

From the above mentioned results, it is concluded that, the mixes prepraed with NA, stone dust as filler and VG 40 bitumen offered highest resistance to moisture susceptibility and maximum resilent modulus value as compared to the other mixes. However from Marshall stability point of view, the samples made with RCA, stone dust and VG 40 as binder, resulted maximum values compared to other mixes. However in general, satifactory values are observed when RCA and stone dust have been used in the bituminous mixes.

References

Mills-Beale, J., & You, Z. (2010). The mechanical properties of asphalt mixtures with recycled concrete aggregates. Construction and Building Materials, 24(3), 230-235.

Giri, J. P., Panda, M., & Sahoo, U. C. (2018). Performance of Bituminous Mixes Containing Emulsion-Treated Recycled Concrete Aggregates. Journal of Materials in Civil Engineering, 30(4), 04018052.

Motter, J. S., Miranda, L. F. R., & Bernucci, L. L. B. (2015). Performance of hot mix asphalt concrete produced with coarse recycled concrete aggregate. Journal of Materials in Civil Engineering, 27(11), 04015030.

Impact Of Mode Shift Behaviour with the Introduction Of Metro rail Service From Nagole To Miyapur Corridor In Hyderabad City

S Praveen[1*], S Ramesh Kumar[2], M Kumar[3]

[1] *Assistant Professor, CVR College of Engineering,Rangareddy, India;*

[2] *Associate Professor, MVSR Engineering College, Hyderabad, India*

[3] *Professor, University College of Engineering, Osmania Universty, Hyderabad-7, India*

* *e-mail: samarthipraveen@gmail.com*

Introduction

Hyderabad city is a large metropolitan city in India with a huge population. The traffic and transportation problems are continuously rising over the past several years. As the existing transportation modes – buses and Para transit– are already saturated, there is a need for developing Mass Rapid Transit System (MRTS). Recognizing this, the Hyderabad Metro Rail Project (HMRP) was planned to be developed as a Public-Private Partnership (PPP) model in India. In order to cope with ever increasing traffic and growing travel demand a Mass Rapid Transport system (MRTS) that involves a 72 km elevated metro project was proposed for Hyderabad city. The study area selected for the project is Hyderabad city covering all metro rail phases and routes.

The Nested Logit model was developed to determine the mode choice behaviour of travellers in the Hyderabad City. Pendyala (1998) studied that the most important factors affecting mode choice are short-distance trips, trip conducted within daily urban systems, traveller's socio economic characteristics. Koppleman and Wen (2001) used the Logit model for the mode choice between the public transport and private transport system for Atlanta. Parameters which were used in the model were travel time and travel cost. Rajat Rastogi and K.V.Krishna Rao (2003) analyzed the travel characteristics of commuters accessing the transit stations in Mumbai, India, with a view to identify the policies that can improve the conditions of the transit access environment.

ALOGIT software is used to find out the mode choice behaviour by incorporating variables and best fit mode choice model developed. The travel behaviour of Hyderabad users has been studied through model.

Methodology

- To determine the travel behavior and travel information through collection of data like destination, cost of the trip, trip purpose and mode choice attributes, etc., on existing mode choices in Hyderabad city.
- To identify the factors influencing the modes chosen in Hyderabad area.
- To develop the relation between modes and factors influencing the modes in the model.
- To develop the Logit model for existing mode choices and also by incorporating Metrorail as a new mode choice using ALOGIT software.
- To perform Statistical tests
- To validate the models.

Results and Concluding Remarks

Table 1. Statistics for Nested Logit Model for Hyderabad City

Variable	Coefficient Estimates	Variables
Travel time	$-0.1454*10^{-4}$(-0.5)	Generic
Travel cost	-0.8855*10-4(-5.4)	Generic
Age	-0.2797(-3.5)	Para-transit
Vehicle ownership	-0.3847(-3.5)	Self-car
Household income	-0.4680*10-1(-0.7)	RTC bus
Purpose of trip	-3.151(-5.4)	MMTS

Householdsize	-0.3998(-4.5)	2 wheeler
Occupation	0.2197(5.8)	Metro rail
	Structural Parameters	
Logsum	0.9207	
L(0)	-1364.0830	
L(θ)	-925.2132	
ρ 2 with respect to zero	0.2952	
ρ 2 with respect to constant	-0.0392	
Sample Size	701	

L (0): Likelihood value with Zero Coefficients
L (θ): Likelihood value at Convergence
L(c): Likelihood value with constant Coefficients
ρ2: Rho- squares statistics
Values in parenthesis are t-statistics

The results of statistical analysis are good. Log likelihood values showed good convergence. Logsum value is between o and 1, thus indicating the best fit nest for which the private transit i.e., 2-wheeler, self-car, hired car and Para transit are grouped under a single First nest. In public transit i.e., RTC bus & MMTS are grouped under Second nest and Metro rail is taken in a separate nest. The negative sign of age for para-transit indicates that as the age increases the preference to choose para-transit as their mode will decrease. The negative sign of Household size for 2-wheeler indicates as household size increases, the preference to choose 2-wheeler decreases. The positive sign for vehicle ownership of self-car mode indicates people prefer self-car. The purpose of trip indicates negative value for MMTS, which reveals that, the purpose of choosing MMTS as mode choice for work, study and leisure etc is decreasing and it is significant. The negative sign of household income indicates that as household income increases the preferences to choose RTC bus in Hyderabad city decreases and it is more significant. The occupations of people with higher order are willing to prefer metro rail. Rho-square values indicate towards goodness-of-fit of the model.

Conclusions:

1. The results would be helpful in understanding and predicting the individual decisions so as to forecast the travel demand and need of metro rail as a mode choice and model was developed.
2. Mode choice is influenced by the factors such as age, vehicle ownership, household size, household income, purpose of trip and occupation etc.
3. Majority of the Hyderabad users (73%) are using public transport to reach their destinations rather than private transport
4. It is clear from study, that maximum of the Hyderabad users (about 63%) are willing to use metro rail as mode choice in Hyderabad city.
5. Most of the Hyderabad users are willing to use metro rail as mode choice due to the significant variables such as less travel time, comfort, safety, affordability and eco-friendly nature of travel.

References

1. Koppelman, F. S. and Bhat, C. (2006). "A self instructing course in mode choice modelling"; Multinomial and Nested logit models Prepared for U.S. Department of Transportation Federal Transit Administration.
2. Pendyala, R.M. (1998), "Causal Analysis in Travel Behaviour Research: A Cautionary Note", *Travel Behaviour Research: Updating the State of Play*, Elsevier Science Publishers, B.V., The Netherlands, 35-48.
3. Rajat Rastogi and K.V.Krishna Rao (2003), "Travel characteristics of commuters accessing the transit stations in Mumbai, India", Journal of Transportation Engineering, Volume 129 Issue 6 - November 2003.
4. S.Ramesh Kumar, M.Kumar, C.S.V.Subrahmanya Kumar & S Praveen (2018), "A Case Study on Mode Choice Analysis in Hyderabad City", International Journal of Emerging Technology and Advanced Engineering, Volume 2, Special Issue 5, 238-243.
5. Subbarao, S. A. (2013) "Analysis of Household Activity and Travel Behaviour-A case of Mumbai Metropolitan Region", International Journal of Emerging Technology and Advance Engineering , 3 (1), 98-109.

Geotechnical Engineering

Compaction and Strength Characteristics of Eggshell Powder and Flyash Based Geopolymer

P. Shekhawat[1*], R. M. Singh[2], G. Sharma[1]
[1] *Malaviya National Institute of Technology Jaipur, Jaipur, India*
[2] *University of Surrey, Guildford, United Kingdom*
* *e-mail: 2016rce9504@mnit.ac.in*

Introduction

Geopolymers are basically polymers produced from aluminosilicate-bearing raw materials viz. Flyash, calcined clay, slag etc. These raw materials are often called precursors in the process of geopolymerization. Geopolymerization is basically a polycondensation reaction which involves the dissolution of alumina and silica from a precursor in the presence of some alkaline activator solution forming tetrahedrally coordinated cementitious materials. Geopolymer can be deemed as a substitute to conventional Portland cement and soil stabilized with Portland cement for utilizing as a pavement material. The manufacturing of cement consumes high energy and emits a large amount of carbon dioxide into the atmosphere. For geopolymerization, silica, alumina, and calcium oxide-rich farm waste eggshell powder and industrial waste flyash can be used as a precursor. Silica, alumina, and calcium oxide from the precursors form C-S-H, C-(N)-A-S-H gel after dissolution which improves the mechanical strength of the geopolymer. Utilization of such wastes in construction promotes sustainable development.

The maximum dry density of any composite is an important factor which later affects the mechanical strength of the composite. The maximum dry density of geopolymer is reported to be influenced by specific gravity of precursor materials, whereas the optimum activator content is reported to be dependent upon availability of moisture in liquid alkaline activator. For example, a decrease in maximum dry density was found for compacted clay-flyash geopolymer when compared to maximum dry density of compacted clay and compacted clay-cement, because of lower specific gravity of flyash (Phetchuay et al. 2014; Sukmak et al. 2015). The unconfined compressive strength is also an important factor to assess the mechanical strength of any composite in compression. The strength of FA-based geopolymer has been reported to increase with the increase in activator ratio Na_2SiO_3/NaOH (Detphan & Chindaprasirt 2009; Sukmak et al. 2013; Morsy et al. 2014).

This study aimed to find out the compaction and strength behaviour of alkaline activated eggshell powder and flyash based geopolymer. The optimum geopolymer composites found in the research have the potential to be employed in sub-base/subgrade application as a sustainable construction material made from waste-by-product. Since no literature is found till date about the inclusion of eggshell powder in geopolymer, this study is a novel approach.

Materials and Methods

In this study, eggshell powder (ESP) and flyash (FA) are used as precursor and mixture of sodium silicate (Na_2SiO_3) and sodium hydroxide (NaOH), with fixed activator ratio Na_2SiO_3/NaOH as 2 is used as liquid alkaline activator. The concentration of NaOH was fixed as 10M. Three different precursor ratios 30ESP:70FA, 50ESP:50FA and 70ESP:30FA were adopted for this study. The geopolymer composites were cured at ambient room temperature for 7 days, and 28 days. Some basic physical properties tests for raw precursor materials were performed. Modified proctor and unconfined compressive strength tests were conducted to analyze the dry density behaviour of the geopolymer with activator content along with mechanical strength of different precursor combination in compression, respectively.

Results and Concluding Remarks

Optimum activator content (OAC) of the precursor combinations was assessed at which maximum dry densities were achieved. Highest MDD, among all precursors' ratios, was achieved for geopolymer mix

70ESP:30FA. Maximum dry densities were then compared with required dry density for fill material in embankment construction.

Unconfined compressive strength was assessed for 7 and 28 days cured geopolymer composites having different precursor combination. Highest compressive strength, among all precursor combinations, was achieved for geopolymer mix 50ESP:50FA. Unconfined compressive strengths were then compared with different codal provisions of pavement construction for application purpose.

References

Detphan, S. & Chindaprasirt, P., 2009. Preparation of fly ash and rice husk ash geopolymer. *International Journal of Minerals, Metallurgy and Materials*, 16(6), pp.720–726.

Morsy, M.S. et al., 2014. Effect of Sodium Silicate to Sodium Hydroxide Ratios on Strength and Microstructure of Fly Ash Geopolymer Binder. *Arabian Journal for Science and Engineering*, 39(6), pp.4333–4339.

Phetchuay, C. et al., 2014. Calcium carbide residue: Alkaline activator for clay-fly ash geopolymer. *Construction and Building Materials*, 69, pp.285–294.

Sukmak, P. et al., 2015. Sulfate Resistance of Clay-Portland Cement and Clay High-calcium Fly Ash Geopolymer. *Journal of Materials in Civil Engineering*, 27(5), pp.1–11.

Sukmak, P., Horpibulsuk, S. & Shen, S.L., 2013. Strength development in clay-fly ash geopolymer. *Construction and Building Materials*, 40, pp.566–574.

Application of similitude rules in geo-technique and foundation engineering: Shake table test with scaled down model pile*[Paper ID:GT-005]*

M.K.Pradhan[1 a*], Shuvodeep Chakaraborty[1], G.R.Reddy[1], K.Srinivas[1]

[1] *CED,BARC,Trombay,Mumbau, 400085*
[2] *RSSD,BARC,Trombay,Mumbau, 400085*
[a]*Homi Bhabah National Institude, Anushaktinagar, Mumbai, 400094*
** e-mail: mpradhan@barc.gov.in*

Abstract

The trend for laboratory model testing under static and dynamic loading conditions is increasing day by day to simulate the response of a real world structure.In the present study shaking table test has been performed in 1G gravitational field with scaled down model pile to study the responses of pile foundation with compliance to similitude rules. The applicability of similitude law for shaking table test with non-liquefiable and liquefiable soil is also discussed. Similarity relationship for static condition on pile foundation under lateral load with significant deformation is also presented briefly. The fundamental dimensions such as length, mass etc. are selected for designing the scaled down model with similarity scale. The practical application of this study is very useful in carrying shake table test of a model scaled down pile foundation. A case study of practical application of similitude rule for conducting shake table test is also presented.

Introduction

Similitude also known as laws of similarity describes how to design a model and carry out model test that resembles the real size structures. In civil engineering full scale test is very expensive, time-consuming and sometimes not feasible. Hence, it is preferably performed through a scaled down model test. The steps for transferring the model test results to generate expected output of prototype are presented in Figure 1. Rocha[] had developed similitude model for the soil for static condition under 1g gravitational field. Similitude of soil structure under dynamic loading was studied by Kagwa by using ratios of forces [3]. Nonlinear dynamic response of ground was examined using Buckingham's theorem by Kokusho et al. [4]. Iai [2] had derived the similarity law for dynamic shaking table test in 1g gravitational field.Iai et al. [3] had proposed two stage scaling relationship called generalized scaling relationship for centrifuge test.

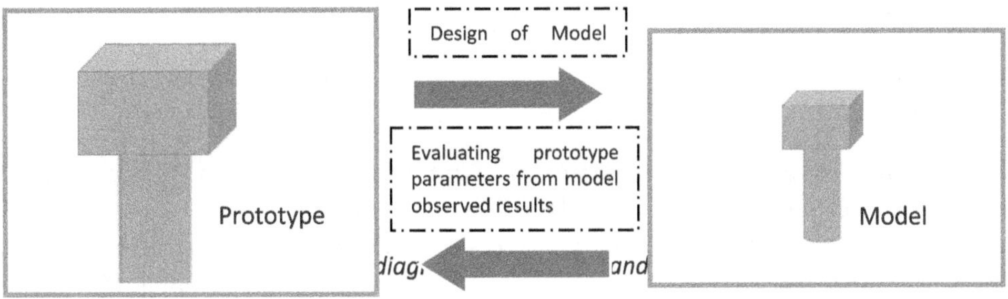

A case study shake table test under 1g gravitational field

Shake table test is performed to study the responses of model pile foundations with raft subjected to sinusoidal base excitation. The structural components of the model pile foundation with raft are scaled down for shake table test. After designing the prototype pile foundation, different dimensions and parameters for model pile foundation is obtained based on similitude rules with suitable scaling factors. The responses of the model pile and the soil-pile interactions, including the inertial and kinematic actions on the model pile, under shakings for liquefied and non-liquefied soil conditions are obtained. The results are further analysed and correlated to predict prototype structure results based on similitude rules with scaling factors for both liquefied and non-liquefied soil conditions as given by Iai [2].

In the present case study aluminium circular hollow section pipes is used as scaled down model pile for reinforced cement concrete (RCC) solid section. List of parameters used in this study are length, diameter, stress, modulus of elasticity etc. In piled raft combination system, piles are made with hollow circular aluminium pipe with 1.20mm thickness.The outer diameter of the model pile is to be determined which corresponds to 520 mm diameter of reinforced Concrete pile with M25 grade. Tests were carried out in a shake table varying input parameters such as input frequency and input ground accelerations.

Results and Concluding Remarks

From the similitude rule of pile under lateral loading the relation between model and prototype is obtained and is given in Eq. 1 [1].

$$n_E n_I = n_G n_L^4 \tag{1}$$

Where, n_E, n_I, n_G, n_L are scale factor for elastic modulus of pile, moment of inertia of pile, shear modulus of soil and length of pile.Moment of Inertia of Model Pile (hollow Aluminium pipe) with outer diameter $D_o$mm, thickness (t) 1.20mm and mean diameter (D_m) the moment of inertia can be calculated as: $I_m = \frac{\pi}{8} t {D_m}^3 = \frac{\pi}{8} * 1.2 * (D_0 - 1.2)^3$. The Moment of Inertia of Prototype Pile (Solid concrete pipe) of diameter D is: $I_P = \frac{\pi}{64} D^4 = \frac{\pi}{64} * 520^4$.

$$n_E = \frac{YoungsModulusofaluminummodelpile}{YoungsModulusofconcreteprototypepile} = \frac{69}{25}, n_I = \frac{MomentofInertiaofaluminummodelpile}{MomentofInertiaofprototypepile}$$

$$n_G = \frac{Shear\ stiffnessofmodel\ soil}{Shear\ stiffness\ of\ prototype\ soil} = 1, n_L^4 = \frac{(Lengthofaluminummodelpile)^4}{(Lengthofconcreteprototypepile)^4} = \frac{0.75^4}{15^4}$$

Putting the values of scale factors in Eq. 1, the outer diameter of model pile, D_o=27mm

Table 1. Model pile to prototype pile conversion

Sl. No.	Property	Prototype/ Actual	Model/ Shake table Test	Remark
1	Material	Concrete	Aluminium	
2	Diameter (mm)	520	27	Hollow pipe of 1.2 mm thick
3	Length (m)	15.00	0.75	
4	Density (Kg/m^3)	2400	2700	
5	Young's Modulus (GPa)	25.8	69	
6	Poisons Ratio	0.15	0.334	
7	Shear Modulus (GPa)	11.2	26	

Hence aluminium pipe of outer diameter 27mmwith thickness of 1.2mm and 0.75m length considered in present experiments can be scaled of actual/prototype pile for a concrete pile of 520mm diameter and 15m length. The scale factors are also used for corresponding interpretation of the results of the prototype. This study will definitely attract the researchers todevelop the similitude relations for shake table testing of model pile foundation or similar kind of geotechnical problem.

Acknowledgments: Author is gratefully acknowledged academic and research support provided by institution, HBNI and BARC.

References

1. D. Muir Wood (2004) Geotechnical Modelling, Spon Press, London
2. Iai, S. (1989) Similitude for shaking table tests on soil-structure-fluid model in 1g gravitational field, Soils and Foundations, 29(1), pp105-118
3. Kagwa, T. (1978) On the similitude in model vibration tests of earth structures, Proceedings of Japan Society of Civil Engineers, No. 275, pp. 69-77
4. Kokusho,T., Iwatate T. (1979) Scaled model tests and numerical analyses on nonlinear dynamic response of soft grounds, , Proceedings of Japan Society of Civil Engineers, No. 285, pp. 57-67

Numerical Analysis of Group of Spin Fin Piles under Different Loading Conditions

A. I.Dhatrak[1], N. G. Tale[2], S. W. Thakare[3]
[1]*Dean Academics, Government College of Engineering, Amravati, India.*
[2]*PG Student, Department of Civil Engineering, Government College of Engineering, Amravati, India.*
[3]*Associate Professor, Department of Civil Engineering, Government College of Engineering, Amravati, India.*
E-mail: anantdhatrak@rediffmail.com, nishatale41@gmail.com, sanjay.thakare1964@gmail.com

Introduction

A variety of structures similar to marine dolphins, dock-fendering systems, tower foundations, inundated platforms, and abutments are installed on deep foundations, which are subjected to uplift loads. Monopiles are widely used to support offshore and onshore wind turbines.Unlike conventional piles, monopiles generally have large diameters ranging from 1 to 6 m and these foundations are generally suitable for water depth up to 35 m. Improvement in the pile capacity can be achieved by providing fins near the top or bottom portion of the monopiles.A Spin Fin pile is a modified pile that has four plates welded along the length of traditional monopile at 90° to each other. It is an emerging form of pile foundation that is capable of resisting large lateral and uplift displacement.

Babu*et al.* (2018)[4] carried out numerical analysis on Lateral Load Response of Fin Piles. The behaviour of regular pile and fin piles with different relative densities of sand, fin orientations, fin numbers and position were investigated. It was concluded that, at higher fin length, star fin piles carried more lateral load followed by straight and diagonal fin piles. Fins placed near the pile top provided more lateral resistance than those placed near the pile bottom.Peng*et al.* (2010)[3] carried out numerical analysis on laterally loaded fin piles. A 3D computer simulation of laterally loaded fin piles was presented to explore the effect of fin dimensions on their load bearing capacity in sand. A fin pile had the optimum fin efficiency when the fin length is half the pile length. Nasr*et al.* (2013)[1] carried out experimental investigation on laterally loaded finned piles in sand. The tests and analysis were carried out by varying the length, width, shape of the fins and type of pile. They concluded that ultimate lateral load improvement depended greatly on length of the fins and increased significantly to the value of $L_F/L_P = 0.4$ and Pile with rectangular fins was more effective in improving the lateral behaviour of piles.

Materials and Methods

A three-dimensional finite element model was established in order to analyse the behaviour of group of conventional and Spin fin pile. The computations were carried out using MIDAS GTS 3D finite element software. A group of three, four, five and six piles are considered for analysis. Mohr-Coulomb model was assigned to soil and piles were considered as elastic material. The relative density of sand was varied. For lateral loading, fins were provided at top portion of pile and for uplift and vertical loading,fins were provided at bottom portion of pile.

Results and Concluding Remarks

Analyses were carried by varying the number of piles in group and relative densities of sand. The ultimate load capacity was found from load settlement curves. The percentage increase in load carrying capacities of spin fin pile as compared to conventional pile for loose sand condition for lateral, uplift and vertical loading are presented below in figure 1.

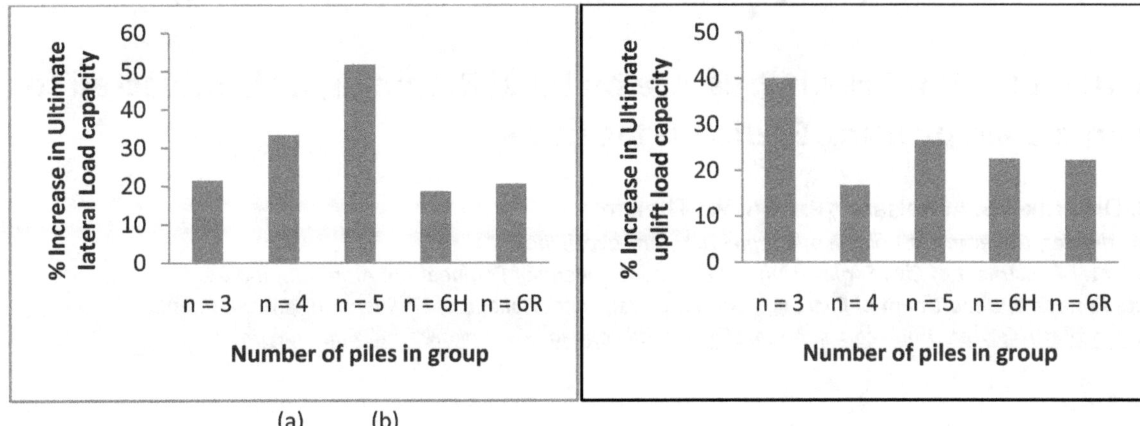

(a) (b)

(c)

Figure 1.Percentage increase in ultimate load capacities of Spin Fin Pile in loose sand for (a) lateral loading, (b) uplift loading, (c) vertical loading.

Based on the results of the present study, the following conclusions are drawn:

1. The percentage increase in ultimate load capacity of group of six piles in hexagonal arrangement is maximum in loose sand condition as compared to conventional pile group. (upto 22 %)
2. The percentage increase in ultimate lateral load capacity of group of five piles in pentagonal arrangement as compared to conventional pile group is maximum in loose sand condition. (upto 52 %)
3. The percentage increase in ultimate uplift load capacity of group of three piles in triangular arrangement group is maximum in loose and medium dense sand as compared to conventional pile. (upto 40 %)

References

1. Ahmed M.A. Nasr, (2018) Experimental and theoretical studies of laterally loaded finned piles in sand. Can. Geotech. J. 51: 381–393.
2. IS: 2911 (Part-4)1985 Code of practice for design and construction of pile foundation.
3. J.-R. Peng, M. Rouainia and B.G. Clarke, (2010)Finite element analysis of laterally loaded fin piles. Computers and Structures 88 (2010) 1239–1247.
4. K.V. Babu and B.V.S. Viswanadham, (2018) Numerical Investigations on Lateral Load Response of Fin Piles. Numerical Analysis of Nonlinear Coupled Problems, Sustainable Civil Infrastructures, DOI 10.1007/978-3-319-61905-7_27.
5. W. R. Azzam and A. Z. Elwakil, (2014) Model Study on the Performance of Single-Finned Pile in Sand under Tension Loads. International Journal of Geomechanics, © ASCE, ISSN 1532-3641, 2017.

Reduction of Lateral Earth Pressure on Piled Retaining wall Subjected to Surcharge Load by using Geofoam Inclusion

Dr. A. I. Dhatrak[1], S. A. Belsare[2], Prof. S. W. Thakare[3]
[1]*Dean Academics, Government College of Engineering, Amravati, India.*
[2] *PG Student, Department of Civil Engineering, Government College of Engineering, Amravati, India.*
[3]*Associate Professor, Department of Civil Engineering, Government College of Engineering, Amravati, India.*
E-mail: anantdhatrak@rediffmail.com,sabelsare04@gmail.com,sanjay.thakare1964@gmail.com

Introduction

Piled retaining walls are frequently used when limited space is an issue, or when the soil is relatively soft. Although the walls are narrower than other choices, they are inserted much deeper into the soil to withstand heavy loads.Piled retaining structures are part of several critical infrastructure facilities such as bridges, highways, railways, buildings, underground structure and are constructed to support soil on one of its sides. They need to withstand pressure from retained material and surcharge pressure due to movement of vehicular traffic, loads from foundations of adjacent building on their backfill, man –made and natural dynamic events.It is essential to have proper knowledge of magnitude and distribution of lateral earth pressure on piled retaining wall. If lateral earth pressure on piled retaining wall is higher, then retaining structure become too costly.The geofoam is an effective compressible inclusion as compared to other inclusion used in civil engineering construction because of its multiple functions. Geofoam is a better and cost effective alternative for retaining structure to withstand larger lateral earth pressure.

Pengpeng Ni*et al.* 2012[1] had conducted the experimental analysis on rigid retaining walls with compressible geofoam inclusion considering surcharge load.It was found that the lateral pressure decreased as thickness of geofoam increased. W. R. Azzam*et al.* 2017[2]had conducted the experiment and numerical analysis on axially loaded - piled retaining wall in sandy soil. It was found that as the penetration depth ratio increased, the ultimate capacity of the piled wall also increased. Mustafa *et al.* 2017[3] had conducted the numerical analysis on performanceof rigid retaining walls using EPS geofoam inclusion. It was found that reduction in lateral earth pressure behind the retaining wall was approximately 50 %.V. B. Chauhan*et al.* 2017[4] had conducted the numerical analysis on the assessment of lateral earth pressure reduction using EPS geofoam. It was found that provision of geofoam behind the retaining wall provided a thrust reduction in range of 8-42% for surcharge pressures ranging from 10-50 kPa. Vinil Kumar Gade*et al.* 2018[5] had conducted the experimental study on short and long term performance of EPS geofoam to reduce static and traffic loading induced earth pressure on non – yielding rigid retaining wall. From their study, it was concluded that Percentage of reduction of total lateral thrust on retaining wall increases as the thickness increases.

Materials and Methods

The main objective of the research was to investigate the reduction of lateral earth pressure on piled retaining wall using geofoam inclusion in PLAXIS 2D. The length and diameter was constant throughout the analysis. The analysis was carried out for with and without geofoam inclusion and lateral earth pressure on pile retaining wall was determined. The various parameters those were considered for analysis are various thickness of geofoam and surcharge load. The analyses were carried out for two cases viz.,geofoamprovided upto pile bottom and upto ground level.

Results and Concluding Remarks

Analyses were carried by varying thickness of geofoam and surcharge load. Total earth pressure was found from earth pressure intensity -pile length curve. The variation of total pressure acting on pile retaining wall with respect to Surcharge load is shown in Figure 1. The Percentage of reduction in total earth pressure due to geofoam inclusion for various surcharge load and thickness of geofoamare shown in

Figure 2. Each analysis was performed for two cases viz. geofoam provided upto pile bottom and geofoamprovided upto ground level.

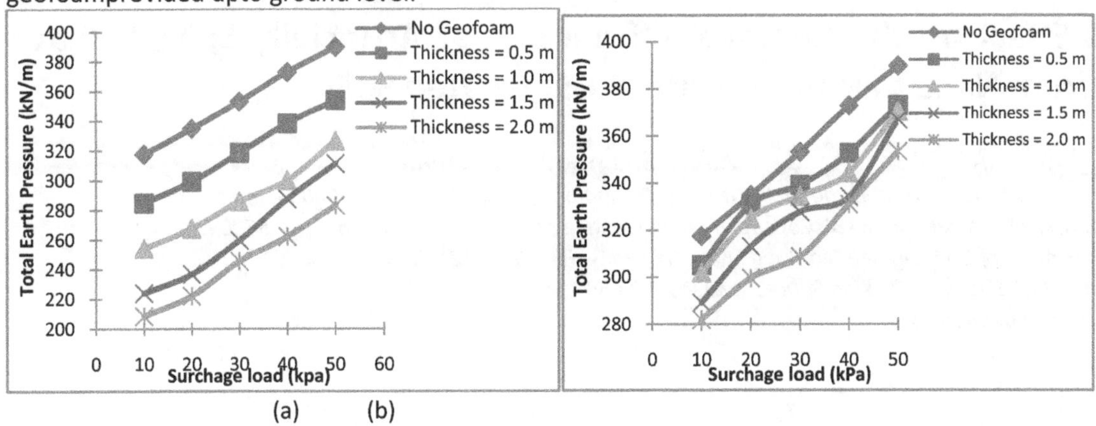

(a) (b)

Figure 1. Variation of total pressure acting on pile retaining wall with respect to Surcharge load (a)Geofoamupto bottom of pile retaining wall, (b) Geofoamupto ground level of pile retaining wall

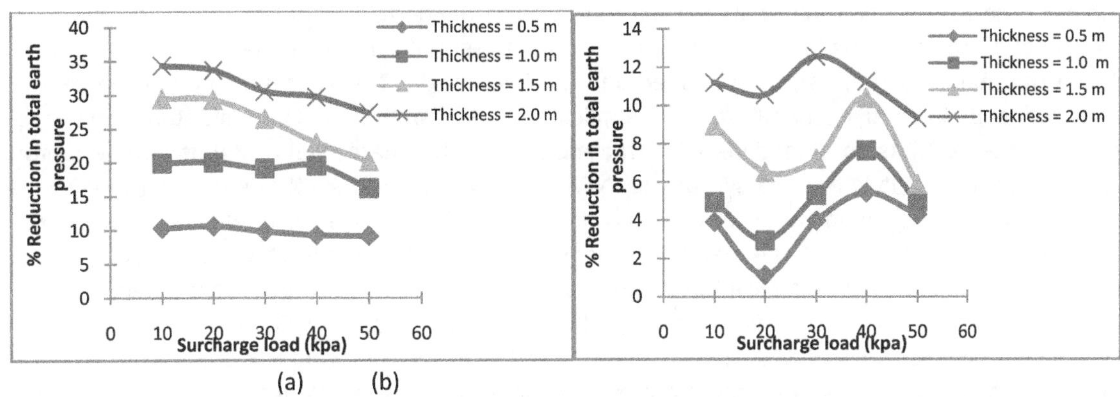

(a) (b)

Figure2. Percentage of reduction in total earth pressure due to geofoam inclusion for various surcharge load and thickness of geofoam(a)Geofoamupto bottom of pile retaining wall, (b) Geofoamupto ground level of pile retaining wall

Based on the results of the present study, the following conclusions are drawn:

1. The percentage reduction in lateral earth pressure is higher when geofoam is provided upto pile bottom as compare to geofoamupto ground level.
2. Lateral earth pressure on piled retaining wall decreases as the thickness of geofoam increases.
3. In case of piled retaining wall with surcharge load condition and geofoamupto pile bottom, percentage reduction in lateral earth pressure is in the range of 9.28% to 34.36% when thickness of geofoam varies from 0.5 m to 2 m.
4. In case of piled retaining wall with surcharge load condition and geofoamupto ground level, percentage reduction in lateral earth pressure is in the range of 1.13% to 12.56% when thickness of geofoam varies from 0.5 m to 2 m.

References

1. Pengpeng Ni, Guoxiong Mei, and Yanlin Zhao (2016) "Displacement - Dependent Earth Pressures on Rigid Retaining Walls with Compressible Geofoam Inclusions: Physical Modeling and Analytical Solutions." *International Journal of Geomechnics*, ASCE, ISSN 1532-3641.
2. W. R. Azzam and A. Z. Elwakil (2017). "Performance of Axially Loaded-Piled Retaining Wall: Experimental and Numerical Analysis." *International Journal of Geomechnics*, ASCE, ISSN 1532-3641.
3. Rashid Mustafa,WasimAkram, SahzadaAman, Mohd.Asif (2017). "Reduction of lateral earth pressure on rigid retaining walls using EPSgeofoam inclusions."*(ICETETSM-2017) IIMT College of Engineering*, Greater Noida, India.
4. V. B. Chauhan& S. M. Dasaka (2017). "Assessment of Lateral Earth Pressure Reduction Using EPS Geofoam-A Numerical Study" *Conference on Numerical Modeling in Geomechanics.*
5. Vinil Kumar Gade, Satyanarayana Murthy Dasaka (2018). "Effect of Long-Term Performance of EPS Geofoam on Lateral Earth Pressures on Retaining Walls." *Springer Nature Singapure Pte Ltd.*

Chelant Effect on the Removal Efficiencies of Artificially Spiked Heavy Metals from Nano Calcium Silicate (NCS) Treated Soils

A.A.B. Moghal [1], S.A.S. Mohammed [2*], M.A.A. Shamrani [3], A. Almajed[4]

[1] *Department of Civil Engineering, National Institute of Technology (NIT), Warangal - 506004, India.*

[2]*Department of Civil Engineering, HKBK College of Engineering ,#22/1,Nagawara, Bengaluru- 560045, India.*

[3]*BRCES, Department of Civil Engineering, King Saud University, Riyadh 11421, KSA.*

[4]*Department of Civil Engineering, King Saud University, Riyadh 11421, KSA.*

** e-mail: abubms@gmail.com*

Introduction

In the past few decades, sorption of heavy metals on different media has become a subject of interest and a major challenge to researchers. Nano technology has now being used in most branches of science and technology. The use of nano materials for soil remediation needs to be studied. Based on the best available techniques for cleaning polluted soils laced with potentially risky elements, immobilizing is thought to be the major economical, hard headed and feasible option. This redressal method decreases pollutant mobility and bio-catalytic mainly with the application of soil amendments. In this study the use of nano calcium silicate (NCS) as an amendment for treatment of contaminated fields and as a material for landfill liners is proposed. It is easy to synthesize, NCS using combustion synthesis method in a short duration compared to other nano metal oxides, and it does not involve use of physical or mechanical techniques. This work aimed to test the stability of nano calcium silicate (NCS) stabilized soils to retain sorbed heavy metals under harsh chemical environments. This study dealt to target proper stabilizer (for each metal ions) in minimizing the removal efficiency. The use of chelant such as EDTA, a weak acid acetic and a strong acid nitric acid and double distilled water are used as desorbing agents, to study which type of stabilizer (and at what dosage) gives maximum removal efficiency and also to evaluate a stabilizer giving the least removal efficiency with aggressive chelant (higher molar), and this can be judged as the best one to treat the contaminated soils in the field (Mallampati et al. 2011).

Materials and Methods

Soil samples were gathered from Al-Ghat which is a kaolinitic soil (a town located 270 km northwest of Riyadh (26° 32' 42'' N, 43° 45' 42'' E)) denoted as soil A and Al-Qatif which is a montmorillonitic soil (a historical coastal oasis area situated on the western shore of the Persian Gulf in the Eastern Province of Saudi Arabia (26° 56⍰ 0⍰ N, 50° 1⍰ 0⍰ E)) denoted as soil B in this paper. NCS was synthesised using combustion synthesis and used in different percentages by weight of soil and denoted as X_1%, X_2%, X_3% having concentration ambit of 0.5, 1.0 and 1.5% respectively. Nitrate salts of copper and zinc were taken as the contaminants with load ratio of 10 and 250 mg/Kg. Four chelating agents were selected for desorption tests two were ligand based agents and the other two were oxidising agents in the form of a strong and weak acids, namely Ethylene di amine tetra acetic acid (EDTA), Di ethylene tri amine penta acetic acid (DTPA), Nitric acid (HNO_3), relying on four molar concentrations (0.01, 0.1, 0.5 and 1.0 M) and a solid to liquid ratio of 1:20, the metal(s) spiked soil was subjected to desorption testing using different chelants. Apart from these four chelants, deionised double distilled water was also used for desorption studies. The percentage removal efficiency was calculated using the following equation

$$\text{Removal efficiency (\%)} = \frac{\text{Contaminant mass in supernatant } (C_L V_L)}{\text{Initial contaminant mass in soil } (C_S M_S)} X\ 100 \quad (1)$$

where C_L and C_S are the concentration of a contaminant in the supernatant in mg/L and soil in mg/Kg respectively, V_L is the volume of supernatant in Litres, and M_S is the dry mass of the soil in Kg.

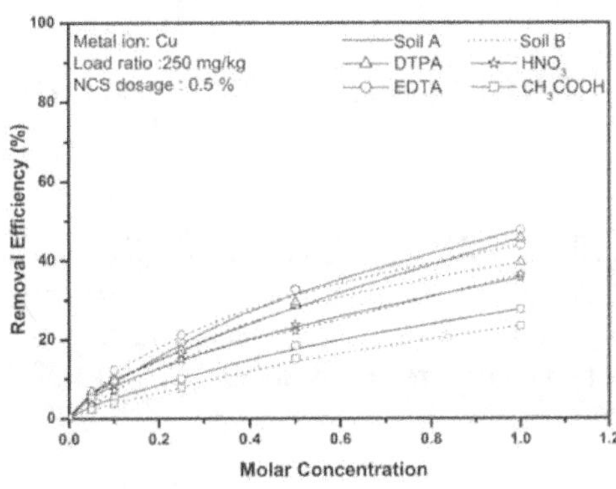

Figure 1. Effect of extracting agents on removal efficiencies if Cu ions on soils.

Results and Concluding Remarks

After conducting exhaustive desorption experiments, it was found that for all the mixtures the removal efficiency for soils was more than 70% and EDTA with 1M concentration gave the maximum removal efficiency. It was found that cationic metal ions were desorbed effectively by EDTA. The order of desorption for metal ions was Zn>Cu. Soils amended with NCS decreased the removal efficiency and encapsulated the metal ions in the interstices of soil NCS mixture and made them immobile and inert, due to changes in redox potential, variation in atomic radius. The solubility product of Cu and Zn played a role in formation of precipitates and chelates. It can be observed in Figure 1 and Table 1 that with the addition of 1.5% NCS the removal efficiency decreased to approximately 40% for both the elements which proves that encapsulation of Cu and Zn has occurred effectively. The following conclusions can be drawn,

1. NCS amended soil retained heavy metals effectively than with only soil, and it may be because of the comportment of colloids in the soil NCS metal solution phase. The dissolution of carbonates leads to considerable widening of pores resulting in higher sorption levels for the 2 metal ions considered in the study.
2. There is an enhancement in immobilization capacity with NCS due to adsorption, entrapment, enclosing/ binding with Calcium and Silica associated salts which are immobile, and are adsorbed by soil coordination process.
3. NCS produces salts which are immobile and with the presence of moisture and CO_2 available in plenty from the surrounding environment, and these have pozzalonic and hydraulic properties similar to cement. This gives an advantage of sprinkling NCS on brown soils without adding water.

Parameters				Maximum Removal Efficiency (%)							
Metal Ion	Extracting Agent	Concentration (Mole)	Load Ratios (mg/kg)	0% NCS		0.5% NCS		1.0 % NCS		1.5% NCS	
				Soil A	Soil B	Soil A	Soil B	Soil A	Soil B	Soil A	Soil B
Cu	EDTA	1	10	82.54	69	71.52	63.4	64.54	51.5	56.47	51.91
		1	250	71.17	71	47.65	43.87	58.4	45.9	39.94	42.19
Zn	EDTA	1	10	88.57	83.67	73.25	65.68	68.54	56.93	61.25	56.19
		1	250	76.54	75.68	54.8	52.26	50.14	44.57	47.94	47.28

Table 1. Maximum removal efficiency of metal ion using different extractants for soils.

Acknowledgments: Funding for this project was provided by the National Plan for Science, Technology and Innovation (MAARIFAH), King Abdul Aziz City for Science and Technology, Kingdom of Saudi Arabia, Award Number (12ENV2583-02), the authors are grateful for this.

References

Mallampati S. R, Mitoma Y, Okuda T, Simion C, Lee B. K (2015) Solvent-free synthesis and application of nano-Fe/Ca/CaO/[PO4]composite for dual separation and immobilization of stable and radioactive cesium in contaminated soils Journal of Hazardous Materials 297(75), 74–82 dx.doi.org/10.1016/j.jhazmat.2015.04.07.

Effect of Enzyme in binding soil grains to improve its geotechnical behaviour

M.A. Lateef [1], S.A.S. Mohammed [1*], A. Almajed[2], A.A.B. Moghal [3]
[1] *Department of Civil Engineering, HKBK College of Engineering ,#22/1,Nagawara, Bengaluru- 560045, India.*
[2]*Department of Civil Engineering, King Saud University, Riyadh 11421, KSA.*
[3] *Department of Civil Engineering, National Institute of Technology (NIT), Warangal - 506004, India.*
**email: abubms@gmail.com*

Introduction

Soil stabilization has become widely adopted process for improving the soil strength and other geotechnical properties. Development of techniques to improvise the soil strength has been researched and implemented to identify their suitability and effectiveness. Use of Enzymes in stabilization of soils has been in practice since recent past. Enzyme induced calcite precipitation (EICP) is a biologically based ground improvement process wherein a treatment solution comprised of urea, calcium chloride, and free urease enzyme is added to the soil. These reagents combine in a reaction sequence comprised of the hydrolysis of urea followed by the precipitation of calcium carbonate, as shown in Equations 1 and 2 below.

$$CO(NH_2)_{2(}aq) + 2H_2O \text{ Urease} \rightarrow CO_3^{2-} + 2NH_4^+ \quad (1)$$

$$Ca^{2+} + CO_3^{2-} \rightarrow CaCO_3 \text{ (s)} \quad (2)$$

The precipitated calcium carbonate may improve the strength of the soil by binding the soil particles together, filling pore spaces, and increasing the roughness of the soil particles, resulting in an increase in the shear strength, stiffness and dilatancy characteristics of soil.

Present study is carried out on two soils (Red soil and Black cotton soil) and their performance on strength characteristics were studied with enzyme dosages. Enzymes are used to boost the reaction process in the soil stabilization. And non-fat milk powder is also used as an additive to test its impact on the soil strength.

Materials and Methods

Soils were collected from two different parts of Karnataka in present study. After the basic tests of the soil were conducted the soil was majorly tested for its unconfined compressive strength (UCS) and the soil column tests were also conducted.

Soil samples were amended with the following ingredients with specific dosage dissolved in deionized water.

- Urea (NH_2 CO NH_2) Molecular weight 60.06 g/mol
- Calcium Chloride ($CaCl_2$ $2H_2O$) Molecular weight 147.02 g/mol
- Non-fat Milk Powder
- Urease Enzyme {Canavalia ensiformis (Jack bean) Type III, powder, 15000-50000 units/g solid}

Both soils (Red and Black cotton soil) were fed with three different enzyme solutions as mentioned below

Solution 1 - 1.0 M Urea+0.67 M Calcium Chloride+3g/l Urease Enzyme
Solution 2 - 1.0 M Urea+0.67 M Calcium Chloride+3g/l Urease Enzyme+4g/l non-fat milk powder
Solution 3 - 0.37 M Urea+0.25 M Calcium chloride+0.85g/l Enzyme+4g/l non-fat milk powder

UCS tests were conducted on samples of the soils prepared using Solution 1, Solution 2 and Solution 3 after a curing period of 7, 14 and 21 days. Soil column tests were also conducted by preparing a soil column with a diameter of 60 mm and a height of 120 mm. Firstly the desired weight of oven dried soil was mixed with deionized water equivalent to optimum moisture content by weighing of soil and was gently placed in the soil column made of polythene material, after the soil was filled in the columns solution 1, solution 2 and solution 3 were poured on the surface of the soil column and were made to percolate in the soil for a duration of 7 days, once the soil absorbed the solutions, bottom of the soil columns were pricked with needles to drain the excess liquid and once the liquid got drained, the soil samples were removed and were subjected to surface drying. The dried samples were then tested further.

Results and Concluding Remarks

It was observed that solution 2 depicted better performance in improving the strength characteristics when UCS tests were performed on Red soil and black soil. Solution 3 showed maximum results for red soil when soil columns were tested whereas solution 1 improved the strength of black cotton soil to the maximum. The increase in strength of the UCS value for red soil treated with solution 2 and cured for 14 days when compared to the raw red soil was found to increase from 0.07 MPa to 0.44 MPa. And when black cotton soil sample was treated with solution 2 and cured for 21 days UCS value increased to 0.22 MPa when compared to raw black cotton soil which gave strength of 0.013 MPa.

Further SEM and EDAX data of the treated soil samples gave a better understanding of $CaCO_3$ precipitation. It was found that with the addition of enzyme there was an increase of 47% in calcium ion concentration as per EDAX. Similarly soil digestion studies gave a value of 4-21% increase in concentration of calcium ions. Thus it proves that calcium precipitation has occurred. It can also be concluded that use of enzymes in stabilization makes the soil develop strength and overall effect of each solution i.e. Solution 1, 2 and 3 were found to perform better when compared to the untreated soil. Precipitation of $CaCO_3$ was better due to the use of non-fat milk powder in the solution. UCS test results can be observed in Figure 1 which gives an understanding that the use of solution 2 depicts gain in strength for red soil (refer Solution2R) as well as for black cotton soil (refer Solution2B).As per Almajed et al. 2018, the mechanism of strength formation in EICP-treated specimens is not governed only by the amount of precipitation but is also influenced by the method of preparation and particle packing. Hence percolation method seems to have enhanced the strength as the enzyme and other constituents had ample time to undergo reaction to form precipitates and provided sufficient inter particle binding throughout the column. Initially the UCS testing gave sufficient increase in strength but ultimately there was a decrease as can be seen in figure 1 after 21 days and was likely due to flushing of organic matter and ammonium chloride salt precipitates from the treated soil. Hence this suggests that always EICP treated specimens should be thoroughly rinsed with water prior to measuring its mechanical properties.

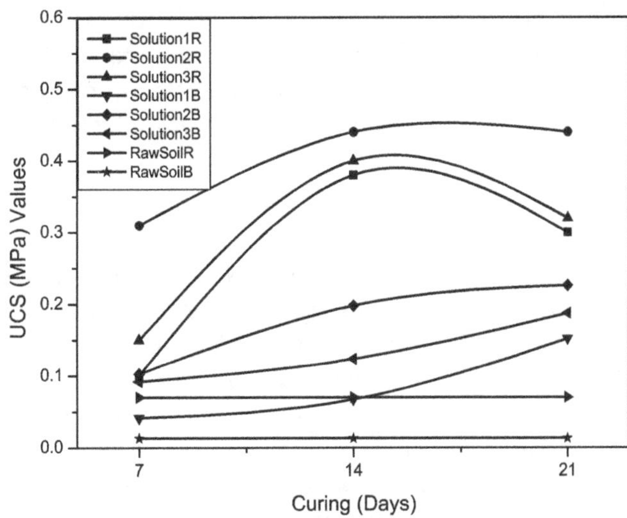

Figure 1. Variation of UCS Values for soils treated with Solution 1, 2 and 3

References

Almajed A, Hamed K T, Edward K (2018) Baseline Investigation on Enzyme-Induced Calcium Carbonate Precipitation. Journal of Geotechnical and Geoenvironmental Engineering, 144(11), 0401808 1–11. https://doi.org/10.1061/(ASCE)GT.1943-5606.0001973

Impact of Sewage water intrusion on Bearing Capacity and Dewatering by ElectroKinetic Remediation

M V PARIKSHITH[1], C S DARSHAN[1], K RAJATH[1]
[1]*Civil Engineering Department, The National Institute Of Engineering, Mysuru, India*
** e-mail: parikshith.mv@gmail.com*

Introduction

One of the stark realities of 21st century is that the demands for soil usage are increasing by several billions, while the uncontaminated soil available to meet the demands is insufficient due to contamination of soil by industrialization and urbanization activities. There are many remediation techniques such as bioremediation, Phytoremediation, thermal and physical treatments. Out of which Electrokinetic method is rapid, less expensive and does not produce any noise. Many researchers have carried studies on all remediation methods including the ElectroKinetic remediation (EKR) by involving the construction of practical model. Al-Hamaiedh et al. 2011 studied the effect of oil contamination on soil properties and the feasibility of EKR to treat the oil contaminated soil by passing 1-4 mA of current using electro osmotic cell and obtained an oil removal efficiency of 30-50%. The objectives developed from a few surveys of specialist's works are as per the following, to determine the basic geotechnical properties of the soil sample taken for study which characterize the soil sample and to study the changes in soil shear property due to the intrusion of sewage water. To remediate the polluted soil by EKR method by developing lab scale model and determining the efficiency of this method by comparing the polluted soil properties with treated soil properties and to see whether the method is feasible without any negative effects. Impact of pollutant on bearing capacity of soil is also illustrated according to IS code by considering an instance.

Electrokinetic remediation method is still in assessment in all research facilities since soil pollutant interaction cannot be known effectively. ElectroKinetic Remediation is a process in which a low-voltage direct current electric field is applied across section of contaminated soil by introducing electrodes into the soil to move the contaminants. The principle of Electro kinetic remediation is similar to a battery. After electrodes are introduced and connected to a source of electric current, molecules dissociate into ions and moves towards oppositely charged electrodes. This dissociated ion attracts the opposite charged end of dipole free molecule and thus this method also acts as dewatering method. This technique requires external processing fluid as medium for conduction of current through the soil, but the pollutant chosen for this study itself act as conducting medium.

Materials and Methods

The EKS model apparatus consists of tank made of non-conductive material which is of acrylic glass to prevent short circuiting. The dimensions of the tank are 0.4mx0.3mx0.2m and thin copper plates are used as electrodes. A 12V battery is used as source for applying low voltage direct current for treatment of contaminated soil. The tank's one face is made detachable for easy removal of soil after each trail and its one half of the base plate is provided with holes near cathode for the drainage of waste water.
These are the following steps involved in the study:

- Polluted area is investigated and the spread of the contaminant (sewage water) in that area is sighted. Pure Soil sample is collected from a certain distance (90ft) apart from the polluted place to avoid pollutant. Top soil is not considered for the experimentation, it is removed to certain extent 30cm which is obtained from literature review.
- Both insitu density and field moisture content are determined in accordance with IS 2720 to stimulate the field condition in the laboratory during the different test trails.
- Basic geotechnical properties of the soil sample is determined and shear parameters of the uncontaminated soil sample is determined by stimulating the field condition.

- Sewage water sampling is done from sewage farm after screening process to avoid floating particles. Storage measures have been taken to prevent sewage water from photo degradation using plastic containers to store sewage water. BOD and COD of sewage water are analysed.
- Soil sample is contaminated with different proportions of sewage water (10%, 15%, 20% and 25%) is tested for shear parameters which are compared with the shear parameters of uncontaminated soil and the 25% of sewage water is considered for further experimentation because the area primarily investigated has spread of sewage water of same weightage and more percentage leads to more water content which serves as conducting medium for current.
- Preparation of lab scale model and treating the manually polluted soil (25% sewage water) which is compacted to field density by introducing electrodes and completing the circuit connections. The polluted soil sample is treated for different electrode spacing (10cm and 15cm). Shear parameters of treated soil samples is determined and compared it with shear parameters of uncontaminated and field condition soil samples.
- The direct current and voltage is sensed by using multimeter at both the terminals to ensure the working of EKR. Water content near two electrodes is found after 24hrs for both the spacing of electrodes to know the flow percentage of water from cathode to anode and to theoretically justify the shear parameters changes. Effect of contaminant on soil bearing capacity of soil is determined by calculating bearing capacity for 2m x 2m square footing for all conditions using IS-6403 (1981) and efficiency of EKR is determined.

Results and Concluding Remarks

Analysis of basic geotechnical properties of soil shows that soil sample collected has sand fraction of 70% and fines of 20%. The value of coefficient of uniformity and coefficient of curvature obtained is 32.8 and 1.58, shows that it is well graded non uniform soil. The specific gravity value of 2.64 also confirms the large fraction of sand particles. The permeability value of field density compacted soil obtained by passing sewage water is in the order of 10^{-3} which urges the removal of pollutant immediately, because in the real scenario it might reaches the ground water. The shear parameters for field condition was 28.5 kpa cohesion (c) and 38.6^0 angle of friction (Ø) gets reduced to 0 kpa (c) and 30.3^0 (Ø) on contamination with sewage water (25% by weight of soil). As sewage water is added, it forms a thin film around soil particles which reduces cohesion (c) and enhances lubrication which in turn reduces the value of (Ø). The remediation increases c and Ø value to 14.3 kpa and 34.2^0 for 15 cm spacing and 11.3 kpa and 32.6^0 for 20cm spacing respectively. The difference in the treatment extent for different spacing between electrodes is because as the length to be travelled by current increases, resistance also increases which causes reduction in the effect of current. The difference in water content of soil determined near two electrodes will justify the reduction of c-Ø values after contamination. Efficiency of this method calculated based on the bearing capacity values which improved from the value 330.4 kN/m^2 to 821.2 kN/m^2 is 59% for 0.3 amp current range. Hence this method is feasible for removal of pollutant by dissociating the water molecule into ions which moves towards oppositely charged electrodes. This dissociated ion attracts the opposite charged end of dipole water molecule and serves as dewatering method in this study helps to improve bearing capacity which plays a major role in footing designs. Efficiency of the method can be increased by supplying current of range more than 1 milliampere/cm^2 and by decreasing the spacing between electrodes appropriately.

References

Al-Hamaiedh, Husam Damen, Omer Nawaf Maaitah (2011) Treatment of oil polluted soil using electrochemical method Alexandria Engineering Journal 50.1: 105-110.

IS-6403 (1981), Code of Practice for Determination of Bearing Capacity of Shallow Foundations (First revision). Bureau of Indian Standards, New Delhi.

SP 36-Part 2(1988), Compendium of Indian standards on Soil Engineering, Bureau of Indian standards. New Delhi.

Punmia BCe, Ashok Kumar Jain (2005) Soil mechanics and foundations. Laxmi Publications (P) LTD, Delhi.

Effectiveness of copper slag blended with lime on the strength of fine-grained subgrade soil

G.R. Harish[1*], T.N.Rajendra[1]
[1] *Department of Civil Engineering, New Horizon College of Engineering, Bangalore, India*
** e-mail:gr.harish1@gmail.com*

Introduction

Weak soils are causing severe damages to the structures such as buildings and pavements built over them. If the pavement is constructed over weak/soft subgrades, it will undergo large deformation owing to lower load carrying capacity of weak soils and severe volume change owing to swell – shrink behaviour of soft soils. This phenomenon will result in rutting mode of premature failure requiring immediate strengthening or reconstruction of the whole structure. Thus, for safe design such weak/soft soils need to be improved before construction. Stabilization is one of the conventional and widely used method to strengthen the weak subgrade soils. The soil behaves differently when it is added with other materials such as lime, fly ash, cement, etc. Several researchers have investigated the strength characteristics of stabilized soil with various stabilizers (Athanasopoulou 2014, Ahmed et al. 2013, Arino and Mobasher 1999, Gupta et al. 2012, Goraj et al. 2003, Ramdas et al. 2011,) Therefore, it is important to understand the mechanism involved with the enhanced behaviour of the soil under stabilized condition.

In this study, an attempt is made is to understand the strength characteristics of fine-grained soil when stabilized with copper slag and lime through laboratory study.

Materials and Methods

Fine grained soil was collected from Jigani, Bangalore, Karnataka. The soil was tested for its engineering properties and all the test were carried out as per relevant IS codes. The soil consists of 5% sand, 53% silt and 42% of clay. The specific gravity of soil is 2.66 and it possesses a plasticity index of 18%. The soil is classified as A-7-6 as per HRB classification system. The maximum dry unit weight of the soil range from 16.18 to 15.06kN/m^3 and optimum moisture content range from 19.2% to 22% respectively under standard and modified Proctor condition. The soil possesses an unconfined compressive strength of 133kPa under unsoaked condition and the value reduces to 50kPa under soaked condition. It possesses a free swell index of 50% and is highly expansive in nature. The soil possesses a CBR of 4% under unsoaked condition and upon soaking for 96 hours, the value reduced to less than 2%. Based on the MoRT&H (2013) requirements of subgrade, the selected soil is considered as weak soil and needs improvement if it is intended to be used in the pavement construction. For improving the selected soil, copper slag and lime were considered.

The lime fixation was carried out through pH test. The optimum lime percentage was found out to be 6%. The copper slag content was varied up to 20% with an increment of 5%. Compaction characteristics of the soil mixed with 6% lime and varying percentages of copper slag was investigated. A series of unconfined compressive strength and CBR tests were carried out in the laboratory under unsoaked and soaked conditions. The stabilized UCS samples were cured for varying curing periods (3, 7, 14 and 28 days). CBR test specimens were prepared with 6% of lime and optimum copper slag content and were cured for 7 days before testing. A minimum of three samples were tested for each test condition.

Results and Concluding Remarks

From the study it was observed that there is no substantial change in the compaction characteristics. Table 1 and 2 shows the variation of unconfined compressive strength of fine grained soil treated with constant lime and varying percentage of copper slag for different curing period under unsoaked and soaked conditions respectively. The unconfined compressive strength was found to increase with the addition of copper slag (i.e., 5%) and further addition of copper slag resulted in reduced compressive strength. The compressive strength of the soil increased by 1.9, 3.7, 6.7 and 9.3 times in comparison with the unsoaked strength under unstabilized condition. Similarly, the compressive strength of the soil increased by 4.2, 8.2, 15.7 and 19.7 times in comparison with the soaked strength under unstabilized condition.

The CBR value of the soil increased from 4% (unstabilized) to 14% (stabilized) and less than 2% (unstabilized) to 9% (stabilized) respectively under unsoaked and soaked conditions.

Table 1. UCS of copper slag Stabilized Silty soil and 6% lime (unsoaked condition)

Curing period (days)	UCS value in unsoaked condition (kPa)				
	Fly ash content (%)				
	0	5	10	15	20
0	133	258	195	180	160
7	133	489	340	260	210
14	133	891	560	440	390
28	133	1241	780	560	480

Table2. UCS of copper slag Stabilized Silty soil and 6% lime (soaked condition)

Curing period (days)	UCS value in unsoaked condition (kPa)				
	Fly ash content (%)				
	0	5	10	15	20
0	50	210	150	130	130
7	50	410	290	210	210
14	50	786	460	380	350
28	50	986	680	500	480

Based on the test results, the following conclusions are drawn.

- The addition of 5% copper slag with 6% lime yields maximum strength enhancement of the selected soil.
- The unconfined compressive strength of the soil increase with increased curing period and the improvement range from 1.9 to 9.3 and 4.2 to 19.7 times respectively under unsoaked and soaked conditions.
- CBR of the soil increased by 3.5 and 6 times respectively under unsoaked and soaked conditions resulting in reduced pavement thickness, improved durability and long lasting pavements.

References

Athanasopoulou A (2014) Addition of lime and fly ash to improve highway subgrade soils. Journal of Materials in Civil Engineering 26(4): 773-775.

Arino AM, Mobasher B (1999) Effect of ground copper slag on strength and toughness of cementitious mixes. ACI Materials Journal 96(1):68–73.

Bulbul Ahmed, Md. Abdul Alim and Md. Abu Sayeed (2013) Improvements of soil strength using cement and lime admixtures. Earth Science 2(6): 139-144.

Gorai B, Jana RK, Premchand (2003) Characteristics and utilization of copper slag - A review. Resources Conservation and Recycling, 39: 299–313.

Gupta RC, Blessen Skariah Thomas, Prachi Gupta, Lintu Rajan, Dayanand Thagriya (2012) An experimental study of clayey soil stabilized by copper slag. International Journal of Structural and Civil Engineering Research, 1(1): 110 – 119.

TL Ramdas, N Darga Kumar, G Yesuratnam (2011) Geotechnical characteristics of three expansive soils treated with lime and fly ash. International Journal of Earth Sciences and Engineering, 4: 46-49.

Enhancement of sub grade soil strength using activated fly ash and fiber reinforcement

A. Malathi Narra[1], B.Monica Seles[2]
[1] *Assistant Professor,Civil Engineering Department,VR Siddhrtha engineering college,vijayawada*
[2] *M.tech Student, Geotechnical Engineering,VR Siddhartha Engineering College,vijayawada*
malathin76@gmail.com, monikaseles064@gmail.com

ABSTRACT
Black cotton soil deposits cover twenty percent area in India and are associated with geotechnical problems such as low strength and excessive differential settlement. Construction of pavements or any other engineering structures on this type of soil affects the performance due to their inferior characteristics whenever comes in contact with water. In view of this, the present investigation aimed at evaluating the individual and combined effects of fly ash and randomly oriented fiber on the behaviour of soil. Different proportions of fly ash were added to the soil in the presence of 10M of potassium hydroxide as alkaline activator and 0.5% of synthetic fiber as reinforcement. A series of unconfined compressive strength tests, California bearing ratio tests and Oedometer tests were performed on natural as well as alkali activated fly ash treated soils to investigate the engineering behaviour. Results showed significant enhancement of strength and reduction in swelling pressure upon activation.

1. INTRODUCTION
Black cotton soils are made generally of clay minerals of the Montmorallonite group. Montmorallonite mineral is shaped by an octahedral sheet is stuck between two tetrahedral sheets to form a three-layer element. Many of these elements are stacked one above the other to form a particle. These soils are swell, get smooth nature and lose their strength whenever it contacts with water, and shrinks become brittle nature when it dries, these properties rise difficult issues while constructions. Stabilization of expansive soils without revealing them from construction area is an effective technique for road development. Soil stabilization is changing of geotechnical or physical properties of soil by adding additives. Different types of additives like fly ash, lime cement have been used to improve the quality of soils; Thermal power plants are the main source of fly ash. In India 100 tonnes of fly ash produced from thermal power plants due to coal combustion for power generation per year .The by product coal combustion is fly ash and it is a waste material. Fly ash is significantly uses for soil stabilization and number of researches have been conducted by investigators and analyze the feasibility of fly ash on soil properties (**Error! Reference source not found.-Error! Reference source not found.**).

Reinforcement is one of the method for improve of sub grade soil. Kumar et al. (2006) studied that the effectiveness of crimped and plain polyester fibers inclusion on compressive strength of clay. The results demonstrate that compression strength varied with the addition of various percentages of fiber. Kumar et al. (2016) contemplated that increasing of fiber content up to 0.25% decreased the swell pressure. Kumar et al. (2013) reported that fly ash fiber reinforcement increased the strength of soil (Choudhary 2014; Moghal et al. 2017; Sudhakaran and Sharma 2018). In this present study synthetic fibers are used as reinforcement. Blending of fiber and fly ash to the soil slightly improves the load carrying capacity of sub grade. This may not sufficient for black cotton soil for sub grade application. For more improvement of strength and decrease in swelling there must be some need for bond between additives and soil.

For bonding between fly ash, fiber and soils, commonly cement and lime are have long been utilized as primary binders. The main disadvantage of cement and lime stabilization, when they used as additives in soil stabilization they react with soil and produces carbon dioxide in to the environments.

2. MATERIALS

2.1 Soil:-

Soil is collected from construction site at a depth of 1m near Kankipadu region; Krishna district Andhra Pradesh. The collected soil is dried and pulverized for conducting physical properties tests as per the required sieve sizes of a test. These tests are conducted as per the ASTM; Physical properties.

2.2 Fly ash:-

Fly ash is procured from VTPS thermal power plant Ibrahimpatnam near Vijayawada, Andhra Pradesh. Fly ash is the finely divided residue that obtained from the combustion of pulverized coal from thermal power plants and other industries. Presently, over 20 million metric tons (22 million tons) of fly ash are used annually in a diffusion of engineering applications.

2.3 Polyester fibers:-

Polyester fibers are manmade fibers. These are manufactured by using chemical synthesis; ester is the main group in their main chain. In this investigation polyester fiber are used as reinforcement. These are alkaline, acid and chemical resistant materials.

2.4 Potassium hydroxide

Potassium hydroxide or caustic potash is extremely highly base; they form strong alkaline solution when they dissolve in water. Potassium hydroxide is in pellets form, molecular weight is 56.1056 gm/mol. The alkaline solution is prepared by dissolving the required amount of KOH pellets in to the distilled water and kept the solution for cooling, and the ph value of this alkaline solution is 12.

4. Results and discussions:

Unconfined compression strength increased 95kpa to 200kpa increased with only fly ash up to 30% , CBR values increased from 1.3 % to 4.4% ,by the addition of 0.5% of fiber as reinforcement to the same preposition of fly ash UCS and CBR values increased from 95kpa to 250kpa and 1.3% to 4.4%. However, in case of alkaline activated fly ash and reinforced fiber specimens lead to much greater improvement in soil strength, which had not fly ash and fiber binders. Potassium hydroxide used in this present study for preparing alkaline solution because of its high efficiency alkaline activation.

Mixing of 10M potassium hydroxide alkaline solution in preparation of testing soil samples, improves the cementitious bond between aluminium and silica particles. This resulted in densification and high stiffness of alkaline activated soil. Alkaline activated fly ash and fiber reinforced soil specimens testing results shows that there is high improvement in soil strength. The unconfined compression strength and CBR test results improved up to 800kpa and 30% for 7 days curing, 1200kpa and 84% for 28 days curing.

5. References

1. Kumar, p and singh S.P., 2008 Fiber reinforced fly ash subbases in rural roads. Journal of Transportation Engineering, 134(4), pp.171-180

2. Moghal, A.A.B., Chittoori, B.C and Basha, B.M., 2018. Effect of fibre reinforcement on CBR behaviour of lime-blended expansive soils: reliability approach. Road Materials and Pavement Design, 19(3), PP.690-709

3. Phanikumar, B.R and Sharma, R.S., 2007.Volume change behaviour of fly ash stabilized clays. Journal of materials in Civil Engineering , 19(1),pp.67-74.

4. Phanikumar, B.R and Sharma, R.S., 2004. Effect of fly ash on engineering properties of expansive soils. Journal of Geotechnical Engineering, 130(7), pp.764-767.

Sonic Integrity Tests for Under-reamed piles – A Case Study

K. M. Kouzer[1*], P. G. Dileep Kumar[2]
[1] *Associate Professor, Department of Civil Engineering, Govt. Engineering College Kozhikode, Kerala, India*
[2] *Assistant Professor, Department of Civil Engineering, Govt. Engineering College Kozhikode, Kerala, India*
** e-mail: kouzer@gmail.com*

Introduction

Sonic Integrity Testing (SIT) is widely used measurement method to check the integrity of a concrete pile after installation. Sonic Integrity Testing is based upon the principle of sending a low strain shock wave from the Pile head through the pile shaft and registration of the response. The reflections of the Pile toe and/or change in pile diameter give information on the integrity of the pile shaft. Sonic Integrity testing does not provide any indications of the load bearing capacity of the pile foundations, but only provides information on the condition of the structural integrity of concrete piles. SIT provides quick and inexpensive results compared to core drilling, inspection by excavation, and load tests, which are time consuming and costly. Pile response is displayed immediately on screen and can be printed or stored on hard disk for further analysis. The aim of analyzing SIT measurement is to quantify the size of possible defects or irregularities that have arisen in the pile shaft during installation or driving. Measurements indicate whether the piles will be capable of passing the load on the pile head to the bearing layers.

Practical application for non-destructive integrity testing of deep foundations was first developed in 1970s in Europe. A number of research works carried out on Pile Integrity Testing (Koten and Middendorp 1980; Schaap and Vos 1984; Rausche et al 1992; Brückl et al 2010) and with the development of digital recording and processing of results, advanced data processing became possible and the range of application was greatly extended. Effect of Age on Bored Concrete piles were studied (Brückl et al 2010). Shallow anomalies of pile foundations were detected by Pile Integrity Tests with three dimensional effects (Chow et al 2003) and with the help of tip reflections (Chai and Phoon 2013). Defects of Bored piles were studied by using low strain integrity tests (Thilakasiri 2016). In this paper, a case study of Sonic Integrity Tests is presented to study the defects of under reamed piles.

Materials and Methods

Under reamed piles of a hostel building at Kozhikode was constructed more than 15 years back and the work was stopped due to technical problems. Now it is required to verify the quality of the pile foundations before continuing the further work. Hence it was proposed to conduct Pile Sonic Integrity Tests on all piles to investigate the depth and quality of piles. The bulb diameter is two times the pile diameter. Lengths of the piles are mentioned as hard stratum.

The equipment used for the test is SIT+ manufactured by Profound VB, Netherlands and the velocity signals were analyzed using Profound SIT v7.95 software. The test is conducted based on IS:14893 (2001). When a hammer loads a pile foundation (during driving or testing), a disturbance, modelled as an incident wave, is introduced to the pile. Because of discontinuities in the pile and interactions with the surrounding soil, reflected wave are also introduced, travelling in the opposite direction to the incident wave. The incident waves and reflected wave, and any subsequent reflections, interfere throughout the length of the Pile. Defects such as an inclusion, hole, or crack are reflected in the same direction as the hammer blow. A bulge or increase in cross section is reflected in the opposite direction. Thus in the reflective wave showed SIT testing, one can reveal defects after installation or after driving. Some defects cannot be detected by SIT testing, as small inclusion, local loss of cover and debris etc. The acceleration signal is converted to velocity and presented on screen as a function of time. All results are easily stored for use in reports. The velocity graph of an undamaged pile consists of a very large peak at impact. A smaller peak, corresponding to the pile toe, can often be distinguished. Soil changes will show immediate reflections.

Results and Concluding Remarks

Sonic Integrity tests were conducted on 60 piles having diameters varying from 300 mm (46 Nos.) and 375 mm (14 Nos.). An input value of stress wave velocity is taken as 3800 m/s. The stress wave velocity in a concrete varies between 3600 m/s and 4400 m/s depending on composition and age. After testing the piles, the stress wave velocity is readjusted in range 3600 m/s to 4400 m/s in post-processing stage to obtain the value of wave velocity, which corresponds, with the known length of the piles. SIT signals with adjusted wave velocity and tested length of the piles are recorded for all piles. As concrete is a heterogeneous material, no specific value can be assigned for stress wave velocity through it, though it is assumed to vary in the range mentioned above.

After hitting the Pile head with a hammer, a shock wave will propagate through the concrete upto the toe of the pile. The shock wave will reflect (partially) and will propagate back to the pile head at which it is detected by the sensor. The detection of this reflected hammer blow is called the toe-reflex. The time interval between the initial hammer blow and the toe-reflex is equal to the time required for a shock wave to propagate through the concrete twice (down and up). The toe reflection of a pile is observed as a positive or negative velocity pulse for a sound pile and the pile length is estimated as 8.5m from pile top (Figure 1). The defects at the middle of the piles are noted by observing the positive reflections in between two negative reflections [Figure 1(b)] and observed that there is a reduction in impedance around 5.0m depth below pile top which shows the decrease in cross sectional area of the pile and it may due to necking or presence of less stiff material in piles. The bulb portion of the under-reamed piles can also be verified by analyzing the velocity graphs. The increase in impedance shows increase in cross sectional area of the pile which will indicate the presence of bulb. The location of the bulb is estimated as 7m from pile top for Pile no. 4 [Figure 1(a)], whereas the location of bulb for defective pile could not be identified so clearly [Figure 1(b)]. Based on the observations on all the piles, the approximate length of piles, location of bulb and defects are estimated. However, the size and quality of the bulb portion cannot be ascertained by Sonic Integrity Test and hence additional load tests are required for confirming the axial capacity of piles.

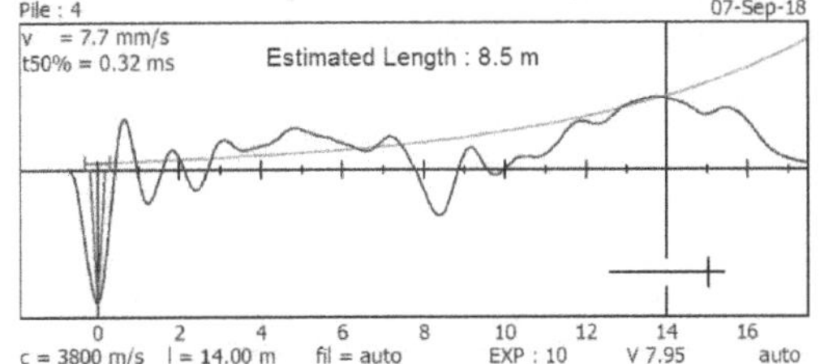

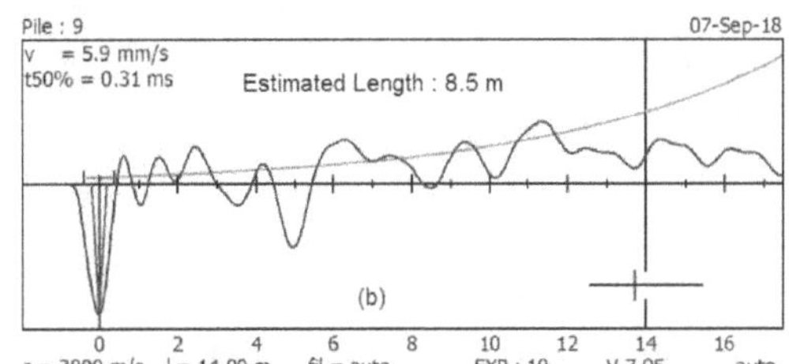

Figure 1. Sonic Integrity Test Results for (a) Sound Pile (b) Defective Pile

References

Brückl J, Wang X T, Wu W (2010) Field tests on effect of concrete age on pile integrity testing, Rivista Italiana Di Geotecnica 1: 27-34.

Chai H Y and Phoon K K (2013) Detection of Shallow Anomalies in Pile Integrity Testing, International Journal of Geomechanics (ASCE) 13(5) : 672-677.

Chow Y K, Phoon K K, Chow W F and Wong K Y (2003) Low strain integrity testing of piles: Three dimensional effects, Journal of Geotechnical and Geoenvironmental Engineering (ASCE), 129(11) :1057–1062.

IS:14893 (2001) Indian Standard Non-Destructive Integrity Testing of Piles (NDT) - Guidelines, Bureau of Indian Standards.

Koten H Van and Middendorp P (1980) Equipment for Integrity Testing and Bearing capacity of piles, International Seminar on the Application of stress-wave theory on piles, Stockholm, Sweden, 69-76.

Rausche F, Likins G E and Shen R K (1992) Pile integrity testing and analysis, Proc. 4th Int. Conf. on the Application of Stress Wave Theory to Piles, Balkema, Rotterdam, Netherlands, 613–617.

Schaap L and Vos J de (1984) The Sonic Pile Test Recorder and Its Application, Application of Stress-Wave Theory on Piles, Proceedings, Second International Conference, Rotterdam.

Thilakasiri H S (2006) Interpretation of Pile Integrity Test (PIT) Results, Annual Transactions of IESL, The Institution of Engineers, Sri Lanka, 78-84.

A STUDY ON THERMAL CONDUCTIVITY OF COARSE AND FINE GRAINED SOILS

Suravarapu Sairam [1], Kadali Srinivas [2]
[1] *Post Graduate Student, Civil Department, VNR Vignana Jyothi Institute of Engineering and Technology, Hyderabad, India, e-mail: s.sairam0106@gmail.com*
[2] *Associate Professor, Civil Department, VNR Vignana Jyothi Institute of Engineering and Technology, Hyderabad, India, srinivasciv@gmail.com (Corresponding author)*

Introduction

Thermal conductivity of soil is an important parameter in assessing the heat transfer phenomena of soils in many of the Geotechnical applications such as ground heat exchangers, foundations of current transfer stations, rocket launching pads, underground tunnels, nuclear wastes dumps and deep repositories. Measurement of thermal conductivity using a single probe is the most widely used method based on transient heat method. Thermal conductivity of soils depends upon density, water content, grain size distribution and mineralogy of soils.

Researchers have worked on the relation between thermal conductivity and density, water content, void ratio, saturation and grain size of the soil and developed theoretical and empirical models to obtain thermal conductivity easily and accurately (De Vries 1963, Kersten 1949, Gangadhara rao and Singh 1999, Zhang 2017). In this study, an empirical model has been developed to predict thermal conductivity from Specific surface area (SSA) and cation exchange capacity (CEC) of soils and also other properties of the soils are considered since properties like density and water content can majorly influence thermal conductivity of the soils.

Materials and Methods

In this study, seven different types of soils are used in which three are sands namely coarse-medium-fine sands designated as CS, MS, and FS respectively and two are commercially available soils sodium bentonite and Kaolin designated as BT and WC respectively and two naturally occurring soils black cotton soil and moorum designated as BC and MR respectively are used for the analysis. Thermal conductivity values of these soils were obtained in the lab by using a single thermal probe (ASTM D 5334-08, Gangadhara rao and Singh 1999) at different water contents and density and also at constant density in dry condition. Regression analysis was performed to develop the correlations between thermal conductivity and density, water content, void ratio, saturation and grain size of the soils. Further analysis was carried out to ascertain the influence of SSA (EGME method, Arnepalli et.al. 2008) and CEC on thermal conductivity of soils. The values of SSA and CEC of the soils used in the study are presented in Table 1

Table 1. SSA and CEC values of the soils used in the study.

Soil	CS	MS	FS	BC	MR	BT	WC
SSA (m^2/g)	0.6	0.9	2.4	188	70	393	13
CEC (meq./100g)	5.65	6.52	6.96	117.83	89.13	133.48	17.39

Results and Concluding Remarks

The variation of thermal conductivity with specific surface area at a constant density of 11.1 KN/m^3 and at 0% water content for sands and other soils used in the study is shown in Figure 1 and Figure 2.

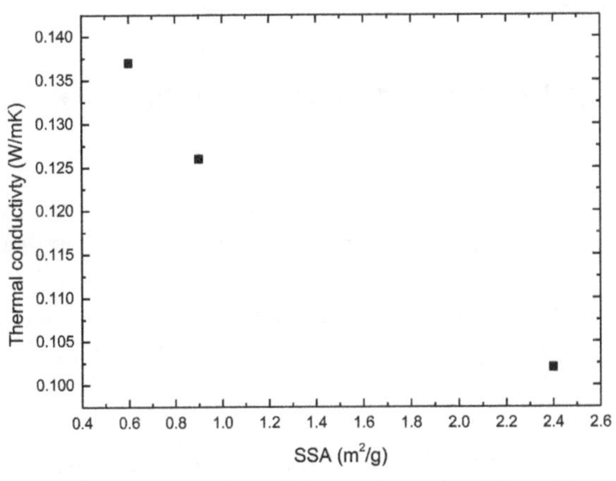

Figure 2. Variation of thermal conductivity with SSA for sands

Figure 2. Variation of thermal conductivity with SSA for fine grained soils

From Figure 1 and Figure 2 it can be observed that with increase in specific surface area thermal conductivity of the soil decreases except for WC due to the presence of passive mineral kaolinite.
The equation to predict thermal conductivity from SSA for sands is given below

$$K = -0.0404(SSA)^2 + 0.1406(SSA) + 0.0322 \quad (1)$$

The equation to predict thermal conductivity from SSA for fine grained soils is given below

$$K = 0.0007(SSA)^3 - 0.01(SSA)^2 + 0.001x + 0.0354 \quad (2)$$

The equation to predict thermal conductivity from CEC for sands is given below

$$K = 0.0268(CEC) - 0.0495 \quad (3)$$

The equation to predict thermal conductivity from CEC for fine grained soils is given below

$$K = -0.0018(CEC)^3 + 0.033(CEC)^2 - 0.0027(CEC) + 0.0802 \quad (4)$$

Where K = Thermal conductivity in W/mK
The above four equations can be used for sands and fine grained soils to determine thermal conductivity values accurately by using SSA and CEC at constant density and water content.

References

1. Arnepalli, D.N., Shanthakumar, S., Hanumantha, Rao, B. and Singh, D.N. (2008). "Comparison of methods for determining specific-surface area of fine-grained soils." Geotech Geol Eng, Vol. 26, pp. 121-132.
2. ASTM International. 2008. D 5334-08 Standard Test Method for Determination of Thermal Conductivity of Soil and Soft Rock by Thermal Needle Probe Procedure. ASTM International.
3. De Vries D. A (1963) Thermal properties of soils. Physics of the plant environment. New York: John Wiley & Sons: 210-235.
4. Gangadhara Rao MVBB, Singh DN (1999) A generalized relationship to estimate thermal resistivity of soils. Can Geotech J 36(4):767-773.
5. Kersten MS (1949) Laboratory research for the determination of the thermal properties of soils. Bulletin No. 28. Minneapolis, MN: University of Minnesota Engineering Experiment Station.
6. Zhang N and Wang Z (2017) Review of soil thermal conductivity and predictive models. International Journal of Thermal Sciences 117: 172-183

Examining the Applicability of Micro-pile for Bridge Foundation in Difficult Terrain Using Pile Load Test

Shubham Srivastava[1*], Mohd. Zain[2], Rajesh Kumar[3]
[1] *P.hD Scholar, Department of Civil Engineering IIT BHU Varanasi, U.P.*
[b] *P.hD Shri Ramswaroop Memorial University, Tindola, Barabanki, Lucknow*
[c] *Professor, Department of Civil Engineering IIT BHU Varanasi, U.P.*
* *e-mail: shubham.rs.civ18@itbhu.ac.in*

Abstract - Construction of foundation of bridges in remote hilly terrain, in adverse climatic condition, is very difficult due to the limited working period and unavailability of transportation of machinery and equipment. In such situations where regular pile foundation is not feasible, micro-piles prove to be better solution. The study deals with the conditions in the field, field testing of micro-piles and their applicability as bridge foundation. In remote areas and difficult hilly terrains there are several problems faced during construction of bridges. The hilly atmosphere has oxygen deficiency, negative temperatures, chilling cold, snowfall, frozen water bodies, lack of adequate transportation facilities and roads are some of the innumerable problems faced during construction. A number of environmental conditions allow only a limited period for construction which is hardly four-five hours a day. Considering the terrain and hydraulic factors at such sites there is limited option of foundation for the bridges. Usually such terrain has cobbles/gravels in the river bed. The seepage rate is likely to be very high. The high seepage rate requires heavy de-watering works while undertaking construction of substructure. Also such heavy concreting will be time consuming. Since, the time slot for construction is limited, regular procedures do not work.

Based on scour criteria, for such heavy loads as on bridges deep foundations are suitable but boulders, cobbles in such area do not allow sinking of well foundation, leaving us with only one option – piles. However, due to limited transportation facilities, roadways, it is not possible to transport such heavy rigs to such remote and inaccessible areas. Therefore, micro-piles are the most suitable option for bridge foundation considering all the above factors and limitations. A micro-pile is a small-diameter drilled and grouted non-displacement pile that is typically reinforced. It is constructed by drilling a borehole, placing steel reinforcement, and grouting the hole and can bear axial loads as well as lateral loads. They are suitable and can be installed in restrictive areas and in all soil types and ground conditions. In India, the use of micro-piles has been restricted only to slope stabilization in hydro projects and earth retaining structures and not in bridge foundation and therefore, it will be a new area for future endeavours.

Keywords: Micro-pile, Field load-settlement test, Bridge Foundation.

SUB-SURFACE INVESTIGATION

A case study for Leh - Ladakh region was taken up. The sub-surface investigation (SSI) was carried out by bore holes of 40m depth each at the planned abutments and intermediate pier locations for the SSI. It was planned for two test piles at each location to be driven up to a depth of 20m and tested for vertical static load which will be twice the design load or up-to the pre-mature failure limit. Results obtained from SSI and micro pile testing shall be utilized for the design and subsequent construction of micro pile foundation for design of appropriate substructure and superstructure.

The specifications of micro pile consist of 273 mm outer diameter steel casing with M-35 grade cement grout filled inside at a pressure of 4 to 5 bar, re-drilled to place four nos. 32 mm diameter longitudinal bars with a centralizer to maintain spacing. This micro pile shall be tested for static load as per IS 2219 Part IV. The bore holes were to be drilled. Standard penetration test (SPT) to be conducted in bore hole at the depth of every three meters or refusal. The collection of disturbed and un-disturbed soil sample for lab test was to be undertaken. And results were to be analyzed for type of soil strata, water table & safe allowable bearing capacity.

Tests Conducted:

Field tests

(a) Boring at the proposed site to ascertain the type of soil strata at the requisite depth and collection of soil samples, both disturbed and undisturbed by boring tools.

(b) SPT conducted for measuring the penetration resistance of the soil, which is measure of its bearing capacity.

(c) Depth of Water Table was to be determined.

Laboratory test

(a) Bulk Density, Moisture content was determined for samples collected in field.

(b) Particle size analysis was carried out as per IS 2720 (Part-4).

(c) Atterberg's Limits were determined as per IS 2720 Part- 5 1985.

Results and Conclusion

The results of pre-production pile load test have been listed below. However, the micro-pile with length 21m, outer diameter = 273mm, casing thickness 8mm, given stiffness factor T = 1.3 is flexible and slender.

- Safe lateral load capacity of micro-pile is found to be 1.8 Ton and 11 Ton acting alone and in a group respectively
- The vertical load carrying capacity of pile is 32 Ton in normal conditions and 40 Ton in seismic condition.
- The worst vertical load expected on a pile is 23.4 Ton which is smaller than 32 Ton, hence safe.
- The micro-piles have been planned as flexible compression member and to take large lateral load and bending moment, the outer peripheral piles have been battered at 10^0 with vertical.

The problems owing to high altitude, remoteness, weather constraints and other difficulties as explained earlier can be overcome by micro-piles. Application of micro-pile for bridge foundation in submergence in river is debatable issue, as IRC78-2014 Section 9 Pile foundation in its clause 709.1.7

enumerates that minimum allowable diameter of bored pile should be 1000 mm for river water zone and 750 mm for bridge on land such as viaducts and ROBs etc. However, IRC: SP109-2015 permits use of small diameter pile in special circumstances. Based upon the success results, the design of micro-pile foundation and sub structure would be carried out. It is expected that this work would be considered as a distinguishing milestone in the era of bridge construction in similar difficult situations, repair and rehabilitation of endangered structures.

References

1. Abd Elaziz, A.Y & El Naggar, M.H. (2012) Performance of Hollow Bar Micro-piles under Axial and Lateral Loads in Cohesive Soils.
2. Benett, J.K. and Hothem, (2010) Hollow bar Micro-piles for Settlement Control in Soft Clays.
3. Bishop, J.A., et al. (2006) Class I and Class II Micro-piles with Hollow Bar Reinforcement Load Tests and Performance Measurements.
4. Bruce, D.A., et al. (1997) Micro-piles, The State of Practice Part I.
5. Cadden, A.J.,et al. (2004) Micro-piles: Recent Advances and Future Trends.
6. Cavey, J.K., et al. (2000) Observations of Mini-pile Performance under Cyclic Loading Conditions.
7. Elkasabgy,M.,and El Naggar,M.H. (2007) Finite Element Analysis of Axial Capacity of Micro-piles.
8. Federal Highway Administration (FHWA 2005) Micro-piles Design and construction Mc Lean VA: US department of Transportation.
9. Gomez, J.E., Rodriguez et al. (2007) Hollow Core bar Micro-pile –Installation, Testing and Interpolation of Design Parameter.
10. Han, J., Ye, S. (2006) A Field Study of Behavior of a Foundation Underpinned by Micro-piles.
11. Holman, T.P. and Tuozzolo, T.J. (2006) Advanced Interpretation of Instrumented Micro-piles Load Tests.
12. International Building Code (IBC) (2006).
13. IRC SP 109 (Indian code on Micro-pile).
14. IS 2911 Part I to IV.
15. Poulos, H.G, & Davis, E.H. (1980) Pile Foundation Analysis and Design. John Wiley and Sons.
16. Richards, T.D. and Rothbauer, M.J. (2004) Lateral Load on in Micro-piles.

Response of fly ash-bentonite based landfill liner to chemical solutions

G Suneel Kumar[1], Ede Lakshmi Suresh Babu[2] and Rabi Narayan Behera[3*]

[1] *Ph.D. Student, Department of Civil Engineering, National Institute of Technology Rourkela, Odisha, India, 769008.*

[2] *M.Tech Student, Department of Civil Engineering, National Institute of Technology Rourkela, Odisha, India, 769008.*

[3] *Assistant Professor, Department of Civil Engineering, National Institute of Technology Rourkela, Odisha, India, 769008.*

* *e-mail: rnbehera82@gmail.com*

Introduction

Hydraulic barriers are a part of lining systems which are responsible for preventing a potential pollutant present in the waste containment from migrating to surface water, groundwater and thus polluting them. Natural clays, calcium or sodium-based bentonite mixed with specific geo-materials are widely used in the construction of hydraulic barriers such as liners and covers due to their low hydraulic conductivity (i.e. $k \leq 10^{-7}$ cm/sec). However, the long term interaction of landfill liner with leachate leads to the change in its permeability. Thus, addressing the need to study the interaction of leachate with liner material.

In the present study, Sodium bentonite (low sodium content) instead of calcium bentonite was used as the hydraulic conductivity of Na-bentonite is approximately five times lower than that of the Ca-bentonite at similar void ratios (Mesri and Olson 1971). Chlorides and Sulphates are widely distributed in nature and are common constituents in the leachate (Arasan and Yetimoglu 2006; Yilmaz et al. 2008), the chemical properties of the landfill leachate listed by Kayabali et al. 1998, also showed larger amounts of Cl^-, Na^+, K^+ ions which are exchangeable. Therefore, chlorides and sulphates of different cations such as Sodium, Calcium, and Iron etc. are chosen for the permeation process, which best represents the in-situ conditions in the laboratory. The preference for replacement is the lyotropic series, which is Li^+, Na^+, K^+, Rb^+, Cs^+, Mg^{2+}, Ca^{2+}, Ba^{2+}, Cu^{2+}, Al^{3+}, Fe^{3+} (Sposito 1981; McBride 1994). Because Na^+ is at the lower end of the lyotropic series, Na-bentonites are more prone to cation exchange when permeated with solutions containing divalent or trivalent ions (Sposito 1981).

Therefore, in the present study, bentonite mixed with fly ash (FA-B) was permeated with different chemical solutions and their chemical compatibility as a liner material was assessed.

Materials and Methodology

Cohesionless material includes Fly ash collected from Aditya Alumina Ltd. situated in Lapanga town of Sambalpur district, Odisha. Cohesive material includes commercially available sodium based bentonite in the local market.

The basic geotechnical characteristics of the materials were found out by performing different laboratory tests (Table 1). In order to assess the chemical characteristics, permeability tests are carried out using a pressure-permeameter, under different heads of 15m, 25m, 35m and 45m of water respectively, on different trial mixes of fly ash and bentonite with different chemicals.

The bentonite content in the fly ash-bentonite mixture is taken up to 15% @ 5% increment starting from 5%. The permeant solutions used include NaCl, NaOH, HCl, $CaCl_2$, H_2SO_4, and $FeCl_3$, which are chosen to best represent the acidic, basic, neutral nature of the landfill leachate as well as to better portray the effect of monovalent, divalent, trivalent cations present in the in-situ landfill leachate.

Results and Concluding Remarks

Table 1. Geotechnical properties of fly ash and bentonite

Property	Fly ash	Bentonite
Specific Gravity	2.08	2.74
Liquid limit, *LL* (%)	-	268
Plastic limit, *PL* (%)	-	54
Linear Shrinkage, L_s	-	29
Plasticity Index, *PI*	-	214

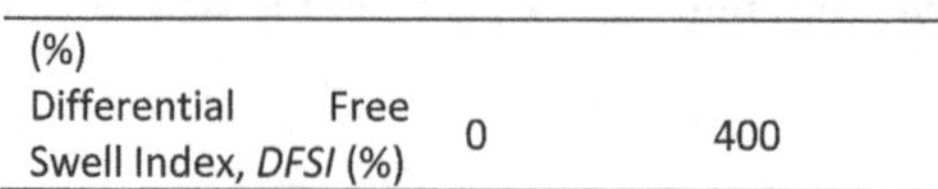

(%)		
Differential Free Swell Index, *DFSI* (%)	0	400

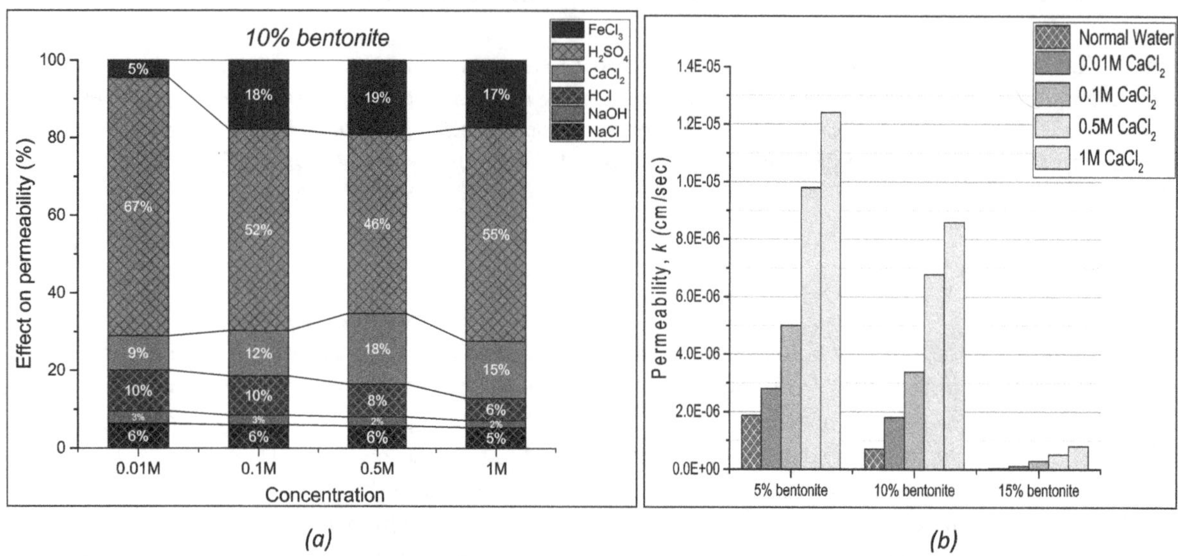

(a) *(b)*

Figure 2. (a) Effect of concentration and type of permeant solution on permeability of FA-B mixture for 10% bentonite (b) Effect of bentonite content and $CaCl_2$ concentration on permeability of FA-B mixture

The basic criteria to be fulfilled by a liner material include the strength criteria (≥0.2MPa) and the permeability criteria ($k \leq 10^{-7}$ cm/sec). Based on the test results of fly ash and bentonite mixtures (i.e. 85:15, 90:10 and 95:05) with six types of permeating solutions (i.e. NaCl, NaOH, HCl, $CaCl_2$, H_2SO_4 and $FeCl_3$) of five different concentrations (i.e. 0, 0.01M, 0.1M, 0.5M and 1M), the following major inferences were drawn. However, the strength characteristics of FA-B mixtures as assessed by Alla et al. 2017 increases with the ageing of liner material due to increasing overburden. Therefore, the permeability characteristics were given priority in this work.

- Irrespective of the nature of the permeant solution (acidic/basic/neutral), the hydraulic conductivity of all the fly ash-bentonite mixtures increased with an increase in chemical concentrations (Figure 2(b)).
- The net repulsive forces were better supressed by multivalent ions than the monovalent ions, resulting in the flocculation of bentonite particles leading to higher permeability (Figure 2(a)).
- The effect of valence of cation is more at higher concentrations than at lower concentrations i.e., with increase in cation valence, the hydraulic conductivity increases and has negligible effect when solutions were dilute (≤0.01M).

References

Alla V, Sasmal SK, Behera RN, Patra CR. Development of alternate liner material by blending fly ash, local soil and bentonite. In: Proceedings of Indian Geotechnical Conference GeoNEst, 14-16 December 2017, IIT Guwahati, India, 95

Arasan S, Yetimoglu T (2006) Effect of leachate components on the consistency limits of clay liners. In: 11th National Soil Mechanic and Foundation Engineering Congress, Trabzon, Turkey (pp. 439-445)

Kayabali K, Kezer H (1998) Testing the ability of bentonite-amended natural zeolite (clinoptinolite) to remove heavy metals from liquid waste. Environmental Geology 34(2-3):95-102

McBride MB. Environmental chemistry of soils. Oxford University Press. New York; 1994.

Mesri G, Olson RE (1971) Mechanisms controlling the permeability of clays. Clays and Clay Minerals 19:151-158

Sposito G. The thermodynamics of soil solutions. Oxford University Press; 1981.

Yılmaz G, Yetimoglu T, Arasan S (2008) Hydraulic conductivity of compacted clay liners permeated with inorganic salt solutions. Waste Management & Research 26(5):464-473

Utilization of Fly Ash-Bentonite Mixture as an Alternate Liner Material

Vamsi Alla[1], Rishab Das[2], Rabi Narayan Behera[3*]
[1] *PhD candidate, Civil Engineering Department, NIT Rourkela,Rourkela, India*
[2] *M.Tech student, Civil Engineering Department, NIT Rourkela,Rourkela, India*
** e-mail: rnbehera82@gmail.com*

Introduction

A landfill liner should be a low permeable barrier which prevents a pollutant in a waste containment system from entering surface water, ground water and thus polluting them. The performance of the lining system is greatly affected by its hydraulic conductivity (Daniel 1987, 1990). Highly expansive clays such as bentonites (because of their low hydraulic conductivity, $k \leq 10^{-9}$ m/sec) mixed with certain geo-materials are used for the construction of hydraulic barriers. Bentonites consist of montmorillonite minerals; they are highly plastic in nature. Though bentonites are low permeable, they exhibit very low strength and very high volume change with variations in water content which is not acceptable in the case of hydraulic barriers. Compacted sand-bentonite mixtures are widely used as a liner material in municipal and hazardous waste containment (Akgun et al, 2015; Al-Rawas et al, 2006). In spite of its low hydraulic conductivity, the usage of sand-bentonite mixtures as liner material is decreasing now-a-days due to the non-availability of sand and growing environmental concerns related to its dredging. Therefore, there is an urgent need to find some alternate material for the construction of liners. This forms the scope to explore the usage of waste material like fly ash as an alternate liner material to sand. In this present work, the geotechnical investigation has been carried out to study the hydraulic conductivity, unconfined compressive strength, plasticity, and swelling characteristics of fly ash bentonite mixtures with bentonite content 5, 10 and 15 percent by dry weight respectively and suggested suitable fly ash bentonite mix considering the above criteria.

Materials and Methodology

The materials used in this present study are fly ash and bentonite. The fly ash used in this present study was taken from Aditya Alumina Ltd. situated in Lapanga town in Sambalpur district, Odisha. The bentonite used in this study is sodium based bentonite which is commercially available in local market.

Experimental procedure: In this study, both physical and geotechnical properties of materials used were determined. Fly ash and bentonite mixtures with dry weight percentages of bentonite of 5, 10 and 15% were prepared i.e., FA+B (95:5; 90:10; 85:15). The liquid limit and plastic limit of above mentioned trial mixes were determined. Swelling characteristics for these trial mixes were obtained by conducting differential free swell index test. Optimum moisture content and maximum dry density were determined for these samples at four compaction energies of 355, 592, 1296 and 2700 kJ/m^3 respectively. Then, the hydraulic conductivity of these sample mixes compacted at their OMC and MDD are determined by falling head method. The effect of bentonite content and compaction effort on the strength of fly ash-bentonite mixes were determined by conducting unconfined compressive strength test.

Results and Concluding Remarks

The effect of bentonite content on various characteristics of the fly ash- bentonite mixtures such as plasticity, swelling, compaction, hydraulic conductivity and strength were evaluated and are presented here. All the fly ash- bentonite trail mixes are satisfying the plasticity criteria and it was observed that the addition of bentonite to fly ash increases plasticity index of the mixture. This may be due to the fact that bentonite being a very high plastic soil when mixed with non-plastic fly ash, it imparts plasticity to fly ash-bentonite mixture. The increase of plasticity in fly ash-bentonite mixes is considered good in reducing the hydraulic conductivity ($\leq 1 \times 10^{-9}$m/sec). In 1994, Benson et al. conducted regression analysis for laboratory test results of 67 landfill sites data and suggested that, hydraulic conductivity ≤ 1 × 10^{-9} m/sec can be achieved if the liquid limit ≥ 20%, the plasticity index ≥ 7%. In 2002, Qian et al. has achieved the hydraulic conductivity of 1 × 10^{-9} m/sec with *PI* > 10% and mix containing 15-20% of silt/clay size particles. Differential free swell test results showed that addition of fly ash to bentonite has decreased the swelling behaviour of bentonite. After 95% fly ash addition to the bentonite, its free swell index got reduced to 6.7%. The decrease in swelling is due to the replacement of plastic fines of bentonite by non-plastic fines of fly ash (Phanikumar et al, 2007). The addition of fly ash to bentonite causes flocculation and cementation to take place and thus reducing the swelling characteristics of fly ash-bentonite mixes. Four different compaction energies such as standard Proctor (SP, 592 kJ/m^3), reduced standard Proctor (RSP, 355 kJ/m^3), modified Proctor (MP, 2700 kJ/m^3) and reduced modified Proctor (RMP, 1296 kJ/m^3) tests were carried out for three fly ash- bentonite mixtures (95:5; 90:10; 85:15). The results were analysed to determine the effect of bentonite content and compaction energy on hydraulic conductivity and unconfined compressive strength of above mentioned fly ash-bentonite mixtures and the same are presented in Figures 1 and 2 respectively.

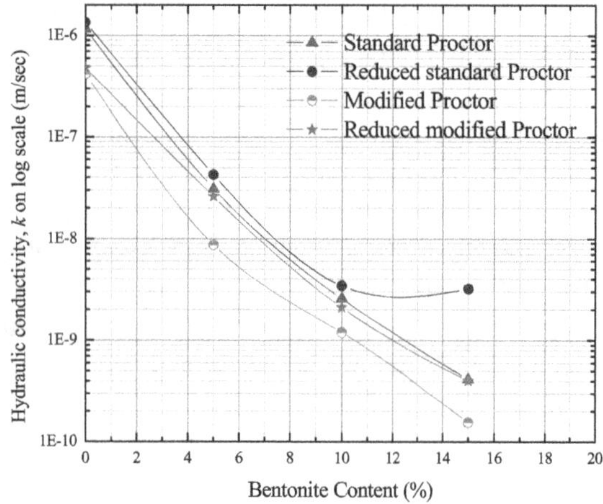

Figure 1. Variation of hydraulic conductivity (k) with bentonite content

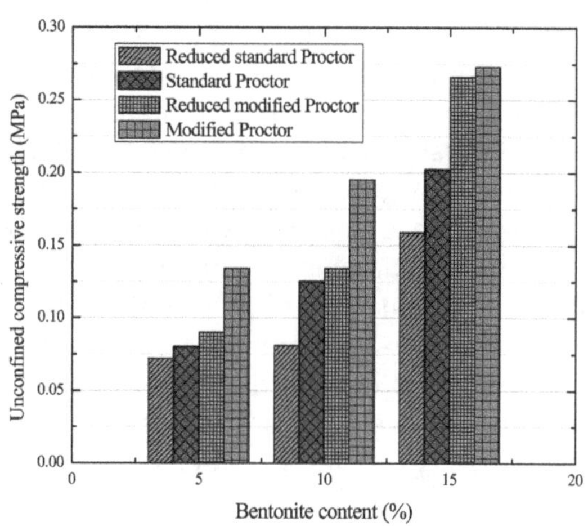

Figure 2. Variation of UCS with bentonite content

By comparing the experimental results with acceptable criteria for hydraulic conductivity ($k \leq 10^{-9}$ m/sec), unconfined compressive strength ($\sigma > 0.2$ MPa), plasticity characteristics (*LL* ≥ 20% and *PI* ≥ 7%), and swelling characteristics, the following fly ash-bentonite mixtures are suggested as liner material in waste containment system. They are:

- FA+B (85+15) mixture compacted under modified Proctor compaction
- FA+B (85+15) mixture compacted under reduced modified Proctor compaction
- FA+B (85+15) mixture compacted under standard Proctor compaction.

References

Akgun H, Ada M, Kockar M K (2015). Performance assessment of a bentonite–sand mixture for nuclear waste isolation at the potential Akkuyu Nuclear Waste Disposal Site, southern Turkey. Environ. Earth Sci., 73, 6101–6116.

Al-Rawas A A, Mohamedzein Y E, Al-Shabibi A S, Al-Katheiri S (2006). Sand–Attapulgite Clay Mixtures as a Landfill Liner. Geotechnical and Geological Engineering, 24, 1365–1383.

Daniel D (1987). Earthen liners for land disposal facilities. In Geotechnical Practice for Waste Disposal "87, GSP No. 13, ASCE. 21-39.

Daniel D (1990). Summary review of construction quality control for earthen liners. In Waste Containment Systems: Construction, Regulation, and Performance, GSP No. 26, ASCE, R. Bonaparte, ed. 175-189.

Phanikumar B R, Sharma R S (2007). Volume Change Behaviour of Fly Ash-Stabilized Clays. American Society of Civil Engineers, 10.1061/ASCE0899-1561200719:167.

Water Resources Engineering

Urban Flood Inundation Mapping using HEC-RAS – Case Study

V Swathi [1], K. Srinivasa Raju [2*], V Anirudh Reddy [1]
[1,2] *Department of Civil Engineering, BITS Pilani Hyderabad Campus, Hyderabad, India 500 078*
[2]*Professor*
** e-mail: ksraju@hyderabad.bits-pilani.ac.in*

Introduction

Urban floods have emerged as natural hazard across metropolitan cities of India. As infiltration of rainwater is hindered, most of the rainfall is converted to runoff. However, shrinking of water bodies, clogging of storm drains deplete their flood carrying capacity (Vijayalakshmi and Babu 2011). The vulnerability of people in low-lying areas and vicinity of water bodies is one of the major concerns. Hyderabad, India has witnessed major urban floods in 1908, 2000 and 2008 affecting life and property (Ahmed et al. 2013). This necessitates identifying flood-prone areas and extent of flooding in the city (Kafle et al. 2007). Numerous researchers modelled various aspects of flood vulnerability and hazard mapping (Tate et al. 2002). Most of the models require huge amount of data pertaining to cross-sections of water bodies, hydrologic parameters, land use data, dimensions of storm drains and invert elevations etc. The present study proposes a methodology to develop flood inundation maps for major water bodies in a data scarce environment for an urban catchment in Hyderabad, India. This provides information about the areas that are likely to be vulnerable to floods for probable inflow scenarios to Hussain Sagar Lake.

Materials and Methods

The Greater Hyderabad Municipal Corporation (GHMC) divided city into sixteen storm water zones based on topography. Among them, zones 12 and 13 are the prime urbanised areas of the city. Layouts of storm water network of these zones were acquired from GHMC. The storm water network layout was overlaid on the Digital Elevation Model (DEM) of 30m (NRSC 2017). Hydrologic Engineering Center-River Analysis System (HEC-RAS) 1D was chosen for its capability to model under scarce data without compromising on quality of the flood inundation mapping. HEC-RAS was appropriate to model urban flooding as the model is built on St. Venant shallow water equations of natural channels (Brunner 2002). The stream center line, flow paths, bank lines and cross section of Hussain Sagar Lake and other ponds were digitised in ArcMap. After the layers were created, the cross-section and geometric details were generated in HEC-RAS. Flood inundation was modelled for steady and unsteady flow conditions, for a subcritical flow regime. In steady flow analysis, critical depth was assigned as the upstream condition. In unsteady flow, inflow hydrograph was assigned as upstream. Normal depth was downstream condition at both the flow conditions, for all the scenarios. Discharge from catchment was calculated using rational method. The outcomes of the model include water surface elevations in the lake. The results of HEC-RAS were further converted to a layer format, to visualise on Google Earth.

Results and Concluding Remarks

From HEC-RAS results, the inundation depths: 0, 0-0.5 m, 0.5-1.5 m, 1.5 m (and above) were classified as No flooding (N), Low (L), Medium (M) and High (H) flooding areas. It is assumed that all the water bodies on upstream of Hussain Sagar Lake are at full capacities. For an inflow of 142 cumecs, the flood depth in the lake ranged from 0.5 to 1.5 m at steady flow and 0.7 to 1.9 m at unsteady flow. The storm water network and water bodies were able to cater for 113 cumecs, areas under N and L are 78% and 21%. For 142 cumecs inflow N, L and M areas were 65 %, 26% and 9%. For 170 cumecs, L, M and H were 30%, 56% and 13%; whereas for 226 cumecs L, M and H were 22%, 60% and 17%.

Steady flow analysis under predicted the flood extent by 10-20% compared to unsteady flow. This is due to the difference in stage-discharge calculations. Steady flow analysis calculates the stage-discharge using kinematic wave and unsteady flow analysis uses dynamic wave. Another difference between steady and unsteady flow is the effects of in-channel and offline storage on flood attenuation wave. Steady flow does not account for them.

Applicability of HEC-RAS was explored for an urban catchment. Responses of water bodies, storm water network for inflow scenarios to Hussain Sagar Lake were analysed at steady and unsteady flow. The outcomes include flood depth in the lake and flood extent. Steady flow analysis under-simulated flood extent by 10-20% compared to unsteady flow. The existing storm water network and water bodies could convey the discharges for 142 and 113 cumecs, but were insufficient for 170 and 226 cumecs. The analysis was based on data availability, few assumptions and limitations of HEC-RAS.

Acknowledgments: This work is supported by Information Technology Research Academy (ITRA), Government of India under, ITRA-water grant ITRA/15(68)/water/IUFM/01.

References

Ahmed, Z., Rao, D. R. M., Reddy, K. R. M., & Raj, Y. E. (2013). Urban flooding-case study of Hyderabad. Global Journal of Engineering, Design and Technology 2(4): 63-66.

Brunner, G. W. (2002). HEC-RAS river analysis system: User's manual. US Army Corps of Engineers, Institute for Water Resources, Hydrologic Engineering Center.

Kafle, T. P., Hazarika, M. K., Karki, S., Shrestha, R., Sharma, S., & Samarakoon, L. (2007, May). Basin scale rainfall-runoff modelling for flood forecasts. In Proceedings of the 5th Annual Mekong Flood Forum, Ho Chi Minh City, Vietnam (pp. 17-18).

NRSC (2017) Digital elevation model. http://bhuva n.nrsc.gov.in/bhuvan_links .php. Accessed on May 2017.

Tate, E. C., Maidment, D. R., Olivera, F., & Anderson, D. J. (2002). Creating a terrain model for floodplain mapping. Journal of Hydrologic Engineering 7(2): 100-108.

Vijayalakshmi, D. P., & Babu, K. J. (2011). Floodplain modelling materials and methodology. International Journal on Transportation and Urban Development 1(1): 12.

Entropy-based spatiotemporal patterns of the daily rainfall variability in the Krishna river basin, India

G.Ravi Kumar[1], R.Maheswaran[1*]
[1] *Department of Civil Engineering, MVGR College of Engineering, Vizianagaram, India*
** e-mail: maheswaran27@yahoo.co.in*

Introduction

Rainfall is the primary component of the hydrological cycle, describes the transfer of mass and energy from the earth to the atmosphere and vice versa (Brunsell, 2010). The spatiotemporal variability of rainfall has a large impact on agricultural productivity (Jin & Wang, 2017). Categorization of most probable water resources helps in identifying the potential regions and it is based on the analysis of spatial and temporal variability of historical rainfall in terms of amount and intensity and it is essential for water resources planning and management (Kawachi et al., 2001). There are fairly minimal studies on the spatial and temporal variability of rainfall for the Krishna river basin using 0.25^0x0.25^0 daily rainfall gridded data and still, there is a gap of identifying the grids with potential water resources availability.

Objectives of the study are:

1. To investigate the rainfall variability for monthly, seasonal and annual time series and to identify the seasons contribution to the annual variability and months contribution the season variability
2. To investigate the Intra-annual distribution of monthly rainfall amount's over-a-year using apportionment entropy
3. To develop a correlation between the mean apportionment entropy and mean annual rainfall and to make a comparative assessment of the potential availability of water resources.

Materials and Methods

Entropy-based Metrics

Shannon entropy concept of measure of uncertainty is used in this study to determine the rainfall variability in terms of space and time. The advantages and limitations of entropy derived metrics are briefly described in (Singh, 1997).

The generalized equation defining entropy(H)(C.E.shannon 1948), is as shown in Eq.(1).

$$H(X) = -\sum_{i=1}^{N} P(x_i)\log_2[p(x_i)] \tag{1}$$

Where P :{ p 1, p 2....pN} is the probability distribution of X, N is the sample size. –Log2P(xi) is the measure of uncertainty about the event occurring with probably P(xi). H (X) is the entropy of X :{ x 1, x 2, xn}, and H(X) is expressed in terms of bits. For a deterministic variable, if the entropy H(X) is zero then it has minimum uncertainty / Maximum certainty, about the probability that it will take on a certain value is 1, and the probabilities of all other alternative values are zero. On the other side if all the events are equally likely then the distribution is highly uncertain and it is equal to logN.

Marginal Entropy (ME)

The marginal entropy H(X) measures the degree of uncertainty about the random variable X with the probability distribution P(x) and it's calculated from a single historical time series and it gives variability contained within the entire length of time scale. In the present study, to understand the spatial distribution of variability based on the amount of rainfall over the 113 years, rainfall time series of different time scales (annual, seasonal, monthly) were considered. (A. K. Mishra et al., 2009)The mathematical expression of ME is given by Eq.(2).

$$H(X) = -\sum_{i=1}^{113} P(x_i)\log_2[p(x_i)] \tag{2}$$

Apportionment Entropy (AE)

To study the variability of the amount of rainfall of months within year Apportionment entropy is used (Maruyama et al., 2005). Considering the amount of rainfall (ri) in a month, (i = 1, 2, 3 . . . 12) of a year to the total amount of rainfall in that year (R), the likeliness will be ri/R, denoted by pi. Probabilities for a particular grid point are expressed in discrete form by taking into account all the aggregate rainfall available at that grid point and their likeliness. Finally, using Shannon entropy, AE can be calculated from Eq. (3).

$$AE(X) = -\sum_{i=1}^{R} P(x_i)\log_2[p(x_i)] \tag{3}$$

Data

The analysis is carried out using daily rainfall (mm/day) gridded data developed by (Pai et al., 2014) with a horizontal resolution of 0.25° × 0.25° grid, the temporal resolution of one day from the period 1901–2013.

Results and Concluding Remarks

1. Distinct spatial patterns were detected for monthly and seasonal time series.
2. There is an increase in the variability of annual precipitation amount from south to north of the Krishna river basin indicating the spread of the rainfall is high in the south when compared to north of the basin (see Fig.2).
3. Coupling mean annual rainfall (see Fig.1) with the apportionment entropy leads to improved categorization of rainfall addressing rainfall amount as well as its variability within a year temporally.

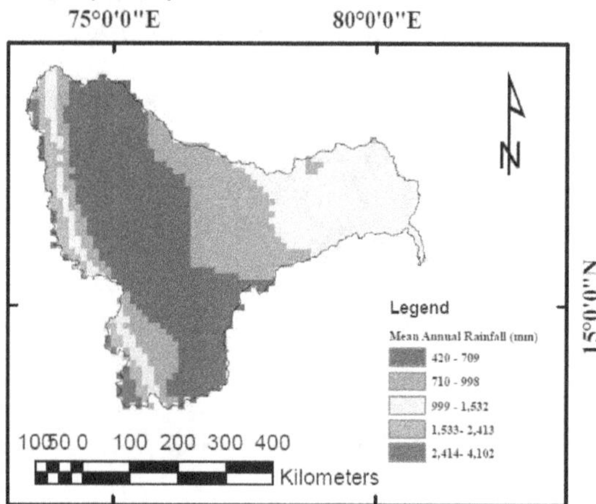

Fig.1: Spatial distribution of mean annual rainfall covering a period of 113 years

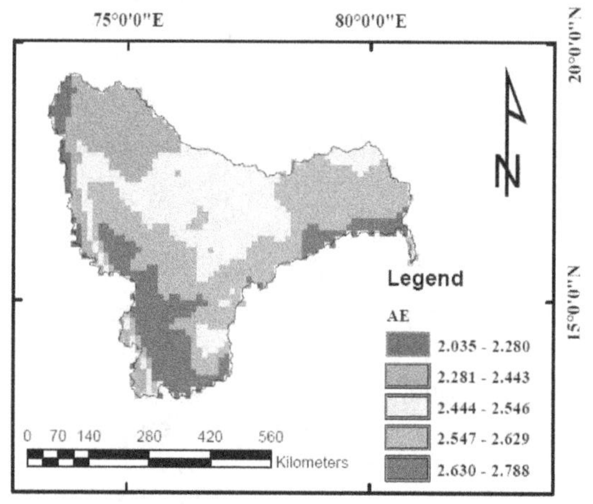

Fig.2: Spatial distribution of monthly rainfall variability based on Apportionment entropy

Acknowledgements: This research is funded by SERB, India through the ECRA/16/1721 held by Dr. Maheswaran Rathinasamy.

References

Brunell, N. A. (2010). A multiscale information theory approach to assess spatial-temporal variability of daily precipitation. Journal of Hydrology, 385,165–172. https://doi.org/10.1016/j.jhydrol.2010.02.016

C.E.Shannon. (1948).The mathematical theory of communications, I and II.Bell System Technical Journal, 27(1), 379–423. https://doi.org/10.1002/j.1538-7305.1948.tb01338.x

Jin, Q., & Wang, C. (2017).A revival of Indian summer monsoon rainfall since 2002.Nature Climate Change, 7(8), 587–594. https://doi.org/10.1038/NCLIMATE3348

Kawachi, T., Maruyama, T., & Singh, V. P. (2001). Rainfall entropy for delineation of water resources zones in Japan. Journal of Hydrology, 246(1–4), 36–44. https://doi.org/10.1016/S0022-1694(01)00355-9

Maruyama, T., Kawachi, T., & Singh, V. P. (2005).Entropy-based assessment and clustering of potential water resources availability.Journal of Hydrology, 309(1–4), 104–113. https://doi.org/10.1016/j.jhydrol.2004.11.020

Mishra, A. K., Özger, M., & Singh, V. P. (2009).An entropy-based investigation into the variability of precipitation.Journal of Hydrology, 370(1–4), 139–154. https://doi.org/10.1016/j.jhydrol.2009.03.006

Effects of land use and land cover changes on surface runoff of a semi-urban catchment

Joygopal Jena[1*], S. Nath[2], D. Satpathy[2]
1 Departemnt of Civil Engineering, GITA, Bhubaneswar, India
2 Department of Civil Engineering, CET, Bhubaneswar, India
** e-mail: jenajoygopal@gmail.com*

Introduction

Land use and land cover (LULC) of a watershed serve as a key factor that influences watershed hydrology. Direct runoff in a catchment depends on the LULC, soil type and rainfall characteristics. Urbanization changes its LULC drastically, where the impermeable surface increases considerably for which infiltration of rain water reduces thereby increases peak discharge and runoff volume.

Remote sensing has been a primary source of creating LULC data, which is used here. Application of remotely sensed data made possible to study the changes in land cover in less time, at low cost and with better accuracy in association with GIS that provides suitable platform for data analysis, update and retrieval (Kachhwala, 1985, Chilar, 2000, J. S Rawat, M. Kumar et al., 2015).

Watershed is an ideal unit for planning and management of land and water resources (Gajbhiye and Sharma, 2012, 2013; Gajbhiye et al., 2014(d)). Several methods are available for estimation of surface runoff from rainfall. The curve number method (SCS-CN Method) is simple and the most popular, which has been adopted here. CN model is used where constraints like slope, land cover, type of soil, area of watershed etc. are considered for runoff estimation. (Balvanshi, and Tiwari, 2014).

In this study, the changes of runoff pattern for a catchment in the sub-urban area of Bhubaneswar City in Odisha state has been compared considering changes of LULC at different periods. GIS is utilized for study of the land use and land over. The changes of runoff pattern for the said catchment are corelated to changes of LULC between the periods 2005 and 2012.

Materials and Methods

Study Area: The present study area covers Gangua-Jhumka watershed, situated in outskirt of Bhubaneswar the capital city of Odisha state lies between 20º16'23.268" N and 21º15'20.0" N latitude and 85º15'00"E and 85º44'52.695"E longitude. The Gangua nala is the major tributary of Kuakhai River originates from village Gadakan and meets the river Daya. The river Jhumka is another major drain flow through the area. Based on the land use land cover pattern the classification for lands and area covered is stated in Table 1.

Table 1. Land Classification of the study Area

Type of land	Agricultural land	Built up land	Forest land	Water body	Waste land	Wet land
% of Area in 2005	39.37%	24.37%)	23.24%	2.83%	10.15%	0.166%
% of Area in 2012	32.39%	32.71%	22.48%	1.528%	8.905%	2.21%

Data used: Most of the data used in this study are obtained through satellite imageries. Ancillary data such as topographic maps, aerial photo and Land-Sat images for the year 2005-06 and 2011-12 are obtained from Orissa Space Application Centre (ORSAC), Bhubaneswar, Odisha. The details of the images used are presented in the Table 2. Topo sheets bearing number H/11, H/15, 73 H/12 and H/16 are used in this work. The rainfall Data for year 2005-06 and 2011-12 collected from the Department of Water Resources (DoWR), Govt. of Odisha.

Table 2 Information about satellite image

Satellite Image	Spatial Resolution	Acquisition Period	Source	Path & Row
LISS-III	23.5m	2005-06 (Rabi-Kharif)	NRSC	Column-106, Row-58
LISS-III	23.5m	2011-12 (Rabi-Kharif)	NRSC	Column-106, Row-58

Methodology: The following steps are taken up. (1) Initially a base map was prepared from all Topo sheets covering the study area. Various thematic maps such as, *LULC map, Soil map, Contour map, Elevation map, Slope map* are prepared using Arc GIS 10.3. (2) The *LULC map* was generated with the help of satellite data using unsupervised classification. The classification of the image was performed by using supervised classification in which, spectral classes are grouped first, based on the numerical information of the data, and are then matched. The masked satellite images for the year 2005-06 and 2011-12, are used to analyse the land coverings. (3) The contour map of the study area has been prepared from the toposheet of the scale 1:50,000 with a suitable contour interval. Using the GIS platform i.e. Arch GIS 10.3 Software the contour lines are digitized on the topo sheet. (4) The *contour map* is used to generate a Digital Elevation Model (DEM), which consists of an optimal array of round elevations at regularly spaced intervals. The DEM is developed by 3D analysis of contours and Triangulated Irregular Network (TIN). (5) Generation of CN map is the most important step. To create the CN map, the soil map and land use map were uploaded to the Arc GIS. The soil map and land use map were selected for intersection, after intersection a map with new polygon representing the merged soil-land map. In both the time period 2005-06 and 2011-12 there are six land use land cover classes and four hydrological soil group presents in the study area. (6) The runoff characteristics and soil erosion of the command area are controlled by the degree of slope. Using the contours, the slope map was prepared. (7) Finally, estimation of Runoff and Runoff depth are done respectively, by SCS Runoff Curve Number Method and SCS Model.

Results and Concluding Remarks

Various maps as discussed are prepared presented in full paper. Some typical maps are shown below.

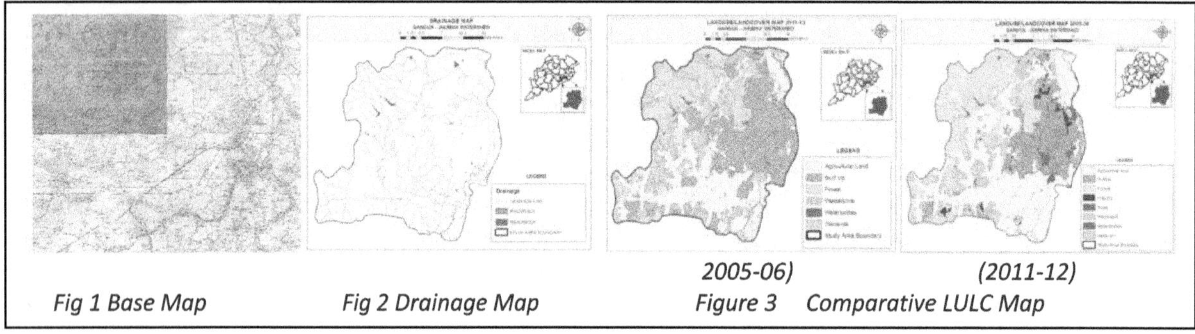

Fig 1 Base Map *Fig 2 Drainage Map* *2005-06) (2011-12)* *Figure 3 Comparative LULC Map*

Rainfall data, calculated CN value are used to estimate the required runoff. Runoff for individual the years 2005-06 and 2011-12 is calculated using SCS-CN method and presented in Table 3.

Table 3 Runoff Depth Results

	Rainfall (mm)	Weighted CN	S (mm)	Q (mm)
2005-06	1452	68.27	118.05	1307.613
2011-12	1579	68.41	117.29	1434.80

Mapping and monitoring of land use/land cover is important for many management and planning activities and considered as an important element for understanding the earth and its whole system. The present study shows how well LU/LC classification and its change analysis of the year 2005-06 and 2011-12 of the study area can be easily carried out using Remote Sensing and GIS technology.

References

Balvanshi A, and Tiwari H.L., (2014) "A Comprehensive Review of Runoff Estimation by the Curve Number Method", International Journal of Innovative Research and Science, Engineering and Technology, Vol.3, Issue 11, pp. 17480-17485, November

Gajbhiye, S and Sharma, S.K., (2012) Land Use and Land Cover change detection of Indra river watershed through Remote Sensing using Multi-Temporal satellite data, International Journal of Geomatics and Geosciences, vol.3, no. 1, pp. 89-96

Gajbhiye, S. (2015), Estimation of Surface Runoff Using Remote Sensing and Geographical Information System, International Journal of u-and e-Service, Science and TechnologyVol.8, No.4 pp.113-122.

Capacity Assessment of Dnyanganga Reservoir, Maharashtra by Geospatial Technology for Sustainable Water Resource Management.

P. K. Kedar[1*] S.A. Gaikwad[2] M. M. Kulkarni[3] S. D. Bhagat[4] S. K. Ghanekar[5]

1. Assistant Engineer Grade II, Maharashtra Engineering Research Institute, Nashik 4
2. Sub Divisional Engineer, Maharashtra Engineering Research Institute, Nashik 4
3. Executive Engineer, Maharashtra Engineering Research Institute, Nashik 4
4. Superintending Engineer, Maharashtra Engineering Research Institute, Nashik 4
5. Director General, Maharashtra Engineering Research Institute, Nashik 4

E mail: -eerecwrd@gmail.com

Introduction

One of the essential inputs required for effective water planning of reservoir is assessment of its present storage capacity. It is therefore essential for the irrigation manager to know the quantum of water available in the live storage zone. Remote sensing technique for reservoir sedimentation surveys is essentially based on mapping of water-spread areas at the time of satellite over pass. It uses the fact that water-spread area of the reservoir reduces with the sedimentation at different levels. The parameters namely water-spread area and the elevation information are used to calculate the volume of water stored between different levels. These capacity values are then compared with the previously calculated capacity values to find out change in capacity between different levels.

The Maharashtra Engineering Research Institute (MERI), the State's Water Resources Department, has done substantial work in the field of reservoir capacity assessment. This institute has carried out more than 350 surveys of capacity assessment of Maharashtra and India .Present paper deals with study of Dnyanganga reservoir by satellite remote sensing technique has been conducted after 46 (1971-2017) years of its first impounding.

Data (Field and Satellite)

The Dnyanganga reservoir lies between Latitude 20': 22': 15.00" N to 20':26': 16.00" N and Longitude 76': 39': 45.00" E to 76':35':44.76" E. The reservoir was constructed on confluence of river Dnyanganga, near village Geru matargaon in Khamgaon taluka of Buldhana district. Total catchment area of the reservoir is 200.000 sqkm. The designed gross storage capacity of the reservoir is 36.264 $Mm^3$and live storage capacity is 33.930 Mm^3.The designed dead storage capacity is 2.335 Mm3.The reservoir was first impounded in the year 1971.Presently due to scanty rainfall and higher demand of water for domestic purpose arises problem of equitable distribution of water hence capacity assessment of present storage was necessary for irrigation manager. Hence the hydraulic data of reservoir such as Elevation Area Capacity table, Controlling levels, area under submergence were taken up from the field officials.

Satellite Data includes the Resourcesat 1(P6) LISS III Nine images and Resourcesat 2 LISS III five images with a resolution of 23.5 m were analyzed for this study.

Methodology

The basic approach is to find out the water-spread area from satellite data for different water levels between MDDL to FRL. The difference between aerial spread of water between current year and earlier years is the areal reduction at these levels. The methodology for estimation of live storage capacity of reservoir using remote sensing consists of following major tasks.

1) Digital data base creation,
2) Estimation of water-spread area,
3) Calculation of reservoir capacity,
4) Comparison of result with previous surveys,
5) Estimation of live capacity loss due to sedimentation.

National Remote Sensing Centre website was browsed and a list of cloud free dates of IRS P6 (with LISS III sensor) and Resourcesat2 (with LISS III sensor) satellite pass over Dnyanganga reservoir was prepared for the period between Year 2013 and 2018. The reservoir levels and corresponding water

Spread areas for dates of satellite pass were obtained from the field office. The selection of the satellite images was done after studying the draw down pattern of the lake levels, selected satellite data in Geo-referenced mode was procured from the NRSC Hyderabad.

Water Spread Area Extraction-Area extraction is done by either Normalized Difference Water Index (NDWI) or by unsupervised classification method. For Dnyanganga reservoir, unsupervised classifications out puts were generated for specific scene and range of unsupervised classification for water body delineation was noted for respective scene. From analysis, WSA around FRL is observed as 2.954 Mm^2 and WSA around MDDL is observed as 0.651 Mm2.then computation of reservoir capacity at different elevations has been done using following prismoidal formula.

$$V = \frac{\mathbf{H}}{\mathbf{3}} * \left(\mathbf{A_1} + \mathbf{A_2} + \sqrt{\mathbf{A_1} \times \mathbf{A_2}}\right)$$

Where, V = Reservoir capacity between two successive elevations h_1 and h_2

H = Elevation difference ($h_2 - h_1$)

A_1 and A_2 are areas of reservoir water spread at elevation h_1 and h_2.

Concluding Remark and Comments

1. The Revised live Capacity is found out to be 25.337 Mm^3 against Design Capacity of 30.831 Mm^3
2. The total loss of Reservoir Capacity due to the sediment deposition during the period of 46 years (1971-2017) is Estimated as 5.494 Mm^3 hence Annual % loss of capacity is 0.390 % .
3. As the half life of the reservoir is almost over, it has lost about 18 % of its useful storage. Hence, it may be predicted that in another 50 years and when the economical life of the reservoir will be over, it may have lost the capacity about 35 % which will definitely affect water planning.
4. Average rate of Siltation is found out to be 5.972 Ha-m / 100 km^2 / year which on higher side than the designed silt rate which is 3.57 Ha-m / 100 km^2 / year. But it is on safer side than the mentioned average rate of siltation of reservoirs for Narmada and Tapi basin which is 17.42 Ha-m/ 100 km^2 / year as per Compendium on Silting of Reservoirs in India Central Water Commission, MOWR, New Delhi 2015.
5. Periodical monitoring of siltation helps keeping the content table updated. As remote sensing technique is a cost effective and fairly accurate it is being used for such surveys by Maharashtra Engineering Research Institute, Nashik.

References

- CWC (2015), Compendium of silting of reservoir in India, Technical report on silting of reservoir in India, WS & RS directorate, Central Water Commission, New Delhi.
- Various sedimentation surveys conducted by Maharashtra Engineering Research Institute, Nashik.

Effective Water Resources Management by using Web-Enabled Sensors and Communication Networks

Dr. A. G. Matani
Professor- Mechanical Engineering, Government College of Engineering, Amravati – 444 604 [M.S.] India ,
Email: ashokgm333@rediffmail.com, dragmatani@gmail.com

Abstract

One of the key factors for improving the water use efficiency and conservation would be the establishment of a framework for efficient and dynamic management of database for water through use of latest technologies including ICT tools. Sensors and Big Data gathering allow real time monitoring of water quantity and quality, precision irrigation, smart leakage detection, enabling better planning and decision-making. Advanced Artificial Intelligence and Geographic Information Systems are transforming large-scale measurement; while internet and wireless technologies, intelligent decision support systems and other innovations are impacting across the water sector. Web-enabled sensors and communication networks provide an opportunity for water stakeholders to obtain information in near real time about physical and environmental variables such as temperature, soil moisture levels and rainfall. Smart metering technologies can also provide individuals, businesses and water companies with information in near real time about their own water use, thus raising awareness about usage, locating leakages and offering better control over water demand. This paper highlights the latest developments in optimizing water resources utilization by using state of the art Information technology applications in various parts of the world.

Key words: Rivers, lakes, dams and reservoirs 3D mapping, urban environment and impact assessment, urban hydrology, urban management and modelling.

NASA providing training program on water resources and disaster management

The launch of several Earth Observation (EO) sensors from advanced satellites provides world-wide continuous measurements on various hydrological components which are essential input data for hydrological modelling. The data gaps due to lack of on-the-ground monitoring of water resources around the world are now available using satellite acquisition. Thus, satellite products and sophisticated computational techniques for the management of water can play an important role in present and future of water resources. The satellite remote sensing for hydrological applications includes, but not limited to rainfall (Global Precipitation Measurements (GPM) and Tropical Rainfall Measuring Mission (TRMM); Soil moisture (Soil Moisture Active Passive (SMAP) and Soil Moisture Ocean Salinity (SMOS); Actual Evapotranspiration (Surface Energy Balance System); Mapping Evapotranspiration with Internalized Calibration (METRIC) and Surface Energy Balance Algorithm for Land (SEBAL); Groundwater level monitoring by Gravity Recovery and Climate Experiment (GRACE). Using satellite data and GIS, water bodies such as rivers, lakes, dams and reservoirs can be mapped in 3D. The spatial water availability maps can be generated. The concerned authorities can use the information for identifying the sites or regions that need effective protection and management and decisions can be made regarding the sustainable management of water resources in the identified regions. The GIS can be used effectively for this purpose to combine different hydro geological themes objectively and analyze those systematically for demarcating the potential zone. There are several urban applications where satellite based remotely sensed data are being applied, namely; urban sprawl / urban growth trends, mapping and monitoring land use / land cover, urban change detection and updating, urban utility and infrastructure planning, urban land use zoning, urban environment and impact assessment, urban hydrology, urban management and modelling.

Drone Technology is aiding data collection for crop breeding in Africa

Preliminary results of a study shows that using drone technology could cut labor and costs spent in collecting data for maize breeding by at least 10%. The International Maize and Wheat Improvement Center (CIMMYT) in Southern Africa have adopted the use of unmanned aerial vehicles (UAVs) -- drones to collect data as a critical part of a breeding programme. UAVs facilitated the collection of instant data gathering and that drones are able to collect data from 1,000 plots in 10 minutes or less a task that might take 8 hours to do so manually. In this way, drone application reduces labor and costs of data collection. CIMMYT started using drones in 2013, and to date UAVs are used in maize breeding in Eastern and Southern Africa, Latin America and Asia.

Nano Ganesh – a revolutionary ICT tool for farm irrigation

Nano Ganesh developed by Shri Santosh Ostwal is an electronic modern simple and low-cost solution in e-irrigation that empowers farmers to control water pumps with the help of a mobile phone. After installing the Nano Ganesh unit at the pump end, a farmer can switch it on or off with the help of a mobile phone from any distance. His phone also displays the availability of the power supply at the pump end as well as on/off status. Hence, farmers are not needed to physically visit sometimes hazardous pump sites in remote locations all the time.

ICT applications in city water supply systems controlling water losses

City water supply systems in most of the urban local bodies (ULBs) in India have an inefficient water distribution with significant losses and UFWs. One of the challenges in managing the leakages, theft, inefficient customer billing, operation & maintenance, asset management is the lack of real-time data/information at the decision making level. By using ICT like sensors, automated communication, control system in industrial process would provide substantial opportunity in reducing the specific water consumption, thus also providing co-benefits on water charges and power consumption. Smart sensors like Smart Levee, water quality / water level sensors, GIS, automatic weather station, weather prediction model, etc can enhance the early warning system during extreme events such as flash flooding etc. besides also managing infrastructure and reservoir systems. ICT can also help in scheduling the optimal time of irrigation, remote management of irrigation activities, and optimal water use for agriculture, which helps in preventing damage due to drought stress or over irrigational practices.

Conclusions

The success of smart water management depends not only upon improving the ICT technology itself; it depends upon expert knowledge and the collaboration of multiple stakeholders, including from water, industry, urban planning and environment. The use of fast growing mobile-based networks and Apps allowing for rapid, reliable decisions on monitoring, acquiring and processing real-time data on water level, rainfall, runoff, water quality and leakage detection, needs to be promoted. It is now up to all of us to use technologies and data wisely and to work together to better inform operational, maintenance and planning decisions in water management. Using drone technology could cut labour and costs spent in collecting data for maize breeding by at least ten per cent. With increased demand for better seeds to adapt to changing climate, breeders have turned to unmanned aerial vehicles (UAVs) - drones for precise gathering of data from the field to enable more efficient maize breeding in most of Southern Africa. The use of drones to collect data may be an efficient way if you look at large acreages.

Estimation of Rainfall of Nizamabad District Using Autoregressive Model

B. Ramakrishna[1], CH. Rama Rao[2], CH. Sanjeev[3]

[1] *Assistant Professor, Department oF Civil Engineering, Rajiv Gandhi University of Knoweldge Technologies, Basar, India e-mail:bhukyaramakrishna@gmail.com*

[2 3] *Under Graduate Students ,Department of Civil Engineering , Rajiv Gandhi University of Knoweldge Technologies, Basar, India . e-mail: ramaraoiiit301@gmail.com*

Introduction

Earth is very special because it has much amount of water; the total earth surface covers 71% of water. In this 96.5% is salt water, this is found in the ocean only just 3.5% of water on the earth is fresh water. In most of the this freshwater 68% is glaciers and icecaps, one third is in ground water and remaining 2% of water is in river. Rainfall is the only source for all these. High intensive rainfall cause floods and low rainfall causes drought. The economy of the country like India depends on agriculture. If we estimate the occurrence and intensity of rainfall on any area we can give proper planning for agriculture also this forecasting of rainfall is of vital importance for flood caution, allocation of domestic and irrigation water in drought seasons that is why the estimation of rainfall is compulsory. Different models have been proposed to estimate rainfall; it may be conceptual or mathematical model. A conceptual model is very complicated that is why everyone is looking for a mathematical modeling, which concentrates only on the relation between different hydrological parameters. Autoregressive models (AR), conventional time series models such as Thomas-fiering model, Autoregressive moving average (ARMA) models, Autoregressive integrated moving average (ARIMA) model, Autoregressive moving average with exogenous inputs (ARMAX) have been applied by many researches in their studies, as they predict reasonable accurate results. Rai et.al (2007) proposed about stochastic time series model for prediction of annual rainfall and runoff in Mahshara watershedof lower Gomati catchment. Shaonlee Chakrabortyn et.al (2015) proposed the development of time series Autoregreesive model for prediction of rainfall and runoff in kelo watershed Chhattisgarh.The present study was conducted with the prime *objective* to develop a *stochastic autoregressive time series model to estimate rainfall of Nizamabad district*. With the data available appropriate model has been selected, Autoregressive time series model is generated and validity of the model is checked. It has been found that AR (2) model is appropriate for the estimation of rainfall in Nizamabad district.

Materials and Methods

The average monthly rainfall data of Nizamabad has been collected since 1900 to 2002.Order of the model is identified by using Autocorrelation and Partial autocorrelation *(Kottegoda and Horder, (1980))*[3] .The steps involved in modelling are Preliminary analysis , Identification of order of the model , Estimation of Autoregressive parameters *(Box and Jenkins, 1976)*[1] and checking for the validity of the proposed model. The schematic representation of methodology used in the time series modeling is shown in figure 1. The autocorrelation functions and partial autocorrelation functions were determined for the 95% probability limits .The autocorrelation function and partial autocorrelation functions with 95% probability limits up to 20 lags of the series (lag k) were computed and the autoregressive model of second order AR(2) was selected for further analysis.

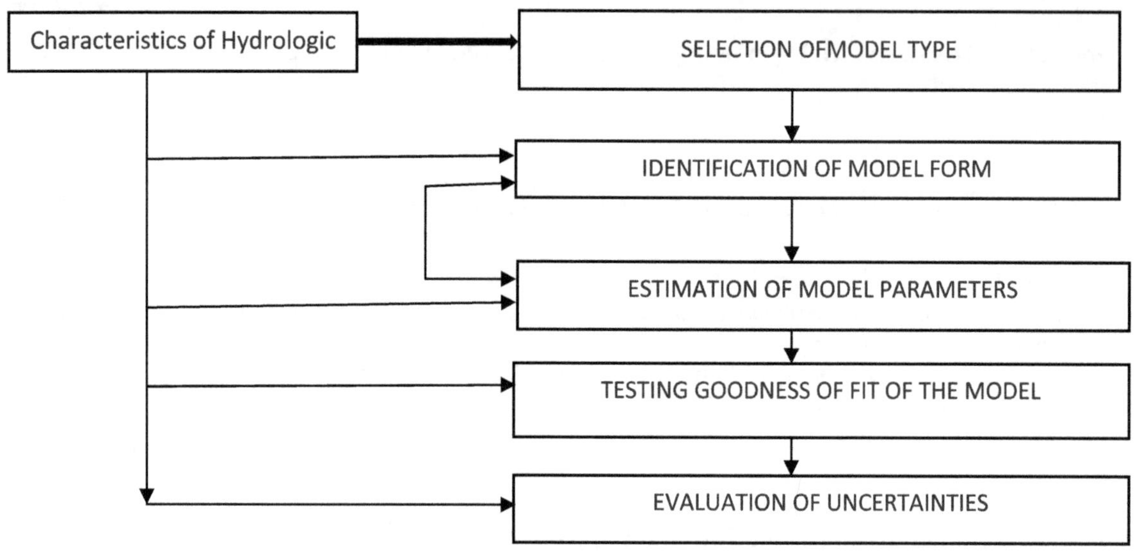

Figure 1. Schematic representation of methodology used in the time series modeling

Results and Conclusion

The main objective of the study is to develop a stochastic time series model, for the generation of annual rainfall of Nizamabad district. The annual rainfall data of the district from the year 1901 to 2002 was collected and used for the development of model. Autoregressive (AR) models proposed by *Box and Jenkins* of orders 1 and 2 were tried for annual rainfall series and different parameters were estimated by the general recursive formula proposed by *Kottegoda* . AR(2) model is was provided accurate results show in figure 2., This model is tried to develop to establish water conservation structures in order to mitigate the problem of water scarcity during summer season in Nizamabad district.

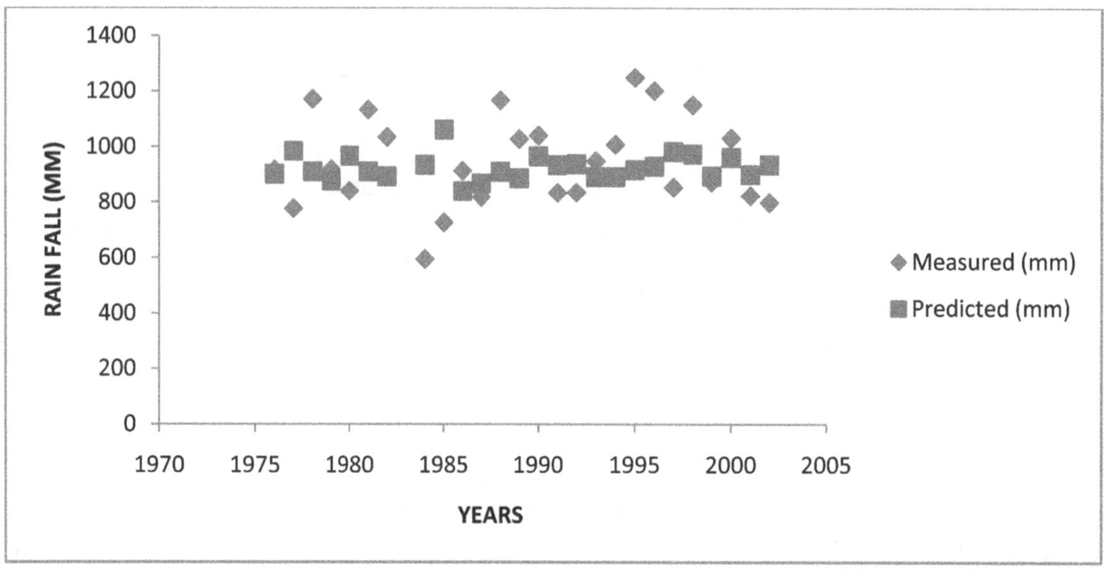

Figure 1. Comparison of measured and predicted series for rainfall

References

1. *Box, G.E.P. and Jenkins, G., (1970).* "Time series Analysis, Forecasting and control".First Ed. Holden – Day Inc., San Francisco
2. *Shaonlee Chakraborty, D.M. Denis and A. Sherring*. Development of Time Series Autoregressive Model for prediction of rainfall and runoff in Kelo Watershed Chhattisgarh ISSN: 2319-1120 /V2N2: 153-163 © IJAEST
3. *Kottegoda N.T, Horder M.A, (1980).* Daily flow model rainfall occurs using pulse and transferfunction. J. Hydrology, 47: 215-234

Application of non-destructive evaluation techniques for analysing drip emitter clogging patterns

Venkata Ramamohan Ramachandrula[1*] and Ramamohan Reddy Kasa[2]
[1] *Executive Director, Water and Livelihoods Foundation, Tarnaka, Secunderabad-500017, Telangana*
[2] *Professor, Centre for Water Resources (IST), JNTUH, Kukatpally, Hyderabad-500085, Telangana, India*
* *e-mail: ramamohan.ramachandrula@gmail.com*

Introduction

Drip irrigation systems are pressurised irrigation systems that deliver irrigation water to the plants without distribution and evaporation losses. Normally, a pressure of 1.5-2.0 kg/cm^2 (15 to 20 m pressure head) is maintained in the drip system so as to deliver water uniformly across the agriculture land. Emitter are an important component in the drip irrigation systems which deliver water to plants at atmospheric pressure after necessary dissipation of pressure. The emitter surfaces are provided with labyrinth flow paths which help in dissipation of pressure head while water passes through them. Clogging of drip emitter paths is a serious problem and is generally classified as physical, chemical and biological clogging (Bucks et al, 1979). Clogging of emitters results in reduced and non-uniform discharge across the agricultural land.

Since the emitters, particularly in-line emitters, are enclosed by the Poly Ethylene (PE) lateral pipes, the emitter surface is not visible to naked eye and very difficult to physically dissect without disturbing the deposited foreign materials. Non-Destructive Testing and Evaluation (NDTE) is a very useful branch of science whose applications are well-known in detection of defects in industrial products and medical diagnosis of internal human organs. Recent studies demonstrate applications of NDTE in soil drainage properties and water flow through porous media (Nguyen and Indraratna, 2019).

Applications of NDTE to understand drip emitter clogging patterns is an emerging area explored by very few researchers. Yanfang et al (2015 & 2018) analysed the chemical clogging patterns of drip emitters in laboratory settings by physical dissection followed by Field Emission Scanning Electron Microscope (FESEM). Current study focusses on study of chemical clogging of emitters due to excessive hardness in irrigation water. Application of various non-destructive methods, such as, X-ray radiography, Scanning Electron Microscope (SEM) scanning and Computed Tomography (CT) scanning was tried on four emitter samples. These emitter samples were collected from the farmers, after they used the drip systems for two to five years. Such a study is useful in assessing the role of geometry and orientation of labyrinth flow paths of the emitters on the clogging intensity. Comparison and merits of using these methods are summarised as major findings.

Materials and Methods

Normally drip emitters are of two types, viz., cylindrical and flat. Different drip irrigation companies in India use cylindrical and flat emitters with varying types of labyrinth flow paths on the emitters. In the current study, four drip emitters, all of cylindrical type, were chosen for NDTE tests using different methods. The four samples were collected from three villages in Yadadri Bhongir district, Telangana State in India during April 2019. Simultaneously, water samples were collected from the source wells and hardness and other chemical parameters were analysed.

The five emitter samples collected from farmers' fields were subjected to different NDTE tests, such as, X-ray scanning, Scanning Electron Microscope (SEM) and Computed Tomography (CT) scanning. The X-ray scanning was done on Fuji FDR D-EVO general X-ray system used for medical purposes in hospitals. While the X-rays easily passed through the low-density polyethylene outer tube, the inner emitter cylinder was clearly visible in the radiographic images. But, the $CaCO_3$ deposits could not be distinguished clearly from the emitter surface.

A SEM, unlike X-ray machines, allows study of exposed surfaces by magnification by several times to the original size. Electron beams emitted will fall on the exposed surfaces of the sample and return electrons are captured by the detectors. The surfaces are characterized by the absorption and reflection of electrons by different materials. The SEM equipment of Oxford Instruments (INCA Penta FETx3) make available at Advanced Analytical Laboratory, Andhra University was used for this purpose. The SEM analysis provided the magnified view of the outlet areas and sectional views that show the interface of both outer PE pipe and emitter. But, the SEM analysis could not penetrate the outer PE lateral pipe to expose the emitter clogging patterns.

Finally, the Procon make X-ray Computed Tomography (CT) mini equipment, available at the Nuclear Engineering and Technology programme, Mechanical Engineering, IIT Kanpur, was used to scan the emitter samples. The CT mini, with a resolution capacity of 28 micro meters, provided not only radiographic images but also hundreds of sectional 2D images on XY, YZ and XZ planes. Reconstruction of 3D images using these 2D images is also attempted using proprietary Avizo Fire software.

Results and Conclusion

The purpose of this study is to explore the applicability and utility of various NDTE methods, such as, X-ray scanning, SEM exposure and CT scanning to the problem of understanding the emitter clogging patterns.

Out of the three NDTE methods used, CT scanning offered clear advantage in terms of magnification to 28 micro meters resolution as well as ability to expose the physical and chemical deposits on the emitter surface. Both simple X-ray and CT scanner – which measure X-ray absorption properties of the exposed objects - have limitation in differentiating the outer PE material and that of emitter, which have closer X-ray absorption properties. The CT mini scanner offers several 2D sectional views in any desired plane and leaves the possibility of 3D image reconstruction and advanced numerical processing.

Whereas, SEM equipment has the ability to zoom to the magnification 200 micro meters but only the exposed surfaces. Both SEM and CT scanning, which apply different techniques of identifying physical materials, seems to useful in characterizing the physical and chemical deposits and cross-verifying results. However, this forms the future scope of work as far as the current study is concerned.

Acknowledgments: The authors acknowledge and thank DST-PURSE programme, Advanced Analytical Laboratory, Andhra University Visakhapatnam for facilitating emitter scanning using SEM equipment and Dr. Prabhat Munshi, Dept. of Mechanical Engineering, IIT Kanpur for guiding on use of CT mini equipment and Avizo software.

References

Bucks, D.A, Nakayama, F.S. & Gilbert, R.G (1979) Trickle irrigation water quality and preventive maintenance. Agric. Water Manage., 2: 149-162. https://doi.org/10.1016/0378-3774(79)90028-3

Liu Yanfang, Li Dan, Wu Pute, Zhang Lin, Zhu Delan, Chen Junying (2018) Clogging characteristic of different emitters in drip irrigation with hard water. Transactions of the Chinese Society of Agricultural Engineering. 34(3): 96-102. https://doi.org/10.11975/j.issn.1002-6819.2018.03.013

Liu Yanfang, Wu Pute, Zhu Delan, Zhang Lin & Chen Junying (2015) Effect of water hardness on emitter clogging of drip irrigation. Transactions of the Chinese Society of Agricultural Engineering, 31(20): 95-100. https://doi.org/0.11975/j.issn.1002-6819.2015.20.014

Nguyen T T and Indraratna B (2019) Micro-CT Scanning to Examine Soil Clogging Behavior of Natural Fiber Drains. J. Geotech. Geoenviron. Eng., 145(9), 04019037, pp.1-16. https://doi.org/10.1061/(ASCE)GT.1943-5606.0002065

DEVELOPMENT OF FLOOD FORECASTING MODEL FOR GODAVARI SUBBASIN USING MIKE 11

SANDEEP NERELLA[1], MUSKE SRUJAN TEJA[2], P.RAJA SEKHAR[3]

[1,2] Research Scholoar In Civil Engineering Department, University College Of Engineering,Hyderabad, India.

[3] Professor University College Of Engineering,Hyderabad, India.

srujanteja12@ gmail.com, sandeepnerella092@gmail.com

Introduction

Hydrological modeling of large river catchments has become a challenging task for engineers due to its complexity in collecting and handling of data such as rainfall, gauge-discharge data, and topographic and hydraulic parameters. A flood forecasting system may include all or some parts of the following three basic elements: (i) a rainfall forecasting model (ii) a rainfall-runoff forecasting model, (iii) a flood routing model. The ability to provide reliable forecast of river stages for a short period following the storm is of great importance in planning proper actions during flood event. This study was taken up to develop a mathematical model for flood forecasting system in Godavari basin area. The developed model has been calibrated based on recent years i.e. 2012, 2013 and validated for the year 2014.The main aim was to estimate the model parameters using statistical data and the comparison of simulated water levels and discharges with actual observations.

Methodology

In this study, an attempt has been made to develop a flood forecasting model for the Godavari basin using MIKE 11. The approach includes rainfall runoff modelling using GIS inputs, hydrodynamic flow routing, calibration and validation of the model with three years of observed field discharge data. This work describes a forecasting system recently developed in coordination with the Central Water Commission (CWC) for the river Godavari. The system consists of two model concepts i.e. a numerical, calibrated model consisting of a hydrological part (MIKE 11 - NAM), hydraulic part (MIKE 11-HD) and a flood forecasting system. The Flood forecasting model comprises of following modules: a) Rainfall-runoff module (RR), b) Hydro-dynamic module (HD), c) Flood Forecasting module (FF).

Results and discussion

a) Results for water level

The developed model has been applied to Bhadrachalam Station data of Godavari river basin. The simulated hydrograph of water levels at Bhadrachalam station has been compared with the actual water levels at the same station.

The Figure 1 show the comparison between the actual water levels versus simulated water levels at Bhadrachalam station for the years 2012 and 2013. The peak water levels of the simulated hydrographs were observed to be 47.15m and 46.50 m on 22nd August, 7th September, 2012 respectively.

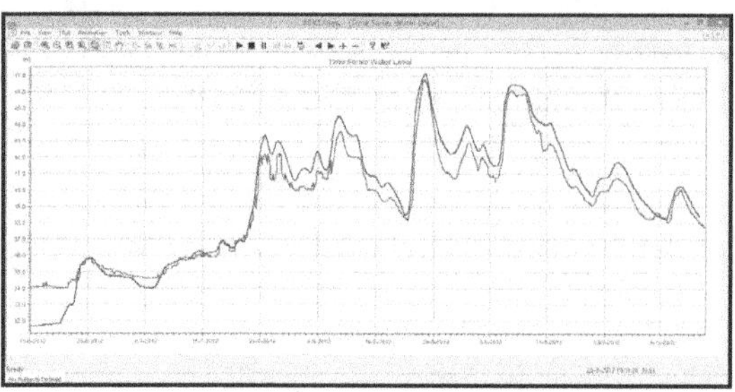
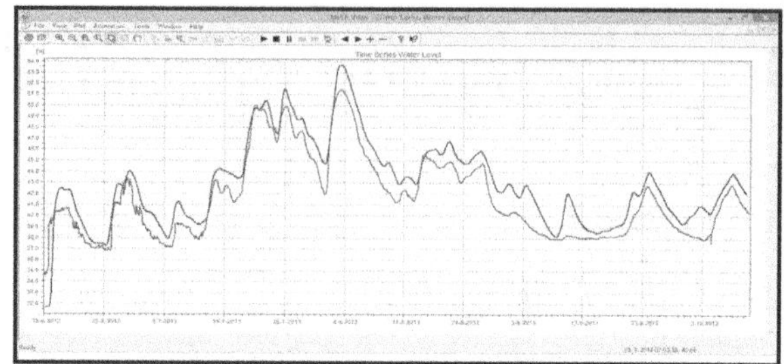

Figure 1: Comparison of water levels at Badrachalam for the year 2012 and 2013

The statistical analysis of Coefficient of Correlation (R), Coefficient of Determination (R2), Volume of Error (VE) and Peak Error (PE) was done for the Bhadrachalam station data of 2012 and 2013. From the regression analysis, the, Coefficient of Determination was found to be 0.973 and volume of error and peak error are calculated as 0.002 and -0.04 respectively. The negative value of the peak error indicates that the simulated value exceeds observed value,the Coefficient of Determination was found to be 0.909 and volume of error and peak error are calculated as 0.443 and –0.585 respectively. The negative value of the peak error indicates that the simulated value exceeds observed value.

b) Results for discharge

Discharge graphs were compared only at Bhadrachalam, flood forecast station. During high floods, discharges are not observed but are compared based on the statistical data. The simulated discharge versus time scale graph (Blue color) was plotted along with the actual discharge versus time scale graph (Red Color).

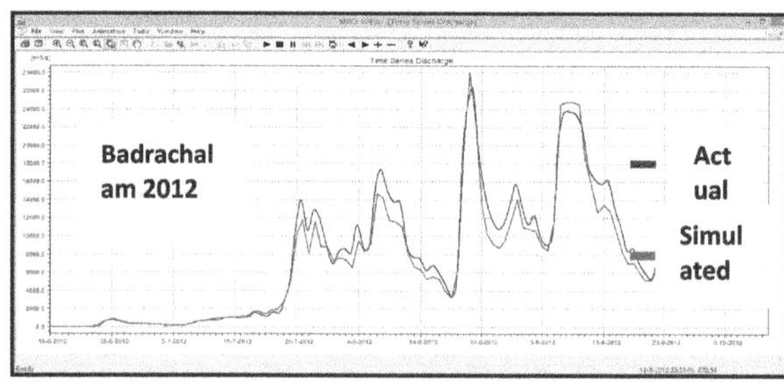

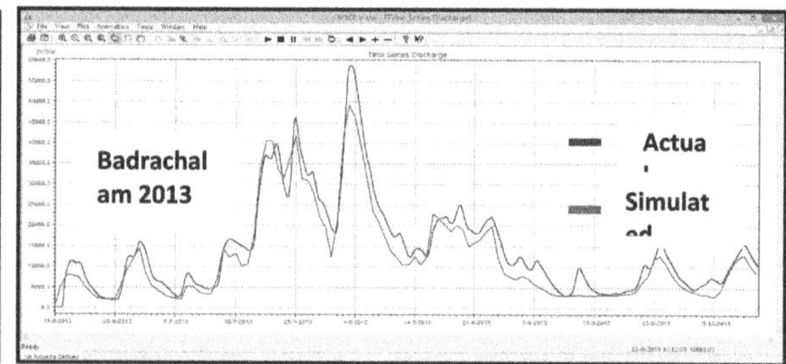

Figure 2: Comparison of Discharge at Bhadrachalam for the year 2012 and 2013

Figure 3 , shows the comparison of actual discharge with the simulated discharge levels at Bhadrachalam for the year 2012 and 2013. They were found to be in good agreement with each other for both the years.

c) Flood Forecasting Results

Flood forecasting studies were carried out for Perur to Bhadrachalam stretch and forecasts were given at Bhadrachalam. The reason for selecting Bhadrachalam for issuing forecasts was due to the fact that, it is a flood prone area.

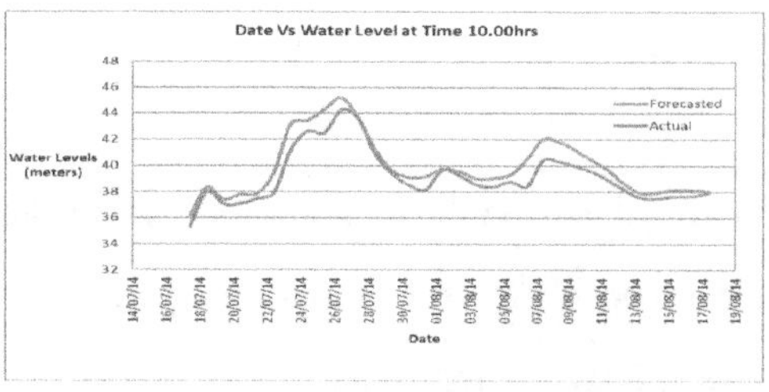

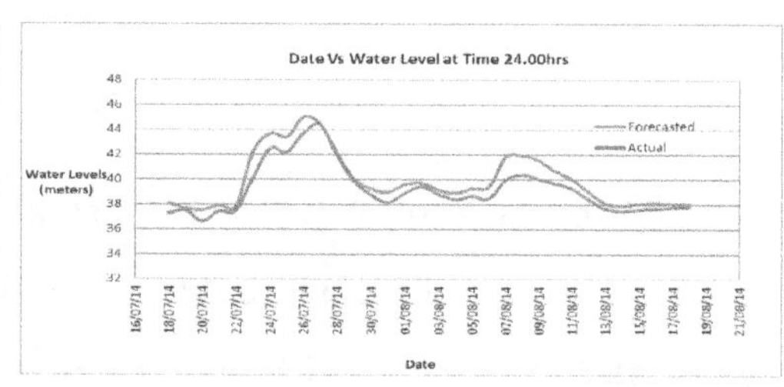

Figure 3: Comparison between Actual versus Forecasted water levels of Bhadrachalam at 10:00 hrs. and 24 hrs.

Comparison between Actual versus Forecasted water levels of Bhadrachalam at 10:00 hrs for the month of July, 2014 is shown in figure 5 .The peak water levels was observed as 44.28 m and forecasted level was observed to be 45.167 m elevation on 26th July, 2014. The Error between the forecasted and actual was about 1.96%. Comparison of actual water level and forecasted water level at Bhadrachalam station at 24 hrs for the month of July, 2014 is shown in figure 6. The peak water level was observed to be 43.77m and forecasted water level was 45.038 m. The Error between the forecasted and actual was about 1.46%

The following are the conclusions from the study:

- A hydrometric, flood forecasting and telemetry network has been developed in Godavari basin. Operations of both simple and complex river systems are analyzed using Mike11.
- The model parameters using statistical data and comparison of simulated water levels and discharges with actual observations are estimated. By Simulation studies have been conducted and the simulated result shows that the computed hydrographs match well with the observed hydrographs.
- The coefficient of determination during the years 2012, 2013, 2014 at Bhadrachalam were found to be respectively 0.973, 0.909, and 0.985.
- The peak error value was observed to be -0.04, 0.585,-0.019 respectively for the years 2012, 2013, 2014.
- The volume errors as observed from the plots at Bhadrachalam for the years 2012, 2013, 2014 were calculated as 0.002, 0.443, and 0.020 respectively.
- From the Flood Forecasting studies carried out at Bhadrachalam for 10hrs the peak water level was observed at 44.28 m and forecasted level equal to 45.167 m. The observed error was 1.96% same, also for 24 hrs the peak water level is observed at 43.77 m and forecasting level was observed at 45.035m, thus showing an error of only 1.46%.

References

1. Abbott, M. B., and Ionescu, F. (1967)."On the numerical computation of nearly-horizontal.flows." Journal of Hydraulic Research., S, 97~117.
2. Agrawal, S. K., Kharya, A. K., and Gupta, P. K. (2006)."Flood Forecasting in River Yamuna - A Mathematical Model Approach." Indian Disaster Management Congress.,29-30.
3. Ammentorp, H. C., Jergensen, G. H., and Kalken, T. (1998). "FLOOD WATCH - A GIS based decision support system." Proc. of the third Int. Conf. on Hydro informatics.,24-26.

Overland flow simulation using artificial neural networks (ANNs)

Ankit Chakravarti[1], Nitin Joshi [2] and Himanshu Panjiar [3]
[1] Department of Civil Engineering, K. L. Deemed to be University, Guntur District-522502, Vijayawada, Andhra Pradesh, India
[2]Department of Civil Engineering, Indian Institute of Technology, Jimmu, J& K, India
[3] Department of metallurgical & Material Science, Jimma University, Ethiopia, East Africa
**e-mail: (Corresponding Author) ankitcae@gmail.com*

Introduction

The overland flow can be describe as a thin sheet of flow found before irregularities cause a gathering of the runoff into discrete stream channels. The overland flow is generated as a result of rainfall in excess of the saturated hydraulic conductivity of the soil or by saturation of the soil surfaces. The overland flow process plays a vital role in the hydrological cycle also is one of the most complex hydrological phenomena to comprehend, due to its tremendous temporal and spatial variability of the basin characteristics and rainfall patters (Kumar et al. 2005). A number of models i.e. artificial neural networks, physically based, black box and conceptual models have been used to simulate the complex hydrological processes such as Overland flow (Poff et al. 1996; Abrahart et al. 1999; Dawson et al. 2001, Rajurkar *et al.*, 2002; Riad et al. 2004, de Voa et al. 2005; Jain & Srinivasulu, 2006; Wu et al. 2010; Chen et al., 2013, Chakravarti et al 2012, 2014 & 2015) shown to be one of the most promising tools in hydrology (ASCE Task Committee, 2000 and Kalteh et al. 2008;). However, due to its complexity and spatio-temporal variation, a few models can accurately simulate this highly non-linear process to forecast future river discharge, which are required for safe and economical aspects of hydrologic and hydraulic engineering design and water management purposes.

The main objective of the present study is to conduct laboratory experiment for the generation of overland flow data using rainfall simulator. For the validation this observed data, a model is establish for estimating observed runoff data using ANN technique. The networks were trained and tested using data that represent different characteristics of the watershed and rainfall patterns. The sensitivity of the network performance to the content and length of the calibration data was examined using various training data sets. Finally, suggestions are made concerning necessary refinements to the existing ANN prior to transfer to operational use.

Materials and Methods

1. Rainfall simulator

The experimental study was performed using rainfall simulator for overland flow i.e. rainfall runoff process. The rainfall simulator enables to demonstrate some of the major physical processes found in hydrology, including rainfall-runoff i.e. overland flow; the abstraction of ground water by wells etc.

2. *Simulated Rainfall Patterns*

The simulated rainfall pattern is used having rainfall intensity of 90 mm/hr having two different overland plane slopes of 1% and 2% were used. From the experimental data it is seen that for a given rainfall intensity the time to peak reduces with increase in the slope of overland plane.

3. Modeling Approach Using ANNs

The ANN model for overland flow analysis was developed using MATLAB. We collected 586 data point for overland flow at experimental model of catchment for 6 events at various slope. These recorded data was used for model development by considering various input variables (slope, rainfall intensity and rainfall duration) and output overland flow discharge, the data base used for ANN model generation. For developing ANN model, the available data were separated as 70% for training, 15% for testing and 15% for validation.

Results and Concluding Remarks

The behavior of model during training, testing and validating was performed which shows its capability to predict the process input output relation.

The training process of ANN model was terminated when the overall error on the testing dataset was minimal. The main function of the training process is to reach an optimal solution based on some performance measurement such as overall error, coefficient of determination known as R value. The validation sets are usually used to select the best performing network model. In this paper, the ANN was the optimal at 1 million iterations with 5 hidden nodes. It was found that that training (R = 0.99) and validation (R = 0.98) phases gives good agreement with coefficient of determination value.

The comparison of observed and predicted overland flow hydrograph using ANN model as shown in Fig. 1 and Fig. 2, which reveals that the model used to predict the laboratory overland flow data is highly efficient for the rising, equilibrium discharge and recession limb of the hydrograph using manning's roughness coefficient is equals to 0.028, for all the events with Nash–Sutcliffe efficiency greater than 95% which is also useful for decision making in the area of water resources management and planning, flood forecasting etc.

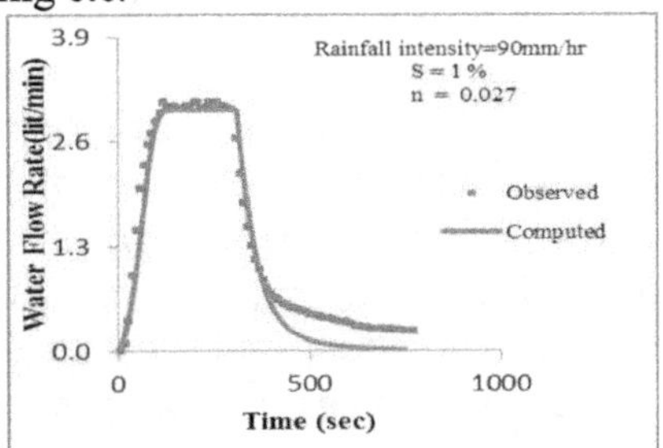

Figure 1 Comparison of observed and computed hydrograph for rainfall intensity 90mm/hr at 1% slope of the plane

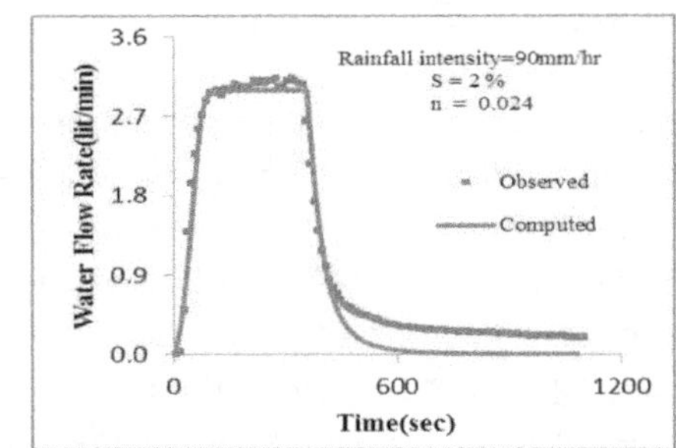

Figure 2 Comparison of observed and computed hydrograph for rainfall intensity 90mm/hr at 2% slope of the plane

References

Abrahart, R.J., See, L.M., Kneale, P.E., 1999. Using pruning algorithms and genetic algorithms to optimise network architectures and forecasting inputs in a neural network rainfall–runoff model. Journal of Hydroinformatics 1, 103–114.

Ankit Chakravarti, Nitin Joshi and Himanshu Panjiar, 2015. "Rainfall runoff modeling using artificial neural networks." *Indian Journal of Science and Technology, Vol. 8(14), pp 1-7*

Ankit Chakravarti and M. K. Jain (2014). "Experimental investigation and modeling of rainfall runoff process." *Indian Journal of Science and Technology, Vol. 7(12), pp 2096-2106,* ISSN – 0974-5645.

Ankit Chakravarti, M.K.Jain and Kapil Rohila (2012). "Experimental investigation of rainfall runoff process." *SWAT International conference in Department of Civil Engineering, IIT Delhi-India. 18-20 July, PP: 91*

ASCE Task Committee on Application of Artificial Neural Networks in Hydrology. 2000a. Artificial neural networks in hydrology. I: preliminary concepts. *J. Hydrol. Eng. ASCE,* 5(2):115-123.

ASCE Task Committee on Application of Artificial Neural Networks in Hydrology. 2000b. Artificial neural networks in hydrology. II: hydrologic applications. *J. Hydrol. Eng. ASCE,* 5(2):124-137.

Kalteh, A.M. 2008. Rainfall-runoff modelling using artificial neural networks (ANNs): modelling and understanding. *Caspian J. Env. Sci.,* 6(1):53-58.

Dawson, C.W. & Wilby, R.L. 2001. Hydrological modelling using artificial neural networks. *Prog. Phys. Geog.* 25(1):80-108.

de Vos, N.J., Rientjes, T.H.M., 2005. Constraints of artificial neural networks for rainfall–runoff modeling: trade-offs and model evaluation. Hydrology and Earth System Sciences 9, 111–126.

Jain, A. & Srinivasulu, S. 2006. Integrated approach to model decomposed flow hydrograph using artificial neural network and conceptual techniques. *J. Hydrol.* 317:291-306.

Kumar, A.R.S., Sudheer, K.P., Jain, S.K., Agarwal, P.K., 2005. Rainfall–runoff modeling using artificial neural networks: comparison of network types. Hydrological Processes 19 (6), 1277–1291.

Rajurkar M P, Kothyari U.C. and Chaube U.C. 2002. Artificial neural networks for daily rainfall–runoff modelling; *J. Hydrol. Sci.* 47:865–877.

Riad, S., Mania, J., Bouchaou, L., Najjar, Y. 2004. Rainfall runoff model using an artificial neural network approach. *Mathematical and computer Modelling,* 40:839-846.

Wu, C.L., Chau, K.W., Fan, C., 2010. Prediction of rainfall time series using modular artificial neural networks coupled with data-pre-processing techniques. Journal of Hydrology 389 (1-2), 146–167.

Performance of flexible mattress as protection for local scour around spur dikes

Lisha Sunny[1*], K.P. Indulekha[2]
[1] *Student, Department of Water Resources, College of Engineering Trivandrum*
[2] *Assistant Professor, College of Engineering Trivandrum*
* *e-mail: lishasunnythuruthel@gmail.com*

Introduction

Local scour is a major challenge experienced by structures constructed in river leading to their failure. There is an immediate change in the flow pattern around the structures. A number of experimental studies on local scour around spur dikes were reported in the past years which led to an overall understanding of the topic. Elawady et.al(2001) conducted experiments on submerged spur and proposed the relation of flow depth, spur length and height with scour area. Corrado et.al (2008) investigated the effect of spur length, spacing, height along with the main hydraulic parameters and proposed design equations to predict spur failure in terms of previous set of variables.

Throughout the world, there are many reported cases of failure of water structures, the primary reason being scouring. One of the most popular methods of stabilizing the riverbed and channel is the use of series of spur dikes. Scouring at the tip of the spur dikes is a major challenge. Two main types of protective measures given around spur include bed armouring and flow altering structures. Bed armouring structures increases the anti-scouring capability of the bed, for example, riprap, flexible mattress and geobags. Flow altering structures alters the flow patterns and tends to protect the spur. Parker et.al(1998) carried out experiments on flexible mattress as a protection around bridge pier and proposed design recommendations for flexible mattress. Huang et.al (2018) conducted experiments examining the performance of flexible mattress in resisting the scour around spur and proposed empirical formula for optimum flexible mattress width around the spur.

This paper presents a study on the erosion pattern around spur dike in mobile bed. This study also extends to series of spur dikes. The effect of flexible mattress width, spur dike length, spacing between spur dikes in the development of scour were investigated.

Materials and Methods

The experiment was conducted in a rectangular flume of fixed slope of dimensions 9.7m x 0.6m x 0.45m in Coastal Engineering lab, College of Engineering Thiruvananthapuram. The test section was of length 3.5m and width 0.6m. To avoid artificially induced end effects caused by sudden flow impact occurring at either end of the test section, gravels were distributed immediately upstream and downstream of test section. The bed consists of 10cm height throughout the test section.

The first spur dike was placed at a distance of 1.25m from the upstream of test section. Spur dikes were made up of loose gravels of size 10 to 12mm. The flexible mattresses were made up of concrete blocks of size 24 x 24 x 6mm placed uniformly with a 2mm gap on a textile cloth. The experiments were performed in a submerged condition. The stability of spur dikes during test condition was observed visually. Erosion was measured at 5cm interval along the width of flume and 25cm interval along the length of test section. Variables investigated (summarised in Table 1) include discharge, spur length, spacing between spur, flexible mattress width.

Fig 1 Flexible mattress laid around the spur dike

Results and Concluding Remarks

The experiments were conducted to study the influence of flexible mattress to prevent scour around spur dikes. Cases for single spur dike and series of three spur dikes were examined. When no protection was given, the spur dike was observed to be unstable. When the discharge was doubled, erosion around spur dike was observed to increase by nearly 50%. When experiments were done with a single spur dike, results showed that when width of flexible mattress is 1.1 to 1.3 times the length of flexible mattress, spur dike obtained stability and erosion depth decreased by 60%. When the width of spur dikes increased, the eroding area around spur dikes increased.

When experiments were done with series of three spur dikes, it was observed that as the spacing between spur dikes increased, the mutual influence of spur dike was observed to decrease and erosion increased in the area between spur dikes. As the spacing increases the width of flexible mattress provided to attain stability increased. When the spacing between the spur dikes were above 3.5 times the length of spur dikes, the dimensions of flexible mattress remains constant. Thus optimum width of flexible mattress was found.

Acknowledgments: The authors are indebted to the anonymous reviewers for their constructive comments on an earlier version of the manuscript.

References

- Elawady, E., Michiue M., and Hinokidani O. (2001). Movable bed scour around submerged spur-dikes. Journal of Hydraulic Eng. 45 (3): 373– 378.
- Gisonni, C., and W. H. Hager. 2008. "Spur failure in river engineering." J. Hydraul. Eng. 134 (2): 135–145.
- Huang, Wei. Maggie Creed, Fei Chen, Huaihan Liu, and Aixing Ma. (2018). Scour around Submerged Spur Dikes with Flexible Mattress Protection. Journal of Waterway, ASCE. 144(5): 1-14.
- Parker, G., C. Toro-Escobar, and R. L. Voigt, Jr. (1998). "Countermeasures to protect bridge piers from scour". NCHRP 24-7.Vol 1 pp: 15-19.

Risk Assessment of Flooding and Riverbank Erosion of Vamanapuram River Basin

Merin Jude[1], Indulekha K.P.[2]

[1] *Student, Department of Civil Engineering, College of Engineering Trivandrum*

[2] *Assistant Professor, Department of Civil Engineering, College of Engineering Trivandrum*

* *e-mail: merinjude92@gmail.com*

Introduction

Flood is the most devastating, widespread and frequent natural hazard of the world. In most cases this is due to the heavy rainfall for long days on the upstream highlands. Riverbank erosion is a significant problem worldwide and it is associated with an extensive loss of land along the course of the river. Vamanapuram River basin experiences flash floods. The short duration intense rainfall, insufficient drainage etc. create conditions leading to flood. Flood proofing and identifying the flood prone areas are more important in managing the flood. Extensive deforestation in the upstream portions of the river has also accelerated soil erosion, while clay and sand mining has affected bank stability.

The objectives of this study are to prepare the flood inundation map of Vamanapuram river basin for 100 year flood, to identify different road stretches affected by the 100 year flood, to detect bankline migration and quantify the erosion and deposition over a period of 50 years.

Materials and Methods

The materials used for the study include Survey of India toposheet, Landsat imageries, ASTER GDEM, gauge station data and flood discharge data of the study area, landuse map, soil map, road network map and the census data.

The following is the methodology adopted for the study:

- The Vamanapuram watershed was delineated from ASTER GDEM using ArcGIS software.
- The waterbodies were extracted from Landsat Images by creating the NDWI images.
- Flood frequency analysis was done. The annual maximum stage data for the river for a period of years were arranged in the descending order and the plotting position recurrence interval was then calculated using the Weibull formula.
- A flood inundation model was prepared by using Hydraulic models like HEC-RAS 4.0 and HEC-GeoRAS 4.0. The 100 year flood discharge at different reaches were given and steady flow analysis was conducted and the output was exported to ArcGIS where a flood inundation map corresponding to the 100 year flood profile was created.
- The 100 year flood profile was overlaid on the road network map to identify the road stretches which will be affected during a 100 year flood.
- From the Survey of India Toposheets and the NDWI images of Landsat, the banklines of the river for different years were digitised. These were then overlaid one above the other to identify the channel planform change. Gridwise assessment of the erosion and deposition area along the river course was done.

Results and Concluding Remarks

Flood inundation mapping was done using HEC-GeoRAS. Fig 1 shows the areas submerged during the event of a 100 year flood and the road stretches that would be affected by it. This inundated layer obtained from the flood inundation mapping using HEC-GeoRAS was then overlaid over road network map of the study area and the road stretches at risk were identified.

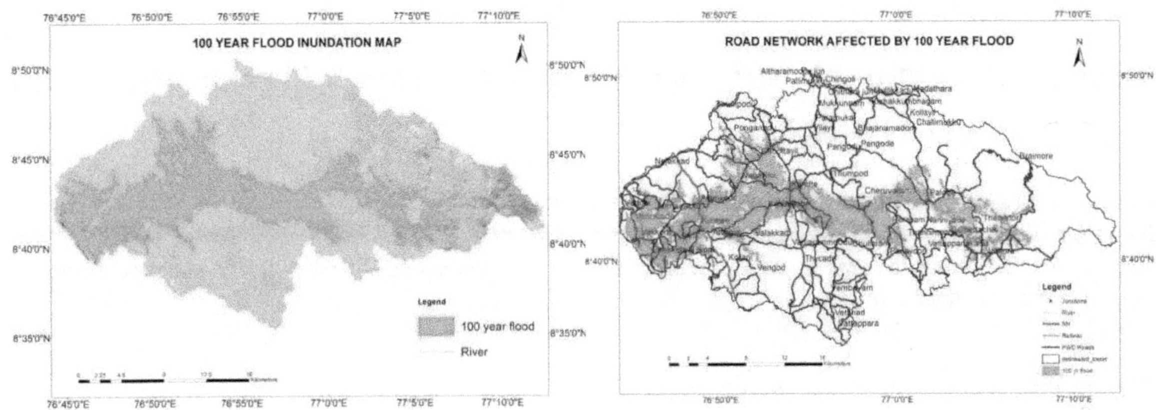

Figure.1. The areas submerged during the event of a 100 year flood and the road networks affected

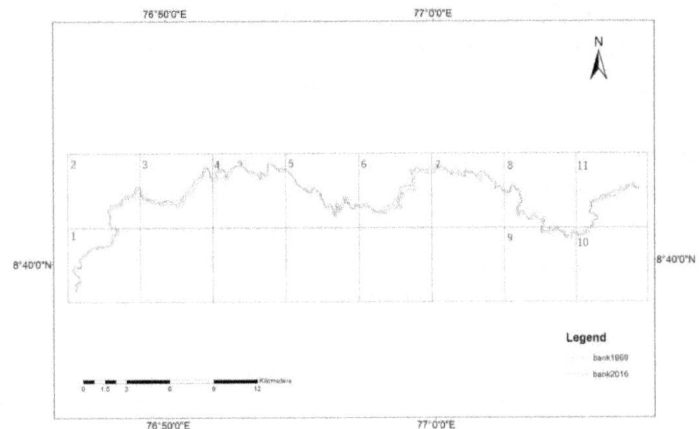

Figure.2. Gridwise Assessment of Channel Planform Change.

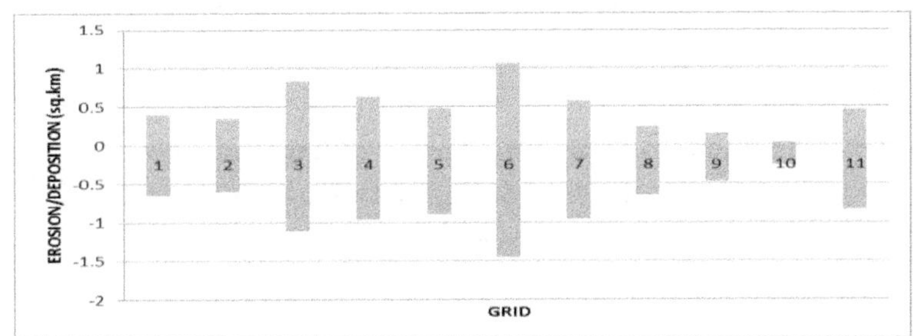

Figure.3. Gridwise change in land area (sq.km) over 1967-2016

The river bankline of 1967 and 2016 was delineated from SOI Toposheets and Landsat Images respectively. A gridwise assessment was done as shown in Figure 2 to analyse the factors for erosion and deposition in each grid. Figure 3 shows the gridwise change in land area from 1967-2016. This can be used to identify the areas which have more erosion and these areas could be prioritised. In short, the GIS and remote sensing technologies can be effectively used to predict the extent of flooding and to quantify riverbank erosion.

References

Ajin, R. ,S., R. R. Krishnamurthy, M. Jayaprakash and Vinod. P. G.(2017) Flood hazard assessment of Vamanapuram River Basin, Kerala, India: An approach using Remote Sensing & GIS techniques. Advances in Applied Science Research, vol. 4, pp. 263-274.

Dabojani, D., Deb Mithun, Kar Kanak Kant (2013).River Change Detection and Bankline Erosion Recognition using Remote Sensing and GIS. Forum Geographic, Vol 13, Issue 1, pp 25-30.

Holistic Assessment of Abstractions from Precipitation using Programming Language

Aman Srivastava[1]*, Pennan Chinnasamy [1], Leena Khadke[2]
[1] *Centre for Technology Alternatives for Rural Areas, Indian Institute of Technology Bombay, Powai, Mumbai, India*
[2] *School of Earth, Ocean and Climate Sciences, Indian Institute of Technology Bhubaneswar, Bhubaneswar, India*
** e-mail: amansrivastava013@iitb.ac.in*

Introduction

In India, the production through agriculture is especially dependent on rainfall as more than 60% of population is relying on agriculture (C Sanjay et al., 2018). As per the global statistics, the India and China are consuming around 40% water for irrigation (Gleick et al., 2000, Kehui xu et al., 2010). The major concern of the research community is to optimize the usage of water obtained as precipitation. Monsoon season is boon for agriculture but after the end of season the stored water is essential. The storage of water by arresting the runoff and the rainfall is the current requirement for agriculture and other uses, hence runoff is indicator for gain of water whereas water distributed/abstracted in various forms like evaporation, evapotranspiration (Raz-Yaseef et al., 2012), interception, infiltration, surface detention, and storage are considered as the losses in rainfall. The rainfall-runoff process is a complicated non-linear outcome of above hydrologic parameters (Ozgur kisi, 2012) and the analysis of such parameters is based on the catchment characteristics and factors affecting on it. The bulk of empirical formulae had been suggested by researchers for estimation of these parameters. The parameters associated with rainfall are calibrated manually using the data obtained from the field. The task of data loading into the empirical relations is time consuming, tedious and demands experience. Obtaining the data from the field and analysing the same in the laboratory increases the efforts and delays the work. To address this, an attempt has been taken in this paper to develop computer-based programming for estimation of precipitation in the given catchment along with the abstractions of the precipitation in the same catchment.

Materials and Methods

The program has been developed using server-side scripting language called PHP (Hypertext Pre-Processor that earlier stood for Personal Home Page). Using this, a web application has been created where all the parameters required to be calculated are available in menu.

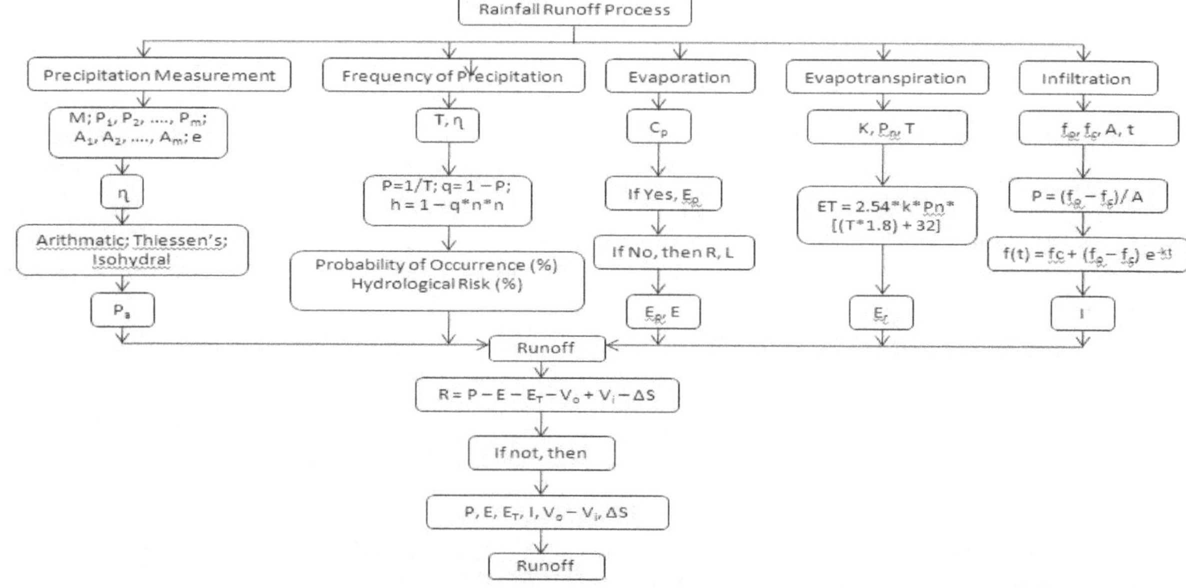

Figure 1. Working Software Model of Rainfall-Runoff Process

The development of program was done with the simplified equations of arithmetic, statistics, and theory of probability and empirical equations associated with (from Garg S.K., 2013) rainfall-runoff process. These equations were generalised with most favorable and generic values of related factors for calculation of parameters affecting the catchment hydrology. The program started with the calculation of magnitude of precipitation using arithmetic, Thiessen's polygon and Isohyetal methods. This step was followed by frequency measurement of precipitation using probability. The remaining parameters like evaporation was calculated based on the type of pan evaporimeter used; the evapotranspiration was estimated using Blaney- Criddle and infiltration was calculated using Horton's equation. In program itself, the runoff from the catchment was derived using water budget equation. The water stored over the surface after precipitation could also be provided in the program. The results obtained from model were validated with manually solved problem. The flowchart encoded in program is as shown in Figure 1.

Results and Concluding Remarks

The output was compared with observed data (*software validation*) provided by the standard specification or by solving one problem manually and compared the same by solving the problem using the program. The comparative analysis of the result obtained from these two approaches was found to be the same. The parameters calculated by manually and via programs are showing minimum error. The percentage error found in mean precipitation by Thiessen's method is 0.041%. Also, the hydrological risk in both the calculation is found to be 0.01%. These errors can be taken as approximation in field for calculations. As this was an empirical cum numerical approach the output obtained from this program will vary if the characteristics considered here had another formula with regards to field conditions. Therefore, the software application was limited to give result for only considered conditions

Table 1. Results obtained from manual calculations and developed program

Parameter	Units	Manual calculation results	Program results
Number of rain gauge required	-	12	12
Mean Precipitation by Thiessen's Method	cm	131.16	131.1057
Probability of occurrence of rainfall		12.5%	12.5%
Hydrological Risk		48.709%	48.7031%
Depth of water evaporated	m	0.00632	0.00632
Potential Evapotranspiration	m	10.09	10.0919
Rate of Infiltration	mm/hr	1.63333	1.633377
Runoff	m	119.5175	119.517

The direct beneficiaries of this productive methodology are the site engineers, as they will obtain results in the field itself. The overall idea of the paper was to avoid manual calculation by providing a computer-based program for water budgeting in the given catchment. The future scope of this paper would be to develop an android application for improved usage of this proposed program.

References

Garg S K (2013). Hydrology and water resources engineering. 23rd edition, Khanna Publisher, Delhi.

Gleick, P. H., & Cain, N. L. (2004). *The world's water 2004-2005: the biennial report on freshwater resources.* Island Press, London.

Kisi, O., Shiri, J., & Tombul, M. (2013). Modeling rainfall-runoff process using soft computing techniques. *Computers & Geosciences, 51,* 108-117. https://doi.org/10.1016/j.cageo.2012.07.001

Raz-Yaseef, N., Yakir, D., Schiller, G., & Cohen, S. (2012). Dynamics of evapotranspiration partitioning in a semi-arid forest as affected by temporal rainfall patterns. *Agricultural and Forest Meteorology, 157,* 77-85. https://doi.org/10.1016/j.agrformet.2012.01.015.

Sanjay, C., Bhasker, P., Damodare, S. L., & Abhale, A. P. (2018). Statistical analysis of seasonal rainfall variability in Nasik district by using GIS interpolation. *JOURNAL of Pharmacognosy and Phytochemistry, 7*(4), 2072-2077.

Xu, K., Milliman, J. D., & Xu, H. (2010). Temporal trend of precipitation and runoff in major Chinese Rivers since 1951. *Global and Planetary Change, 73*(3-4), 219-232. https://doi.org/10.1016/j.gloplacha.2010.07.002.

Estimating parameters associated with well hydraulics by developing groundwater software model

Aman Srivastava[1*], Pennan Chinnasamy [1], Leena Khadke[2]
[1] Centre for Technology Alternatives for Rural Areas, Indian Institute of Technology Bombay, Powai, Mumbai, India
[2] School of Earth, Ocean and Climate Sciences, Indian Institute of Technology Bhubaneswar, Bhubaneswar, India
* e-mail: amansrivastava013@iitb.ac.in

Introduction

The past research shows a lots of model studies over groundwater hydrology. From the past few decades, the groundwater hydrology is shifting towards modeling perspective to increase evaluation efficiency (Stephen, 2017). The issues like declination in groundwater level and the increased susceptibility of surface water is attracting the attention of the researchers towards focused study on groundwater hydrology (Asano. 2016). The study of well hydraulics plays a major role in groundwater management and is mainly concern with the structure and composition of earth material as the 65% of geographical area is covered by hard rock (Saraf et al. 2004). The aquifer found as the most saturated formation of earth which not only stores water but also yield in sufficient quantity (Dingman, 2015). Basically, the confined and unconfined aquifer exhibits the steady and unsteady state of groundwater flow. This results into the determination of the permeability and discharge through aquifer (K Subramanya 2013). The broad idea of the present work is to avoid manual calculations by providing a computer-based program in order to simplify the conception of well hydraulics. The C language has been used for the same. The site engineers are the direct beneficiaries for this adopted productive methodology as instead of collecting data from the field and analysing it in the laboratory, the same can be achieved on the field itself.

Materials and Methods

The development of program has been done using C language. The menu in the program has covered the parameters like discharge, recuperation of well, well efficiency, well loss, and specific capacity of well. The flowchart encoded in program is as shown in Figure 1. As the inputs are provided to the program, it will execute the calculations as the sequence of parameters encoded in it. The provided inputs into model will simply execute the output. The fundamental equations embedded into programming were assumed the most accepted and generalised values for its simplification. The model can be calibrated with field observations. The model validation was done by comparing the model output with observed data provided by standard specifications. The results obtained for a problem using program was also compared with the results obtained by manually solving the problem.

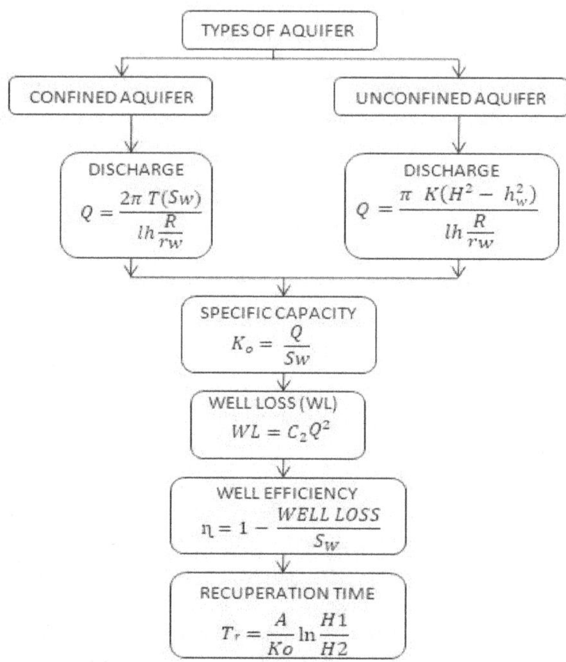

Figure 1: Groundwater Software Model

Results and Concluding Remarks

The groundwater software model developed in this paper provided analytical cum mathematical approach of understanding the well hydraulics using computer-based programming. The mathematical interdependencies between the parameters resulted into formula (*modeling*). For this, famous empirical formulae proposed by several researchers were used in the development of the program. When the parameters of the delinquent area (input) was applied over the model (*model calibration*), the program interpreted the data (processing) and relieved the information into the obligatory formula already defined in the program which produced one of the results (output) for the input data. The output was compared with observed data (*model validation*) provided by the standard specification like that of the report of *Groundwater Resources Estimation Committee, Ministry of Water Resources, Government of India* etc. As this was an empirical cum numerical approach of estimating the properties of well, hence the output obtained from this model would vary if the characteristics considered here have another formula (empirical or numerical) with respect to some other field conditions. Therefore, the model was limited to give result of only those conditions for which the corresponding formula is being considered in the currentprogram.

The comparison between model output and manual calculations are showing the deviation in range of –0.5% to +0.5%. For example, the value of well efficiency obtained through model is 99.99% and through manual calculation is 99.57%. The outputs for other parameters are shown in table. The output for confined aquifer has also beencalculated.

Table 1. Results obtained from manual calculations and developed program for unconfined aquifer

Characteristics/ Type of Aquifer	Units	Unconfined aquifer by program	Unconfined aquifer by manual calculations
Diameter of Well (Dw)	m	0.45	0.45
Width of Aquifer (B / H)	m	30	30
Radius of Influence (R)	m	300	300
Drawdown at Well (Sw)	m	3	3
Permeability of Soil (K)	m/s	2.3148×10^{-4}	2.3148×10^{-4}
Depth of Depression of Water in Well (H1)	m	4	4
Depth of Recuperation of Water in Well (H2)	m	0.4	0.4
Results			
Discharge	m^3/s	0.015938	0.01593
Specific Capacity	m^2/s	0.00531	0.00531
Well Efficiency		99.57%	98.59%
Recuperation Time	sec	68.96	68.9296

Hence, it can be concluded that the methodology of understanding the conceptions of well hydraulics provided an interdisciplinary approach to the researchers to link the field level analysis on wells with computer-based programming. The future scope of this paper would be to transmute this program into an android application for improved use of this contemporary model.

References

Asano, T. (Ed.). (2016). *Artificial recharge of groundwater*. Elsevier. Butterworth Publishers, London.

Stephen A. Thompson (2017). Hydrology for water management. CRC Press, London. https://doi.org/10.1201/9780203751435

S. Lawrence Dingman (2015). Physical hydrology.Waveland Press, inc. Long Grove, Illinois.

Gorelick, S. M. (1983). A review of distributed parameter groundwater management modeling methods. *Water Resources Research*, *19*(2), 305-319. https://doi.org/10.1029/WR019i002p00305.

Jones L., Willis R., & Yeh W. W. G. (1987). Optimal control of nonlinear groundwater hydraulics using differential dynamic programming. *Water Resources Research*, *23*(11), 2097-2106. https://doi.org/10.1029/WR023i011p02097

Saraf, A. K., Choudhury, P. R., Roy, B., Sarma, B., Vijay, S., & Choudhury, S. (2004). GIS based surface hydrological modelling in identification of groundwater recharge zones. *International Journal of Remote Sensing*, *25*(24), 5759-5770. https://doi.org/10.1080/0143116042000274096.

K Subramanya (2013). Engineering hydrology. Tata McGraw-Hill Education, Delhi.

2D numerical modelling of short-term morphological change in meander bends

K.P. Indulekha[1], R.L. Abhijith[2*],
[1]*Assistant Professor, Depatment of civil engineering, College of Engineering Trivandrum*
[2]*PG Student, Department of Civil Engineering, College of Engineering Trivandrum*
** e-mail: abhijith2555@gmail.com*

Introduction

Prediction of morphological changes in meander bends is important to assess the performance of river-training works and channel navigability. With use of a groyne field, both sufficient navigable depth improvement and bank erosion protection are possible in meander bends. The hydraulic performance of a groyne field depends mainly on its layout and dimensions and local hydraulic and morphological conditions. To optimize groyne field design, either a physical or a numerical river model study is required Several studies have been carried out to predict groyne field morphological changes in a river reach using physical river models (e.g., Klingeman et al. 1984; Kuhnle et al. 1999). On the other hand, several 2D river models have been used by many investigators for hydrodynamic and morphological simulation. Abad et al. 2008 and Papanicolaou et al. 2010 well validated the two and three-dimensional numerical models with groyne field performance in straight or meandering physical river model studies for river bank protection or flow diversion. Lotsari et al. (2014) studied morphological changes in the braided lower Tana River using numerical simulations. Very recently Karmaker and Dutta, 2016 conducted a study to predict the short-term morphological change in large braided river using 2D numerical model.

Based on the physical model experiments carried out for the typical meandering section of Vamanapuram river, this study developed a MIKE 21 C numerical model to predict the short-term morphological changes in the selected section, under different groyne field layouts, for the better understanding of the flow, scour and sediment deposition patterns associated with implementing the groynes.

Materials and Methods

A physical model of a typical section of Vamanapuram River was constructed in the Coastal Engineering Laboratory at the College of Engineering Trivandrum (Figure 1). Experiments were conducted with and without groynes to identify the critical regions of erosion. For each set of experiments velocity is measured at 14 different sections of physical model using ADV instrument. The eroded bed levels after experiments were measured at the different points in 14 sections. Different layout designs for a series of groynes were tested to determine the optimal design for the given situation, in terms of the projection lengths of the groynes, the spacing between the groynes as a factor of the projection length (S/L ratio), and the orientation of the groynes with regard to a constant discharge condition. A mathematical model, MIKE 21 C is used to simulate the physical model numerically. Simulated flow depth and velocity were calibrated with the observed data from the physical model, and predicted bed levels were validated with observed bed levels, under no groyne condition and with groyne condition for 45, 90,135 degree orientation respectively for a S/L ratio 2.5 Later calibrated numerical model is used to evaluate the behaviour of model morphology, under all possible orientation and S/L ratio of the groynes that are difficult to carry out by physical model experiments.

Results and Concluding Remarks

This study reveals that a 2D morphological model with appropriate sediment functions and boundary conditions can be used for adequately predicting short-term morphological changes in a meander bends.

From the 7 sediment transport formula available in the numerical model it was found that Engelund and Hansen formula with 90% bed load and 10 % suspended load is optimum for the present situation. Calibrated numerical model showed the same trend of erosion as in the physical model. Numerical model successfully simulate the effect of groyne field with 45, 90,135 degree orientation (Figure 2). From the results obtained from the physical and numerical models, it was found that groynes with a S/L ratio 2.5 with 10 cm projection is optimal. Higher S/L ratio makes more erosion near the groyne tip and in some situations model became unstable in higher S/L ratio.

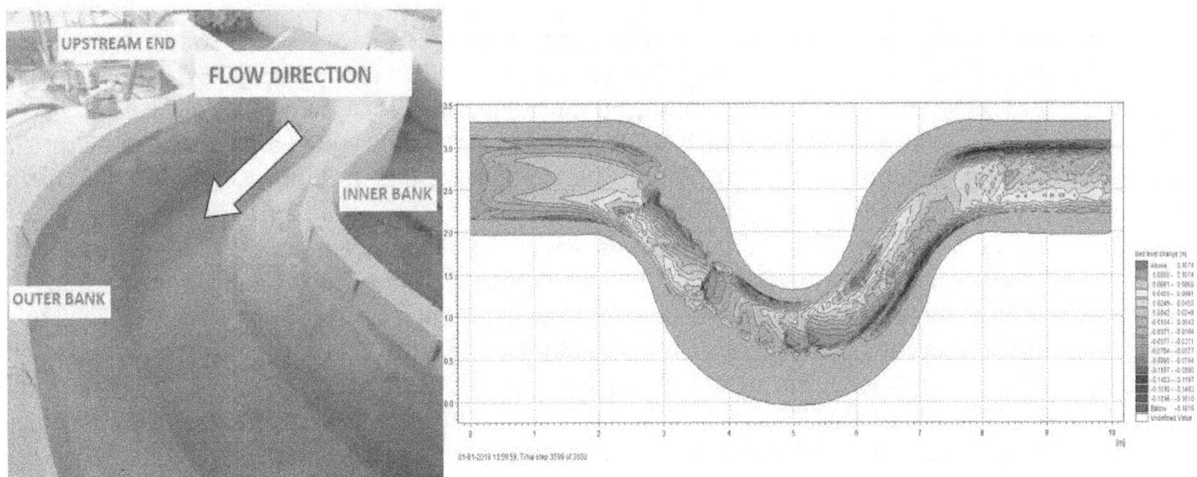

Figure 1. physical model *Figure 2. bed level change simulated for 45 degree groyne with S/L ratio 2.5*

Acknowledgments:

The authors are thankful to the Kerala State Council for Science, Technology and Environment, Kerala, India (Grant No. ETP/1566/2016/ KSCSTE) for providing financial assistance for physical modelling under Engineering Technology Programme (ETP).The authors are also grateful to DHI organization for providing evaluation licence of MIKE 21 C software and full support during the project.

References

Abad, J. D., Rhoads, B. L., Guneralp, I., and Garcia, M. H. (2008). “Flow structure at different stages in a meander-bend with bendway weirs.” J. Hydr. Eng., 10.1061/(ASCE)0733-9429(2008)134:8(1052), 1052–1063.

Klingeman, P. C., Kehe, S. M., and Owusu, Y. A. (1984). “Streambank erosion protection and channel scour manipulation using rockfill dikes and gabions.” Technical Rep. WRRI-98, Water Resources Research Institute, Oregon State Univ., Corvallis

Kuhnle,R.A.,Alonso,C.V.,and Shields,F.D.(1999).“Geometry of scour holes associated with 90° spur dikes.” J. Hydr. Eng., 10.1061/(ASCE) 0733-9429(1999)125:9(972), 972–978

Lotsari, E.,Wainwright,D.,Corner,G.D.,Alho,P.,andKayhko,J.(2014).“Surveyed and modelled one-year morphodynamics in the braided lower Tana River.” Hydrol. Processes, 28(4), 2685–2716.

Papanicolaou, A. N., Elhakeem, M., Dermisis, D., and Young, N. (2011). “Evaluation of the Missouri River shallow water habitat using a 2D-hydrodynamic model.” River Res. Appl., 27(2), 157–167.

Tapas Karmaker and Dutta (2016) “Prediction of Short-Term Morphological Change in Large Braided River Using 2D Numerical Model” J. Hydraul. Eng.,ASCE 10.1061,pp 56-62.

Sub Watershed Prioritization Based on Morphometric Characteristics Using Fuzzy Logic

B. Naga Malleswara Rao
Professor, Department of Civil Engineering, CVR College of Engineering, Hyderabad, India
e-mail: bnmrao@gmail.com

Introduction

The prioritizations of watersheds are carried out to take appropriate soil and water conservation measures. Some of the works done by various researchers in the fields of soil loss estimation and watershed prioritization have been studied and used for the analysis of the present work (Suresh et al., 2004; Ratnam et al., 2005; Kalin and Hantush, 2009; Pandey et al., 2009; Niraula et al., 2011; Pai et al., 2011). Geographic Information Systems (GIS) and remote sensing (RS) techniques were useful spatial technologies for morphometric characterization of subwatersheds (Aher et al., 2010; Rao et al., 2011). In recent research studies on prioritization of watersheds using multi-criteria evaluation through fuzzy analytical hierarchy process (Aher, P. D., 2013). A fuzzy prioritization model is developed in the study to incorporate the uncertainty in estimating the complex array of critical factors drawing from soils, land use, groundwater depth, slope etc. Sub watersheds are classified using a fuzzy linguistic triplet (poor, average and good). In the present study four land use variable model is used to prioritize sub watersheds for watershed developmental activity. This study attempts to delineate the sub watersheds as poor, average and good and its spatial distribution for associated land use types in Ghanpur watershed.

Materials and Methods

The watershed taken up in the present study is situated in Ghanpur mandal which is in the western part of Warangal district and is geographically located between North latitude 17^0 7′ 9″ and $17^0$9′48″ East longitude $79^0$1′52″ and $79^0$4′37″.The survey of India topo sheets 56 O/5 and 56 O/2 cover the entire watershed area.

Prioritization of Sub-watersheds by Morphometric Analysis

The parameters required for the prioritization are obtained from the GIS database of the study area. The parameters are calculated in GIS platform. All these parameters are normalized for using them in the prioritization depending on the relationship among the parameters.

Fuzzification: choosing the membership functions

Fuzzification consists of computing the fuzzy membership degree (FMD) of the selected parameters pertaining to the fuzzy sets indicating 'poor', 'average' and 'good' for prioritization. The membership function of a fuzzy set, usually expressed as $f_A(x)$, defines how the grade of membership of x in A is determined. Several suitable functions which can be easily adapted to specific requirements can be used for defining flexible membership grades. Trapezoidal, Gaussian and sigmoid functions with three parameters are the commonly used membership functions in many engineering studies. The parameters of these membership functions can be appropriately chosen to get either a S- or bell shape. These two shapes seem to be intuitively reasonable shapes for many applications (Sasikala et al., 1996; Jiang, 1996; Mitra et al., 1998). For these reasons, bell-shaped and S- Shaped functions are adopted in this study. For the sigmoid function, the following equation is used:

$$\mu = \cos^2 \alpha \quad (1)$$

Where, in the case of a monotonically decreasing function:

$$\alpha = \frac{(x - point\,c)}{(point\,d - Point\,c)} * \frac{\pi}{2} \quad (2)$$

When x < point C, $\mu = 1$

In the case of a monotonically increasing function:

$$\alpha = \left[1 - \frac{(x - point\,a)}{(point\,b - Point\,a)}\right] * \frac{\pi}{2} \quad (3)$$

When x > point b, $\mu = 1$

Using Eqs (1), (2) and (3) the membership functions for slope gradient, percentage of vegetation cover, type of soils and groundwater depth (GWD) are implemented. The values of inflection points of membership function are chosen based on discussion held with the local engineers involved in prioritizing the sub watersheds.

Model Development

For the objective function of prioritization of sub-watersheds, seven morphometric evaluation variables were decided as criteria's.

Sub Watershed Rank $\mu_{SWR}(x)$ = f [soils, slope, GWD, land use] (4)

Eq. 4 is expressed as a fuzzy model. Eq. 4 requires knowledge of the fuzzy membership degrees that X adopts for the selected land use elements in the fuzzy sets poor, average and good suitability, which are obtained by fuzzification. Thus, $\mu_{SWR}(x) = \max[\min\{\mu_{PS}(x), \mu_{PSl}(x), \mu_{PGW.}(x), \mu_{PLu}(x)\}$

$$\min\{\mu_{GS}(x), \mu_{GSl}(x), \mu_{GGW.}(x), \mu_{GLu}(x)\}$$

$$\min\{\mu_{AS}(x), \mu_{ASl}(x), \mu_{AGW.}(x), \mu_{ALu}(x)\}] \quad (5)$$

where, SWR: Sub Watershed Rank

P: Poor, G: Good, A: Average and S: Soils, Sl: Slope, GW: Groundwater, Lu: land Use

Results and Concluding Remarks

By integrating the four land use variable models for sub watershed prioritization Figure 1 is obtained. One land use factor alone cannot properly characterize prioritization and thus a good number of watershed variables would provide more reliable results. Models are built up using four, five and six parameters at a time, applying Eq. 5. The present model incorporating the percentage of vegetation cover (Lu), slope (Sl), groundwater depth (GW) and percentage of loam soils (S) to perform the analysis.

Table 1. Prioritization rankings of the sub-watersheds.

	Sub watersheds					
	4D1C8u	4D1C8v	4D1C8w	4D1C8x	4D1C8y	4D1C8z
Fuzzy Membership Degrees	0.020	0.443	0.128	0.382	0.017	0.152
Prioritization Ranks	**5**	**1**	**4**	**2**	**6**	**3**

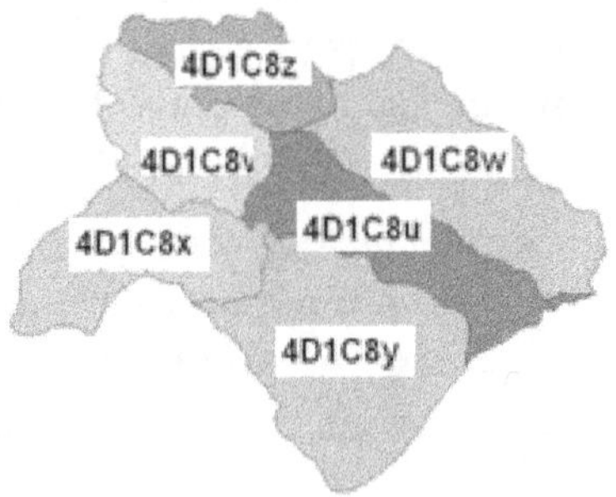

Figure 1. *Prioritization rankings of the sub-watersheds*

References

Aher, P. D., K. K. Singh, and H. C. Sharma (2010) Morphometric characterization of Gagar watershed for management planning. In 23rd Convention of Agricultural Engineers and National Seminar. Rahuri, India: Mahatma Phule Agril. University, 6-7 February.

Kalin, L., and M. Hantush (2009) An auxiliary method to reduce potential adverse impacts of projected land developments: subwatershed prioritization. EnvironmentalManagement, 43(2): 311-325.

Niraula, R., L. Kalin, R. Wang, and P. Srivastava (2011) Determining Nutrient and Sediment Critical Source Areas with S WAT: Effect of Lumped Calibration. Transactions of the ASABE, 55(1): 137-147.

Pandey, A., V. M. Chowdary, B. C. Mal, and M. Billib (2009) Application of the WEPP model for prioritization and evaluation of best management practices in an Indian watershed. Hydrological Processes, 23(21): 2997-3005.

Pai, N., D. Saraswat, and M. Daniels (2011) Identifying priority subwatersheds in the Illinois river drainage area in Arkansas watershed using a distributed modeling approach. Transactions of the ASABE, 54(6): 2181-2196.

Analysis of Precipitation for Improve Watershed to Krishna River

J Gopala Rao[*1], Gunwant Sharma[2], Sudhir Kumar[3]
[1]Ph.D. Scholar, Department of Civil Engineering, MNIT Jaipur, Rajasthan, India
[2]&3Professor, Department of Civil Engineering,
[*]Corresponding author. Email: gopalarao1992@gmail.com

Introduction

The present study is an attempt to evaluate the spatial and temporal rainfall variability of surrounds Krishna basin and Rainfall water Contribution to Krishna River from Andhra Pradesh (AP) and Telangana Areas in Sothern India. The Krishna River is the fourth-biggest river in terms of water inflows and river basin area in India. After the Ganga, Godavari and Brahmaputra (MWR et al. 2017, 2018). Hence the proper understanding of rainfall pattern and its trends may help water resources development and to make decisions for the developmental activities of that place. The annual normal rainfall of AP is 940 mm. during the North-East Monsoon (October to December).

The spatial and temporal variability of rainfall at Seonath sub-basin in Chhattisgarh State (India) for 49 years (1960-2008). Showed the results a downward trend in annual and seasonal rainfalls by using nonparametric tests and Spearman's rho test (parametric) to detect the trend (Chakraborty et al. 2013) and (Sharma et al. 2018) others used same techniques for different time steps.

Materials and Methods

The present study carries 0.25° × 0.25° gridded data set of Annual rainfall from the period 1994 to 2013 (20 Years) was used for the spatial and temporal variability and trend analyses in the present study shown in Figure 1 a). These data are obtained from the Indian Meteorological Department (IMD), Pune. Therefore, Gridded data sets of rainfall and temperature have been used in many hydrological and climatological studies worldwide (Dash et al. 2013), for hydro-climatic forecasting with ARIMA (Sidiq et al. 2018), climate attribution studies and climate model performance assessments. The following methods of Kendal's Rank Correlation, Sen's slope estimator, the percentage change in rainfall, Mann-Whitney Pettitt's Test, Euclidean clustering method, ARIMA were applied to the present study.

Kendall Rank Correlation (τ): the numerical value of "tau" (Kendall et al. 1970) and it was written as τ=S/D. Where S is the Kendall score, D is the maximum possible value of S. **ARIMA Model:** The equation (1) for the simplest basic ARIMA (p, d, q) model. Where, AR: p = order of the autoregressive part, I: d = degree of differencing involved and MA: q = order of the moving average part

$$Y_t = C + Y_{t-1} + Y_{t-2} + \cdots \ldots + Y_{t-p} + e_t - e_{t-1} - e_{t-2} - \cdots \ldots - e_{t-p} \quad (1)$$

Results and Concluding Remarks

Detailed trend analysis for annual rainfall was carried out from 1994 to 2013 using Non-parametric tests. It is important to check the autocorrelation properties of the data before deciding on the selection of trend detection test. Based on the Mann-Kendall test on rainfall data, the trend analysis results for all the watersheds have presented in Table 1. If the p-value is less than the significance level α = 0.05, the null hypothesis H_0 is rejected. Rejecting H_0 indicates that there is a trend in the time Series while accepting H_0 indicates no trend was detected.

Table 1. Time series values of Annual rainfall at different Watersheds

Watershed	Cluster Variable	Mean	Mann-Kendall Test			Sen's slope (mm/Yr)	% Change in Rainfall	Pettitt's Test	
			Kendal's tau	P-value	α			t	P-value
W1	C1	698.228	-0.100	0.559	0.1	-4.687	-13.4267	2009	0.479
	C2	700.124	0.100	0.559	0.1	2.248	6.420802	2004	0.643

W2	C11	722.888	-0.016	0.948	0.1	-0.826	-2.28404	2012	0.982
	C22	751.313	0.037	0.846	0.1	0.699	1.860483	1994	0.982
W3	C111	790.048	-0.016	0.948	0.1	-0.438	-1.10894	2008	0.932
	C222	840.720	-0.026	0.897	0.1	-0.453	-1.07768	2011	0.947

Table 2. Summary of each Parameter ARIMA and Selection Criteria (AIC) for Annual Rainfall

Watershed	Cluster Var.	Mean	S.D	Best Model (p,d,q)	SSE	MSE	RMSE	AIC
W-1	C1	698.2	175.0	(1,1,0)	635308.200	35294.900	187.800	246.184
	C2	700.1	150.0	(2,1,0)	246042.320	13669.010	116.910	237.770
W-2	C11	722.8	196.4	(1,1,0)	473278.170	26293.200	162.150	243.270
	C22	751.3	166.5	(1,1,0)	378358.700	278083373.000	144.900	240.315
W-3	C111	790.0	168.6	(1,1,0)	428781.000	23821.100	154.300	241.517
	C222	840.7	186.7	(1,1,0)	569132.500	31618.400	177.800	247.178

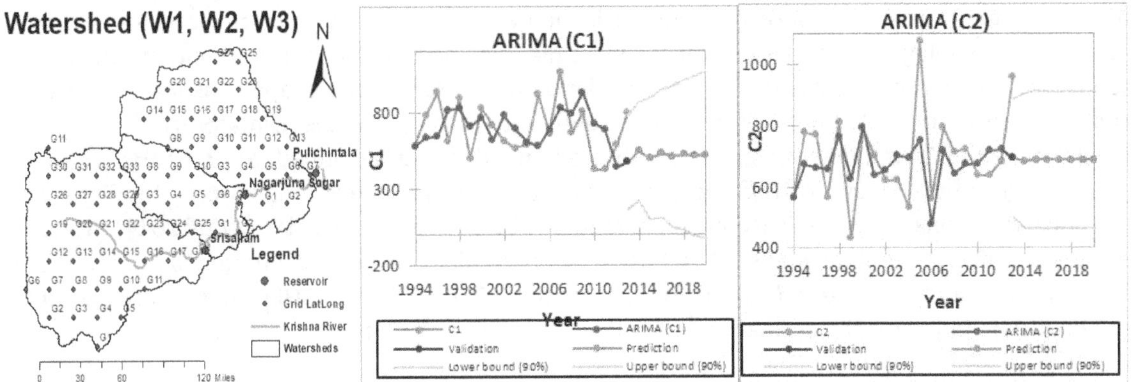

Figure 1. a) Geographic locations of Watersheds (W1, W2, and W3), b) and c) are trends of C1 and C2

In those plots, C1 and C111 are the significant decreasing trend with magnitude shown in Table 1. Similarly, C2, C11, C22, and C222 are the significant increasing trend with magnitude shown in Table 1. One cannot reject the null hypothesis H_0. From ARIMA, the forecasting trend is in increasing in C1 variable and C2 variable have straight cum randomly increasing trend, Table 2 showing the reference results of the C1 and C2 variable. It may be concluded that the annual rainfall in 3 watersheds, out of 3 watersheds, W1 (C2) is getting good value. Here forecasted next 7 years by using 20 years data. This model is considered appropriate to predict the annual rainfall for the upcoming years to assist decision-makers to establish priorities for water demand, storage with the distribution.

Acknowledgment: We acknowledge the Indian Meteorological Department (IMD) in Pune for providing useful meteorological gridded datasets for this research work.

References

Chakraborty S, Pandey RP, Chaube UC, Mishra SK (2013) Trend and variability analysis of rainfall series at Seonath River Basin, Chhattisgarh (India). Journal of Hydrology, 247

Dash SK, Saraswat V, Panda SK, Sharma N (2013) A study of changes in rainfall and temperature patterns at four cities and corresponding meteorological subdivisions over coastal regions of India. Glob. Planet. Chang: 10.1016/j.gloplacha.2013.06.004

Kendall MG (1970) Rank Correlation Methods. 2nd Ed., and New York: Hafner.

Ministry Of Water Resources, River Development and Ganga Rejuvenation (2017-18) Annual Report. New Delhi, http://mowr.gov.in/sites/default/files/Annual_Report_MoWR_2017-18.pdf

Sharma S, Ranjeet, Dubey P, Mirdha IS, Sikarwar RS (2018) Precipitation Trend Analysis by Mann-Kendall Test of Different Districts of Malwa Agroclimatic Zone. Environment and Ecology 36 (2A), 664-671

Sidiq M (2018) Forecasting Rainfall with Time Series Model. IOP Conf. Series: Materials Science and Engineering, 407, DOI:10.1088/1757-899X/407/1/012154

Monthly reference evapotranspiration estimation using ANN model for Surathkal,Karnataka

M.Sarika[1*], A.Mahesha[2]
[1] *Research Scholar,* [2] *Professor ,Department of applied Mechanics and Hydraulics ,NITK,Surathkal,Karnataka,India*
** sarikacivil@gmail.com*

Introduction

Evapotranspiration is one of the major components of the hydrologic cycle. Accurate estimation of evapotranspiration is essential in many studies such as hydrologic water balance, irrigation scheduling, and efficient water resource planning and management. Irrigation futures aim to identify an appropriate method for the calculation of reference crop evapotranspiration. The ET_0 can be measured by the lysimeter or estimated indirectly using the climatological data. The measurement of ET_0 is a time consuming method and need carefully planned experiments.

Most of the ET_0 methods giving good results comparable with Penman Monteith ET_0 may be selected at regional level for reasonable estimation of $ET_{0.}$(K.Chandrasekhar Reddy,2014).In recent decades, artificial neural networks (ANN) has shown their high ability for estimating parameters and modelling the complex systems(Nazanin Abrishami et.al,2018). ANN and ANFIS models are identified as beneficial models for the estimation of ET_0 (Sivash Gavili et.al,2018.

Materials and Methods

The daily values of maximum and minimum temperature, relative humidity,sunshine hour, wind speed data collected for past 8 years 2001 to 2008 for the surathkal. From this monthly average values of all the parameters calculated for all 8 years. The average monthly values for all the 8 years calculated by using the CROPWAT software. This data is used as observed evapotranspiration data.

Imported the both input,Target files to the MATLAB workspace.Here out of 100% data , 70% data is selected for training the NN model. 15% data used for Validation. 15% data used for Testing.The neural network is trained with given input and target datas. Here MLP is used. In MLP network Back propagation learning alogoritham is used. Input data is feed forward through the network optimize the weights between neurons.Adjustment of the weights is done by back propagation of the error during training phase.The entire process was carried out using MATLAB.

Results and Concluding Remarks

Performace indices of ANN model

Time series	Slope of the scatter plot		Intercept of the scatter plot		Coefficient of determinion R^2		RMSE	
	Training	Testing	Training	Testing	Training	Testing	Training	Testing
Monthly	0.97	1.1	0.1	-0.3	0.98	0.94	0.17	0.18

Conclusions:

The nearly unit slope and zero intercept of scatter plots indicate that ANN model ETo comparable with that of CROPWAT (FAO PM).High value of R^2 ,low value of RMSE indicate satisfactory performance of the models.The proposed model may therefore used for the ETo estimation in the selected region with degree of accuracy.

References

Nazanin Abrishami,Ali Reza Sepaskhah,Mohammad Hossein Shahrokhnia (2018).Theoritical and Applied Climatology, http://doi.org/10.1007//s00704-018-2418-4

Walison B.Alves,Glauco DE S.Rolim,Lucas E.DE O.Aparecido,(2017),Journal of the Association of agricultural Engineering,116-1125. http://doi.org/10.1590/1809-4430-Eng.Agric.v37n6p1116-1125/2017.

K.Chandrasekhar reddy,(2014),Artificial neural network models for estimating reference evapotranspiration in rajendranagar region,International journal of civil engineering and technology,ISSN 0976-6306,PP.195-201.

Laaboudi A, Mouhouche B, Draoui B (2012) Neural network approach to reference evapotranspiration modeling from limited climatic data in arid regions. Inter J Biometeor 56:831–841.

Kumar M, Raghuwanshi NS, Singh R (2011) Artificial neural networks approach in evapotranspiration modeling: a review. Irrig Sci 29:11–25.

AHP-TOPSIS Method Approach for Priority Analysis of a Small Hydro Power Project

Shilpesh C. Rana[1]*, Jayantilal N. Patel[2]
[1]Civil Engineering Department, SVNIT Surat, Surat, Gujarat, INDIA
[2]Civil Engineering Department, SVNIT Surat, Surat, Gujarat, INDIA
* e-mail: scrana-ced@msubaroda.ac.in

ABSTRACT

When the different alternatives are available, it becomes difficult to choose the best sequence of alternatives. In that situation AHP-TOPSIS Method Approach helps deciding the best sequence of alternatives. The objective of this paper is to carry out priory analysis of small hydro power projects. For analysis a case study of Sakarda branch canal is taken. AHP-TOPSIS method has been used for analysis. Three criteria rated power, project cost and distance of the power house to the grid line and four canal fall locations at chainages 11214 m, 14022 m, 16550 m & 18234 m are on Sakarda branch canal considered as alternatives. The project cost was calculated by designing hydro power components (using Indian Standards Guideline) and applying actual market rates. Distance of the power house from the grid line obtained using 'Map Distance Ruler Lite' application. With the AHP method consistency of decision criteria matrix has been checked and Priority of the small hydro power project decided using TOPSIS method. Based on the results of AHP-TOPSIS method it is concluded that the priorities for the location of small hydro power project are: at chainages 18234 m, 14022 m, 11214 m, , 16550 m.

Keywords: *AHP; Main Canal; Small Hydro Power Project; TOPSIS method*

Introduction

There are 25 branch canals off taking from Narmada main canal phase-I (Narmada main canal chainage-0 to 144.5 km.) out of which one branch canal (Sakarda) have been taken for research. Four small hydro power plant locations (at chainages 11214 m, 13242 m, 14022 m, 16550m & 18234m) have been identified for the priority analysis. In this research work, an approach to find priority of small hydro power house location is developed using Analytic Hierarchy Process (AHP) and TOPSIS Method. Location of Sakarda branch canal is shown in Figure 1.

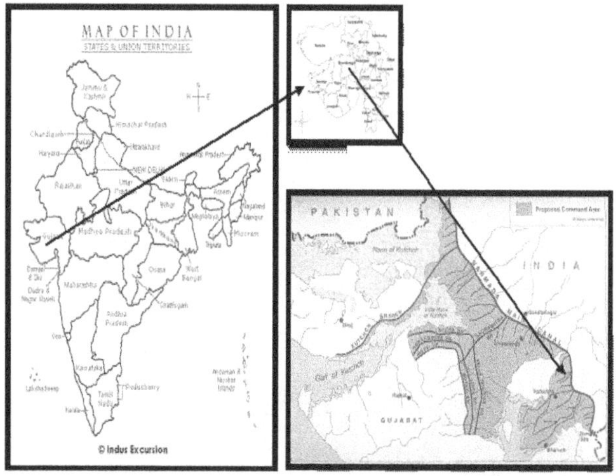

Figure 1. Location of Sakarda Branch Canal

Methodology

For the priority analysis small hydro power plants AHP and TOPSIS method has been used.

Analytical hierarchy process (AHP) method

The analytical hierarchy process method was developed by Saaty (1977, 1980, 1982, 1995) . The method employs the following steps to solve a problem (Ioan et al. 2017):

Step 1: Identifying the problem.

Step 2: Establishing the decision-making criteria

Step 3: Establishing the decision-making alternatives.

Step 4: Determining relative weight of criteria by comparing the criteria in pairs

Step 5: Normalising the comparisons between criteria.

Step 6: Determining the consistency factor of the decision criteria matrix.

When determining the consistency relation, one considers the following rule: if $CR< 0.10$, then the matrix is consistent, meaning that the vector of the weights is well-determined.

TOPSIS method

The method employs the following steps.

Step 1: Establishment of the decision matrix

Step 2: Normalization of decision matrix.

Step 3: Determining the weighted decision matrix.

Step 4: Identify the Positive and Negative Ideal Solution.

Step 5: In this step the relative closeness of particular alternative to the ideal solution is calculated.

Step 6: Ranking CI values in decreasing order to compare relatively better performances.

Results and Concluding Remarks

In this case study, rated power and project cost are considered to be the most important criterion. As results of AHP method, the decision criteria matrix is consistent. Using TOPSIS method, relative closeness to the ideal solution values $Ci = S'i / (S_i^- + S_i^+)$ has been calculated, which are as shown in Table 1.

Table 1. Alternative wise relative closeness to the ideal solution values

Alternatives	Ci = S'i / (Si- +Si+)
A1	0.31
A2	0.62
A3	0.25
A4	0.70

Decreasing ordered relative closeness values (Ci) are 0.31, 0.62, 0.25 and 0.70 for alternatives (A_4-A_2-A_1-A_3). Hence out of four alternatives (at chainages chainages 11214 m, 14022 m, 16550 m & 18234 m) the best alternative for the location of small hydro power project is fourth at chainage 18234m followed by second, first and third. The proposed approach can be used to prioritize small hydro power plant locations when different alternatives are available. The analysis in this research paper reveals that the waitages plays an important role for selection of small hydro power project location.

References

Ioan A, Gheorghe B, Ioan G, George N & Florin G (2017) Choosing the optimal technology to rehabilitate the pipes in water distribution systems using the AHP method. Energy Procedia 112: 19-26.

Performance of permeable concrete for ground water recharge in HBTU campus: Kanpur

Himanshu Tripathi[1] and Dr. Sunil Kumar[2*]
1. *Department of Civil Engineering, HBTU Kanpur, Uttar Pradesh*
**2. Professor in Department of Civil Engineering, HBTU Kanpur, Uttar Pradesh*

*E-mail–*himtripathi18@gmail.com

Introduction

Today's scenario in Kanpur, particularly in the urban area, continuous urbanization along with the population growth, the land is being covered with impermeable surfaces such as residential, commercial buildings, highways, etc. Pervious concrete is one of the most effective pavement materials to address a number of important environmental issues, such as recharging groundwater and reducing storm water runoff thereby improving the groundwater quality. To this end, the present study focuses on, a) the strength and permeability of permeable concrete, b) Investigating the performance characteristics of permeable concrete such as compressive strength, tensile strength, permeability coefficient, void content for groundwater recharge.

Materials and Methods

Experiments were carried out in Concrete laboratory of HBTU Kanpur. Portland-pozzolana cement was selected as per IS 1489 (PART-1), coarse aggregate size 12.5 mm and 6.25mm and fine aggregate were used in the tests. Methods of tests for aggregates of concrete were carried out as per IS: 2386 (Part III). 24 cubes of size 10*10 cm and 12 cylinders of 15*15 cm were casted with 12 cube and 6 cylinders in a group and the size of coarse aggregate used in the test was 12.5 mm and 6.25 mm respectively. The concrete mix ratio was 1:1.5:3. The water - cement ratio used in the mix was 0.42. A chemical admixture was added to the concrete mix for enhancing the compressive strength of permeable concrete. The test specimens were immersed in water in curing tanks after 24 hours of casting for 28 days.

Compressive strength of the specimen shall be calculated by dividing the maximum compressive load taken by specimen by its cross- sectional area.

$$\text{Compressive strength} = \frac{compressive\ load}{cross\text{-}sectional\ area\ of\ the\ specimen}$$

The void content was determined according to the ASTM C1688[5]. Total percentage of voids present by volume in a specimen was given according to the formulae-

$$\text{Void content (V) \%} = \frac{T-D}{T} * 100$$

$$T = \frac{M}{V}$$

Where,
V= void content (%).
T= theoretical density. of air free basis (Kg/m^3)
D= density (Kg/m^3).
M= total mass of all material (Kg).
V= sum of absolute volume component ingredients in the batch (m^3).

The coefficient of permeability was determined in the laboratory by a permeameter. The test were carried out by subjecting the concrete specimen of known dimensions contained in a specially designed cell, to a known hydrostatic pressure from one side, and measuring the quantity of water percolating through it during a given interval of time and computing the coefficient of permeability by the formulae:

$$K = \frac{Q}{A * T * H/_L}$$

Where,

K= coefficient of permeability in cm/sec.

Q= Quantity of water in mm^3 percolating over a certain a period after the steady state has been reached.

T= Time in sec over which Q is measured.

A= Area of specimen $cm^{2.}$

H/L= Ratio of the pressure head to thickness of specimen (both expressed in same units).

Figure 1. Cube for compressive test

Figure 2. Cylinder for permeability test

Figure 3. Permeability Apparatus

Results and Concluding Remarks

The compressive strength and tensile strength tests were conducted on the mixes obtained. Darcy's law was used for calculating the mechanical properties. The 28-day strength and permeability coefficient were obtained as 23.3 N/mm^2 and 0.197 cm/sec respectively. Mechanical properties of permeable concrete were found very well. The proper study and implementation of porous concrete at large scale concrete floors and open spaces can simultaneously cater to the need of aesthetics and preservation.

References

1. *M.I. Gambhir (2018) Concrete Technology. Tata McGraw Hill Education Private Limited, Delhi.*
2. *Tennis P, Leming M, Akers D. Pervious concrete pavements. Skokie, IL: Portland Cement Association; 2004.*
3. *Wanielista M, Chopra M. Performance assessment of Portland cement pervious pavements: report 2 of 4: construction and maintenance assessment of pervious concrete pavements. 2007.*

Quality and quantity analysis of water resources in part of Hirakud command area addressing water logging and quality issue for canal irrigation

B. Pradhan[1], P.C. Swain[2*]

[1] Department of Civil Engineering, Veer Surendra Sai University of Technology, Burla 768018

[2] Professor, Department of Civil Engineering, Veer Surendra Sai University of Technology, Burla 768018

** e-mail: biswajitpradhan030@gmail.com, prakashchswain@gmail.com*

Introduction

Canal water is mainly used for irrigation purpose in the study area, which is a part of Hirakud irrigation project. Canal irrigation network from Hirakud multipurpose project causes water logging problems in the head reach of the canals. In this research, a simple framework is developed and implemented to show the interactions between surface water and groundwater in a canal irrigated area for which average runoff, recharge, and groundwater table fluctuation of study area has been considered. The model is generated using VISUAL MODFLOW based on canal irrigated area of Hirakud irrigation Project of Sambalpur, Odisha. Quality analysis of water is done for the available water resources which helps to improve the productivity of farmland and helps in efficient water management. The quantitative study is done using VISUAL MODFLOW for groundwater modelling and CROPWAT to find crop water requirements for each crops. Quality analysis of water is an important aspect to increase the productivity of the crops, so an analysis of the important water quality parameters is done. Finally a quantitative and qualitative analysis of available water resources of the command area is proposed for the benefit of the farmers. This will not only help in reducing the waterlogging, but also can irrigate the command area in the tail reach of the canals. Application of Differential Global Positioning System (DGPS) to collect the data with high accuracy makes this research even more unique.

Materials and Methods

Waterlogging is the main issue in the head reach of the canal. Generally waterlogging occurs due to over irrigation and rise in the groundwater level in the head reach of the canal. To analyse the groundwater flow direction, groundwater modelling is done using visual Modflow 2000. In the visual Modflow 2000 a sum total of 11 no of pumping wells are proposed near to the water logged area to solve the water logging issues. A detailed work is carried out for the extraction of water from the waterlogged area to be supplied to the deficit areas by the help of groundwater modelling technique in visual ModFlow. The positions of pumping wells near the water logged area and no of pumping wells proposed with their optimum pumping capacity. Similarly CROPWAT is used to find out the water requirement for individual crops. After finding out the crop water requirement for each crop per year and no of wells proposed with their optimum pumping capacity, a proper water supply management can be done to avoid water logging issue in the head reach of the canal.

The Chaunrpur canal has three major waterlogged areas and an amount of 800, 920 and 810 cubic meter per day can be pumped out. So in this way Chaunrpur canal can contribute 2530 cubic meter per day of water in a year. The Senhapali consist of 5 waterlogged areas where water can be pumped out. The amount of pumping rate has been derived as 700, 920, 940, 776 and 850 cubic meter per day and the Parmanpur canal commands have 3 water logged area and water can be pumped out is 780, 720,930 m/dSay respectively. So pumping well installed near Sehnapali canal, Chaunrpur canal and Parmanpur canal commands can contribute up to 3334410 cubic meter of water in a year. The total amount of water required for all crops to be irrigated in Chaunrpur, Senhapali and Parmanpur canal command is found out to be 691862600m^3 as per CROPWAT. This amount of water is calculated for single cropping season. As a whole year consist of two cropping season in the study area, so one year will need double the amount of the water that is required for a single cropping season. So a total amount of 179884250 m^3 of water is needed in a year from the study area. As previously derived that the proposed pumping well can contribute up to 3334410 m^3 of water, hence the rest amount of water can be fulfilled by canal water.

The proposed plan can contribute to 24.09 % of water for irrigation purpose which was is being wasted and increasing as groundwater level near the canals. The water contributed by the pumping well will save the canal water which can be used in the tail reach of canal command.
The quality analysis shows almost all the parameters like Sodium Absorption Ratio (SAR)., Permeability Index (P.I), Percent sodium (Na^+), Calcium to Magnesium ratio (Ca^{2+}/ Mg^2), Kelley's index (KI), Magnesium Hazard (MH) / Magnesium adsorption ratio (MAR), Residual sodium carbonate (RSC),Potential salinity (PS) etc. satisfies the criteria for irrigation, hence the available water in the study area is suitable for irrigation. Further few parameters are not satisfying due to the wastage of industries in the study area and that has to be taken care off to maintain the quality of water.
The results of quality analysis of water is also found acceptable for irrigation. That quality of water should be maintained properly so that the wastage from industries should not mix up with the irrigation water in the study area.

Results and Concluding Remarks

The groundwater flow modelling is done using Visual Modflow and Quality analysis of water is undertaken to figure out the suitability of available water for irrigation. One of the objectives of this work is to solve the water logging issue by providing a quantitative assessment of canal water. This work will help the farmers to improve the yield by the efficient management of available water resources in the study area.
Based on the experimentation, the following conclusions are arrived at: The analysis reveals that the groundwater table remains at higher RL in the command areas of Parmanpur distributary of Sason main canal, compared to the Senhapalli and Chaurnpur distributary as per the survey in DGPS instrument. The investigation reveals that out of total water requirement of 13837250 m^3 per year, 3334410 m^3 can be supplied from pumping well installed near the water logged areas which is 24.09 % of total water, which can be supplied to the command area at the tail reach. This will not only help in reducing the waterlogging; but also can irrigate the command area in the tail reach of the canals, where canal water is much less than design discharge. The quality analysis shows that almost all the parameters are within the acceptable range for irrigation.

References

Samantaray, S., Rath, A., and Swain, P. C. (2017). "Conjunctive use ofgroundwater and surfacewater in a part of hirakudcommand area."International Journal of Engineering and Technology (IJET),ISSN (Print) : 2319-8613, ISSN (Online) : 0975-4024,DOI: 10.21817/ijet/2017/v9i4/170904039,Vol 9 No 4 Aug-Sep 2017.

Ravikumar P. and Somashekar R.K. Geochemistry of groundwater, Markandeya River Basin,Science Press and Institute of Geochemistry, CAS and Springer-Verlag Berlin Heidelberg (2011) Chin.J.Geochem. (2011)30:051–074 DOI: 10.1007/s11631-011-0486-6.

Kumar, S., Pavelic, P. George, B., Venugopal, K., Nawarathna, B. (2013). "Integrated modeling framework to evaluate conjunctiveuse options in a canal irrigated area."Journal of Irrigation and DrainageEngineering, Vol. 139, No. 9, September 1, 2013. © ASCE, ISSN0733-9437/2013/9-766-774

Basagaoglu, H. and Marino, M. A. (1999). "Joint management of surface and ground water supplies." Ground water, Vol. 37, No. 2, Page 214-222.

Khadri, S. F. R., and Pande, C. (2016). "Ground water flow modelling for calibrating steady state usingMODFLOW software: a case study of Mahesh River basin, India."Model. Earth Syst. Environ. (2016) 2:39,DOI 10.1007/s40808-015-0049-7.@Springer International Publishing Switzerland 2016.

Chakravorty, B., Pandey, N. G. and Kumar, S. (2014). "Effect of conjuctive use on waterlogging in lower Gandak basin of bihar." International Journal of Scientific Engineering and Technology, ISSN: 2277- 1581, 18-20 Dec 2014.

Suspended sediment concentration measurements using acoustic dopplervelocimeter

K.P. Indulekha[1] , Sebin xavier[2], Aiswarya Prabhu[2]
[1]*Assistant Professor, College of Engineering Trivandrum*
[2] *Student, Department of Civil Engineering, College of Engineering Trivandrum*
** e-mail: indulekhakp@gmail.com*

Introduction

The measurement of suspended sediment concentration (SSC) is an important aspect of many stream, river and coastal water studies. The conventional sampling methods used are intrusive, labour intensive and expensive. A surrogate method for the estimation of SSC is experimentally tested in this study using Acoustic Doppler Velocimeter (ADV). ADV is an instrument used to estimate velocity of flowing fluid based on Doppler shift principle. Studies conducted by Elci et al., (2008) have shown that, Signal to noise ratio (SNR) value obtained as subsidiary along with velocity data is a representation of sediment particle in the flowing fluid. Recent studies conducted by Mehmet Ozturk (2017) also showed the sensitivity of the acoustic signals towards measured SSC, sediment size and size distribution. Salehi and Strom (2009) examines the relationship between suspended sediment concentration and the signal to noise ratio recorded by a Nortek Vectrino Doppler velocimeter for two velocimeter acoustic output power levels.

In the present work, an experimental study is conducted in a flume with external sediment supply to correlate the SNR reading with suspended sediment concentration (SSC) and Linear equations are developed to obtain SSC from SNR reading for different soil.

Materials and Methods

The experimental work was done in a rectangular open channel flume in the Coastal Engineering Lab, College of Engineering Trivandrum. The flume was operated as a feedback system where the water was recirculated by centrifugal pump system.The experiment is done in a flume using a 16 MHz Sontek micro ADV placed at different locations along the length of the flume. There was 2 sets of experiments done, one with the erodible bed, i.e., flume bed filled with river sand in order to simulate the natural river flow condition. And the other set without erodible bed in which sediment was supplied externally.The signal strength readings received by an ADV was in the form of SNR readings.
In order to correlate the SNR readings with sediment concentration, conventional method was used by sampling technique. Water sample was collected by pumped sampling mechanism using a burette with bulb arrangement into a 100 ml sampling bottles to test for the suspended sediment concentration. Filtration technique with Whatman 40 grade filter paper was used to find out sediment concentration in ppm (mg/l).In order to study the effect of particle size on the SNR reading, experiment is done with four types of soil (sand, laterite, marine and kaolinite clay) whose mean size (D_{50}) varies from 0.61 mm to 0.011mm. Sieve analysis was done to find the particle size distribution of fine sand and laterite soil. For marine clay and kaolinite clay hydrometer analysis was conducted to obtain the particle size distribution curve since their particle size is smaller than 75 μm.And a multivariate regression model using a statistical modelling tool called Statistical Package for Social Sciences (SPSS) with general applicability is developed to obtain SSC from SNR, coefficient of uniformity (C_u), coefficient of curvature (C_c) and mean size (D_{50}) as parameters.

Results and Concluding Remarks

A methodology was developed and applied to allow the use of an ADV for the measurement of suspended sediment concentration (SSC) in a hydraulic flume. ADV measurements were taken along the flume with importance given to obtaining signal to noise ratio(SNR) and Results showed that, output of the ADV (SNR) varies proportional to the sediment concentration in the flowing water. It also showed the dependence of the soil particle moving along the flow, since SNR value corresponding to same concentration was observed to be different for change in the soil (Figure 1). Correlation obtained and linear equation obtained for each soil type is given in TABLE 1.

A prediction equation relating SSC measurements with SNR readings and particle size parameters of the suspended soil were obtained with a good correlation of 0.894. Thus, this technique of using ADV may provide reasonably accurate estimates of SSC under favourable circumstances. Moreover, it gives SSC estimates concurrent with velocity measurements.

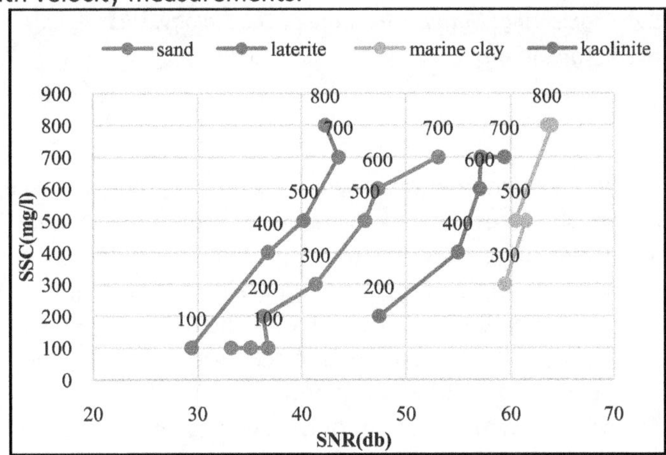

Figure 1 Variation of SNR against SSC for different soil for 10cm sampling depth

TABLE 1 Pearson correlation obtained and linear equation developed.

Soil type	Pearson correlation, R^2	Linear equation developed
Sand	0.656	SSC = 28.781*SNR-872.398
Laterite	0..718	SSC = 52.614*SNR-1587.994
Kaolinite	0..845	SSC = 113.051*SNR-6312.608
Marine	0.784	SSC = 18.517*SNR – 519.958

Acknowledgments: The authors are thankful to the Kerala State Council for Science, Technology and Environment, Kerala, India (Grant No. ETP/1566/2016/ KSCSTE) for providing financial assistance for research work under Engineering Technology Programme (ETP).

References

- Mehmet Ozturk(2017), Sediment Size Effects in Acoustic Doppler Velocimeter-Derived Estimates of Suspended Sediment Concentration, *Water*.
- M. Salehi and K. B. Strom, (2009), Suspended Sediment Concentrations Measurements of Muddy Sediments with an ADV, World Environmental and Water Resources Congress 2009 @ ASCE, pp. 3572 – 3578.
- SebnemElçi, Ramazan Aydın, and Paul A. Work (2008), Estimation of suspended sediment concentration in rivers using acoustic methods, Environmental Monitoring Assessment, 159, pp. 255 - 265.

ANALYSIS OF WATER COLUMN SEPARATION PHENOMENON IN PUMPING MAIN

G. Deepchand[1*], E.Venkata Rathnam[2]
[1] *M.Tech, Department of Civil Engineering, National Institute of Technology, Warangal.*
[2] *Professor, Department of Civil Engineering, National Institute of Technology, Warangal.*
** e-mail: deepchandgundapaneni@gmail.com*

Introduction

Surge or hydraulic transient is a phenomenon in which a pumping main is susceptible to heavy damage due to extremely high or low pressures caused by rapid changes in discharge through the main. The changes in discharge may be due to starting or stopping of pump, opening or closing of valve, power failure, single pump failure (when multiple pumps are in parallel operation).In some cases, during power failure a cavity forms at the peak along the alignment due to the difference in outflow velocity and inflow velocity. After sometime, the cavity collapses causing a shock wave to propagate throughout the length of pipe. This phenomenon is called water column separation.

Generally during design phase of pumping main, only power failure condition is considered to analyse the main for surge pressures. But, it is necessary to check the water column separation occurrence also to provide adequate protection system. In this study, a pumping main is analysed for both power failure condition and water column separation and a protection system is provided to mitigate surge effects. The main objectives of this study are

1) To conduct surge analysis for the transmission main for possible failure cases like power failure and water column separation.

2) To design proper safety devices and to locate them along the transmission main to minimise the surge effects.

Materials and Methods

In this study, a water transmission main of 800 mm diameter DI class K9 pipe is analysed for surge pressures. Two pumps of $0.22m^3/s$ discharge each are pumping water from Chalivagu reservoir to water treatment plant near Koglvai village in Warangal district. The pump details, pumping main details and alignment details of main are collected.

Using SAP2 software, surge analysis was done for power failure and water column separation conditions without providing any protection devices. Based on the results, a protection system is adopted and analysis was done by providing the adopted system. If the results are not within permissible limits, the protection system is upgraded or new protection devices are employed for surge analysis until an adequate and economical system is reached.

Results and Concluding Remarks

Surge analysis was conducted for the transmission main without any protection devices for power failure condition. The maximum pressure of 86.45 m was occurred at 90 m chainage which is less than 1.5 times the working pressure. The minimum pressure of -37.74 m was occurred at 14580 m chainage which is far less than -10 m (permissible vapour pressure). Hence, protection is required to mitigate downsurge.

Surge analysis for water column separation is done by selecting some possible locations for occurrence of water column separation and by assuming that water column separation is occurring at different combinations of these locations and surge analysis is done for all these combinations. The maximum pressure of 111.1 m was occurring at 90 m chainage when water column separation was occurring at 6600 m and 12690 m simultaneously. As the maximum pressure is less than 1.5 times of working pressure, there is no need to provide any devices against upsurge.

To mitigate downsurge, one way surge tank was provided as primary protection and several trails were done by providing OSTs of different sizes at different locations and also with air cushion valves. After several trails, the final protection system adopted consists of

a) One way surge tank of 3 m diameter, 2m storage height with a staging height of 10m with a connecting pipe of 250 mm diameter to connect the tank to the main is provided at chainage 12690 m.

b) Three air cushion valves with size of 200 mm are provided at chainages 3240m, 6600m and 10590 m.

c) A stand pipe of 300 mm diameter with top level of pipe as 258.277 m at a chainage of 15660m.

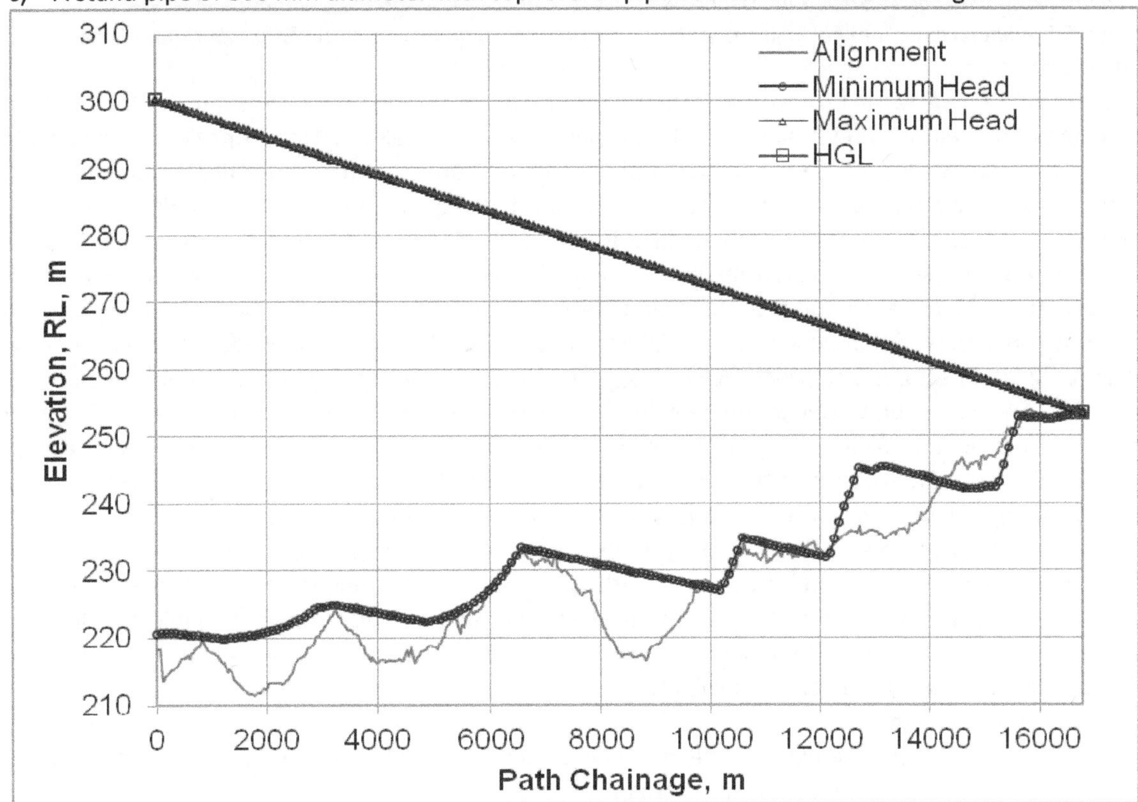

Figure 1.Maximum and minimum piezometric heads with adopted protection system.

By providing the adopted protection system, the maximum pressure is maintained at the working level i.e. the maximum piezometric head is limited to HGL and the minimum pressure is reduced from -37.74 m at 14580m chainage to -5.72 m at 15270m chainage.

References

- Akpan, P.U., Jones, S., Eke, M.N., and Yeung, H. (2015). "Modelling and transient simulation of water flow in pipelines using WANDA transient software." *Ain shams engineering journal, Elsevier publications*, 8, 457-466. http://dx.doi.org/10.1016/j.asej.2015.09.006
- Chaudhry, M.H. (2014). *Applied hydraulic transients, 3rd ed.*, Litton Educational/Van Nostrand Reinhold, New York.
- Lokesh Kumar, M., Phanindra Vital Reddy, G., and Adarsh Kumar, A. (2015). "Hydraulic and surge analysis in a pipeline network using Pipeline Studio." *International journal of engineering research and technology*, 4(2), 41-48.
- Sridharan, K., and Raghuveer Rao, P. (2013). "Surge analysis program version (SAP2) User manual and Design guidelines manual."
- Streeter, V.L., and Wylie, E.B. (1967). *Hydraulic transients*, McGraw hill, New York.

Management Aspects of Surface Irrigation-Yield Loss Due to Water Logging in Godavari Basin-India

A.DhavaleswarRao Bhandaru[1] B.Dr. RVRK Chalam[2] C.Sunitha Chilaka[3] D.Reshma.Tabassum

[1]*Research Scholar/Vagdevi College of Engineering/JNT University/Hyderabad-TS-India*
[2]*Principal & Professor/K.L.Engineering College/K.L.Deemed to be University-Guntur/Hyderabad-TS-India*
[3]*Executive-Engineer/Irrigation Department/Government of Telangana-Karimanagar-TS-India*
[4]*Deputy Executive-Engineer/Irrigation Department/ Government of Telangana-Karimanagar-TS-India*
[1]*eswar54@yahoo.com* [2]*dr.chalam@gmail.com* [3]*chsunitha81@gmail.com* [4]*reshma.tabu@gmail.com*

Introduction

The Sri Rama Sagar Project was constructed across river Godavari in Pochampad(V) to irrigate a total command Area of 0.67 mha.(Administrator-cum-CE-SRSP et.al.1981) Irrigation projects are specifically designed for increasing the crop production and yet the very projects may cause degradation and loss of production and land resources through soil salinization, if adequate provisions for drainage and salinity control are not made. An intensification of agricultural production on a global scale is necessary in order to secure the food supply for an increasing world population (Bruno Glaser et.al2002). The Irrigation potential created under SRS project is about 0.67million hectares and it has fallen to 0.242 million hectares at present. It was observed that an area of 42,729 Ha. under G6-Godavari basin was identified as waterlogged and about 55,180 hectares was prone to water logging. Un controlled application of water in the command area of the major irrigation project generally lead to deleterious consequences like water logging, salinity and alkalinity problems. Saline Water is one of the most common pollution in fresh ground water (David Keith Tod et.al 1980).On account of this fact, water-logged area has been increasing in G6-Godavari Basin Excess salts, regardless of composition; generally keep the soil clays in a flocculated state.

Since the problems of water logging and soil salinity were diagnosed in recent years, the several studies were conducted by visiting the sites and collecting the soil samples and tested in laboratories. However, there is a decreasing trend in crop yields per hectare due to water logging salinity/alkalinity in Sri Ram Sagar Project Command. The results obtained suggest that about 39% of area is under water logging condition and affected by Salinity and alkalinity.

The main objective is

1) To examine the magnitude and socio-economic consequences of water logging and salinity due to canal irrigation.
2) To examine the production loss, growth of weeds, infestation of diseases etc. due to, mismanagement of surface irrigation.
3) To suggest appropriate measures for overcoming the problems of water logging and soil salinity.
4) To suggest Leeching and drainage arrangements.

Hence this problem is selected for suggesting Remedial Measures.

Materials and Methods

In General, water quality along the Corridors is good and complies with CPCB Surface Water Quality norms. Ground water is a major source of domestic as well as agricultural water supply in the area. Details of water quality features are obtained. Application of more irrigation water than required by crops is practiced in G6-Godavari basin. Water logging and soil salinity are the two major problems affecting the agricultural productivity and sometime becomes too severe to take it out from economic crop production (Central Soil Institute.et.al2002). In India, vast areas have been brought under irrigation without provision of drainage component. Even though the water quality is good, large scale irrigation has led larger scale water logging and wide spread salinity in the canal command area.

In this research Method the change in soil profile has been studied. In addition, excess use of fertilizers like N,P &K during the years 2015 to 2018 has been increased from 69469 tons to 88520 tons.The case study of Sri Rama Sagar Project indicates that water logging is caused by the interaction of large number of factors such as ground water recharge, drainage, surface irrigation, cropping patterns, ground water with drawl for irrigation, soil characteristics, seepage, excess use of fertilizers from field channels and distributaries. In command area of Sri Rama Sagar Project observation wells indicated waterlogged and non-waterlogged areas in the same command.

Site investigation and site surveys were revealed the information like topographic details, soil profile, salinity/alkalinity, EC, pH and soil Texture in villages of Sri Rama Sagar Project Command.

For delineating water logged area, dug wells are located in the command and depth of water table is observed season wise post monsoon depth of water table are generally considered for delineating water logged area.

It is seen that the expected yield of rice stages at 3.5 t/ha. and the average yield is 2.6 t/ha. Water logging and salinity have shown an increasing trend in many of the developing countries(Kasuhan Kitilia et.al2004). Water is even a scarce resource in these areas and even then loss of productivity is caused by over use, misuse and mismanagement of water.

Results and Concluding Remarks

The soils were analysied for chemical properties as per procedures out lined by Jackson, 1973.The Ec of experimental area varied from 0.05 to 4.4mmhos/cm and the pH varied from 7.6 to 8.6 . The Ec value decreased is slightly alkaline soils and increased for Saline Soils.

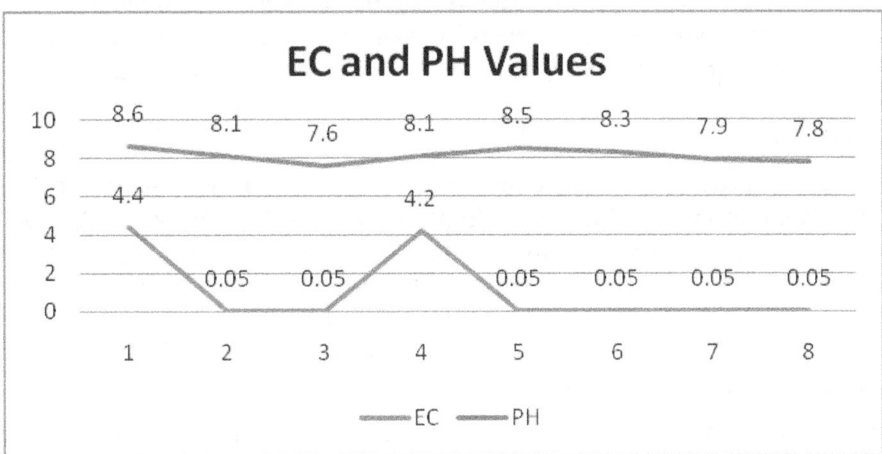

Figure 1. Trend Showing the variation in Ec & pH Values

Soil pH is one of the most important parameters which influence plant growth. The pH and EC are arrived. The research is based on data collection for last 5 years for 6 different crops. Arrive loss in yield per each crop after which financial evaluation is to be done.

The main Object of this research is to suggest leaching and arranging the pattern of drainage by conducting several field and laboratory tests. Similarly, for the removal of excess water, installation of surface and sub surface drainage is an important regulatory measure (Ravaneet Saini, Gopala Krishnaan,NK.Tuli,Gurmeet Kaur et al.2013) drainage followed by leaching with good quality water removes excess salts from the root zones, of the crop. An efficient drainage system is essential for the quick disposal of the storm water and excess irrigation water(Prof.Bancy Mati et.al 2014).Wter logging is a common problem of management. It is observed that the action taken by Irrigation department is not effective.**Hence,research is done to suggest methods of leaching and adopting different types of drainage systems as a pre-requisite, at the stage of planning canal irrigation projects itself in order to avoid huge social costs due to water logging and salinity.**

References

1) Administrator-cum-C.E: A Manaual on Sri Ramasagar Project Command Area-Published in1987
2) Bruno Glaser.2002Biology and Fertility of Soils-June 2002, Volume 35, Issue 4.
3) David Keith Todd: Ground Water Hydrology- Vol.1980,ISBN9971-51-173-8
4) Kasahun Kitila, Ayub Jalde, Mekonnen Workina : Evaluation and Characterization of Soil Salinity ISSN: 2376- 8053
5) Central Soil Salinity Research Institute-India: Methodology for Identification of Water logging and Soil salinity
6) Ravneet Saini, Gopal Krishan, NK Tuli and GurmeetKaur:: Water Assessment of Water logging and Salinity in Punjb
7) Prof. Bancy Mati: How You Can Prevent/Reduce Occurrence of Water logging in Agricultural Lands 2014

Spatial trends in ground water recharge with respect to landuse and precipitation in Wardha sub-basin, India

R. Deshpande, M Sargaonkar, A. Sharma*, R. Kadaverugu

[1]*CSIR-National Environmental Engineering Research Institute, Nagpur -440020, India*
** e-mail: a_sharma@neeri.res.in*

Introduction

The ground water is one of the severely stressed source of water, which is unscrupulously extracted for agricultural, domestic and industrial use in India. Declining ground water is becoming a serious concern for policy makers and has potential to affect human wellbeing. Changes in landuse and urbanization is further exerting pressure on ground water, especially where the recharge potential is minimum. In this context we have analyzed historic changes in ground water levels between pre and post-monsoon, and aimed to understand its relation with the landuse and precipitation in the Wardha sub-basin (3721.73 km2), India.

Materials and Methods

Ground water data from around 90-95 bore-wells were analyzed during the period 2010-2013. Spatial statistical methods like Inverse Distance Weight (IDW) and Kriging were utilized through R and ArcGIS software in order to extrapolate the ground water level profiles over the study area. Difference in water levels over seasons (either recharged or depleted) is correlated with the spatial rain-fall pattern and land use land cover information.

Results and Concluding Remarks

Krigging shows better interpolation of ground water level than IDW method, while validated with observed data. In general, trends of ground water uneven recharged/depletion is observed where due to anthropogenic activities are dominant in comparison with other land cover. The sudden rise in ground water in the event of rainfall is also observed in areas with history of drought. The percentage of recharge area is also growing over time from 5% to 10 % during the study period. However, trends of uneven recharged/depletion substantiate the need for urgent intervention in conserving the ground water resources for sustained human consumption and also summarizes the historic changes taking place in the light of climate variability.

Acknowledgments: Authors thank the Director, CSIR-NEERI for providing all the support for carrying out this work.

References

Machiwal, D., Singh, P. K., Yadav, K. K., 2016. Estimating aquifer properties and distributed groundwater recharge in a hard-rock catchment of Udaipur, India. Appl Water Science. https://doi.org/10.1007/s13201-016-0462-8

Pebesma, E., 2019. The meuse data set: a brief tutorial for the gstat R package, NIST/SEMATECH e-Handbook of Statistical Methods. http://www.itl.nist.gov/div898/handbook/

Comparison Between Remote Sensing Based Drought Indices for Telangana Region

Soumyasree Dixit[1*], Mahesh Ramesh Tapas[2], Syed Tayyaba[2], K. V. Jayakumar[3]
[1]PhD Scholar NITW,[2] MTech Students NITW,[3] Dean IRAA NITW
** e-mail: dixitsoumya1993@gmail.com*

Introduction

The grievousness of Drought is becoming a paramount issue among the extreme events in the region of Telangana due to poor and dawdle monsoon, very high temperature, and inadequate water resources. The satellite-based drought indices may be applicable on a regional scale to help decision-makers in taking timely and appropriate actions. The Various Remotely Sensed Drought indices available are Enhanced Vegetation Index (EVI), Evaporation Stress Index (ESI), NDVI, Temperature Condition Index (TCI), Vegetation Condition Index (VCI), Vegetation Drought Response Index (Veg DRI), Vegetation Health Index (VHI), Water requirement Satisfaction Index (WRSI), Normalized Difference Water Index (NDWI), Soil adjusted water index (SAVI), Soil Water (SW). The drought in May month for the Warangal Region (both Rural and Urban) is computed using Normalized Difference Vegetation Index (NDVI), Normalized Difference Water Index (NDWI), Vegetation Condition Index (VCI), and Soil Water (SW), from the year 2014 to 2019 for the month of May. The objective of this study is to classify images unsupervised and compare drought indices by calculating the percentage of area under each Drought Category.

Materials and Methods

The Landsat eight images of Warangal are downloaded from USGS NASA for path 82 and row 235. The image is processed in Qgis 3.8.0 to find NDVI, NDWI, VCI, SW. The output raster image is classified unsupervised in ERDAS imagine 2015.

The total area of Warangal District (both urban and rural) is classified into five different drought categories. The percentage area coming under each drought category with various drought indices are computed and compared.

Normalized Difference Vegetation Index (NDVI) calculates vegetation by measuring the difference between Wavelengths which vegetation strongly reflects and vegetation absorbs.

$$NDVI = (NIR-RED) / (NIR+RED)$$

Normalized Difference Water Index (NDWI) indicates water content in plants and soil and is determined by calculating analogy with Normalized Difference Vegetation Index (NDVI).

$$NDWI = (NIR-SWIR) / (NIR+SWIR)$$

Vegetation Condition Index (VCI) anchor on the influence of drought on vegetation and can provide information on the arrival, duration, and severity of drought by considering vegetation changes and comparing them with previous values.

$$VCI= (NDVI- NDVI_{MIN}) / (NDVI_{MAX} - NDVI_{MIN})$$

The Reflectance of the soil decreases with increase in Soil Water (SW) for all Wavelengths.

$$SW= NIR\ Band/ Blue\ Band$$

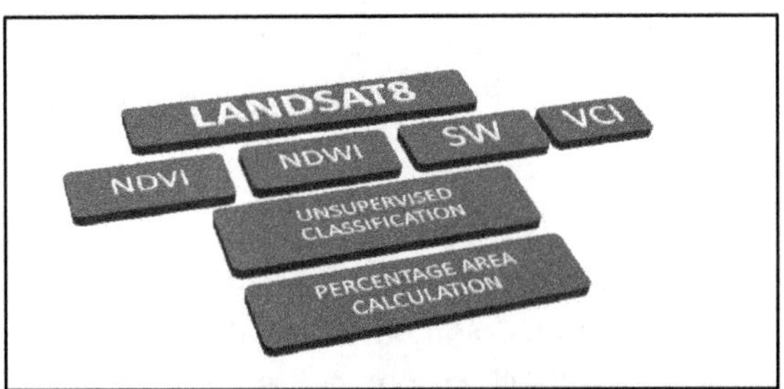

Figure 1. Flowchart of Methodology.

Results and Concluding Remarks

The maximum, minimum values of NDVI, NDWI, SW, and VCI is discussed in Table 1. The Percentage of the area under each drought category for the years 2014 to 2019 is discussed in Table 2. The Minimum value of NDVI was observed in the year 2016. The Maximum value of NDVI was observed in the year 2014. The Minimum value of NDWI was observed in the year 2015. The Maximum value of NDWI was observed in the year 2019. The Minimum value of SW was observed in the year 2019. The Maximum value of SW was observed in the year 2015. The Maximum percentage of Extreme Drought was found in the year 2017.

Table 1. Maximum and Minimum values of Drought Indices

		2014	***2015***	***2016***	***2017***	***2018***	***2019***
NDVI	***Maximum***	*0.465925*	*0.460922*	*0.410087*	*0.460858*	*0.413698*	*0.431791*
	Minimum	*-0.0684027*	*-0.0564994*	*-0.127692*	*-0.0946926*	*-0.0197515*	*-0.0708879*
NDWI	***Maximum***	*0.318672*	*0.316489*	*0.247844*	*0.293753*	*0.279692*	*0.320453*
	Minimum	*-0.211752*	*-0.565197*	*-0.164865*	*-0.185156*	*-0.4474*	*-0.176433*
SW	***Maximum***	*2.27812*	*2.32896*	*2.25028*	*2.39653*	*2.14751*	*2.27001*
	Minimum	*0.77635*	*0.817597*	*0.875172*	*0.801198*	*0.838527*	*0.75181*

Table 2. Percentage Area and Drought Category

YEAR	*DROUGHT INDEX*	*EXTREME DRY*	*DRY*	*NORMAL*	*WET*	*EXTREME WET*
2014	NDVI	43.81093	15.49146	20.84126	13.04571	6.810316
	NDWI	43.81093	19.54231	18.79485	11.47588	6.366482
	VCI	43.81093	15.49179	20.84126	13.04577	6.810316
	SW	43.81093	12.68223	17.67645	18.49968	7.3307
2015	NDVI	43.81	18.32	18.62	12.90	6.35
	NDWI	43.81	18.79	19.32	12.77	5.30
	VCI	43.81	18.32	18.62	12.90	6.35
	SW	43.81	14.18	17.69	15.77	8.54
2016	NDVI	43.81	19.28	15.33	14.02	7.56
	NDWI	43.81	15.34	20.26	12.25	8.32
	VCI	43.81	14.02	19.28	15.33	7.56
	SW	43.81	20.32	15.33	12.30	8.23
2017	NDVI	43.81093	14.43144	21.55199	15.49018	4.715137
	NDWI	43.81093	17.66989	20.54002	14.03157	3.94003
	VCI	43.81093	14.4307	21.55281	15.49039	4.715137
	SW	43.81093	13.9115	18.80404	13.51555	9.958086
2018	NDVI	43.81093	17.4982	19.25025	12.77672	6.663694
	NDWI	43.81093	18.29748	19.5041	12.1665	6.210021
	VCI	43.81093	17.49804	19.25058	12.77676	6.663694
	SW	43.81093	14.34689	17.40538	15.29835	9.138506
2019	NDVI	43.81	16.39	20.21	15.22	4.36
	NDWI	43.81	18.51	20.20	14.30	3.54
	VCI	43.81	16.31	20.21	15.22	4.36
	SW	43.81	14.18	18.78	14.91	8.31

References

Eriyagama, N., V. Smakhtin and N. Gamage, 2009: Mapping Drought Patterns and Impacts: a Global Perspective. IWMI Research Report No. 133. Colombo, International Water Management Institute, http://www.iwmi.cgiar.org/Publications/IWMI_Research_Reports/PDF/PUB133/RR133.pdf.

The Analysis of Non-stationarity and Evolution of Droughts

Das Subhadarsini, N.V.Umamahesh

Department of Civil Engineering, National Institute of Technology, Warangal, India
e-mail: das.darsini1992@gmail.com

Introduction

Drought is the costliest natural hazard which destroys human life and property. The devastating effects of drought and its increasing trend in drought intensity and severity have led to research extensively for better understanding, monitoring and prediction of drought. There are different types of droughts based on different hydrological variables and economic conditions. The name of the four main droughts (with their respective drought indices) are like meteorological drought (SPI-Standardized Precipitation Index), agricultural drought (SSI-Standardized Soil moisture Index), hydrological drought (SRI-Standardized Runoff Index) with other types of droughts such as Ground Water drought (SGI- Standardized Groundwater Index) and socioeconomic drought.

There are also some modified Standardized Indices i.e. SPEI (standardized precipitation-evapotranspiration index) and the Self-calibrated Palmer Drought Severity Index (SC-PDSI) which were compared with SPI and a correlation was found among the drought indices(Wang et.al 2017). Because of this correlation between different drought indices, there is a propagation from meteorological drought to agricultural and hydrological drought. Kwon et.al (2016) explored the time-varying joint return periods for drought duration and severity, to assess how the frequency of a significant drought has varied over time to provide a context for the changing climate and drought severity in the region and to investigate relationships between the large-scale climate indices such as the ENSO and PDO on the drought patterns. H. Varikoden et.al(2015),by analyzing droughts in India, suggested that droughts associated with El Niño events brought severe drought conditions over WG region, and there was no considerable difference in cumulative rainfall associated with the two types of droughts(El Niño and non- El Niño) over central India. Jitendra Singh et.al (2016) found the effect of urbanization in changing the character of ISMR(Indian Summer Monsoon Rainfall) extremes by performing a nonstationary frequency analysis. Xiang Zhang et.al (2017) quantified the relationship and evolutionary process among these four types of droughts with impact on crop yield in India and interestingly found there was no time lag between these four kinds of drought, except for the evolution from meteorological to vegetation drought by reconstructing the time series of droughts from 1981 to 2013.

The objective of the study is to find out the reasons of non-stationarity in climate change in India and different non-stationary analysis for drought prediction and its impact on agricultural productivity in India.

Materials and Methods

There are different types of non-stationary analysis for the prediction of different types of droughts. A nonstationary, multivariate, Bayesian copula model for drought severity and duration is developed and applied to explore the time-varying joint return periods for drought duration and severity, to assess how the frequency of a unusual drought has varied over time to provide a context for the changing climate and drought severity in the region (Han Kwon et.al 2016). A spatial multi-linearregression approach was used for quantifyingthe contributions of decadal PET and precipitationvariations to drought duration and intensity (Shanlei Sun et.al 2016). Under non-stationarity, Generalized Additive Models for Location, Scale and Shape (GAMLSS) algorithm was used to fit a time-varying location parameter of lognormal distribution with the initial values ($\alpha 0$) of the traditional Reconnaissance Drought Index (RDI) to establish NRDI for drought monitoring in a climate change(Javad Bazrafshan et.al 2018). And as there is no worldwide accepted parametric distribution for meteorological and hydrologic variables, a nonparametric multivariate Standardised drought index (NMSDI) indicates the reliability and effectiveness by showing the variations of developed NMSDI is well consistent with those of 1-month SPI and SSI (Yuelu Zhu et.al 2015). The modified Mann-Kendall test trend method, Rescaled Range (R/S) analysis, Morlet wavelet analysis were used to capture the trend of severity and duration of historical drought, persistence of drought, to calculate the period of dry and wet condition, to calculate the joint return period of two typical scenarios respectively (Shengzhi Huang et.al 2014).

Results and Concluding Remarks

By analysing about different hydrological variables causing different types of droughts, it is concluded that the non-stationarity in climate change plays the main role in drought condition. From some research papers, it was found that in India, the non-stationarity in climate change may be occurred due to urbanisation, human intervention and due to some climatic indices i.e. ENSO, IOD. Because of this non-stationarity in climate change, the occurrence of unusual drought events causes bad impact on agricultural productivity which is the root of the financial source in India. About 72% of the drought years in India are caused by the influence of Pacific Ocean as it is found from historical years (13 of the 18 years) that most of the drought conditions of ISMR are associated with El Niño.

So for better improvement in drought prediction, monitoring and making drought policy to face any extreme drought condition, it is needed to analyze about the direct and tele-connected impacts of the climate indices i.e. ENSO, IOD on the drought condition by selecting the best non-stationary analysis and best drought index.

References

Bazrafshan, J., & Hejabi, S. (2018). A Non-Stationary Reconnaissance Drought Index (NRDI) for Drought Monitoring in a Changing Climate. *Water Resources Management*, *32*(8), 2611–2624. https://doi.org/10.1007/s11269-018-1947-z

Huang, S., Huang, Q., Chang, J., Chen, Y., Xing, L., & Xie, Y. (2015). Copulas-Based Drought Evolution Characteristics and Risk Evaluation in a Typical Arid and Semi-Arid Region. *Water Resources Management*, *29*(5), 1489–1503. https://doi.org/10.1007/s11269-014-0889-3

Kwon, H. H., Lall, U., & Kim, S. J. (2016). The unusual 2013–2015 drought in South Korea in the context of a multicentury precipitation record: Inferences from a nonstationary, multivariate, Bayesian copula model. *Geophysical Research Letters*, *43*(16), 8534–8544. https://doi.org/10.1002/2016GL070270

Salvi, K., & Ghosh, S. (2016). Projections of Extreme Dry and Wet Spells in the 21 st Century India Using Stationary and Non-stationary Standardized Precipitation Indices. *Climatic Change*, *139*(3–4), 667–681. https://doi.org/10.1007/s10584-016-1824-9

Singh, J., Vittal, H., Karmakar, S., Ghosh, S., & Niyogi, D. (2016). *Urbanization causes nonstationarity in Indian Summer Monsoon Rainfall extremes*. 269–277. https://doi.org/10.1002/2016GL071238.Received

Sun, S., Chen, H., Ju, W., Wang, G., Sun, G., Huang, J., ... Yan, G. (2017). On the coupling between precipitation and potential evapotranspiration: contributions to decadal drought anomalies in the Southwest China. *Climate Dynamics*, *48*(11–12), 3779–3797. https://doi.org/10.1007/s00382-016-3302-5

Varikoden, H., Revadekar, J. V., Choudhary, Y., & Preethi, B. (2015). Droughts of Indian summer monsoon associated with El Niño and Non-El Niño years. *International Journal of Climatology*, *35*(8), 1916–1925. https://doi.org/10.1002/joc.4097

Wang, H., Pan, Y., & Chen, Y. (2017). Comparison of three drought indices and their evolutionary characteristics in the arid region of northwestern China. *Atmospheric Science Letters*, *18*(3), 132–139. https://doi.org/10.1002/asl.735

Wang, W., Ertsen, M. W., Svoboda, M. D., & Hafeez, M. (2016). Propagation of Drought: From Meteorological Drought to Agricultural and Hydrological Drought. *Advances in Meteorology*, *2016*(4), 1–5. https://doi.org/10.1155/2016/6547209

Zhang, X., Obringer, R., Wei, C., Chen, N., & Niyogi, D. (2017). Droughts in India from 1981 to 2013 and Implications to Wheat Production. *Scientific Reports*, *7*(February), 1–12. https://doi.org/10.1038/srep44552

Zhu, Y., Chang, J., Huang, S., & Huang, Q. (2015). Characteristics of integrated droughts based on a nonparametric standardized drought index in the Yellow River Basin, China. *Hydrology Research*. https://doi.org/10.2166/nh.2015.287

USE OF LINEAR PROGRAMMING TO DETERMINE OPTIMAL CROPPING PATTERN FOR IRRIGATED AND RAIN FED REGIONS OF BIDAR DISTRICT.

Dileep Kumar Kanna[1], M.N. Dandigi[2]
[1]Deparment of Civil Engineering, BKIT,Bhalki, District Bidar Karnataka 585401 India.
[2]Deparment of Civil Engineering, P.D.A.C.Engineering College, Kalburgi.585101 India.
*dilipkanna1@gmail.com

Introduction:

Agriculture sector is the mainstay of Indian economy, contributing about 15 percent of national Gross Domestic Products (GDP)and half of population depends on agriculture and farm related activities. The contribution of agriculture to GDP has declined over the years and at present it is 15 percent. Nevertheless, the importance of agriculture is not likely to decline due to concerns of food, security, employment, rural poverty and availability of wage goods. (Vijay, 2012) The population in India is growing at a rate of 1.4 percent per annum. The country needs to produce more food grains to feed its population which would surpass China's population in 2027. The increase in demand of food grains coupled with limited availability of farm resources need a careful exploration of production possibilities and ways for increasing efficiency of resources on various sizes of farms. (Paroda and Praduman, 2000).The efficiency in agriculture is achieved by optimum utilization of resources (land, labour, capital, irrigation etc.). Optimal allocation of resources is defined as what crops to undertake, how much land to allocate what method of combination of inputs to use on each crop that farm returns are maximum.

Agriculture in India due to inefficiency of allocation, increased input, cost and dwindling profitability of farm products have made it a losing proposition. In view of this it is necessary that the available inputs should be used economically and efficiently since more is efficiency higher is the income. In the present study an attempt has been made to analyze possibilities and prospects of increasing the net farm income by rational resource allocation through optimum production pattern. The objective being to ascertain existing resource pattern, develop optimum combination of resources for different farms and compare with existing pattern. Alsotoanalyzeincome and employment opportunities through reorganization of resources.

Materials and methods:

The study area is Bidar district in Karnataka which was purposively selected for the study. A total of 120 farmers were selected and divided into small and large farmers as per their land holdings. (<2Hectare small and > 2 large farmer). The data on family composition, cropping pattern, resources inventory, and technical coefficient, prices received and paid by farmers were collected from respondents. The mathematical formulation is as shown below:

Maximize $Z = \sum_{i=1}^{n} CjXj$ where j = 1 to n activities

Subjected to $\sum_{j=1}^{n} aijXj_j >= bi$ (I = 1...................... K)

$\sum_{j=1}^{n} aijXj_j <= b_i$ (i = k+1,...... ....m)

$\sum_{j=1}^{n} aijXj_j = b_i$ (I = m+1, ... v)

$X_j, b_i >= 0$

Where, Z = Objective Function, C_j= price of jth activity during kharif and rabi season, X_j= unit of j^{th}production activity during *kharif* and *rabi*,a_{ij} = amount of ith resource required by j^{th} activity, b_i = quantity of ith resource.

The net returns from model were found by deducting variable expenses from gross income. Software package LINGO 17.0 version was used to get optimal values. The variables were: cost of seeds, manure, fertilizer, water, hired human and bullock labour, tractor services, insecticides, pesticides etc. the harvested prices are taken as output and actual market price of input as inputs. The objective function includes sum of year's cash flow. Final amount on objective function is result of production, marketing, borrowing and debt management in the year.

Two different models for small and large farmers were run, one with restricted borrowing termed as S1 and other relaxed borrowing termed as S2. Similarly for large farmers were termed as L1 and L2 respectively. The results were compared with existing pattern in terms of income, employment generation, credit effect on net income etc.

Results and Concluding Remarks :

The optimal solution for two types of farmers under different varying conditions is shown in table1.

Table 1 Net farm returns of cropped area, net crop return per hectare and cropping intensity of small and large farmers

Item	*Existing Farmers*	*Small Farmers S1*	*Small farmers S2*	*Large farmers L1*	*Large farmers L2*
Net farm returns of cropped area (Rs:)	*74620*	*105196*	*113518*	*1156481*	*1771757*
Net farm returns per Hectare (Rs:)	*46931*	*66161*	*71395*	*189898*	*290929*
Cropping intensity in per cent	*132*	*118*	*132*	*110*	*198*

In smallfarmers with restricted capital net farm returns increased by 29 %, net per hectare increased by 29% under S1 model. The same in S2 model (relaxed capital) net income increased by 34% net return per hectare increased 34%. In case of L1 net returns increased by 47% and L2 they increased by 47%. The same in L2 increased by 65% in both net return and per hectare. The impact of credit on net farm returns of small and large farmers were Rs: 8322 and Rs: 615276 respectively which shows model gives large amount of income under relaxed capital. It was noted that credit required was directly related to farm size while credit on income inversely related to farm size.

References :

Aparnati,M., Bhatt,D. 2014. *Linear Programming Model for Optimal Cropping Pattern for Economic Benefits of Mrbc Command Area.*IJIRST international Journal for Innovative Research in Science & Technology.Pp 47-54.

Firzone,J.A.,Coelho,R.D.,Dourado-Neto,D.,Solani.R.,1997. *Linear Programming Model to optimize the water Resource Use in Irrigation Projects; An Application to the Senator Nilo Coelho Project.* Science Agriculture, Piracicaba 1997. Pp 136-148.

Paroda R.S. and Pradumankumar 2000. Food Production and Demand in South Asia. Agricultural Economics Research Review. 13(1): 1-24

Vijay Paul Sharma. 2012. India's agricultural development under the new economic regime policy perspective and strategy for the 12th Five year plan. *Indian Journal of Agricultural Economics.* 67(1): 46-78.

Environmental Engineering

Selective removal of trace hexavalent chromium by new class of redox-active adsorbents

Renuka Verma[1*], Sudipta Sarkar[1]
[1]*Indian Institute of Technology Roorkee, Roorkee Uttarakhand, India*
**e-mail:* renuka.jmi@gmail.com

Introduction

Cr(VI), being a carcinogen, is regulated at concentrations greater than 50µg/L in drinking water. With the possibility of USEPA changing the MCL to 10µg/L, drinking water at many places shall be out of compliance. Although there exists various options for Cr(VI) removal from industrial wastewater, fewer options are available for contaminated drinking water since Cr(VI) is present at trace concentrations in comparison to other competing ions and secondly due to non-specificity of the adsorbent towards target ion. As a result, the treatment process becomes expensive due to the need for frequent replacement of exhausted adsorbent or regeneration. Thus, a material is required which would show high selectivity and removal capacity for such trace concentration of Cr(VI) from contaminated drinking water.

In one recent research project report to California Department of Public Health, it was observed that two weak base anion exchange (WBA) resins, i.e., Duolite A7 and SIR-700 have shown at least an order of magnitude higher capacity than other resins for trace Cr(VI) removal from contaminated groundwater (McGuire et al. 2006). The field results suggested presence of one or more mechanisms along with ion exchange. It is therefore intriguing to find out the mechanism behind such exceptionally high Cr(VI) removal capacity shown by these WBA resins. Motivated by all this, Cr(VI) removal by Duolite A7 was further investigated in the laboratory under different operating conditions and a novel mechanism has been reported for highly efficient chromium removal involving a combination of two mechanisms, ion exchange followed by redox reaction.

Materials and Methods

Five synthetic anion exchange resins (INDION 810, INDION 820, INDION 860, Amberlite IRA 67 and Duolite A7) with different combinations of matrix and functional group were selected for this study. Fixed bed column runs were carried out using epoxy coated glass columns of 11 mm diameter and flow rate was maintained using peristaltic pumps. An initial Cr(VI) concentration of 200 µg/L and a background concentration of 100 mg/L of commonly occurring anions like chloride, bicarbonate and sulphate was kept in the influent. Cr(VI) in the effluent was measured by 1,5-Diphenylcarbazide method (APHA, 2012) and total chromium was measured using ICP-OES instrument. Surface morphology and elemental composition were studied using FE-SEM and EDAX. ATR-FTIR spectroscopy was used to study the changes in the molecular structure of the samples before and after Cr(VI) removal. XPS was also carried out to study the oxidation state of chromium present in the exhausted samples.

Results and Concluding Remarks

The breakthrough curves obtained from fixed bed column study showed that strong base anion exchange resins performed better than weak base anion exchange resins at neutral pH, whereas at slightly acidic pH (pH = 5.0), Duolite A7 outperformed all the resins. For Duoilte A7 resin, breakthrough of 50µg/L (MCL for Cr(VI) in India) was observed around 35,000 BV(Bed Volume), whereas for all other resins, chromium broke through almost instantaneously at less than 1000 BV. Also, the performance of Duolite A7 resin improved with decrease in influent pH. The pH profile of Duolite A7 resin showed that from the beginning of column run, effluent pH was higher than influent pH indicating consumption of H^+ ions during the removal process, with continuous drop in the pH as the column run progressed. Perusals of literature has reported reduction of Cr(VI) to Cr(III) at acidic pH along with consumption of protons. Another breakthrough profile of Duolite A7 was plotted for column running at influent pH 7.0. Once the column was exhausted, it was left undisturbed and restarted after few days. It was observed that upon each restart of the column, there was a significant drop in effluent concentration and breakthrough curve instead of being steeper, became gentler long after restart.

From SEM image, it was observed that virgin Duolite A7 resin is highly porous and has a rough surface morphology, whereas after chromium adsorption, the surface has become very smooth and the visible pores have been filled up. New chromium peaks were observed in the EDAX spectra of exhausted Duolite A7 resin. The XPS spectrum of Cr 2p splits into Cr $2p_{1/2}$ and Cr $2p_{3/2}$ components attributed to spin-orbital coupling. Further, the Cr 2p bands were resolved into five Gaussian peaks, located at binding energy values of 575.46 eV, 576.78 eV, 579.83 eV, 584.47 eV and 586.48 eV which can be attributed to CrN, Cr_2O_3, Cr(VI), metallic Cr and $Cr(OH)_3$, respectively. Thus, XPS result suggested reduction of Cr(VI) to Cr(III) and simultaneous precipitation of Cr(III) into $Cr(OH)_3$. FTIR spectra showed evolution of C=O group in the exhausted samples. This observation leads to a conclusion that there has been oxidation inside the resin.

In the light of above results, we propose a new reactive ion exchange mechanism taking place for the removal of Cr(VI) by Duolite A7 resin. When passed through a WBA column, chromate ions in the solution selectively bind to positively charged functional groups like secondary amine (marked as $-NH_2^+$). The bound Cr(VI) anions thus approach closer to the polymeric matrix and being a strong oxidizing agent, it oxidizes the matrix of the resin. As a result, Cr(VI) anions got reduced to Cr(III) species, vacated the occupied sites and ultimately got precipitated as $Cr(OH)_3$. The vacated functional groups took up further Cr(VI) anions. Thus, the anion exchange sites got automatically regenerated, resulting in a huge capacity for Cr(VI) removal. On the other hand, SBA resins have quaternary amine functionality and thus the steric hindrance caused by bigger size functional groups did not allow chromate ions to approach closer to the matrix. As a result, Cr(VI) could not oxidize the polymeric matrix.

Although, the treated water needs to be characterized with respect to the byproducts of the redox reaction between Cr(VI) and the polymeric matrix, but this process opens the door for the design of new class of redox active agents for trace removal of Cr(VI).

References

APHA, AWWA, WEF (2012), Standard Methods for the examination of Water and Wastewater, 22nd Ed., New York

McGuire, M.J.; Blute, N.K.; Seidel, C.; Qin, G.; Fong, L. (2006) Pilot-Scale Studies of Hexavalent Chromium Removal from Drinking Water. Journal AWWA, 98, 2, 134. https://doi.org/10.1002/j.1551-8833.2006.tb07595.x

Health risk assessment of street vendors at a traffic intersection of Bangalore

Dr. Smaranika Panda[1]
[1] *Department of Civil Engineering, CMR Institute of Technology*
Associate Professor
e-mail: smaranika.p@cmrit.ac.in

Introduction

Air pollution in India is one of the major issues of concern in last few decades. Unplanned urbanisation and rapid growth in various sectors has led to increased pollutant concentration beyond the safe limits in all major cities of India. Off the late, particulate pollution is emerging as one of the biggest problem due to high concentration of associated carcinogenic and toxic elements (Panda and Nagendra, 2018). In order to abate the air pollution, many mitigation measures were implemented by the state and central pollution control boards. Continuous ambient air quality monitoring was carried out in all major cities and towns. However, studies on personal exposure to particulate pollution are limited in India. The amount of personal exposure of particulate is majorly dependent on the location and duration of stay (Babadjouni et al., 2017). Present study focuses on the PM5 exposure of street vendors at a busy traffic intersection of Bangalore. Bangalore has witnessed a rapid growth in last few years. The multidimensional, unexpected growth of the city has increased traffic congestion, small scale industries, and street shops thereby increasing the particulate pollution. Further, the street vendors are getting exposed to the particulate pollution for an entire day. Hence, the present study focuses on health risk assessment of street vendors at an urban hot spot of Bangalore. In this regard, personal exposure monitoring of the street vendors were carried out for 15 days for 8 hour duration using personal sampler (EnviroTech APM 801). The filter samples were analyzed for 15 elements using inductive coupled plasma optical emission spectroscopy (ICP-OES, Make: Optima 5300). Carcinogenic risk factors due to metal exposure were quantified for the street vendors.

Materials and Methods

Study Area: Bangalore city was considered as study site which spreads over 709 KM^2. Kundanahhali traffic intersection (latitude: 12°57′22″N; longitude: 77°42′54″E) which is one of the congested intersections of Bangalore was considered for monitoring. Kundanahalli intersection is a four legged intersection with three 4 lane roads and one 2 lane road. The street vendor sits in a front open shop in the traffic intersection.

Air quality monitoring and chemical characterization of PM5: PM5 monitoring was carried out using personal sampler (EnviroTech APM 801) for 15 days. Total sampling time was 8 hours. The exposure concentrations were quantified gravimetrically by considering the deposited mass and the amount of air flow during the sampling period. The instrument carries air at a flow rate of 2.5 litre/minute which is equivalent to the average human inhalation rate. Chemical characterizations of PM5 for 15 elements (Al, Ba, Ca, Cd, Co, Cr, Fe, K, Mg, Mn, Na, Ni, Pb, Sr, and Zn) were carried out using ICP OES. The elements were extracted using hot acid digestion procedure (IO-3.1) and were analyzed (IO-3.5) based on U.S. EPA methods

Health risk assessment: Health risk assessment was performed for the heavy metals that induce carcinogenic effects (Cr, Ni, Pb, and Cd). Inhalation unit risks for the metals were taken from USEPA.

Results and Concluding Remarks

8 hour average PM_5 exposure concentration for the monitoring period was 537±108 $\mu g/m^3$. Exposure concentration was observed to be exceeding the national ambient air quality standard of PM2.5 (60 $\mu g/m^3$ and PM10 (100 $\mu g/m^3$) for the entire monitoring period (Figure 1). The weekday concentrations were 24%

higher than the weekend concentration. Analyzed 15 elements contribute 15% of total PM_5 exposure concentration (figure 2 and 3). Significant contribution of Mn, Ba, Cu, Zn, Pb from the tailpipe emissions and abrasion of tyres were observed at the traffic intersection. Trace elements such as Ni, Cd and Pb concentrations were observed 2, 4 and 5 times higher than the national ambient standard concentration. The excess cancer risk for street vendors was observed highest for Cr followed by Cd, Pb and Ni.

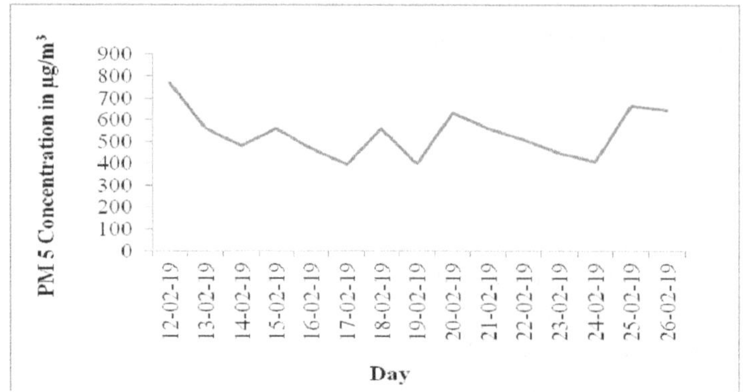

Figure 1. Variation of PM_5 exposure concentration during monitoring campaign

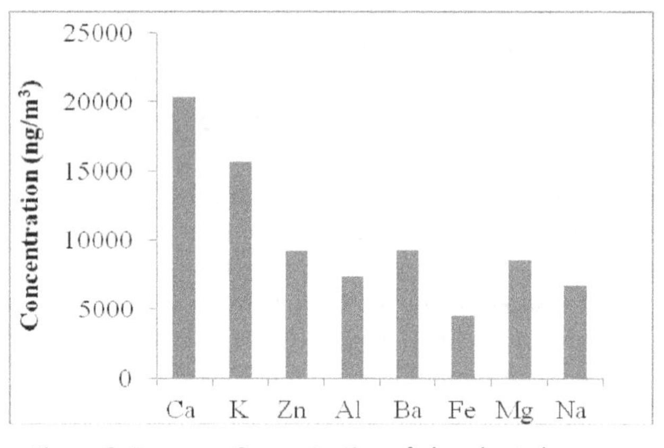

Figure 2: Exposure Concentration of abundant elements (ng/m^3)

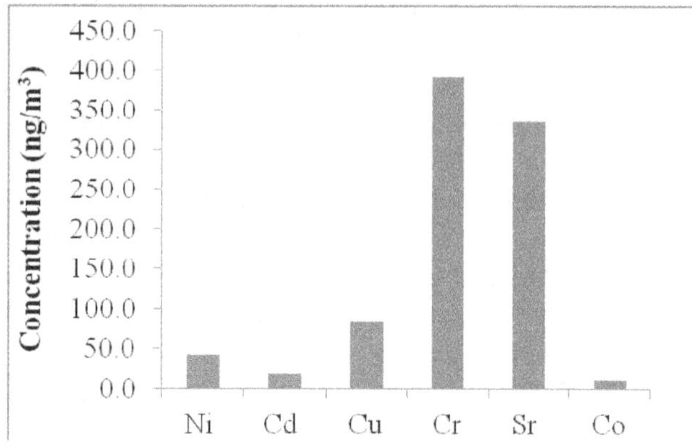

Figure 3: Exposure Concentration of trace elements (ng/m^3)

Table 1. Excess cancer risk of heavy metals for street vendors at traffic intersection

Pollutant	Inhalation Unit Risk (IUR)	Excess cancer risk (10^{-6}) population
Cd	1.8×10^{-3}	33.8
Cr	1.2×10^{-2}	130.8
Ni	2.4×10^{-4}	10.1
Pb	1.2×10^{-5}	26.4

The results of the present study showed alarming pollutant exposure of street vendors at traffic intersections of Bangalore. The obtained results can be used for further epidemiological studies.

Acknowledgments: The author would like to thank KSCST for funding the project

References

Panda S, & Nagendra SM (2018) Chemical and morphological characterization of respirable suspended particulate matter (PM10) and associated heath risk at a critically polluted industrial cluster. Atmospheric Pollution Research, 9(5), 791-803. https://doi.org/10.1016/j.apr.2018.01.011

Babadjouni RM, Hodis DM, Radwanski R, Durazo R, Patel A, Liu Q, and Mack W.J (2017) Clinical effects of air pollution on the central nervous system; a review. Journal of Clinical Neuroscience, 43, pp.16-24. https://doi.org/10.1016/j.jocn.2017.04.028

Screening of substrates for the removal of heavy metals from acid mine drainage (AMD)

A. Shweta Singh*, B. Saswati Chakraborty
Department of Civil Engineering, Indian Institute of Technology Guwahati, Assam, India
** e-mail: shwetasingh@iitg.ac.in*

Introduction

Acid mine drainage (AMD, also referred as acid rock drainage) generated from various active as well abandoned mines pose a serious environmental concern and remains a very challenging industrial wastewater. Owing to its physio-chemical characteristics such as low pH (in case of net acidic AMD), less organic matter (< 10 mg/L) and higher sulfate concentration (1000 – 3000 mg/L) deteriorates the quality of nearby surface streams when disposed of untreated. Low pH (< 3) is the most problematic issue related to AMD generation, as many heavy metals tend to leach out from rocks and remain in the solubilized state, which further adds heavy metal contamination (Colmer et al. 1950). The conventional treatment practiced is the addition of lime to raise the pH and simultaneously precipitating metals out, which is expensive and disposal of hazardous metal-laden sludge is still questionable. Therefore, the utilization of active/passive biological treatment (such as sulfidogenic bioreactors, anaerobic wetlands) has gained attention lately.

In the present study, various waste derived substrates such as cow manure (organic), bamboo chips (cellulosic) and alum sludge (drinking water treatment sludge) were explored for the treatment of AMD. AMD characteristics were similar to the one generated in North-eastern coalfield Ltd (NEC, a subsidiary of Coal India Ltd) in Assam. Metal sorption capacity of different substrates was evaluated at various dosages of batch studies. Thus, the main objective of the study is to identify the potential substrate material based on its pH raising ability and metal removal capacity, which could be applied in bio-remediation of AMD in constructed wetlands.

Materials and Methods

The AMD was directly collected from various coalmines of North-eastern coalfield (Makum) and seasonally characterized. For batch studies, the synthetic AMD was prepared in the laboratory by dissolving known amount of metal sulfate salts in deionized (MilliQ) water, similar to the pollutant concentration of studied coalmines. Bamboo chips were obtained from the local farm units outside IIT Guwahati campus while cow dung was collected from local dairy farms in Amingaon. Alum sludge was procured from the drinking water treatment plant of IIT Guwahati.

Batch studies were carried out with different substrate materials to evaluate its applicability in treating AMD. The metal removal efficiency as well as change in pH were observed under different substrate dose (varying from 0 to 200 g/L). Synthetic AMD had composition of Fe (100 mg/L), Mn (2 mg/L), Al (25 mg/L), Zn (5 mg/L), Co (1 mg/L), Ni (1 mg/L) and Cr (1 mg/L), the pH of AMD was kept 2.50 ± 0.50 (adjusted using 1M H_2SO_4). All batch sorption experiments were performed in 250 mL Erlenmeyer flasks, where the desired dose of the substrate was added in known AMD volume and agitated (120 rpm) at room temperature (27 ± 5°C) for a given contact time period of 24 hours in a horizontal mechanical shaker. Samples were filtered at the end of the agitation period and filtrates were analysed for the change in pH and residual metal concentration. pH was first measured using pH meter (Systronics, India) and then filtrates were preserved by acidifying with HNO_3 to bring the pH < 2 and later analysed for residual metal concentration using Flame Atomic Absorption Spectroscopy (SpectrAA 55B, Varian). The batch experiments were carried out in duplicates. The metal removal efficiency (R, %) was calculated using the following equation (1) :

$$\text{Metal removal (R, \%)} = \{(C_i - C_e)/C_i\} * 100 \quad (1)$$

where C_i and C_e are the initial and final metal ion concentrations, respectively.

Results and Concluding Remarks

The metal precipitation as well as binding of metal ions with the surface of media, depends on the pH of the solution. pH of the solution increased with the increase in the dose for all the selected substrates (Figure 1). An increase in pH up to 5, 7.42 and 7.3 was achieved at highest dose of 200 g/L with alum sludge, cow manure and bamboo chips, respectively. Cow manure was found to be most effective in raising the pH, as reported in earlier literatures (Choudhary and Sheoran 2012). This could be attributed to buffering from bicarbonates and organic acids in manure as well as due to the dissolution of the surface-bound hydroxyl ion from substrate materials (Song et al. 2012). Removal of metals ions increased with simultaneous increase in substrate dose due to the decrease in H^+ on the surface, which results in the less repulsion with metal ions (Table 1) (Namasivayam and Ranganathan, 1995). Higher metal sorption affinity was maximum for Cr followed by Co, Ni, Zn and Fe, and least in Al and Mn, this trend was in accordance with the increase in affinity for higher electronegative metal as described earlier (Kim 2014).

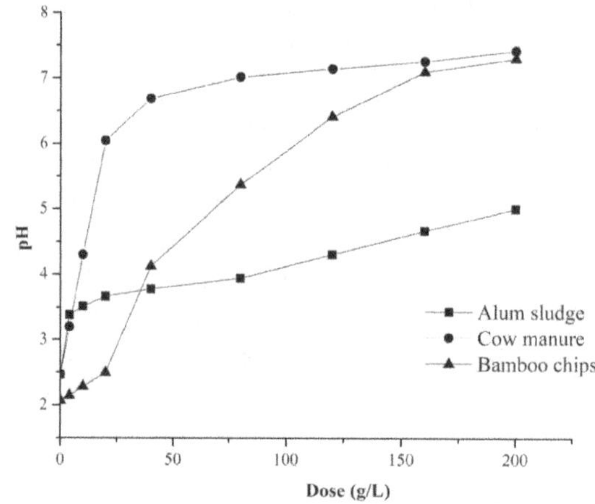

Figure 1. pH Vs substrate dose (g/L).

Table 1. Metal removal efficiency (%) of different substrates under varying dosage.

Substrate	Dose (g/L)	Metal removal (R, %)						
		Fe	Mn	Zn	Al	Co	Ni	Cr
Alum sludge	25	5.149	32.48	45.10	9.621	72.25	87.64	100.0
	50	19.57	49.18	81.87	15.37	85.27	91.28	100.0
	100	50.76	62.25	91.24	21.66	91.61	96.09	100.0
Cow manure	25	100.0	25.61	100.0	43.22	90.95	100.0	100.0
	50	100.0	62.73	100.0	46.80	92.06	100.0	100.0
	100	100.0	72.07	100.0	53.65	92.67	100.0	100.0
Bamboo chips	25	9.924	10.68	30.48	18.97	29.77	23.27	9.304
	50	33.45	41.23	66.18	50.97	40.65	32.26	60.41
	100	48.68	60.04	78.55	62.97	56.29	49.23	67.05

Acknowledgments: The authors would like to acknowledge IIT Guwahati for the infrastructure and lab facilities. We are also grateful to North-eastern coalfields for their assistance during field.

References

Choudhary RP, Sheoran AS (2012) Performance of single substrate in sulphate reducing bioreactor for the treatment of acid mine drainage. Minerals Engineering 39: 29-35. https://doi.org/10.1016/j.mineng.2012.07.005

Colmer AR, Temple KL, Hinkle ME (1950) An iron-oxidizing bacterium from the acid drainage of some bituminous coal mines. Journal of Bacteriology 59(3): 317-328. https://www.ncbi.nlm.nih.gov/pubmed/15436401

Kim MJ (2014) A study on the adsorption characteristics of cadmium and zinc onto acidic and alkaline soils. Environmental Earth Sciences 72(10): 3981-3990. https://link.springer.com/article/10.1007/s12665-014-3287-5

Namasivayam C, Ranganathan K (1995) Removal of Cd (II) from wastewater by adsorption on "waste" Fe (III) Cr (III) hydroxide. Water Research 29(7): 1737-1744. https://doi.org/10.1016/0043-1354(94)00320-7

Song H, Yim GJ, Ji SW, Neculita CM, Hwang T (2012) Pilot-scale passive bioreactors for the treatment of acid mine drainage: Efficiency of mushroom compost vs. mixed substrates for metal removal. Journal of Environmental Management, 111: 150-158. https://doi.org/10.1016/j.jenvman.2012.06.043

Performance analysis of two chamber microbial fuel cell to control the output voltage

P.X Nancy Grace[1*], D. Vasanthi[1], S.Amalraj[2]
[1]*Instrumentation Engineering,Madras Institute of Technology, Anna University, Chennai,Tamilnadu, India.*
[2]*Centre for Environmental Studies, College of Engineering, Anna University, Guindy, Chennai,Tamilnadu, India.*
* *e-mail: nancypeter91@gmail.com*

Introduction

To meet the water & energy demand, lot of technologies are emerging, Microbial Fuel Cell is one of the cost and energy effective methodology to treat the waste water and to produce electricity [1]. MFCs are deliberated as low power, low voltage producer. The great advantage of MFC is the direct conversion of organic waste into electricity by the action of microorganisms. The microorganisms available in the waste water generates the electricity by conducting oxidation reduction reactions, which converts organic energy into electrical energy [2]. Even though MFC produce lower energy, it is from waste, so there is no fuel cost this claims the advantage of MFC compared to chemical fuel cells. It is the only methodology that can produce energy from waste, without any power consumption so MFCs are also suitable for remote locations where the electricity would be barely possible [3] [4] and the places where the human can't enter can be access through the eco robots designed by MFC [5]. MFC has a complex mechanism and interdisciplinary performance so, there are lot of challenges in its modelling and controlling [6]. The single MFC results in voltage, which is fluctuating and that leads to the power output fluctuations, so this proposed work aims to maintain the voltage at a particular constant value [7].

Materials and Methods

Two – Chamber MFC is considered for this proposed work. MFC contains two chambers, anode and cathode chamber. Anode chamber is anaerobic chamber and cathode chamber is aerobic chamber. The two chambers are separated by Proton Exchange Membrane (PEM). The Fuel (Waster water) is an inlet to the anode chamber. The oxygen is an inlet to the cathode chamber. Anode and cathode will be placed in the respective chambers. The microorganisms available in the inlet wastewater attached to the anode. During its respiration it liberates protons (H+) and electrons (e-). The protons transfer done by PEM to the cathode chamber and the electrons are transferred to the cathode through external circuit.

Two-chamber microbial fuel cell model by Zeng is used in this proposed work. The model incorporates Butler-Volmer expressions i.e, fundamental relationship in electrochemical kinetics, biochemical reactions and mass balance and energy balance equations. The modeling is done for both anode and cathode [1] in a LabVIEW software.

Results and Concluding Remarks

Open loop analysis

The Microbial Fuel Cell model equations are simulated in LabVIEW control design & simulation tool box. Acetate oxidation in the anode chamber cause the increase in reaction rate in the anode chamber and the oxygen reduction causes the reaction rate decrease in the cathode chamber. As the reaction rate of anode is increased due to the substrate consumption by the microorganisms, there is a decrease in acetate concentration (C_{AC}) from 1.56 mol m^{-3} to 0.4 mol m^{-3}. As substrate concentration decreases there is an increase in biomass (X) and carbon dioxide (C_{co2}). It should be noted that in cathode chamber dissolved oxygen acts as an acceptors of H^+ ions to form H2O, hence there is a decrease in the concentration of dissolved oxygen (C_{o2}) from 0.308 mol m^{-3} to 0.266 mol m^{-3}. There is an insignificant increase in the cations concentration (M) gets increased with increase in cell current density. It is observed that both are linearly proportional. The microorganisms consumes the substrate and liberates the protons, which leads to increase in protons (H) from 4.56E-05 mol to 0.000501649 mol. It is also observed that the dissolved oxygen reduction leads to small increase in the hydroxyl ions (OH). The anode chamber is acidified and decrease the pH value which cause the increase in over potential (Etta A). Increase in pH of cathode chamber reduced the over potential (Etta c).H^+ ions increase in anode leads the pH of anode chamber to decrease, similarly in cathode chamber slight increase in OH, increase the pH of cathode chamber. Cell voltage is decreased when Cell current density is increased, which leads the changes in power density. It is observed that the maximum power is achieved when cell current density is 7 mA m^{-2} and the corresponding cell voltage is 0.291814 v. Hence, the load is fixed to achieve 7 mA m^{-2}.

Hence from the open loop analysis, it can be concluded that to achieve the maximum power density it is necessary to maintain the cell voltage. Thus the closed loop control of MFC has been carried out.

Closed loop analysis

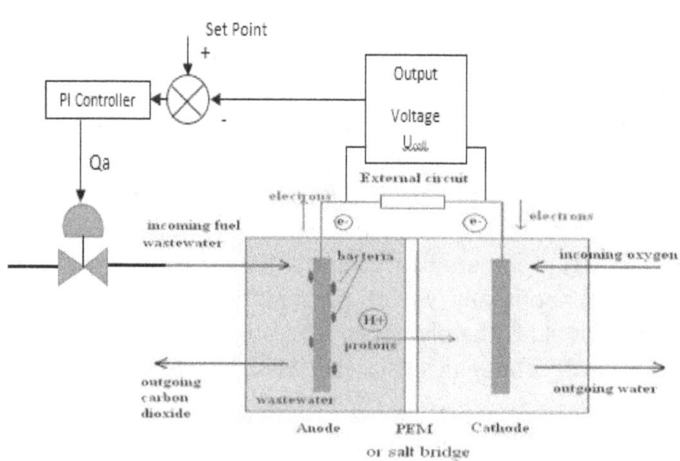

Figure 1. Block diagram of closed loop

Control of output voltage is done by manipulating the inlet flow Q_a. To manipulate the inlet flow to maintain the cell voltage at 0.291814v which gives the highest power, PI controller is implemented. The disturbance which affects the process is concentration of acetate called substrate. Once the disturbance occur at simulation time 520m, the flow is manipulated by the controller, makes the cell voltage to settle in the setpoint. This results in an Integral Square Error of 0.524833. The fluctuations in the cell voltage leads the changes in power density. Hence the PI controller is implemented to maintain the cell voltage which gives the higher power.

References

[1] Yinqzhi Zeng, Yeng Fung Choo, Byung- Hong Kim, Ping Wu (2009), "Modelling and simulation of two- chamber microbial fuel cell" Journal of Power Sources.

[2] Morteza Esfandyari, Mohmmad Ali, Fanaei Reza, Gheshlaghi Mahmood Akhavan Mahdavi (2016), "Mathematical modeling of two-chamber batch microbial fuel cell with pure culture of Shewanella" Chemical Engineering Research & Design.

[3] Carlo Santoro, Catia Arbizzani, Benjamin Erable, Ioannis Ieropoulos (2017), "Microbial Fuel cells: From fundamentals to applications. A review" Journal of Power Sources.

[4] Firas Khaled, Olivier Ondel (2016), "Microbial fuel cells in Waste water Treatment in Building as Potential Solution to Supply Low Power Applications" IEEE International conference on Industrial Technology.

[5] Ioannis Ieropoulos, John Greenman, Chris Melhuish (2003), "Imitating Metabolism: Energy Autonomy in Biologically Inspired Robots" In Proceedings of the AISB '03.

[6] V.B. Oliveira, M.Simoes, L.F.Melo, A.M.F.R pinto (2013), "Overview on the developments of Microbial Fuel Cells" Biochemical Engineering Journal.

[7] Liping fan, Jun Zhang, Xiaolin Shi (2015), "Performance improvement of a Microbial Fuel cell based on Model Predictive Control" International journal of electrochemical science.

Blending methanol as renewable fuel in transportation sector towards controlling air pollution

Dr. Ashok G. Matani
Associate Professor -Mechanical Engineering Department, Government College of Engineering, Amravati- [M.S.] India , E-mail: ashokgm333@rediffmail.com, dragmatani@gmail.com

Abstract

Alternative fuels on internal combustion engine recently has become an attention due to the concern for environmental protection, and needs on reducing dependency on fossil fuels and meeting the current stringent regulation. Alcohol fuel is one of the attractive alternative fuels as it can be produced from renewable resources and is oxygenated. Methanol fuel is a well-known alcohol fuel that can be blended at the lower blending ratio with gasoline and produce better engine operation in spark ignition engine. However, there is a problem related to its methanol properties, especially on its energy content and vapour lock characteristics. Alcohol with higher carbon number such as iso-butanol has high energy content and is able to displace more petroleum gasoline compared to the methanol-gasoline blended fuel. Adding alcohols into gasoline or diesel allows the fuel to have a complete combustion with the present of oxygen which increases its combustion efficiency and reduces greenhouse gas emission. The interest on alcohol as an alternative fuel in the automotive fuel market is expected to grow rapidly in the next decades. In terms of application, Brazil has successfully and widely used ethanol as a fuel for spark ignition engine operation.

Methanol enhancing fuel economy goals

Methanol has a number of physical properties that make it an ideal transportation fuel. For refiners, the use of methanol allows for the expansion of gasoline supply over a greater number of vehicles, and the upgrading of regular gasoline to high premium grades by increasing octane. For automakers, methanol contains oxygen for cleaner fuel combustion, a lower boiling temperature for better fuel vaporization, and a higher blending octane for smoother burning with reduced "knock." This last point has become a critical issue for the automotive industry, which is increasingly recognizing the benefit of using more alcohol – both ethanol and methanol have 109 Research Octane Number (RON) – to boost octane and facilitate higher engine efficiencies to facilitate compliance with CAFE fuel economy goals.

Methanol 15 (M15) in petrol reduces pollution

Methanol is a clean burning drop in fuel which can replace both petrol & diesel in transportation & LPG, wood, kerosene in cooking fuel. It can also replace diesel in Railways, marine sector, generator sets, power generation and methanol based reformers could be the ideal compliment to hybrid and electric mobility. Methanol Economy is the Bridge to the dream of complete hydrogen based fuel systems. Methanol burns efficiently in all internal combustion engines, produces no particulate matter, no soot, almost nil SO_X and NO_X emissions towards near zero pollution. Currently methanol accounts for almost 9% of transport fuel in China. Millions of vehicles running on methanol. China alone produces 65% of world methanol and it uses its coal to produce methanol. Israel, Italy have adopted the methanol 15% blending program with petrol and fast moving towards M85 & M100, Japan, Korea have extensive methanol & di methyl ether usage and Australia has adopted gasoline, ethanol & methanol fuels and blends almost 56% methanol. Methanol has become the choice of fuel in marine sector worldwide and countries like Sweden are at the forefront of usage. Large passenger ships carrying more than 1500 people are already running on 100% methanol. Eleven African and many Caribbean countries have adopted methanol cooking fuel and across the world generator sets and industrial boilers are running on methanol, instead of diesel. Renewable methanol by capturing CO_2 back from the atmosphere is becoming very popular and is seen by the world as the enduring energy solution known to mankind. Methanol is a significant solution to the burning problem of urban pollution worldwide. With slight modifications to existing engines and vehicles structures and fuel distribution infrastructure, 15% of all vehicle fuels can be converted to methanol & di methyl ether (DME). India is planning to implement methanol 15 % blending program with petrol and cost of petrol is expected to come down immediately by 10% and M100 program for buses and trucks is also being implemented in near future.

Methanol in Railways

Indian Railways consumes about 3 billion litres a year and the annual diesel bill is in excess of Rs. 15000 Crores. A methanol locomotive prototype is being implemented by Indian Railways under a grant by Department of Science & Technology and once all 6000 diesel engines are converted to methanol at very minimal cost of less than 1 Crore a engine, the annual diesel bill can be reduced by 50%. Methanol conversion program in railways is complimentary to the goals of electrification in Railways.

Conclusions:

For methanol-blended fuel to actually reduce pollution, proper manufacturing of coal-based methanol must be followed. Petrol supplied in India is already blended with ethanol, however, due to the limited production of ethanol, this is not all across. Petroleum Ministry is improving this instead of building expensive petrol refineries. Methanol-blended automotive fuel petrol and diesel is already in use in a few markets in various blended percentages right from 3 percent to 85 percent. Certain applications, for example some classes of racing cars, run on 100 percent methanol. While the Government had a firm stance on bringing in electric vehicles and even dissuading full hybrids from entering the market, it now seems open to exploring other strategies to reduce pollution.

This paper discusses latest trends in utilizing methanol as a renewable fuel in automotive industries towards air pollution prevention in various parts of the world.

Keywords: Alcohol fuel, iso-butanol, high energy content, better fuel vaporization, higher blending octane, anhydrous ethanol blends.

Molecular simulation applied to membrane water treatment: A review

A. Megha Mohan[1*], B.Pramada.S.K[2]
[1] *Department of Civil Engineering,National Institute of Technology,Calicut,Kerala,India*
[2] *Assistant Professor, Department of Civil Engineering,National Institute of Technology,Calicut,Kerala,India*
* *e-mail: megha_p180115ce@nitc.ac.in*

Introduction

The demand for water for both domestic and industrial use is increasing due to population growth. A variety of contaminants of size ranging from micrometre to tenths of nanometres is present in raw water. Conventional water treatment methods will not be sufficient to bring the quality of water to drinking water standards. Advanced water treatment techniques such as membrane technology, advanced oxidation process, Electrodialysis etc. is required for this. Membrane technology is used for water treatment in order to improve the quality of water. Membrane process uses a physical barrier that may be a porous membrane or filter. It is a versatile technique to separate particles in a fluid. Specially designed membranes with different pore sizes are used to separate particles on the basis of their size and shape with the aid of pressure. Commercially used membranes for removal purpose are microfiltration membranes, ultrafiltration membranes and nanofiltration membranes. These membranes basically differ according to their pore size. The major disadvantage of membrane technology is membrane fouling which occurs due to metal oxides, organic or inorganic colloids, bacteria or other micro-organisms. The efficiency of these membranes can be improved by incorporating functionalised nanomaterials.

The performance of the membranes can be analysed by various characterisation techniques. Physical movement of atoms and molecules can be studied by molecular simulation. In molecular simulation, the atoms and molecules are allowed to interact for fixed period of time, to study the dynamic evolution of the system (Magnus and Erik (2015)). Molecular dynamics is frequently used to analyse the dynamics of atomic level phenomena, such as thin film growth that cannot be observed directly. This method is frequently applied to study the motion of molecules such as proteins and nucleic acids which can be useful for interpreting the results of certain biophysical experiments and for modelling interactions with other molecules. Molecular dynamics can also be used for prediction of protein structure.

Membranes are fabricated by a polymeric base with additives and nanomaterials. By using molecular simulation the macromolecules present in the membrane can be analysed. The analysis of the process in micro level is possible using molecular simulation. In this paper the possibility of applying molecular simulation software in membrane water treatment will be discussed. The various simulation softwares that can be applied to membranes are also discussed.

Materials and Methods

Papers in the period 2009 to 2019 are reviewed in this work. The major keywords used are membrane technology, molecular simulation and membrane molecular simulation. The methods in which the molecular simulation softwares are implemented were reviewed. The feasible softwares for membrane simulation was selected and reviewed.

Results and Concluding Remarks

The most applicable software in membrane water treatment is presented. The advantages of applying molecular simulation in analysing membrane process are also mentioned.

References

Magnus L and Erik L(2015) Aotuomatic GROMACS Topology Generation and Comparisons of force fields for Solvation free energy Calculations. The Journal of Physical chemistry 119:810-823.dx.doi.org/10.1021/jp505332p

Temporal variation of air pollutants concentration over Chennai

Athira. T[1], Agilan. V[2*]

[1] *Research Scholar, Department of Civil engineering, National Institute of Technology Calicut, Kerala, India.*

[2] *Assistant Professor, Department of Civil Engineering, National Institute of Technology Calicut, Kerala, India.*

** e-mail: athira_p180082ce@nitc.ac.in*

Introduction

Air is the vital and most dominant component of the atmosphere. Clean air is a necessity for all the living beings and is considered as a fundamental right. The presence of unwanted substances, particles and gasses in the air makes it harmful and leads to air pollution. Air pollution is significantly potent enough to worsen our living standards and most of the countries are under the threat of air pollution. There are various sources for this pollution among which anthropogenic sources are the predominant ones. The best way to combat air pollution is to make public aware about the severity of air quality in the atmosphere, which enables public to understand severity and importance of protecting air. Therefore, the objective of this study is to interpret the time series air pollutant data and identify the type of pollution prevailing in the study area.

Materials and Methods

Chennai, the capital of Tamilnadu, is chosen as the study area. It is one of the largest and most populated cities in India. The EDGAR v4.3.2 gridded emissions of the air pollutants during the period 1970 to 2012 is used for this study(Crippa *et al.*, 2018). For this study, five air pollutants namely, carbon monoxide, particulate matter (both PM_{10} and $PM_{2.5}$), sulphur oxide and nitrogen oxide are selected. The annual pollutant concentrations for various points of Chennai city are extracted from the EDGAR v4.3.2 gridded dataset. The temporal variation in carbon monoxide, particulate matter (both PM_{10} and $PM_{2.5}$), sulphur oxide and nitrogen oxide are analysed using various statistical methods.

Results and Concluding Remarks

The study results showed an increasing trend in all the air pollutant concentrations over the study area. This increment should be seriously considered as a clear warning. Note that the Chennai's climate has been drastically changed in the last few years and air pollution could be one of the reasons for this change(Sá *et al.*, 2016).

References

Crippa, M. *et al.* (2018) 'Gridded emissions of air pollutants for the period 1970–2012 within EDGAR v4.3.2', *Earth System Science Data*, 10(4), pp. 1987–2013. doi: 10.5194/essd-10-1987-2018.

Sá, E. *et al.* (2016) 'Climate change and pollutant emissions impacts on air quality in 2050 over Portugal', *Atmospheric Environment*. Pergamon, 131, pp. 209–224. doi: 10.1016/J.ATMOSENV.2016.01.040.

A lab scale batch study of free water surface constructed wetland for water hyacinth plantation

A. Nema[1*], K. D. Yadav[1], R. A. Christian[1]
[1] *Civil Engineering Department/SV National Institute of Technology/ Surat, India*
** e-mail: anudeepneman@gmail.com*

Introduction

Most constructed wetlands around the world are still primarily used to treat municipal and domestic wastewaters but treatment of many types of industrial and agricultural wastewaters, stormwater runoff and landfill leachate has recently become also common. Despite the mistrust of many civil engineers and water authorities constructed wetlands have been widely accepted around the world and have become a suitable solution for wastewater treatment. The mechanisms of pollutant removal in constructed wetlands involve an interaction between the bacterial metabolism, plant uptake and accumulation (Osem et al.,2007). The impurities are removed in facultative ponds entirely by natural processes involving both algae and bacteria (Abdel-Raouf et al, 2012). In that order, vegetation is considered as a dominant feature of constructed wetlands and acts as an important biotic factor in the treatment process (Dhote & Dixit, 2009).

Greywater has great potential for reuse due to its availability and its low concentration of pollutants compared with combined household wastewater. Due to the low levels of contaminating pathogens and nitrogen, reuse and cycle of greywater are receiving more and more attention. Recycling of greywater involves installing a system which treats greywater to meet quality standard for non-potable uses. Its importance is to reduce pollutants and minimize the risk of pathogenic transmission. Non-potable applications of greywater reuse include sprinkling irrigation, toilet flushing, laundry, car washing, floor washing, concrete production, etc. Water hyacinth can be used for the treatment of greywater with a constructed wetland system.

Materials and Methods

Water hyacinth collected from the pond near to Bhagwan Mahavir College of Engineering which 2 km away from SVNIT campus. Water hyacinth collected from the pond has an average height of 15 cm. Collected water hyacinth are carefully washed with tap water without making damage to roots. These plants were acclimatised for greywater by increasing the percentage of greywater (50%, 60%, 70%, 80%, 90% and 100%) into the water for 30 days.

Fig.1 Reactor constructed with water hyacinth with batch run for 25 days operation period.

The laboratory set up was operated at batch mode. In the case of batch mode parameters were monitored at 24 hr interval. The batch system starts with an initial greywater volume of 115 L such that the reactor will work up to 25 days, without losing minimum water level in the reactor such that the addition of greywater not required after starting the batch process. The samples have to collect from outlet of the reactor every 24 hr for an operation period of 25 days. Treated samples are collected and tested for pH,

Conductivity, Temperature, BOD3, COD, TS, TSS, TDS and TKN on daily basis. Fig.1. showing reactor constructed with water hyacinth with batch run for 25 days.

Results and Concluding Remarks

The greywater comprised water from wash basins, showers and bathrooms. Greywater from the kitchen was not included. Initial greywater characteristic is shown in table 1 below.

Table 1.Cheratorstics of greywater used in batch process.

Parameters	Unit	Mean value
pH	-	7.32 ± 0.44
Temperature	^0C	27.50 ± 1.28
Electrical conductivity	mS/cm	0.86 ± 0.04
Turbidity	NTU	187.20 ± 16.90
TS	mg/L	608.40 ± 59.35
TSS	mg/L	334 ± 23.72
TDS	mg/L	274.60 ± 62.22
Hardness, as CaCO3	mg/L	180.40 ± 15.56
Alkalinity, as CaCO3	mg/L	210.20 ± 15.35
COD	mg/L	180.20 ± 6.73
BOD3	mg/L	50.60 ± 11.70
Chloride	mg/L	45.80 ± 6.52
TKN	mg/L	3.10 ± 1.21
Oil and grease	mg/L	24.80 ± 4.85

In batch system Conductivity, pH, TSS, and TKN of greywater increase from the initial values after treatment. TS removal occurs only in increased operation time of wetland in the batch system. TDS reduction is because of plant intake and microbial consumption, with 50% removal within 3 days of operation of the batch system. TKN of greywater increases when treated with wetland operated in batch modes water hyacinth reintroduces nutrients to greywater. In the batch system, BOD_3, COD was shows better removal efficiency. The maximum removal efficiency of BOD_3 was 83.3% and of COD was 75% within 3 days of operation of the batch system. Further, the removal efficiency of BOD_3 and COD showed same removal up to the 25th day of operation with no considerable variation. After treatment TS, TSS, TKN values were increasing from the initial value of greywater sample. That showed negative removal efficiency for TS, TSS and TKN in throughout the batch run. Analysing the removal efficiency of the batch system, within 3 days of operation period batch system shows good result in the removal of BOD_3, COD and TDS.

Acknowledgement: Take this humble opportunity to express my deep sense of gratitude to my supervisor Dr. K. D. Yadav and Dr. R. A. Christian, for his valuable guidance, encouragement and constructive criticism during the dissertation work. I also fully acknowledge him for the freedom and unconditional support to complete this dissertation work. I would like to thank all the faculties of the Environmental Engineering Section, SVNIT.

References

Osem, Y., Chen, Y., Levinson, D., & Hadar, Y. (2007). The effects of plant roots on microbial community structure in aerated wastewater-treatment reactors. Ecological Engineering, 29(2), 133-142.

Abdel-Raouf, N., Al-Homaidan, A. A., & Ibraheem, I. B. M. (2012). Microalgae and wastewater treatment. Saudi Journal of Biological Sciences, 19(3), 257-275.

Dhote, S., & Dixit, S. (2009). Water quality improvement through macrophytes-a review. Environmental Monitoring and Assessment, 152(1), 149-153.

An assessment on the effects of propanol as additive with Calophyllum Inophyllum biodiesel-diesel blends

Anant G. Nagpure[1*], W.S.Rathod[1], Supriya Bobade[2]
[1] Department of Mechanical Engineering, Veermata Jijabai Technological Institute, Mumbai, Maharashtra India.
[2] Indian Biodiesel Corporation, Baramati, Maharashtra, India.
* e-mail: nagpureag77@gmail.com

Introduction

Rigorous emission regulations have increased research interest on alternative renewable fuels. The energy demands of the world mainly depend on petroleum-based fuels. About 27% of the energy consumption is fulfilled by petroleum-derived fuels in the transportation sector; this type of fuel should be replaced by biofuels within 2050 [1]. Biofuels exhibit the similar properties like petroleum fuels and can be blended with petroleum fuel. These are bio-degradable, non-toxic, having higher flash point and a better lubricity. However, they have some limitations in fuel properties such as low volatility, high viscosity and pour point. [2] Therefore the fuel properties of the biofuels must be improved before using it in diesel engines. In open literature such investigations can be seen, where focus has been given on biodiesel/diesel blends and the addition of certain alcohols into those blends. As the number of carbons in alcohols increases, they can be more easily blended with diesel and biodiesel. With respect to the number of carbons in atomic structures of higher alcohols, propanol (C_3H_7OH), n-butanol (C_4H_9OH) and 1-pentanol ($C_5H_{11}OH$) have higher cetane numbers, calorific values, viscosities, flame speeds while having lower latent heat of evaporations, ignition temperatures and corrosion risks. Moreover, because these higher alcohols have good solvent capabilities, they can easily be blended with diesel and biodiesel. Some investigations have proven that higher alcohols such as n-butanol and n-pentanol are more effective than lower alcohols and therefore, having more potential in being the next generation biofuels. [3-4]

In this work, as a next generation alternative fuel, the higher alcohol, propanol was blended with diesel and Calophyllum Inophyllum biodiesel and evaluated in a diesel engine to investigate the engine performance, exhaust emission and combustion characteristics. Initially the diesel-biodiesel cream blend was investigated by using B05, B10, B15, B20, B25, B30 blends. The engine tests were conducted at various loads and compression ratios. With this experimentation, the suitable blend found was B20. In further experimentation, propanol in 6%, 12% and 18% by volume was added to diesel-biodiesel blend to make B20P6, B20P12 and B20P18 blends. Brake thermal efficiency(BTE), specific fuel consumption(SFC), hydrocarbon(HC), carbon monoxide(CO), carbon dioxide(CO_2), oxides of nitrogen(NO_X), heat release rate, and cylinder pressure are considered as evaluation criteria. A single cylinder, constant speed direct injection diesel engine was used for exploratory analysis of evaluation criteria at different load conditions.

Materials and Methods

Characterization of test fuels, Engine testing and facilities

The test fuels used were petroleum diesel fuel and Calophyllum biodiesel feedstock, which met the ASTM requirements of fuel properties. All properties were checked at Indian Biodiesel Corporation, Baramati. The fuel tests were carried out on the single cylinder four stroke variable compression ratio C.I. research engine. The engine specifications were: Bore = 87.5mm, Stroke = 110mm, Compression ratio = 12:1 to 18:1, and maximum power output of 3.5 kW at 1500 rpm. The water cooled eddy current dynamometer was used with an electronic exciter for measuring and maintaining the required load. The test setup was equipped with fuel supply system and data acquisition unit. A measuring probe was used to collect the exhaust gas samples at tailpipe of the engine. The collected gaseous emissions were measured using five gas analyzer (AVL). The combustion analysis software (Enginesoft) provided the results of the performance and combustion characteristics.

The entire engine experiments were conducted at a rated speed of 1500 rpm with an injection timing of 27° before Top Dead Center. The engine was allowed to run till the steady state is reached. Then the engine was loaded in terms of 25%, 50%, 75% and 100% at various compression ratios viz. 14:1, 16:1, and 18:1. To generate the base line data, experiments were carried out initially using neat diesel fuel. After recording the base line data, tests were carried out using B20P6, B20P12 and B20P18 blends. The engine performance, emission and combustion characteristics were recorded.

Results and Concluding Remarks

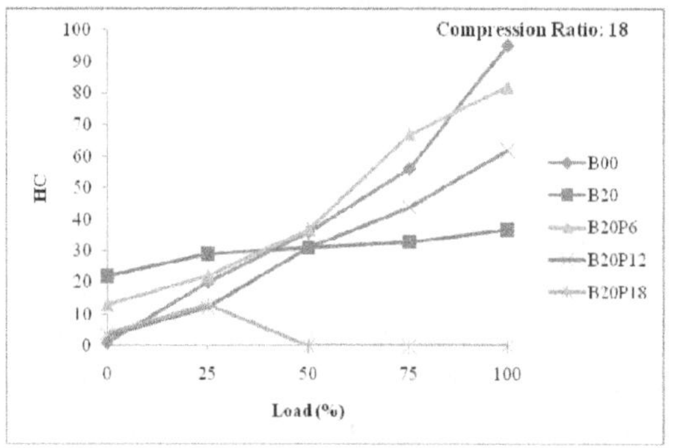

Figure 1: *Variation of Hydrocarbon emissions as a function of engine load.*

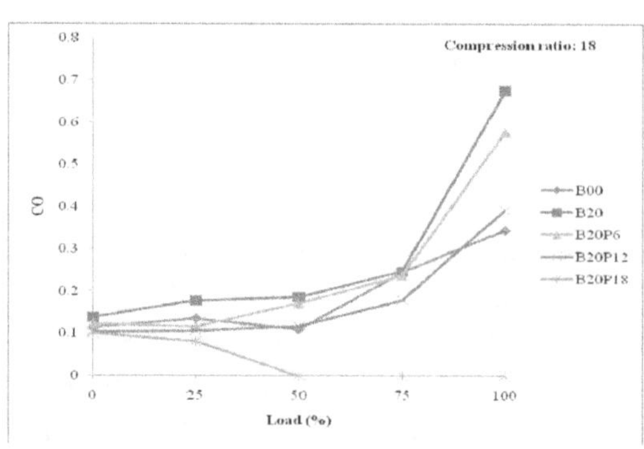

Figure 2: *Variation of Carbon monoxide emissions as a function of engine load.*

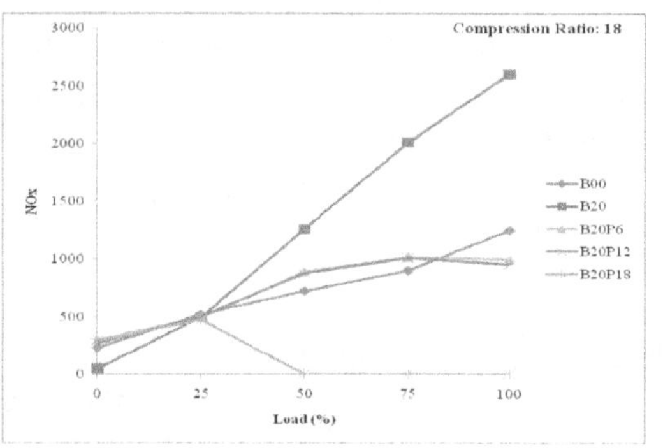

Figure 3: *Variation of Nitrogen oxide emissions as a function of engine load.*

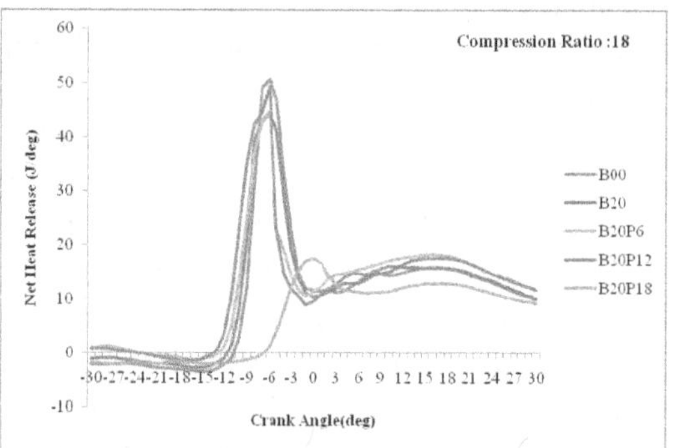

Figure 4: *Variation Heat Release Rate as a function of Crank angle.*

- By using the higher alcohol additive, fuel properties are enhanced. It slightly decreased blends cetane number and flash point.
- B20P6 and B20P12 shown increase in Brake thermal efficiency, but with B20P18 blend engine cannot run at higher load conditions. Experimental evaluation has shown that B20P12 is the optimum value of percentage of additive in cream blends of diesel and biodiesel.
- Better engine performance, improved combustion and considerable reduction in NOx emission is observed. NOx emissions decreased due to higher alcohols have a cooling effect inside the combustion chamber.

References

[1] H.K. Imdadul , H.H. Masjuki , M.A. Kalam , N.W.M. Zulkifli , Abdullah Alabdulkarem, M.M. Rashed , Y.H. Teoh , H.G. How (2016), "Higher alcohol–biodiesel–diesel blends: An approach for improving the performance, emission, and combustion of a light-duty diesel engine", Energy Conversion and Management, 111, 174–185.

[2] Nadir Yilmaz, Alpaslan Atmanli (2017), "Experimental assessment of a diesel engine fuelled with diesel-biodiesel-1-pentanol blends", Fuel 191, pp. 190-197.

[3] Gokhan Tuccar , Tayfun Ozgur, Kadir Aydın (2014), "Effect of diesel–microalgae biodiesel–butanol blends on performance and emissions of diesel engine", Fuel 132, 47–52.

[4] K. Nanthagopal, B.Ashok, B.Sarvanan, Deepam Patel, B. Sudarshan, R. Aaditya R. (2018), "An assessment on the effects of 1- pentanol and 1-butanol as additives with Calophyllum Inophyllum biodiesel", Energy Conversion and Management. 158:70–80.

A Review on Determination of Emerging Environmental Pollutants in Surface water using Chromatography/ Mass Spectrometry Techniques

G.T.N.Veerendra [1*], Dr. B.Kumarvel [2], Dr. M.Mohana Krishna Reddy [3]
[1] *Research Scholar, Department of Civil Engineering, Annamalai University, Chidambaram, Tamil Nadu*
[2] *Associate Professor, Department of Civil Engineering, Annamalai University, Chidambaram, Tamil Nadu*
[3] *Principal Scientist, Analytical Chemistry, CSIR-IICT, Tarnaka, Hyderabad, Telangana*
* *e-mail: tarakaveerendra@gmail.com*

Introduction

Emerging pollutants present in the environment are posing the challenge of new water quality standards. From the past decades, the existence of contaminants was increased in surface water at wide range across the globe and showed a severe threat to both aquatic and human health concern. However those contaminants are not included with present water quality regulations also not in any monitoring programs, the pollutants refer to pharmaceuticals, personal care products (PCP's), industrial effluents, pests of aquaculture, etc. The emerging contaminants prioritised as relentless and bioactive, which cannot be removed by the available conventional wastewater treatment methods nor biodegradable. The long term discharge of effluents with wastewater is addressed to be a hazard, because the contaminants may undergo bioaccumulation and chances of evolving new mixtures in our waters. In specific with low concentrations and diversity in pollutants are treated as the most challenge for detection and analysis, even for water and wastewater processes. The need for trending Analytical Methods and Methodologies associated with High-resolution mass spectrometry and chromatography have significant scope for tracing the potential contaminants of the water environment.

Materials and Methods

The present paper was intended to review and identify the chemically active compounds at various locations of countries like USA, China, South Africa, Bangladesh, Malaysia and some parts of India. In precise, the paper will discuss various techniques and method preparations involved to trace of environmental pollutants using Analytical Equipment like High performance liquid chromatography, Liquid Chromatography/ Mass Spectrometry and Gas Chromatography/ Mass Spectrometry. Besides, Author will also explore about techniques for handling the collected foreign sample, storage procedure, extraction techniques, Standards preparation (RSD -relative standard deviation, LOD -Limit of Detection and LOQ-Limit of Quantification), selection of mobile phase solutions, detector types and Mass library for quantification and determination.

Results and Concluding Remarks

To address all Analytical methods providing prompt and sensitive multi-residue methods proposed by previous authors for the environmental samples and target compounds traced at different locations like Triclosan, Ketoprofen, Malathion, Diazinon, Aldicarb and many. Finally, all traced compounds with average concentrations shall be addressed, and compared with health-based guideline value suggested by WHO will undergo a thorough review.

References

Archer, E.; Petrie, B.; Kasprzyk-Hordern, B.; Wolfaardt, G. 2017. The fate of pharmaceuticals and personal care products (PPCPs), endocrine disrupting contaminants (EDCs), metabolites and illicit drugs in a WWTW and environmental waters. IN: Chemosphere, Vol.174, pp.437-446.

Rai S, Gullapalli MD, Srivastava A, Shaik H, Siddiqui MH, Mudiam MKR., A Rapid Method for the Quantitative Determination of 34 Pesticides in Nonalcoholic Carbonated Beverages Using Liquid-Liquid Extraction Coupled to Dispersive Solid-Phase Cleanup Followed by Gas Chromatography with Tandem Mass Spectrometry. J AOAC Int.

2017 May 1;100(3):624-630. doi: 10.5740/jaoacint.17-0064.
Rashed, M.N. (2013) in Organic Pollutants: Monitoring, Risk and Treatment, M.N. Rashed, Ed., Chapter 7, InTech, http:// dx.doi.org/10.5772/54048
Working Group on Pesticides (2002) Report of the OECD/ FAO Zoning Project, Environment Directorate, ENV/JM/ PEST(2002)15, Organization for Economic Cooperation and Development, Paris, France, http://www.fao.org/fileadmin/ templates/agphome/documents/Pests_Pesticides/JMPR/ENVJM-PEST-2002-15.pdf
Gupta, G.P., Mahapatro, G.K., Puri, S.N., & Ramamurthy, V.V. (2009) in Proceedings of the National Symposium on IPM Strategies to Combat Emerging Pests in the Current Scenario of Climate Change, January 28–30, 2009, Pasighat, Arunachal Pradesh, India, pp 399–412

Mitigation of alkali-induced heave in transformed kaolinitic clays using flyash and GGBS

P.Lakshmi Sruthi[1], P. Hari prasad Reddy[2*]

[1] *Research Scholar, Department of Civil Engineering, National Institute of Technology Warangal, Warangal, India.*

[2] *Associate Professor, Department of Civil Engineering, National Institute of Technology Warangal, Warangal, India.*

* *e-mail: ponnapuhari@gmail.com*

Introduction

Chemically contaminated soil often experiences direct or indirect effects, which even leads to substructure failures. Several case studies related to leakage of chemical contaminants like acids (Assa'ad 1998) and alkalis (Sibley and Vadgama 1986) due to accidental spillage from industries were widely documented. Further, number of studies on abnormal volume change behaviour of alkali-contaminated clays were reported (Rao and Rao 1994; Sivapullaiah and Reddy 2009). From the observed results of induced swelling by alkali, efforts were made by researchers to control alkali induced heave using salt solutions (Sivapullaiah and Manju 2006; Sivapullaiah and Reddy 2010). From the above mentioned studies, it is clear that the role of chemical stabilization using salt solutions to control alkali induced heave is insignificant. Moreover, looking for alternate materials to control alkali induced volume changes, studies were carried out using lime, flyash and GGBS (Sivapullaiah and Manju 2006b; Ghosh et al. 2016). However, the effectiveness of FA and GGBS to control the alkali induced heave in alkali transformed kaolinitic clays (ATKC's) is unrevealed. Thus, the present work focuses on the possibility of using industrial by products like flyash (FA) and ground granulated blast furnace slag (GGBS) for stabilizing the induced heave in alkali transformed kaolinitic clays i.e. red earth (ATRE) and kaolin (ATK).

Materials and Methods

Two kaolinitic clays namely, red earth (RE) and kaolin (K), were selected for the present investigation. Flyash (class F) collected from National Thermal Power Corporation, Telangana and Ground granulated blast furnace slag (GGBS), the by-product of steel industry were selected as heave controlling agents in this study. 10% and 20% of flyash and GGBS were considered as the quantities of stabilizing agents used. Sodium hydroxide solution of 0.1 N and 1 N concentration was considered as a contaminant.

One-dimensional consolidation test was performed according to Puppala et al. (2005) to evaluate the swell potential of stabilized clays. Later, stabilized clay samples were for taken for XRD and SEM analysis.

X-ray diffraction (XRD): Samples were analyzed from 6° to 70°, at a step size of 0.01, using copper k-alpha radiation, PANalytical X-ray Diffractometer.

Scanning electron microscopy (SEM): SEM–TESCAN model microscopy equipment was used for finding out the morphological alterations in the clay samples.

Results

Samples that exhibited high swelling (i.e. ATRE with 0.1 N, 1 N NaOH and ATK with 0.1 N NaOH) were selected for carrying out stabilization studies. Decrease in swelling was noticed with the addition of FA and GGBS. Alkali transformed red earth (ATRE) treated with 20% flyash showed an overall decrease in swelling of about 86% in 0.1 N and 1 N NaOH, whereas 20% GGBS showed a reduction of about 75% in both solution. Furthermore, alkali transformed kaolin (ATK) inundated with 0.1 N NaOH solution showed a reduction in swelling of about 63% and 77% with addition of 10% FA and 10% GGBS respectively (Fig.1). The reduction in swelling with FA and GGBS is attributed to the formation of sodium and calcium based mineral polymers respectively. These polymers bind the partially reacted and unreacted particles restricting the alkali induced swell.

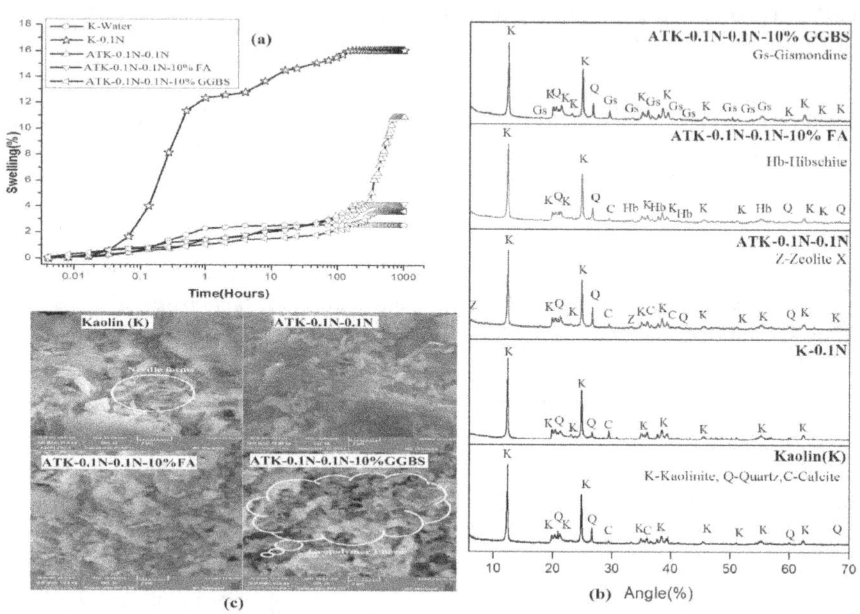

Fig. 1(a) Swelling behavior (b) XRD patterns (c) SEM images of Kaolin (K) and ATK-0.1N-0.1N stabilized with 10% FA and 10% GGBS

Concluding Remarks

- Addition of FA/GGBS to ATKC's effectively reduced the swelling under alkaline conditions
- % FA and GGBS varied with the amount of mineral content, transformation condition and concentration of interacting solution
- 20% fly ash/ GGBS is required for effective control in ATRE w.r.t 0.1 N and 1 N transformation condition.
- Whereas, 10% fly ash/ GGBS is considered to be more effective in controlling the swelling of ATKC w.r.t 0.1 N transformation condition.
- XRD studies revealed that sodium and calcium based geo polymers formed are responsible for the effective reduction of induced swelling in ATKC's.
- SEM studies revealed web or net formations highlighting the mineral polymerization in ATKC's.
- Effective reduction in swelling is partly by replacement of soil with Fly ash/GGBS and mostly by cementation of soil particles by pozzolonic Fly ash/GGBS due to mineral polymerization.

References

Assa'ad A (1998) Differential Upheaval of Phosphoric Acid Storage Tanks in Aqaba Jordan. Journal of Performance of Constructed Facilities 12:71–76. https://doi.org/10.1061/(ASCE)0887-3828(1998)12:2(71)

Ghosh P, Kumar H, Biswas K (2016) FlyAsh and Kaolinite-based Geopolymers: Processing and Assessment of Some Geotechnical Properties. International Journal of Geotechnical Engineering 10(4): 377–386. http://doi.org/10.1080/19386362.2016.1151621

Puppala AJ, Napat I, Rajan, KV (2005) Experimental studies on ettringite-induced heaving in soils. Journal of Geotechnical and Geoenvironmental Engineering 131(3):325–337. https://doi.org/10.1061/(ASCE)1090-0241(2005)131:3(325)

Rao SM, Rao KSS (1994) Ground Heave from Caustic Soda Solution Spillage- A Case Study. Soils and Foundations 34(2): 13–18. http://doi.org/ 10.3208/sandf1972.34.2_13

Sibley MH, Vadgama NJ (1986) Investigation of Ground Heave at ICI Mond Division, Castner-Keller Works, Runcorn. Geological Society, London. Engineering Geology Special Publications 2:367–373.http://doi.org/10.1144/GSL.1986.002.01.62

Sivapullaiah PV, Manju (2006) Ferric Chloride Treatment to Control Alkali Induced Heave in Weathered Red Earth. Geotechnical and Geological Engineering 23(5):1115–1130. http://doi.org/10.1007/s10706-005-1137-7

Sivapullaiah PV, Reddy PHP (2009) FlyAsh to Control Alkali Induced Volume Changes in Soils. Proceedings of the Institution of Civil Engineers – Ground Improvement 162: 167–173. http://doi.org/10.1680/frim.2009.162.4.167

Sivapullaiah PV, Reddy PHP (2010) Potassium Chloride Treatment to Control Alkali Induced Heave in Black Cotton Soil. Geotechnical and Geological Engineering 28(1): 27–36. http://doi.org/10.1007/s10706-009-9274-z

PERFORMANCE OF CEMENTITIOUS BINDERS IN STABILIZATION OF CHROMIUM WASTE IN A LANDFILL

Swapnil Tiwari[1], Dr. Sunil Kumar[2]

[1]Department of Civil Engineering, HBTU, Kanpur, India.

[2]Department of Civil Engineering, HBTU, Kanpur, India.

*email: shyamhari081196@gmail.com, suniljadon@live.in

Introduction

India's 2500 tanneries discharge about 24 million cubic metres of wastewater and about 0.4 million tonnes of hazardous solid waste per annum (CPCB and MoEF, 2003). Most extensively used technique for tannery sludge disposal is dumping it in a sanitary landfill. But the leachate produced in landfill is a point source of soil and underground water contaminants. Leachate of sanitary landfill contains heavy metals such as Chromium, Arsenic, Lead etc., which degrades the quality of soil and underground water. Once the groundwater is contaminated with these pollutants, it is very expensive to treat that water. These toxic metals are potentially responsible for adversely affecting plant and animal's health. Therefore, this study aims to stabilize the chromium laden tannery sludge in a landfill using cementitious binders (cement, lime, fly ash and gypsum). *Palomo and Palacios 2003* had conducted an experimental study on stabilization potential of a cementing matrix, by mixing alkali-activated fly ash, in the presence of hazardous heavy metal chromium and lead, and concluded that presence of chromium suppresses the actuation mechanism of fly ash and, as a result its hardening process due to the formation of sodium chromate. *Rossetti et al. 2002* summarized the optimization of Stabilization/Solidification system for Cr, Pb, Zn and Li in aqueous solution.

Stabilization/Solidification technique is used as waste processing technology for effective disposal of waste, with the objective of safely depositing hazardous industrial waste for disposal. The technology is economical and safe, since metals are converted (or stored as) into highly insoluble salts that do not leach into the groundwater at a remarkable rate. In present study, Stabilization/Solidification technology was applied to dry sludge containing chromium. Twenty mixes of cementitious binders were made for M-10 grade of concrete. The variation in compressive strength and amount of chromium in leachate produced of all mixes was observed.

Materials and Methods

Tannery sludge obtained from common effluent treatment plant Jajmau, Kanpur (U.P.) was used in present study. Locally available PPC cement, sand and stone dust was used with Lime, Gypsum and fly-ash as different cementitious binders for concrete mixture of M-10 grade.

Twenty mixes of different matrices of cementitious binders (Cement, Fly ash, Lime and Gypsum) were made for M-10 grade of concrete (1:3:6). Water/cement ratio for regular concrete was 0.68 and water/(cement+waste) ratio was 0.89. The concrete was partly replaced by weight with obtained sludge in increasing percentage from 0 to 40%. Nine samples (6 cubical+ 3 cylindrical) for each mix are prepared as per IS 516: 1959 for testing (7 and 28 days). Permeability test of concrete as per IS 3085:1965 was conducted after 28 days on cylindrical specimen and leachate was collected.

Colorimetric analysis as per IS 3025 (Part 52):2003 using spectrophotometer at 540nm wavelength was done for leachate samples to determine chromium concentration in leachate produced. Optimization was done for different matrices of concrete.

Results and Concluding Remarks

The leachate produced after permeability test was collected and analysed colorimetrically using spectrophotometer and the result variations (Table 1) are discussed below:

The study shows that the chromium becomes immobilised and leaching of chromium is satisfactory upto 20% replacement of concrete with waste. The values of concentration of chromium in leachate for all the mixes are within the prescribed limit (i.e. 5 mg/l) as per CPCB guidelines for hazardous waste rules, 2016.

The concentration of chromium in leachate of samples was as low as 0.2905 mg/l and as high as 2.3851 mg/l. The optimal result was obtained for Mix 1 with 100% cement, 100% sand and 10% waste replaced by sand with chromium content in leachate – 0.2905 mg/l.

Table 1. Concentration of chromium (VI) in the leachate produced from samples.

S. No.	Mix	Cement In %	Lime In %	Fly-ash In %	Gypsum In %	Sand In %	Stone dust In %	Aggregate		Waste In %	Concentration (mg/l)
								10 mm In %	20 mm In %		
1	MIX 0	100	-	-	-	100	-	50	50	-	-
2	MIX 1	100	-	-	-	100	-	50	50	10	0.2905
3	MIX 2	100	-	-	-	-	100	50	50	10	0.3479
4	MIX 3	80	-	-	20	100	-	50	50	10	0.7838
5	MIX 4	80	-	-	20	-	100	50	50	10	0.8919
6	MIX 5	-	90	10	-	100	-	50	50	10	0.9864
7	MIX 6	-	80	10	10	100	-	50	50	15	1.9459
8	MIX 7	100	-	-	-	-	100	50	50	15	0.6351
9	MIX 8	-	90	10	-	-	100	50	50	15	1.4891
10	MIX 9	80	20	-	-	100	-	50	50	20	0.9054
11	MIX 10	70	10	10	10	-	100	50	50	20	0.7973
12	MIX 11	-	90	-	10	100	-	50	50	20	1.8432
13	MIX 12	80	20	-	-	-	100	50	50	20	0.9358
14	MIX 13	70	10	10	10	100	-	50	50	20	0.8445
15	MIX 14	80	-	20	-	100	-	50	50	30	1.0270
16	MIX 15	100	-	-	-	100	-	50	50	30	1.4392
17	MIX 16	80	-	20	-	-	100	50	50	30	1.7973
18	MIX 17	100	-	-	-	100	-	50	50	35	2.0033
19	MIX 18	100	-	-	-	80	20	50	50	35	2.0202
20	MIX 19	100	-	-	-	100	-	50	50	40	2.3851

References

1. Palomo, A. and Palacios, M., Alkali-activated cementitious materials: alternative matrices for the immobilization of hazardous wastes. Part II. Stabilization of chromium and lead. Cem. Concr. Res., 2003, 33, 289–295.
2. Rossetti, A. V., Di Palma, L. and Medici, F., Assessment of the leaching of metallic elements in the technology of solidification in aqueous solution. Waste Manage., 2002, 22(6), 605–610.
3. CPCB and MoEF (2003) Environmental Management in Selected Industrial Sectors: Status and Need, Central Pollution Control Board and Ministry of Environment and Forests, New Delhi.

Removal of Heavy Metals from Wastewater and Advancement through Zeolites-A Review

G V Sai Krishna, M Chandra Sekhar
Department of Civil Engineering, National Institute of Technology, Warangal, India
e-mail ID: gvsai2110@gmail.com

Introduction

Heavy metals are five times denser than water and naturally occurring components, non-biodegradable, hence persist in the environment. Few of heavy metals are useful for our dietary purposes but most of the heavy metals in their trace concentrations also found to be dangerous.The rapid growth of industrialization, urbanization, and increased living standards are leading to the use of numerous products developed from heavy metals.Wastewaters generated from these processes are posing a great threat to the environment. Heavy metal pollution considered as one of the major environmental concern due to its taxological and carcinogenic nature. This draws towards the removal of heavy metals from wastewater.

Heavy metal removal methods are in effect form past few decades like precipitation, adsorption, ion exchange, membrane filtration,floatation. The chemical precipitations using various chemicals like soda ash, lime, chitosans, peroxides resulted in removal efficiencies varying from partial removal of 70% to complete removal (Quanyuan Chen et al 2018, Junxing Hao 2018,Tiecheng Wang 2019). Adsorption of heavy metal includes various adsorbents like low-cost adsorbents generated from wastes, and effective adsorbents like activated carbon are proven to be prominent in adsorption of heavy metals (S. Ricordel et al 2001,M. Kobya et al 2005). Ion exchange with media prepared for selective metals have shown efficiencies more than 90% (Omid Tavakoli 2017,T.M. Zewail et al 2015, James P. Bezzina et al 2019). Filtrations are suitable for low concentration of heavy metals which yields maximum efficiencies with a decrease in pore size, reverse osmosis is accepted widely due to its efficiency (N. Abdullah et al 2019, Chun-Chun Ye et al 2019, Yifeng Huang et al 2016). Flotation techniques achieve complete removal of heavy metals with the usage of multi-stage cascades along with surfactants (Hongyang Wu et al 2019, MojtabaTaseidifar 2017).

The objective of the study is to review the current scenario of heavy metal removal techniques and to compare them with the advanced adsorption methodologies using zeolites prepared from industrial wastes like fly ash and bottom ash based on the available literature.

Methodology

Fly ash and bottom ashes generated from power plants are rich in $SiO_2$and Al_2O_3which are building blocks of zeolite. Zeolites are hydrated aluminosilicate minerals made from interlinked tetrahedra of alumina (AlO_4) and silica (SiO_4). These tetrahedral units form clusters of polyhedral units with a negative charge to hold a cation in their tetrahedral housing.The research conducted in this area suggested the Characterization of ashes and mineral composition of fly ash are key elements in the synthesis of zeolite using hydrothermal treatment or alkali activation(Maria Visa 2016, Jessica A. Oliveira 2019).From the extensive literature available on the zeolitespreparation and their use as ion exchangersstressed their use as excellent adsorbents of mono, bi, trivalent ions of selective metal or multi-metals(Claudia Belviso 2018).

Results and concluding Remarks

The research carried on the methodologies for removal of heavy metals from wastewaters adsorption has proven efficient and economical due to its availability, regeneration ability. Zeolites are adsorbents prepared from secondary wastes like fly ash and bottom and their ability to adsorb heavy metals makesthem apromising material. The research on the fly ashes carried out with a primary objective of using it as a low-cost adsorbent, succeeded in the development of zeolites from fly ash using alkali activation and hydrothermal treatment. The zeolites synthesis and its use as an adsorbent are proven to be successful but extensive work has to be carried out on the stability of zeolite, leaching phenomenon of zeolite, effects of multivalent cations and their selective capture and recycling opportunities of the exhausted zeolite.

Zeolite generated from fly ash used aslow-cost adsorbents. The review of zeolites as an adsorbent of heavy metals seeks attention in assessing its reusable criteria and effect of various operating parameters. It also highlights the need forstudy on the suitability of final wastes generated as a replacement for building material in less important structures.

References

Belviso, C. (2018). State-of-the-art applications of fly ash from coal and biomass: A focus on zeolite synthesis processes and issues. *Progress in Energy and Combustion Science*, *65*, 109–135. https://doi.org/10.1016/j.pecs.2017.10.004

Bezzina, J. P., Ruder, L. R., Dawson, R., & Ogden, M. D. (2019). Ion exchange removal of Cu(II), Fe(II), Pb(II)and Zn(II)from acid extracted sewage sludge – Resin screening in weak acid media. *Water Research*, (Ii), 257–267. https://doi.org/10.1016/j.watres.2019.04.042

Bratskaya, S. Y., Pestov, A. V., Yatluk, Y. G., & Avramenko, V. A. (2009). Heavy metals removal by flocculation/precipitation using N-(2-carboxyethyl)chitosans. *Colloids and Surfaces A: Physicochemical and Engineering Aspects*, *339*(1–3), 140–144. https://doi.org/10.1016/j.colsurfa.2009.02.013

Chen, Q., Yao, Y., Li, X., Lu, J., Zhou, J., & Huang, Z. (2018). Comparison of heavy metal removals from aqueous solutions by chemical precipitation and characteristics of precipitates. *Journal of Water Process Engineering*, *26*(October), 289–300. https://doi.org/10.1016/j.jwpe.2018.11.003

Hao, J., Ji, L., Li, C., Hu, C., & Wu, K. (2018). Rapid, efficient and economic removal of organic dyes and heavy metals from wastewater by zinc-induced in-situ reduction and precipitation of graphene oxide. *Journal of the Taiwan Institute of Chemical Engineers*, *88*, 137–145. https://doi.org/10.1016/j.jtice.2018.03.045

Joseph, L., Jun, B. M., Flora, J. R. V., Park, C. M., & Yoon, Y. (2019). Removal of heavy metals from water sources in the developing world using low-cost materials: A review. *Chemosphere*, *229*, 142–159. https://doi.org/10.1016/j.chemosphere.2019.04.198

Oliveira, J. A., Cunha, F. A., & Ruotolo, L. A. M. (2019). Synthesis of zeolite from sugarcane bagasse fly ash and its application as a low-cost adsorbent to remove heavy metals. *Journal of Cleaner Production*, *229*, 956–963. https://doi.org/10.1016/j.jclepro.2019.05.069

Ricordel, S., Taha, S., Cisse, I., & Dorange, G. (2001). Heavy metals removal by adsorption onto peanut husks carbon: Characterization, kinetic study and modeling. *Separation and Purification Technology*, *24*(3), 389–401. https://doi.org/10.1016/S1383-5866(01)00139-3

Taseidifar, M., Makavipour, F., Pashley, R. M., & Rahman, A. F. M. M. (2017). Removal of heavy metal ions from water using ion flotation. *Environmental Technology and Innovation*, *8*, 182–190. https://doi.org/10.1016/j.eti.2017.07.002

Visa, M. (2016). Synthesis and characterization of new zeolite materials obtained from fly ash for heavy metals removal in advanced wastewater treatment. *Powder Technology*, *294*, 338–347. https://doi.org/10.1016/j.powtec.2016.02.019

Application of Geo-spatial Method for Mapping of Air Pollutants

Karan singh[1], Deepesh Singh[2]
[1]Department of Civil Engineering, HBTU, Kanpur, India.
[2]Department of Civil Engineering, HBTU, Kanpur, India.
*email: karansingh9194@gmail.com, dr.deepeshsingh@gmail.com

Introduction

Kanpur with an area of about 403.7 km^2 is located in the state of Uttar Pradesh, India and having latitude 26.4499° and longitude 80.3319°. The population of this city was estimated 29.2 lakhs in 2018. The air quality data set of Kanpur city, over a period of three years (2016-2018) was used in the study. These data are monitored regularly each month. Kanpur was named the world's most polluted city in WHO's list. According to the report of World Health Organization, about 90 per cent of world's population is exposed to dangerous levels of pollution and 14 of world's 15 most polluted cities are in India. Kanpur is the most polluted city which came on top. Air pollution have a great impact on vegetation i.e. damage of crops, forests etc, on water bodies i.e. eutrophication in lakes, acid rain effect on lakes, rivers and on human health short term effect i.e. throat infection, allergic reaction, headache, respiratory infection and long term effect i.e. lung cancer, brain damage, liver and kidney damage etc. It is found in many researches that younger children and older people are more effected by air pollution. Air pollution also contribute to the depletion of the ozone layer which protect the earth from ultra violet rays of sun. Some other effect is also a climate change. It is written on international council of clean transport that on two wheeler for compression ignition vehicle 89% of particulate matter is reduced in BHARAT STAGE VI and around 76% of NO_X reduced in BHARAT STAGE VI and on spark ignition 82% and 49% of NO_X and CO is reduced in BHARAT STAGE VI vehicles. In this paper the study has been done in monitoring the 7 stations in Kanpur city as per three year data available in UPPCB website. The study is focuses on the IDW interpolation and monitor the pollutants such as SO_2, NO_2, PM_{10} at all the locations of Kanpur city by raster the shape file in a cell size of 1'(minute).

Materials and Methods

Geo-Spatial analyst tool IDW interpolation of ARC GIS is used to mapping the air pollution. The given data of each year divided as pre monsoon (January to May), monsoon (June to September), post monsoon (October to December). In this study indexes were calculate of all the seasons at all the seven stations which is calculate by taking averages of the pollutants level. ARC GIS with geo-spatial analyst tool IDW interpolation find the pollutants at different location of Kanpur city by raster the shape file of Kanpur city in 1' (minute) cell size. The standards are according to the NAAQs criteria and different color shows different level of pollutants i.e. low, moderate, high, critical. Raster the shape file shows different pollutants level at different locations such as commercial location, industrial location, residential, institutional location were shown. In this study seven stations were taken

STATION	LATITUDE	LONGITUDE
AWAS VIKAS	26.4874°	80.2713°
JAREEB CHOWKI	26.4619°	80.3249°
KIDWAI NAGAR	26.4275°	80.3353°
PANKI	26.4705°	80.2407°
SHASTRI NAGAR	26.4639°	80.3006°
IIT KANPUR	26.51°	80.23°
DADA NAGAR	26.4534°	80.2941°

Results and Concluding Remarks

Mapping of these pollutants shows Post-monsoon air quality was better than monsoon and pre monsoon for all air quality parameters. However, in the study it was found out that winter season has worst air quality. It is observed that PM10 is critical in all the three seasons at all the locations of Kanpur city i.e. monsoon, post-monsoon, pre monsoon. So2 concentration is quite good in all the seasons at all the locations.

MONSOON SEASON

1. NO_2 concentration in IIT Kanpur at latitude 26.51 degree and longitude 80.23 degree low (0-20) due to controlled vehicular movement because students are allowed to ride bicycles rather than any motorbike or cars and at latitude 26.5 degree to 26.6 there is very less movement of bikes and cars due to very congested market area and all other locations is high (40-60) in NO_2 concentrations due to very high vehicular movement and all other sources.
2. SO_2 concentration at all the locations of Kanpur city is low (0-25).
3. PM_{10} concentration in IIT Kanpur at latitude 26.51 degree and longitude 80.23 is high (60-90) and all other locations are very critical i.e. greater than 90 due to vehicles which generates particulate either from direct emission from burning of fuels (especially diesel powered vehicles) or from wear of tyres (vehicle generated air turbulence on road ways).

PRE- MONSOON SEASON

1. NO_2 concentration in IIT Kanpur at latitude 26.51 degree and longitude 80.23 degree is low (0-20) is and at latitude 26.52 degree to 26.62 degree and longitude 79.4 to 80.24 degree is moderate(20-40) and all other locations is high(60-90) in NO_2 concentrations.
2. SO_2 concentration at all the locations of Kanpur city is low (0-25).
3. PM_{10} is very critical more than 200 at all the locations of Kanpur city.

POST- MONSOON SEASON

1. NO_2 concentration in IIT Kanpur at latitude 26.51 degree and longitude 80.23 degree is low (0-20) is and at latitude 26.52 degree to 26.62 degree and longitude 79.4 to 80.24 degree near Nehru Nagar is moderate (20-40) also in latitude 26.44 degree to 26.52 degree and longitude 80.31 degree to 80.32 degree is moderate near Mandhana and all other locations is high(60-90) in NO_2 concentrations.
2. SO_2 concentration at all the locations of Kanpur city is low (0-25).
3. PM_{10} is very critical more than 200 at all the locations of Kanpur city.

References

1. *M. Habil, D. Massey, A. Taneja, Personal and ambient PM2.5 exposure assessment in the city of Agra, Data Brief 6 (2016) 495–502.*

2. *B. Kamarehie, M. Ghaderpoori, A. Jafari, M. Karami, A. Mohammadi, K. Azarshab, A. Ghaderpoury, N. Noorizadeh, Estimation of health effects (morbidity and mortality) attributed to PM10 and PM2.5 exposure using an Air Quality model in Bukan city, from 2015–2016 exposure using air quality model, Environ. Health Eng. Manag. J. 4 (2017) 137–142*

3. *Y.I. Chirino, Y. Sánchez-Pérez, Á.R. Osornio-Vargas, I. Rosas, C.M. García-Cuellar, Sampling and composition of airborne particulate matter (PM10) from two locations of Mexico City, Data Brief 4 (2015) 353–356.*

4. *C.M. Liu, Effect of PM2.5 on AQI in Taiwan, Environ. Model. Softw. 17 (2002) 29–37.*

5 *. A. Zhang, Q. Qi, L. Jiang, F. Zhou, J. Wang, Population exposure to PM2.5 in the urban area of Beijing, PLoS one 8 (2013) e63486. [6] S. Chattopadhyay, S. Gupta, R.N. Saha, Spatial and temporal variation of urban air quality: a GIS approach, J. Environ. Prot. 1 (2010)*

Surface Water Quality Assessment of Ganga River with reference to Water Quality Indices

Shivam Mishra[1], Deepesh Singh[2]
[1]Department of Civil Engineering, HBTU, Kanpur, India.
[2]Department of Civil Engineering, HBTU, Kanpur, India.
*email: shivamm682@gmail.com, dr.deepeshsingh@gmail.com

Introduction

The Ganga basin accounts for a little more than one-fourth (26.3%) of the country's total geographical area and is the biggest river basin in India, covering the entire states of Uttarakhand, Uttar Pradesh (UP), Bihar, Delhi, and parts of Punjab, Haryana, Himachal Pradesh, Rajasthan, Madhya Pradesh, and West Bengal. The river traverses a length of 1450 km in Uttarakhand and Uttar Pradesh while touching the boundary between UP and Bihar for a stretch of 110 km. It then flows through Bihar, more or less covering a distance of 405 km. The length of the river measured along the Bhagirathi and Hugli rivers during its course in West Bengal is about 520 km. Main stem of river Ganga houses a population with high density. In absence of proper sanitation, abstraction of surface and groundwater for irrigation and drinking purposes and partially treated domestic and industrial effluent turns Ganga into a polluted river in the stretch from Kannuaj to trighat in the state of Uttar Pradesh and also makes the water of river Ganga unfit even for bathing purposes. Kanpur is one of the major contributors to the pathetic water quality of river. The study area of Kanpur City (longitude 80°14� E and latitude 26°30� N) is situated at an approximate distance of 435 km SE from Delhi on the right bank of river Ganga and Jajmau industrial area is situated on its downstream side. Large number of industrial establishments (mostly tanneries) at Jajmau has made enormous ramifications upon the habitability in its surroundings. Partially treated industrial effluent coupled with domestic sewage from Kanpur City is discharged into the adjacent river Ganga, resulting in worst quality of water in river Ganga in this part of city. It is estimated that Kanpur generates 450 MLD of sewage every day as well but the existing infrastructure can only treat around 160-170 MLD, remainder goes directly into to river untreated.

Methods and Results

There is effort to compare basic water quality parameters at upstream and downstream side of Kanpur city. Parameters under study are dissolved oxygen, biological oxygen demand, total coli form, and faecal coli form. Data was collected from bithoor site and jajmau site for the year 2016-18. The monitoring results obtained during 2016-18 under National Water Quality Monitoring Programme reflect that organic matter and bacterial population of faecal origin continue to dominate the pollution problem in River Ganga. The major water quality concerns as revealed from the monitoring results are pathogenic pollution as reflected through indicators i.e. Total Coliforms (TC) & Faecal Coliform (FC), organic matter as reflected through Biochemical Oxygen Demand (BOD). Descriptive statistical approach was also applied to analyse the trend and range of data. Major variation in all the considered parameters was observed between bithoor sampling station and jajmau sampling station. CPCB has indentified 764 grossly polluting industries discharging wastewater to main stem of River Ganga (either directly or through drains) and its two important tributaries Kali-east and Ramganga in Uttarakhand, Uttar Pradesh, Bihar and West Bengal. It was observed that water consumed by grossly polluting industries is 1123 MLD. In terms of number industrial units, tannery sector is dominating where as in terms of wastewater generation Pulp & paper sectors dominate followed by chemical and sugar sector. It is also observed that GPI in Bihar generate minimum wastewater (19%) in terms of water consumed whereas GPI in West Bengal generate maximum wastewater 75.5% in terms of water consumed this followed by Uttarakhand (56.7%) and Uttar Pradesh (39%).

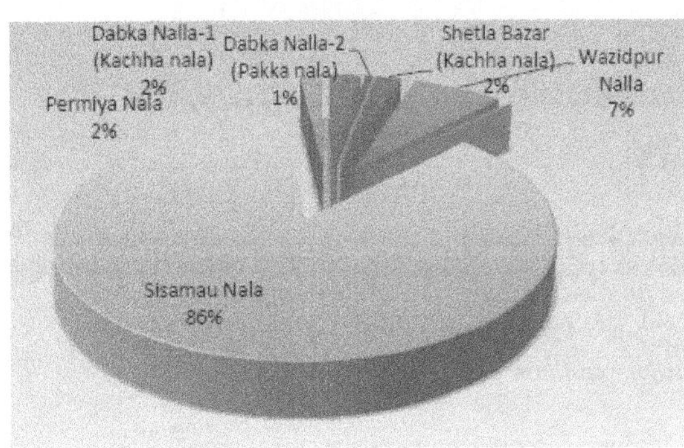

Fig 1: *BOD load distribution of catchment in Kanpur*

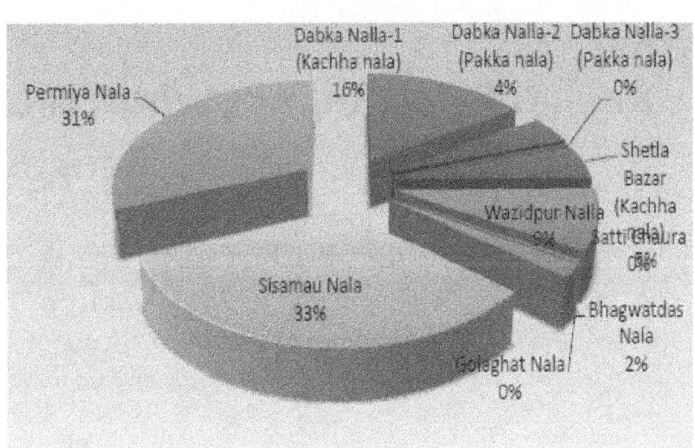

Fig 2: Pie chart showing flow distribution of drains located in Kanpur region

A total of 138 drains were identified, discharging 6087 MLD of wastewater. In Uttarakhand 14 nos. of drains are discharging 440 MLD of industrial and domestic wastewater directly/indirectly to river Ganga. Uttar Pradesh discharges 3289 MLD of industrial and domestic wastewater through 45 drains. In Bihar total 25 no of drains are identified, discharging 579 MLD of wastewater to river Ganga. 1779 MLD of wastewater discharges to river Ganga through 54 drains in West Bengal

Conclusion

Water is one of the most essential natural sources for sustainable life and it is likely to become critically scarce in the coming decades because of the majorly polluted environment due to rapid increase in the population, increasing demands. Especially, River water is one of the Essential drinking water sources entire the world because most of the village areas are situated along the river basins and the people living in villages will use river water as drinking purpose. Uttar Pradesh has 687 grossly polluting industries, finds CPCB. These largely small scale, often illegal units—tanneries, sugar, pulp and paper and chemical—contribute 270 mld of wastewater. But what really matters is the location of the plants. While over 400 tanneries contribute only 8 per cent of the industrial discharge, they spew highly toxic effluent into the river and are located as a cluster near Kanpur. So the concentration of pollution is high. It is alarming that not much is happening to control pollution. In 2013, an inspection of 404 industrial units by CPCB showed that all but 23 did not comply with the law.

References

1. Sharma, Y. (1998). Water Pollution Control - A Guide to the Use of Water Quality Management
2. Singh, B. (2015, June 3). Water quality of river Ganga has an impact on human health: TERI. Varanasi, Uttar Pradesh, India
3. Prakash, M. S. (Jabalpur, 2015, June 3). Survey on Varanasi- Ganga. Retrieved October 9, Trier, 2016, from TERI: http://www.teriin.org/files/press_release_ES_Jabalpur.pdf
4. Vinod Tare, G. R. (August-2013, Page: 17). Ganga River Basin Environment Management Plan: Interim Report. Kanpur, Delhi: Government of India.
5. *Consortium of 7indian IITs, "Ganga river Basin management plan"; 2015.*

Real-Time Ambient Carbon Monoxide Ward–Wise Mapping for Bengaluru City

Rajesh Gopinath[1*], Monica Krishna GM[2], Pratik Kumar Sinha[2], Nitish Kumar[2], Deepak Tripathi[2]
[1] *Faculty, Dept. of Civil Engineering/ BMS Institute of Technology and Management, VTU/ Bengaluru 64, India*
[2] *Student, Dept. of Civil Engineering/ BMS Institute of Technology and Management, VTU/ Bengaluru 64, India*
** e-mail: dr.rajeshgopinathnair@gmail.com*

Introduction

As one among the fastest growing megacity, Bengaluru has sporadically grown in all directions (Dipak DG et al. 2017), with elevated air pollution levels and deteriorating Air Quality index (N Balasubramanya et al. 2017) majorly contributed by Automobile emissions. In this context, Ambient Carbon Monoxide [CO] which is a colourless, odourless, and tasteless gas; is considered to be the most dangerous criteria air pollutant (Aftab Jahan Begum et al. 2017).When compared with all air pollutants; no major exhaustive studies have been carried out in Bengaluru on real-time basis to explore its impending health threats as an inadvertent outcome of rise in petrol engine vehicles at Bengaluru. Most researches on urban outdoor air pollution have focussed mostly on Oxides of Nitrogen and Sulphur (Kanakiya et al. 2015). There has been a dearth of research on CO levels and also the studies that were carried out on CO were restricted to a smaller number of sample size, and also limited by the location attributes (I Suryati et al. 2017; S Harinath et al. 2010). The present study was incepted to plug this gap in research and understand the impending, inadvertent and direct health threats of CO. The major objectives of this research encompasses determination of Temporal variation of Ambient Carbon Monoxide concentration across all wards of Bengaluru using Digital Carbon Monoxide Sensor, followed by Spatial Mapping of Bengaluru for Carbon Monoxide intensity; and Finally delineation of the most critical and non-critical stations.

Materials and Methods

The study envisages monitoring ambient CO levels across all the 198 wards Bengaluru, spread over a period of five months encompassing both winter and summer seasons. The study was carried out with calibrated digital CO meters possessing Stabilized Electrochemical Gas-specific sensor capable of detecting concentrations between 1ppm to 1000ppm. The 11 hours continuous and simultaneous monitoring was engaged at half an hour interval from 7:30 AM to 6:30 PM; with sensor placed at a height of 1.2m to 1.5m from the ground level in accordance with WMO norms. The criteria air pollutant data collection was carried out at junctions/signals, wherein the probability of CO levels being highest was more frequent due to idling conditions of vehicles. The elaborate exercise would result in identification and computation of 'maximum', 'minimum' and 'mean' ambient carbon monoxide intensity levels. The 'exposure-limit' was further compared and consulted with norms of NAAQS, OSHA, EPA, WHO and NIOSH. Subsequently, delineation of the most critical and non-critical places in Bengaluru was achieved; for implementation of corrective measures; and recommendation for preventive measures with respect to town planning.

Results and Concluding Remarks

Startling violations of more than 200 instances have been observed, wherein CO levels were found to thrive beyond 9 ppm, thereby risking the potential of increasing threats of congestive heart failure. Three Violations were found to exceed 15ppm, which can lead to risks such as impaired performance in time discrimination, shortened time to angina response and vigilance decrement. The highest lethal concentration was observed as 18ppm, near Manyata Tech Park.

Also, the major take-away from the study highlights the probable health issues for traffic personnel posted at the junctions. This hence warrants necessary action to inculcate safety and remedial measures ranging from legal actions, to sustainable traffic management measures. As per NAQQS standards out of 198 wards of Bengaluru, 1 ward was in the range of 1-2ppm, 185 were in the range of >2-9ppm and remaining 12 wards were in the high-risk range of >9-14ppm. The same has been showcased in Figure 1. The study concludes with ward-wise specific recommendations, including measures such as 5% in overall wards to be provided with traffic signals, road width to be increased at 2.6% of overall wards, 22.4% of overall wards to enhance green-cover, and dwellers/users from 70% from overall wards to reduce usage of private vehicles and instead to promote usage of public transportation.

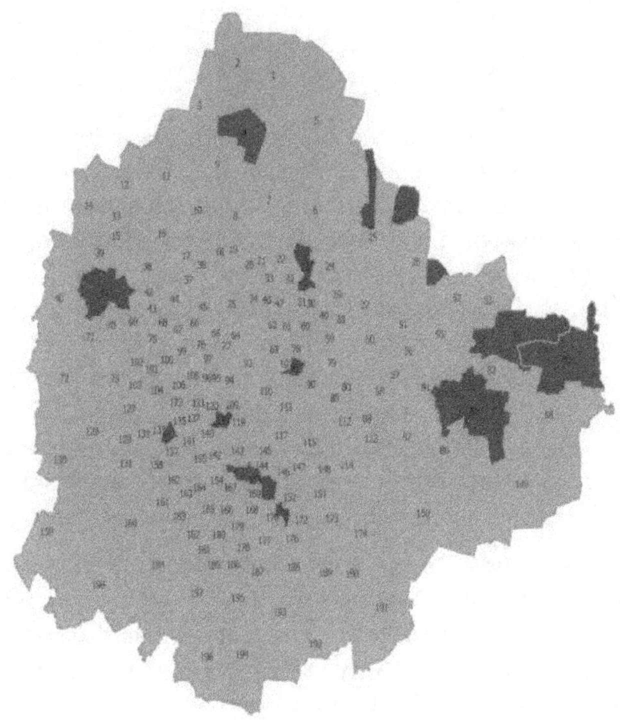

Figure 1. Delineation of 198 observatories of Bengaluru with daily average Carbon Monoxide intensity as per NAAQS.

References

Rajesh Gopinath, Deepak DG BC (2017) A Site-based practical Improvisation for the Analytical Determination of Aspect Ratio. Architecture and Engineering. 2(3):11-20. DOI: 10.23968/2500-0055-2017-2-3-11-20, Эл № ФС77-70026.

N Balasubramanya, Rajesh Gopinath (2017) Historical Trend of Ambient Air Quality Indices at Key Observatories for Major Cities in Karnataka, India. Bulletin of Volgograd State University of Architecture and Civil Engineering, Series: Construction and Architecture. 49(68):189-195. ISSN 1815-4360, PI FS77-19321.

Aftab Jahan Begum, Rajesh Gopinath (2017) Development of Step-Wise Ranking for Indoor Plants as Indoor Air Pollutant Purifiers. ANNALS of Faculty Engineering Hunedoara. 15(4):53-56. ISSN: 1584-2673.

Kanakiya RS, Singh SK, Shah U (2015) GIS application for spatial and temporal analysis of the air pollution in urban area. International Journal of Advanced Remote Sensing and GIS. 4(1):1120-1129. ISSN: 2320-2043.

I Suryati and H Khair (2017) Mapping air quality index of carbon monoxide in Medan city. IOP Conf. Series: Materials Science and Engineering. 180:1-6. DOI: 10.1088/1757-899X/180/012114

S Harinath and Murthy U. (2010) Spatial distribution mapping for air pollution in industrial areas -A case study. Journal of industrial pollution control. 26(2):217-220.

Treatment Technologies Used for Treating Tannery Wastewater Effluent- A Review

Korpe Sneha, P.V. Rao
Department of Civil Engineering, National Institute of Technology, Warangal, India
e-mail: korpesneha11.sk@gmail.com

Introduction

Leather producing Tannery industry is one of the important industries in the world contributing to economic development. But, the fact cannot be denied that the wastewater released by the tanneries is polluting our environment. Tannery waste effluents (TWE) are characterized by their high chemical oxygen demand (COD), biochemical oxygen demand (BOD), total organic carbon (TOC), variation in pH, solid contents, and high turbidity. The effluent released from the industry consists of highly concentrated as well as complex organic matter. Various toxic pollutants such as propylene glycol and oxides, acrylate esters, chromium, cadmium sulphate, tannins, organic and inorganic acids etc. are generally present in tannery wastewater at very high concentrations (Saxena, Rajoriya, Saharan, & George, 2018).These pollutants are proved to be harmful for public health as well as other living beings, if it is not properly treated and released without meeting the standards. It results in water pollution (groundwater and surface water) as well as land pollution. In order to make the tanning process to be sustainable, it is very necessary to adopt sustainable technologies for the production and disposal of tannery effluents.

Various researchers have focussed on treating the tannery wastewater effluent by various physical, chemical as well as biological processes. However treating the effluent by any single process do not prove to be efficient enough, getting the effluent to the discharge standards given by the regulating bodies. The effluent quality can be improved significantly by using more than one process or combination of different processes which will enable reusing the effluent for various purposes such as agriculture or secondary uses. Among the various treatments applied to the tannery effluent, Advanced Oxidation Process (AOPs), hydrodynamic cavitation (HC) process, Coagulation, Electrocoagulation, Electrodialysis etc. are the most used technologies. These techniques can efficiently degrade the complex organic matter present in TWE. These can be used as standalone processes or in combination with conventional treatment process in order to increase the efficiency of removal of the pollutants present in TWE. The present research work aims to review the information regarding the use of various treatment technologies for tannery wastewater effluent, their potential advantages and technical shortcomings.

Materials and Methods

Advanced oxidation processes (AOPs) means application of either advanced oxidation technologies using UV/O_3, O_3/H_2O_2, UV/H_2O_2 or the photo Fenton reaction ($UV/H_2O_2/Fe^{2+}$ or Fe^{3+}). A study reports the comparison among conventional and heterogeneous Fenton oxidation processes in the treatment of tannery wastewater where nZVI particles were produced and used as heterogeneous catalyser in place of Fe(II) in Fenton oxidation (Vilardi, Di Palma, & Verdone, 2018). The principle behind AOP's used for treatment of TWE is the production of highly reactive species having high oxidation potential which helps in degrading the complex organic matter present in TWE. However, AOP's such as application of O_3, UV is comparatively costlier than conventional treatment process.

Hydrodynamic cavitation process is an alternative treatment method used to treat highly contaminated water released from the tanneries. HC describes the process of vaporization, bubble generation and bubble implosion occurring in a flowing liquid as a result of a decrease and subsequent increase in local pressure. Bursting of the bubble/cavities produces OH° radicals which are strong oxidizing agents and are able to degrade the complex structure into simpler form. In addition to that, various chemical reagents having better oxidation potential could also be added in order to increase the removal efficiency. The set-up that is used to perform hydrodynamic cavitation consists of an overhead tank, a centrifugal pump enabling the wastewater to circulate, an orifice or venturi plate in between, to built-up the pressure into the fluid, resulting into cavities producing OH° and pressure valves to maintain the pressure. The removal of contaminants and reduction of COD, TOC, TSS etc. in TWE was carried out using coagulation followed by cavitation as a pre-treatment tool in order to enhance its biodegradability to make it suitable for anaerobic digestion (Saxena et al., 2018). Whereas, HC set-up is designed according to the different types of industrial wastewater treated and contaminants present into it. Though this process is highly efficient in removal of contaminants from wastewater but due to its complex design problem and high cost it is not been in use full-fledged in all industries.

Electrocoagulation/ electrodialysis process is also one of the efficient treatment methods for the removal of COD, NH_3-N, Cr and colour from TWE. The performance analysis of Electrocoagulation/ electrodialysis process was conducted for removal of pollutants from tannery wastewater (Deghles & Kurt, 2016). Effects of current density and electrolysis time were analysed to optimize the electrocoagulation process with aluminium/ iron electrodes by using an integrated EC-ED system composed of a wastewater storage tank, a peristaltic pump, an electrochemical reactor with a set of five pairs of aluminium or iron electrodes, gas separation tank and a sedimentation tank. Due to continuous power supply requirement this method proves to be costly in scaling up at large scales.

Photoelectrocatalytic reduction is also an alternative method which is used for treating TWE, which was carried out for reduction of Cr(VI) by applying electric field to photocatalysis of as-prepared TiO_2 spheres (Zhao et al., 2017). A 500 W high-pressure-mercury lamp was utilized as the ultraviolet (UV) light source, and an applied voltage between the two ends of the reactor was driven by a power supply. This method also requires continuous power supply which makes it costlier.

Results and Concluding Remarks

From the above study we can conclude the following:

- Conventional treatment processes are not enough to treat the highly polluted TWE.
- HC/AOPs/Electrocoagulation etc, can be used as pre-treatment techniques.
- Combination of two or more processes to treat the TWE proves to be more efficient than a single process.
- Residual formation in conventional process is very high and disposal becomes a major challenge whereas, if these pre-treatment strategies are implemented residual formation is less comparatively.

References

Deghles, A., & Kurt, U. (2016). Chemical Engineering and Processing�: Process Intensifi cation Treatment of tannery wastewater by a hybrid electrocoagulation / electrodialysis process. *Chemical Engineering & Processing: Process Intensification*, *104*, 43–50. https://doi.org/10.1016/j.cep.2016.02.009

Saxena, S., Rajoriya, S., Saharan, V. K., & George, S. (2018). An advanced pretreatment strategy involving hydrodynamic and acoustic cavitation along with alum coagulation for the mineralization and biodegradability enhancement of tannery waste effluent. *Ultrasonics Sonochemistry*, *44*(October 2017), 299–309. https://doi.org/10.1016/j.ultsonch.2018.02.035

Vilardi, G., Di Palma, L., & Verdone, N. (2018). On the critical use of zero valent iron nanoparticles and Fenton processes for the treatment of tannery wastewater. *Journal of Water Process Engineering*, *22*(November 2017), 109–122. https://doi.org/10.1016/j.jwpe.2018.01.011

Zhao, Y., Chang, W., Huang, Z., Feng, X., Ma, L., Qi, X., & Li, Z. (2017). Applied Surface Science Enhanced removal of toxic Cr (VI) in tannery wastewater by photoelectrocatalysis with synthetic TiO 2 hollow spheres. *Applied Surface Science*, *405*, 102–110. https://doi.org/10.1016/j.apsusc.2017.01.306

Carbon Sequestration in Imphal West District: A Mitigation to Climate Change

Abdulla Azarudeen A[1*], ThiyamTamphasana Devi[2]
[1]*M.Tech. Student, Department of Civil Engineering, National Institute of Technology, Manipur, Langol-795004.*
[2]*Assistant Professor, Department of Civil Engineering, National Institute of Technology, Manipur.*
* *e-mail:azarcivilian@gmail.com* , thiyam85@gmail.com

Introduction

Carbon sequestration is the long-term carbon storage in plants, soils, geologic formations and the oceans. The IPCC estimated that 1.86 billion tons of carbon is released into the atmosphere annually due to LULC & climate change. The Imphal west district which possesses the large forest area is chosen for the present study since forests enhance carbon sequestration. The land use and the land cover map of Imphal west district is obtained from the maximum livelihood supervised classification. Imphal West district (Manipur state) was classified with six major divisions like; Dense forest, Sparse forest, Scrub/Grass, Cropland, Built-up & Water bodies. The remotely-sensed spectral bands can reveal valuable information such as vegetation structure, state of vegetation cover, photosynthetic capacity, leaf density and distribution, water content in leaves, mineral deficiencies and evidence of parasitic shocks or attacks (Jensen et al., 2007).The InVEST Carbon Storage and Sequestration model is implemented in the study which aggregates the amount of carbon stored in the carbon pools according to the land use maps and classifications and future prediction by using GeoSOS FLUS. This study compared the LULC and the total carbon sequestrated in year 2026 and stored in 1996 and 2016 in Manipur.

Materials and Methods

The preparatory phase of fieldwork has been done before going to the field for study the carbon stocking. During this period the basic information and data were collected related to the literature searching for carbon stocking.

- Satellite imagery: Landsat 8 OLI (Dated 17/01/2016), Landsat 5 TM (11/021996) were downloaded.
- Geological map from Geological Survey of India (GSI).
- Default carbon value from Intergovernmental panel for climate change (IPCC).

Soil map prepared by National Bureau of Soil Survey & Land Use Planning (NBSS&LUP) on 1:500000 scale was used for extraction of physical and chemical properties of soil *viz.* soil depth, soil texture, soil drainage and soil erosion. Default carbon value is prepared by Intergovernmental panel for climate change based upon that holding capacity in different pools like Above ground biomass, Below ground biomass, Soil carbon, Dead wood, Harvest wood products. In this study ESRI ArcGIS 10.2. Software is used for mapping and to predict LULC GeoSOS-FLUS software is used. For calculation and mapping of carbon sequestration InVEST model and Google Earth Pro software are used.

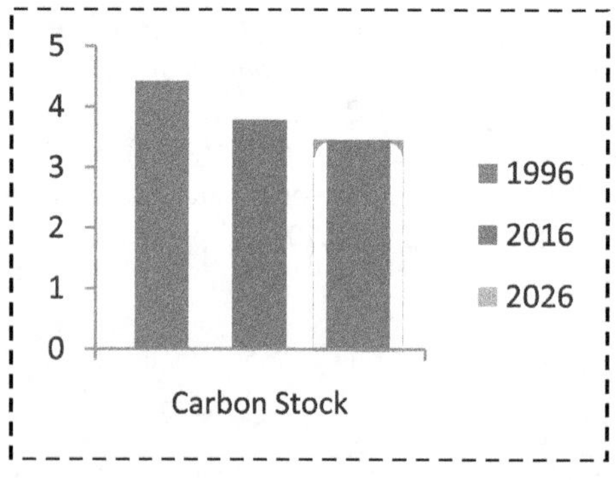

Figure 1. Amount of Carbon Stock (in Kg)

Result and Conclusion

The carbon stocking of the year 1996 for Imphal west district (24.7828° N, 93.8859° E) were observed as 4.48 kg of carbon spread through Imphal west district as Dense forest (12.18%), Sparse forest (14.37%), Crop land (35.16%), Scrub/Grass (25.68%), Built up (12.61%) and Water bodies (0%). Likewise, for the year 2016, it is 3.78 kg of carbon in which Dense forest (10.55%), sparse forest (11.3%), Crop land (39.11%), Scrub/Grass (29.29%), Built up (9.75%) and Water bodies (0%). The Carbon stock is shown in Fig.1.The predicted carbon sequestration for the year 2026 has been calculated as 0.39 kg. It is spread through district with the concentration of minimum -1.8504 mg of C to maximum 0 mg of C at water bodies.

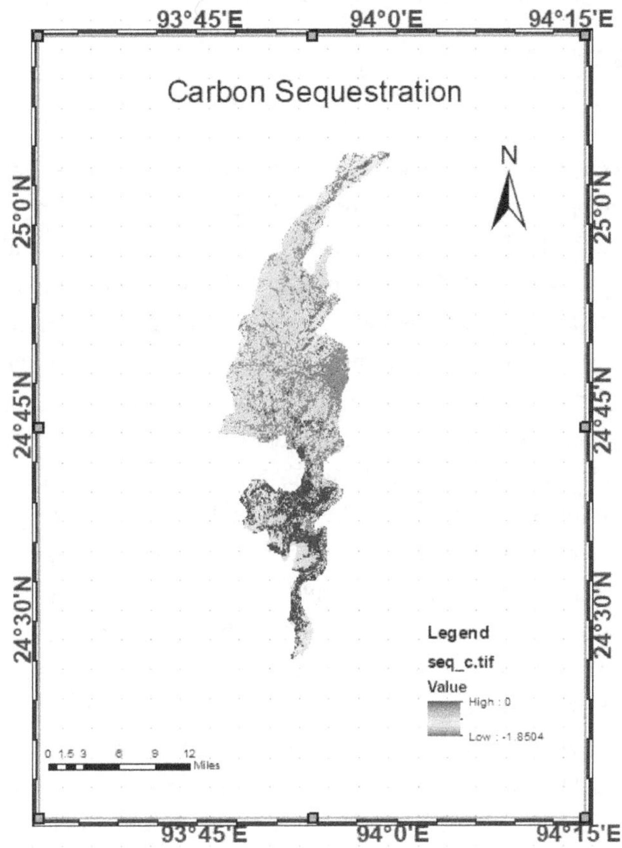

Figure 2. Carbon Sequestration Map

References

Xun Liang, Xiaoping Liu, Li Dan, Zhao Hui, Chen Guangzhao (2018). Urban growth simulation by incorporating planning policies into a CA-based future land-use simulation model. *International Journal of Geographical Information Science*, 32(11): 2294–2316

LU, D. (2001). Estimation of forest stand parameters and application in classification and change detection of forest cover types in the Brazilian Amazon basin. PhD dissertation, Indiana State University, Terre Haute, Indiana, USA

Kuldeep Pareta and Upasana Pareta (2011). Forest carbon management using satellite remote sensing techniques a case study of Sagar district (M.P.), E-International Scientific Research Journal. 3(4):ISSN 2094-1749.

Assessment of Physico-Chemical Parameters of Ground Water Qualities In Aurangabad City, Maharashtra.

Shinde Priyanka S.[1*] B. Prof. R.V. Wanjule[2] C. Principle H.H. Shinde[3]

[1]Research scholar, Department of civil engineering Jawaharlal College of engineering Aurangabad.

[2] Professor, Department of civil engineering Jawaharlal College of engineering Aurangabad.

[3] Principle, Department of civil engineering Jawaharlal College of engineering Aurangabad.

* e-mail:Shindepriyanka121@gmail.com

Introduction

Aurangabad city is rapidly developed due to Jaikwadi Dam at Paithan. It is one of the important source of water for both domestic and non-domestic purpose. Due to high increase in population of city the water demand also increases and all have to face water crises now. In Aurangabad municipal corporation include 18 nearby villages having area 138.5 sq. km and the agricultural practice is totally depends upon ground water and rainfall Water quality depends upon the local geology, human activities like industrial pollution and solid and sewage improperly discharged or disposed. The study area come under the semi-arid and arid zone where there is lack of water demand and hence the monitoring , identifying and taking action against the ground water qualities is essential in Aurangabad city near industrial areas. The total quantity of industrial effluent generated from Chikathana MIDC is 1.5 MLD and 0.5 MLD is domestic effluent. The major cause of ground water pollution is due to possibilities of leakages, seepages, spillage from old ETPs and unscientific disposal of treated effluent on land. The Chikathana and railway station MIDC do not have proper effluent collection and disposal system(Aurangabad MPCB report 2011).In the study area to the left side of sukhna river industrial area is present and to the right side of the river agricultural land is available. Hence the surrounding area of the river and its ground water is polluted and therefore water quality monitoring is very necessary around this area.25 samples are collected from dug well, bore well and hand pumps during post and pre monsoon season of year 2016-17 and 10 physicochemical parameter are analysis standard analytical methods APHA 1992 and IS 3092.The main objective of the present study is **A.** to assess the ground water quality of are in post and pre monsoon seasons with reference to 10 physicochemical parameter. **B.** To study the variation in water quality in study are in pre and post monsoon seasons by graphical analysis. **C.**To classify and asses the ground water in study area for irrigation suitability as per the critical suggested by central soil salinity research institute (CSSRI) India.

Materials and Methods

The preliminary survey is carried out in four villages/ area, namely Pisadevi (upstream part of river sukhna), Brijwadi, Naregao and Chikathana. Total 25 ground water samples are collected from open dug well, bore well and hand pumps along the side of Sukhna River and its surrounding areas Before taking samples the high density polyethylene sample bottles of 1 liter cleaned and soaked it in 10% Nitric acid overnight washed and rinsed with distilled water at the date of sampling and at the sampling site the bottles are rinsed with same Ground water samples before filling. The samples are collected between 9.00 am to 12.00 am in morning. Before taking samples from the bore well, hand pump and dug well water is to be pumped at least 5 minutes and then bottle should fill with that water. All water samples are collected and adequately labeled and preserved by adding 1 ml of HNO_3 and stored at 4° C. the temperature, pH, and TDS are measured immediately at the site. The pH is measured with pH meter and EC and TDS is measured with conductivity meter. The amount of Sodium (Na^+) and Potassium (K^+) are found with the help of Flame photometry. Calcium (Ca^+) and Magnesium (Mg^+) were determined by titration with standard EDTA. NO^{3-} was determined by spectrophotometer. Chloride was determined by standard silver nitrate titration. The samples are taken immediately into the laboratory for analysis under standard method.

Results and Concluding Remarks

In study area pH ranges from 7.5 to 9 during post monsoon season and 7.1 to 8.62 during pre-monsoon, indicating alkaline nature of samples. EC ranges from during post monsoon 900µs/cm to 5250µs/cm with an average 3202µs/cm and during pre-monsoon 1400 µs/cm to 5879 µs/cm with an average 3725µs/cm.Total of ground water samples ranges from 632 mg/l to 5920 mg/l with an average of 1400 mg/l in post monsoon season, similarly in pre monsoon season 552 mg/l to 4112 mg/l with an average 1775 mg/l.. In the study area the nitrate concentration in ground water ranges from 0.1 mg/l to 13.4 mg/l with an average 5.4 mg/l in post monsoon season and 1.6 mg/l to 12 mg/l with an average 5 mg/l in pre monsoon. 100 % ground water samples having nitrate concentration within permissible limits.Calcium is about 52 % of samples are exceeding the permissible limits in post monsoon season and 76% of ground water samples are exceeding limits pre monsoon. Chloride ranges from 158.59 mg/l to 2606 mg/l with an average 685.34 mg/l and 2015.32 mg/l to 3015 mg/l with an average 1075 mg/l in post and pre monsoon season respectively. SAR value in study area were found to range from 5.32 mg/l to 21.35 mg/l with mean 11.53 mg/l. the minimum conc. Is found in sample No. 12 in Naregao in post monsoon and in pre monsoon the minimum conc. found in sample No. 7 in Naregao area. The maximum concentration found in sample no. 13 in post monsoon in Chikalthana area and similarly in pre monsoon the maximum conc. It is found in sample no. 23 in Chikalthana area.

As river water get polluted by treated and partially treated domestic sewage as well as industrial effluents has penetrated through soil and ground water in Misarwadi, Brijwadi, Naregao, Chikalthana area. Also it is influence by quality water recharge. The present study clearly reveals that ground water samples chosen for study area are above the maximum permissible limits expect Nitrate. The suitability of ground water quality for irrigation is determined based on chemical index like Sodium abruption ratio (SAR), Soluble sodium Percentage (SSP) Kelli's Ratio (KR).Hence all sampling station need some degree of treatment before consumption and it also need to protect from getting contaminated. There should be proper awareness program to be held for farmers for the agricultural practices.

References

[1] K. M. Anwar, "ANALYSIS OF GROUNDWATER QUALITY USING STATISTICAL TECHNIQUES⍰: A CASE STUDY OF ALIGARH CITY (INDIA)," vol. 2, no. 5, pp. 100–106, 2014.

[2] I. Standard, *IS 10500-2012 DRINKING WATER-SPECIAFICATION Bureau of INdian Standards Manak Bhavan, 9 Bahadur Shah Zafar marg New Delhi*, 2012th ed., no. May. BIS, New Delhi, India, 2012.

[3] N. Gupta, P. Pandey, and J. Hussain, "Effect of physicochemical and biological parameters on the quality of river water of Narmada, Madhya Pradesh, India," *Water Sci.*, vol. 31, no. 1, pp. 11–23, 2017.

[4] K. C. Nelly and F. Mutua, "Ground Water Quality Assessment Using GIS and Remote Sensing⍰: A Case Study of Juja Location , Kenya," *Am. J. Geogr. Inf. Syst.*, vol. 5, no. 1, pp. 12–23, 2016.

[5] N. S. Rao, "Groundwater quality from a part of Prakasam District , Andhra Pradesh , India," *Appl. Water Sci.*, vol. 8, no. 1, pp. 1–18, 2018.

Review on Application of Sewage sludge as fertilizer

A. Nithin Kumar[1], Dr. M. Chandra Sekhar[2*], M. Sagarika[1]
[1]Department of Civil Engineering, National Institute of Technology, Warangal, India
[2]Professor, Department of Civil Engineering, National Institute of Technology, Warangal, India
* e-mail: anithin09@student.nitw.ac.in

Introduction

Sewage sludge is defined as the leftover secondary product post-treatment of municipal/industrial wastewater . It is a residue of semisolid that results from sedimentation of suspended solids. Generally sewage sludge is found in two types: ***Primary sludge*** that results from trapping of organic matter and suspended solids in sedimentation by gravity, and ***Secondary sludge*** produced from organics degraded by microorganisms. Currently an approximate amount of 38,354 MLD of sewage with equivalent sludge generation is estimated (Gautam et al., 2012). Nutrient potential of sewage in India is to be estimated about 3,50,000 Tonnes/year of Nitrogen, 1,50,000 Tonnes/Year of Phosphorus and 2,00,000 Tonnes/per year of Potassium with an economical value of about 2,600 Millions (Juwarkar et al., 1991). The most prominent methods for sewage sludge disposal are incineration, sanitary landfill, or use for land-based applications including structural soil improvement, soil buffer, and soil amendment (Babatunde and Zhao, 2007). However these methods have some drawbacks as well. In sanitary landfill, the groundwater is polluted by leachate; and the gas from landfill (mainly methane) causes explosion and fire. Additionally, choice of suitable site for sludge landfill is an issue too. For incineration, because of its high moisture content, operating cost is very high. Whereas emissions of toxic air pollutants namely NO_x, SO_2 and dioxins are also problems. Land application of sewage sludge in agriculture may be a more preferred option as it helps in enhancing the essential Organic carbon for soil and nutrients to plants. Researchers round the globe has shown that application of municipal sewage sludge in agriculture has bimproved production and productivity of a wide range of field crops as well as vegetables. Due to presence of high organic carbon in sludge, application to agricultural soil has improved physical and chemical properties and biological activity as well of soil. Thus sewage sludge application to soil enables the recycling of nutrients and may substitute the need for commercial fertilisers. Improper application of sludge in soil may cause disturbance.in the soil properties especially sludge with heavy metals of higher levels such as Cd, Ni, Pb and Zn which may accumulate in plant tissues and can cause food chain contamination (Saha et al., 2015).

Sewage Sludge as an Agricultural Resource

Sewage sludge is rich in organic matter, micro and macro elements that can serve as an efficient fertilizer substituting inorganic fertilizers. Sewage sludge with higher organic matter aids in enhancing the soil properties (Aggelides and Londra, 2000). Sewage sludge consists of several essential micronutrients for plants such as Boron, Chlorine, Copper, Iron, Manganese, Molybdenum, and Zinc: which are not supplied by most inorganic fertilizers (Warman and Termeer, 2005). The primary nutrients in sewage sludge are organic by nature, less soluble in comparison with inorganic fertilizers, releases very slowly. Therefore, sewage sludge can nourish the plants at a slower rate over longer duration with higher efficiency and a low possibilities of groundwater pollution, when applied rate is appropriate (Long, 2001). During the recent past, a number of researches were carried out on application of sewage sludge on cultivation of agricultural crops globally, presented in Table 1.

Table 1. Heavy metals in different crops by the application of sewage sludge

Sl no	Crop	Heavy metals	References
1	Potato	Cd, Cr, Cu, Ni, Pb, Zn	Pakhnenkoa et al. (2009)

2	French Bean	Cd, Cr, Cu, Fe, Pb, Zn	Kumar and Chopra (2014)
3	Cabbage	Cd, Cr, Cu, Ni, Pb, Zn	Ullah and Khan (2015)
4	Sunflower	Cr, Cu, Ni, Zn	Belhaj et al. (2016)
5	Rice grain	As, Cd, Cr, Cu, Fe, Mn, Ni, Pb, Zn	Meena et al. (2016)
6	Sugar cane	Cd, Cr, Cu, Ni, Pb, Zn	Leite Moretti et al. (2016)
7	Tomato	Cd, Cl, Cr, Cu, Fe, Ga, K, N, Mg, Mn, Na, Ni, P, Pb, Zn, SO4	Alghobar and Suresha (2017)
8	wHEAT	Cd, Cu, Fe, Mn, Pb, Zn	Shahbazi et al. (2017)

Concluding Remarks

Sewage sludge is a rich source of plant essential nutrients mainly N, P & K and some of the micronutrients and its application to soil enables the recycling of nutrients and may substitute the need for costly fertilisers. Additionally its application improves soil physical properties (bulk density, porosity, stability of aggregates and water holding capacity) due to its high organic content. It also enhances soil biological activity (microbial biomass carbon and activities of different enzymes) and soil chemical properties (CEC). Production of majority of the agricultural crops has benefitted from the application of sewage sludge. But the major problem associated arises due to rise in heavy metals concentration to soil when improperly applied. Low dosage of sludge application do not cause heavy metals accumulation above the safe limit. It is therefore recommended that prior to application in the soil the dose of sewage sludge must be standardised for a particular crop based on heavy metal concentrations.

References

Aggelides, S. M. and P. A. Londra, P.A. (2000). Effects of Compost Produced from Town Wastes and Sewage Sludge on the Physical Properties of a Loamy and a Clay Soil, Bioresource Technol., **71**, 253.

Alghobar, M.A. and Suresha, A. (2017). Evaluation of metal accumulation in soil and tomatoes irrigated with sewage water from Mysore city, Karnataka, India. Journal of the Saudi Society of Agricultural Sciences, 16: 49-59.

Babatunde, A. O., & Zhao, Y. Q. (2007). Constructive approaches toward water treatment works sludge management: an international review of beneficial reuses. *Critical Reviews in Environmental Science and Technology*, *37*(2), 129-164.

Belhaj, D., Elloumi, N., Jerbi, B., Zouari, M. Abdallah, F.B., Ayadi, H. and Kallel, M. (2016). Effects of sewage sludge fertilizer on heavy metal accumulation and consequent responses of sunflower (Helianthus annuus). Environmental Science and Pollution Research, 23: 20168-20177.

Gautam, R. K., Verma, S., & Islamuddin, N. M. (2018). Sewage Generation and Treatment Status for the Capital City of Uttar Pradesh, India.

Juwarkar, A. S., Juwarkar, A., Deshbhratar, P. B., & Bal, A. S. (1991). Exploitation of nutrient potential of sewage and sludge through land application. *RAPA Report (FAO)*.

Kumar, V. and Chopra, A.K. (2014). Accumulation and Translocation of Metals in Soil and Different Parts of French Bean (*Phaseolus vulgaris L.*) Amended with Sewage Sludge, Bull. Environ. Contam. Toxicol., **92(1)**, 103-108 .

Leite Moretti, S.M. Bertoncini, E.I. Vitti, A.C., Alleoni, L.R.F. and Junior, C.H.A. (2016). Concentration of Cu, Zn, Cr, Ni, Cd, and Pb in soil, sugarcane leaf and juice: residual effect of sewage sludge and organic compost application. Environmental Monitoring and Assessment, 188(3): 163, doi: 10.1007/s10661- 016-5170-1.

Long, K. (2001). The Use of Biosolid (Sewage Sludge) as a Fertilizer/Soil Conditioner on Dairy Pastures, A Review from a Dairy Food Safety Prospective, Biosolids Report.

Meena, R., Datta, S.P., Golui, D., Dwivedi, B.S., and Meena, M.C. (2016). Long-term impact of sewage irrigation on soil properties and assessing risk in relation to transfer of metals to human food chain. Environmental Science and Pollution Research, 23: 14269-14283.

Pakhnenkoa, E. P., Ermakova, A. V. and Ubugunovb, L. L. (2009). Influence of Sewage Sludge from Sludge Beds of Ulan-Ude on the Soil Properties and the Yield and Quality of Potatoes, Moscow University Soil Sci. Bull., **64(4)**, 175.

Saha, S. (2015) Remediation of Heavy Metals Toxicity to Crops in Sewage-Sludge Contaminated Soils, Published PhD thesis, Bidhan Chandra Krishi Viswavidyalaya, West Bengal.

Shahbazi, F., Ghasemi, S., Sodaiezadeh, H., Ayaseh, K. and Ahmad- mahmoodi, R.Z. (2017). The effect of sewage sludge on heavy metal concentrations in wheat plant (Triticum aestivum L.). Environmental Science and Pollution Research, 24:15634-15644.

Ullah, H. and Khan, I. (2015). Effects of sewage water irrigation of cabbage to soil geochemical properties and products safety in periurban Peshawar, Pakistan. Environmental Monitoring and Assessment, 187: 126, doi: 10.1007/s10661-015-4344-6.

Warman, P. R., and Termeer, W. C. (2005). Evaluation of Sewage Sludge, Septic Waste and Sludge Compost Applications to Corn and Forage: Ca, Mg, S, Fe, Mn, Cu, Zn and B Content of Crops and Soils, Biores. Technol., **96(9)**, 1029.

High resolution air quality modelling for sustainable urban planning

Rakesh Kadaverugu[1,*], Asheesh Sharma[1], Chandrasekhar Matli[2,*]
[1]*CSIR-National Environmental Engineering Research Institute, Nagpur -440020, India*
[2]*Department of Civil Engineering, National Institute of Technology, Warangal -506004, India*
** e-mail: r_kadaverugu@neeri.res.in / rakesh927@gmail.com*

Introduction

Air pollution is the established cause for world wide premature deaths. Nearly 6-9 million deaths are attributed to the exposure to toxic ambient air and more than 90% of children across the globe are getting effected. Urban areas are more effected than rural backgrounds, as they are densely populated and having unsustainable use of energy. Understanding of air systems and dispersion dynamics is essential for urban policy planners for effective management and control of harmful impacts. Although the prevailing air quality is measured through ground monitoring stations, these point observations fall short while answering to the questions related dispersion, and cripples the air shed planning ability (Kadaverugu et al., 2019). In order to complement the air quality observations, there is an immense need for augmenting the data with the dispersion models and statistical tools for simulating spatiotemporal contours over urban landscapes.

Materials and Methods

Open source computational fluid dynamics (CFD) tool OpenFOAM (open field operation and manipulation)was used to simulate the dispersion of a conservative pollutant over the 3-dimensional domain of an urban centre, which was developed using the data from open spatial data portal. The developed 3D topography was ingested into the CFD as computational domain and meshed with varying grid resolution to fairly capture the spatial intricacies. Varying boundary conditions were assigned to walls representing buildings, vegetation, water bodies and open spaces in the domain, in order to capture the differential treatment with air flow. Quasi-steady state velocity fields were solved using the Navier-Stokes equations in the domain and coupled with the dispersions of a tracer pollutant to mimic the real life air pollutant transport.

Results and Concluding Remarks

The impact of building geometry and vegetation on wind fields and dispersion ability have been incorporated in the modelling, in order to produce realistic near real time air quality contours of urban centres. Pollution hotspots can be identified in the study area by plugging in real life emission inventory into the study, especially ingesting traffic emissions is a crucial challenge. The framework is a proof of concept for implementing air quality modelling at high resolution and can be expanded with additional modules for tackling atmospheric meteorology and chemistry dynamics for future implementation for urban air shed planning. The method has potential applications in urban green space planning and real time traffic management for sustainable management of air quality. The model will also stand as a best policy planning tool for evaluating the management options in achieving objectives of national clear air programme (NCAP).

Acknowledgments: Authors thank the Director, CSIR-NEERI for providing all the support for carrying out this work.

References

Kadaverugu, R., Sharma, A., Matli, C., Biniwale, R., 2019. High Resolution Urban Air Quality Modeling by Coupling CFD and Mesoscale Models: a Review. Asia-Pacific Journal of Atmospheric Sciences. https://doi.org/10.1007/s13143-019-00110-3

Review on Pretreatment of Ricestraw in anaerobic digestion

M. Sagarika[1], P. Vekateswara Rao[2*], A. Nithin[1]
[1]Research scholar, Department of Civil Engineering, National Institute of Technology Warangal, Warangal, India
[2]Associate professor, Department of Civil Engineering, National Institute of Technology Warangal, Warangal India
* e-mail: smothe@student.nitw.ac.in

Introduction

Agriculture is the major sector in India. India is one of the world's greatest producer of pulses, rice, wheat, corn, sugarcane, spices, etc. In India, over all production is of 686 MT (gross production), out of which, 234.5 MT is the surplus potential, which is 34% of the gross potential (Hiloidhari M. et al. 2014). Out of all lignocellulosic (agricultural) residues, Ricestraw accounts for 154 MT, whereas, its surplus potential is 43 MT (Hiloidhari M. et al. 2014). For using the surplus agricultural residue anaerobic digestion is a feasible technology, which converts organic portion of the biomass into methane, carbon dioxide and digestate in the absence of oxygen. Methane gas can be used as a substitute for the natural gas and digestate can be used as a fertilizer for agricultural purpose as it contains nutrients like phosphorous and nitrogen.

Usually, in anaerobic digestion of cellulosic biomass such as Ricestraw, hydrolysis is the rate limiting step from the four steps hydrolysis, acidogenesis, acetogenesis and methanogenesis (Fu. S. F et al. 2015). which creates the complex structure, that gives the recalcitrance to the hydrolysis in anaerobic digestion (Zhu. J et al. 2010). Lignocellulosic substrates, such as forest waste, crop residues are majorly composed of hemicellulose, cellulose and lignin which are recalcitrant in nature. pretreatment is one of the method to enhance the performance of anaerobic digestion and to elevate the hydrolysis of the recalcitrant matter of the substrate (Shen. S et al. 2014).

Pretreatment

The primary aim of pretreatment technology on lignocellulosic biomass is to change or alleviate the structural and compositional impediments to hydrolysis (Kaur. K, & Phutela. U. G, 2016). There are three types of pretreatments: 1) physical pretreatment(milling, grinding, chipping), 2) chemical pretreatment (acids, alkalis, oxidants), 3) biological pretreatment. The pretreatment technology results in chemical and/or physical changes in the lignocellulosic biomass (Phutela. U. G et al., 2012).

Physical Pretreatment

Physical pretreatment can raise the pore size and accessible surface area of the substrates, and reduce the degree of polymerization and crystallinity of the cellulose in substrates (Harmsen. P. F. H et al., 2010). milling, grinding, chipping are some of the physical pretreatments, milling is the shredding the lignocellulosic biomass into smaller pieces and also creates shearing of the substrate (Hendrik. A. T. W. M, & Zeeman. G. 2009).

Chemical Pretreatment

Chemical pretreatment though alleviating lignin and hemicellulose upgrade the biodegradability of cellulose and to reduce the crystallinity and degree of polymerization of the cellulosic part of the substrate (Behera. S et al, 2014) . NaOH, Calcium hydroxide, ammonia, hydroge peroxide, acid are some of the chemical pretreatments.

Biological Pretreatment

Biological pretreatment has gain attention of all through out the world because of the additional benefits over physical and chemical pretreatments such as less energy needs, surface and reaction specificity, no initiation of the toxic compounds and huge production of desired products (Yuan. X et al, 2012). biological treatment is majorly related with the reaction of fungi such as white, brown, and soft rot fungi (Cianchetta. S et al, 2014).

Combined Pretreatment

Combined pretreatment is experimented for improving the efficiency of the anaerobic system. Combined treatment can be physical-chemical, thermo-chemical, physical-biological, etc.

Concluding Remarks

Pretreatment on Ricestraw is experimented to reduce the recalcitrance of the lignocellulosic substrate for enhancing methane and biogas production. Pretreatment can alleviate the lignin content and results in enhancing the accessible surface area of substrate to microorganisms. Biodegradability of the lignocellulosic substrate is limited to various factors like available surface area, crystallinity of cellulose and lignin content.

References

Hiloidhari, M, Das. D, & Baruah. D. C. (2014) Bioenergy potential from crop residue biomass in India. Renewable and sustainable energy reviews, 32, 504-512. https://doi.org/10.1016/j.rser.2014.01.025

Fu, S. F, Wang. F, Yuan X. Z, Yang, Z. M, Luo. S. J, Wang, C. S, & Guo. R. B. (2015) The thermophilic (55 C) microaerobic pretreatment of corn straw for anaerobic digestion. Bioresource technology, 175, 203-208. https://doi.org/10.1016/j.biortech.2014.10.072

Zhu. J, Wan. C, & Li. Y. (2010) Enhanced solid-state anaerobic digestion of corn stover by alkaline pretreatment. Bioresource technology, 101(19), 7523-7528. https://doi.org/10.1016/j.biortech.2010.04.060

Shen. S, Nges. I. A, Yun. J, & Liu. J. (2014) Pre-treatments for enhanced biochemical methane potential of bamboo waste. Chemical Engineering Journal, 240, 253-259. https://doi.org/10.1016/j.cej.2013.11.075

Kaur. K, & Phutela. U. G. (2016) Enhancement of paddy straw digestibility and biogas production by sodium hydroxide-microwave pretreatment. Renewable Energy, 92, 178-184. https://doi.org/10.1016/j.renene.2016.01.083

Phutela. U. G, Kaur. K, Gangwar.M, & Khullar. N. K. (2012) Effect of Pleurotus florida on paddy straw digestibility and biogas production. International Journal of Life Sciences, 6(1), 14-19. https://doi.org/10.3126/ijls.v6 i1.5550

Harmsen. P. F. H, Huijgen. W, Bermudez. L, & Bakker. R. (2010) Literature review of physical and chemical pretreatment processes for lignocellulosic biomass (No. 1184). Wageningen UR-Food & Biobased Research.

Hendriks. A. T. W. M, & Zeeman. G. (2009) Pretreatments to enhance the digestibility of lignocellulosic biomass. Bioresource technology, 100(1), 10-18. https://doi.org/10.1016/j.biortech.2008.05.027

Behera. S, Arora. R, Nandhagopal. N, & Kumar. S. (2014) Importance of chemical pretreatment for bioconversion of lignocellulosic biomass. Renewable and sustainable energy reviews, 36, 91-106. https://doi.org/10.1016/j.rser.2014.04.047

Yuan. X, Cao. Y, Li. J, Wen. B, Zhu. W, Wang. X, & Cui. Z. (2012) Effect of pretreatment by a microbial consortium on methane production of waste paper and cardboard. Bioresource technology, 118, 281-288. https://doi.org/10.1016/j.biortech.2012.05.058

Cianchetta. S, Di Maggio. B, Burzi. P. L, & Galletti. S. (2014) Evaluation of selected white-rot fungal isolates for improving the sugar yield from wheat straw. Applied biochemistry and biotechnology, 173(2), 609-623. https://doi.org/10.1007/s12010-014-0869-3

Optimization of Process Parameter by RSM for Removal of Methylene Blue Using Stem of Water Hyacinth

Rajnikant Prasad[1*], K.D Yadav[1]
[1] *Department of Civil Engineering, SVNIT, Surat, India*
** e-mail: rajnikantprasad1312@gmail.com*

Introduction

Uses of synthetic dyes have increased in industries like paper, textile, leather, and plastics and hence the effluent from such industries as are of serious concern due to the harmful effect associated on the plants and animals. Textile industries are considered to be one of the most polluting industries in terms of composition and volume of effluent (Blanco et al., 2014). Conventional methods such as coagulation, flocculation, and membrane separation process are not suitable due to its complexity, uneconomical and produce a large amount of sludge (Gadekar and Ahammed, 2019). Among the various methods available, adsorption is found to be the most effective method. The use of activated carbon for adsorption is costly and has difficulty in regeneration, making it impractical for usages in developing countries.

In this study, an investigation was done for the color (MB dye) removal efficiency of water hyacinth stem powder as an alternative to activated carbon using response surface methodology. Water hyacinth (WH) is considered as an invasive plant and it can be used for the adsorption process for color removal this will result not only in the better disposal option for water hyacinth but also for dye effluent. Response surface methodology was used to find the relation between the experimental and observed results. It involves three major steps: experimental design, estimating coefficient in mathematical model and prediction and validation of the model (Sadhukhan et al., 2016).

Materials and Methods

To prepare the adsorbent, WH was collected from Tapi river, Near Jahangirpura, Surat, Gujarat, India. It was washed many times with tap water to remove adhering particles followed by distilled water wash and was kept in sunlight for natural drying for 5 days. The dried material was kept in an oven at 80□C for 24-48 hours followed by grounding and sieving to a particle size of 216 µm. Methylene blue was used as an adsorbate. The stock of 500 mg/L was prepared by dissolving a known amount of dye in distilled water. Adsorption experiment was conducted in a batch system, consisting of a set of 250 mL Erlenmeyer flask with 100 mL of dye solution. The flask was shaken in an orbital shaker at 200 rpm for a specified amount of time. The solution thus obtained was centrifuged at 4000 rpm (Eltek-TC 4100 F) for 10 min and the supernatant was then analyzed at a wavelength of 665 nm under UV-VIS spectrophotometer (Systronics-169). Batch experiments were conducted by varying experimental variables such as pH (2-12), WH dose (0.5-3 g/L), initial dye concentration (25-75 mg/L) and contact time (10-90 min). The percentage of color removal efficiency was calculated as:

$$\text{Colour removal (\%)} = (Co-Ce)/Co \times 100 \qquad (1)$$

where Co and Ce are the initial dye concentration and final concentration after the adsorption process respectively.

Fourier transform infrared spectroscopy (FTIR) (Shimadzu- FTIR-8400S) was performed to identify the binding sites for dye adsorption onto the adsorbent (WH). Scanning electron microscope (Hitachi-S3400N) were carried out to understand the surface morphology and characterization of adsorbent before and after adsorption of MB onto WH.

Experimental design

In this study, experiments were designed using CCD, a subset of RSM for color removal by WH. Total of 30 number of experiments was calculated which includes 16 factorial points, 8 axial points and 6 replicates at the center point. The goodness-of-fit of the applied polynomial model was evaluated using ANOVA and normal plots and residuals.

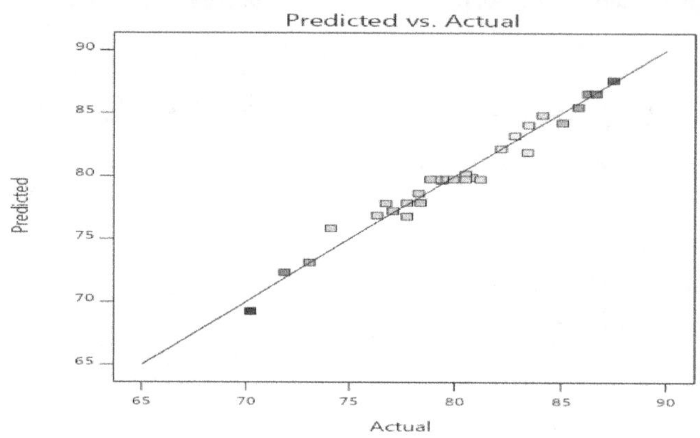

Figure 1. Actual and predicted values for the removal of MB using the stem of water hyacinth.

Results and Concluding Remarks

Fourier transform infrared spectroscopy (FTIR) was performed to detect the binding sites for dye adsorption onto the bio-sorbent. The FTIR spectra of WH before and after adsorption shows peaks at regular interval indicating the complex nature of the water hyacinth stem powder. Scanning electron microscopy shows the presence of uneven surface, pores and fractures surface. The structure clearly signifies the presence of micropores and high surface area. After the adsorption the existence of smooth surface all over the WH conforming the adsorption of dye molecule onto the WH surface.

Batch runs were conducted in CCD model designed experimentally to assess the effect of independent parameter on the response and the results obtained thereon. A polynomial regression model was performed between the response variable and the corresponding coded value (A, B, C, and D) and the best-fitted model equation were obtained as:

Colour removal = $+79.77+3.45A+1.03B+3.04C+1.66D-1.24AB-0.446AC+0.439AD+0.307BC+0.4233BD-1.13CD+3.35A^2-0.893B^3-0.893C^2-1.24D^2$

The maximum removal efficiency observed using the stem powder of water hyacinth was 87.47 %. The results were assessed using predicted R^2 value which was 0.87 and adjusted R^2 value 0.94. From Figure 1 it can be seen that there is a good correlation between the experimental and predicted values. Under optimized conditions, the removal efficiency of MB was 88.47 % obtained under the condition of pH-12, WH dose 2.157 g/L, contact time 90 min and dye concentration 60.43 mg/L from aqueous solution. The predicted results were validated by conducting experiments under given optimum conditions and the result showed the removal of 86.8 ± 2 %. From the above findings, it can be observed that water hyacinth stem powder can be used as an efficient and economical adsorbent for MB dye removal from aqueous solution and the model developed can be used for color removal study.

References

Blanco, J., Torrades, F., Morón, M., Brouta-Agnésa, M., García-Montaño, J., 2014. Photo-Fenton and sequencing batch reactor coupled to photo-Fenton processes for textile wastewater reclamation: Feasibility of reuse in dyeing processes. Chem. Eng. J. 240, 469–475. https://doi.org/10.1016/j.cej.2013.10.101

Gadekar, M.R., Ahammed, M.M., 2019. Modelling dye removal by adsorption onto water treatment residuals using combined response surface methodology-artificial neural network approach. J. Environ. Manage. 231, 241–248. https://doi.org/10.1016/j.jenvman.2018.10.017

Sadhukhan, B., Mondal, N.K., Chattoraj, S., 2016. Optimisation using central composite design (CCD) and the desirability function for sorption of methylene blue from aqueous solution onto Lemna major. Karbala Int. J. Mod. Sci. 2, 145–155. https://doi.org/10.1016/j.kijoms.2016.03.005

SPATIAL AND TEMPORAL VARIATION OF GROUNDWATER QUALITY IN WARANGAL CITY

Prasanta Majee[1], P. Hari Prasad Reddy[2*]
[1] *Research Scholar, Department of Civil Engineering, National Institute of Technology Warangal, Warangal, India.*
[2] *Associate Professor, Department of Civil Engineering, National Institute of Technology Warangal, Warangal, India.*
** e-mail: ponnapuhari@gmail.com*

Introduction:

Water is one of the most valuable resources and ecosystem is greatly dependent on its quality and quantity as it is essential for all the living species on earth. Access to ample and safe water is a basic need for human race. Quality of groundwater is greatly influenced by natural and anthropogenic activity (Kouras et al. 2007). It is mostly deteriorated due to overexploitation, intensive usage of pesticides and fertilizers, ineffective sewage management, rapid urbanisation, mining activities, improper waste management, lack of awareness (Jasmin et al., 2014; Basavarajappa et al., 2015). Assessment of groundwater quality for drinking purpose involves the determination of the chemical composition of groundwater and the remedial measures for the restoration of the quality of water in case of its deterioration demand the identification of possible sources for the contamination of groundwater (Annapoorna et al., 2015). Many researchers (Bodrud-Doza et al., 2016; Selvam et al., 2014) have carried out the studies across the globe to determine the Water Quality Index (WQI). The present study analysed the water samples from 23 different sampling points in and around the Warangal city and determined the Water Quality Index (WQI) and Heavy metal Pollution Index (HPI) to check its suitability for drinking purposes during post monsoon and pre monsoon season.

Materials and Methods

Groundwater samples from twenty-three bore wells were collected from November 2016 to April 2017. From each location, two sets of samples were collected by pre-washed plastic bottles. Water sample meant for metal concentration analysis were acidified with concentrated HNO_3 to decrease its pH value to 2 . Each sample stored at 4°C and analyzed within 3 days from the date of sample collection.
Standards methods (Guide Manual: Water and Wastewater Analysis, CPCB) were used to determine Total Alkalinity (titration with sulphuric acid), Total Hardness (titration with 0.01M EDTA), Chlorides (titration with silver nitrate), Dissolved Oxygen (titration with sodium thiosulphate), Calcium (titration with EDTA). Sulphate and Fluorides were analyzed by using UV Spectrometer. Electrical Conductivity (EC) and pH were determined using pH and EC electrodes. Heavy metals such as Chromium (Cr), Lead (Pb), Nickel (Ni), Iron (Fe), Zinc (Zn) and Copper (Cu) were analyzed using Inductive coupled plasma optical emission spectrometry (ICP-OES).

To determine the potability of groundwater in Warangal city, the Water Quality Index (WQI) was calculated by the following equation (Vasanthavigar et al., 2010):

$$WQI = \sum(W_i \times q_i) = \left[\sum\left(\frac{wi}{\sum_{i=1}^{n} wi}\right) \times \left(\frac{Ci}{Si} \times 100\right)\right]$$

Where W_i is the relative weightage of ith parameter, q_i is the quality rating of ith parameter, C_i is the concentration of ith parameter and S_i is the standard value of ith parameter.

Results

The result showed that the value of different physico-chemical parameters such as total dissolved solids (TDS), Electrical conductivity (EC), Chlorides (Cl^-), Total Hardness, Calcium(Ca^{2+}), Bicarbonates(HCO_3^-), Fluorides(F^-), Dissolved Oxygen(DO) etc were exceeding the WHO limit in most of the samples. The maximum value of fluoride was recorded as 2.59 mg/l, while the minimum was recorded as 0.43 mg/l. The TDS value was exceeded the permissible limit in almost all sampling points.

From the Water Quality Index (WQI) value, it was observed that around 65% of the sample comes under poor water class and only 26% of the samples were suitable for drinking in post monsoon season. In pre-monsoon season 70% of the water samples were found to be in poor water class, 13% were very poor water class and only 17% of the water samples were suitable for drinking purposes.

While in Heavy metal pollution index (HPI), it showed that all the sample points comes under high degree of pollution in both pre-monsoon and post season.

The spatial maps generated by ArcGIS 10.3 indicate that the groundwater quality is very poor in Warangal city and it cannot be used for drinking purposes.

Concluding Remarks

- The water quality index results concluded that most of the water samples are of poor water quality.
- Water quality deteriorates from post monsoon to pre-monsoon. This is due to the depletion of groundwater table. In post monsoon season water quality found to be good compare to pre-monsoon due to the dilution of different minerals because of recharge of aquifer with fresh water during monsoon season.
- Heavy metal pollution index indicates high degree of metal pollution in the study area. This could be because of landfilling, improper sewage disposal of industries, agro-chemical runoff.

References

Annapoorna, H., & Janardhana, M. R. (2015). Assessment of groundwater quality for drinking purpose in rural areas surrounding a defunct copper mine. *Aquatic Procedia, 4*, 685-692.

Basavarajappa, H. T., & Manjunatha, M. C. (2015). Groundwater quality analysis in Precambrian rocks of Chitradurga district, Karnataka, India using Geo-informatics technique. *Aquatic Procedia, 4*, 1354-1365.

Bodrud-Doza, M., Islam, A. T., Ahmed, F., Das, S., Saha, N., & Rahman, M. S. (2016). Characterization of groundwater quality using water evaluation indices, multivariate statistics and geostatistics in central Bangladesh. *Water Science, 30*(1), 19-40.

Jasmin, I., & Mallikarjuna, P. (2014). Physicochemical quality evaluation of groundwater and development of drinking water quality index for Araniar River Basin, Tamil Nadu, India. *Environmental monitoring and assessment, 186*(2), 935-948.

Kouras, A., Katsoyiannis, I., & Voutsa, D. (2007). Distribution of arsenic in groundwater in the area of Chalkidiki, Northern Greece. *Journal of Hazardous materials, 147*(3), 890-899.

Selvam, S., Manimaran, G., Sivasubramanian, P., Balasubramanian, N., & Seshunarayana, T. (2014). GIS-based evaluation of water quality index of groundwater resources around Tuticorin coastal city, South India. *Environmental earth sciences, 71*(6), 2847-2867.

Vasanthavigar, M., Srinivasamoorthy, K., Vijayaragavan, K., Ganthi, R. R., Chidambaram, S., Anandhan, P., Manivannan, R & Vasudevan, S. (2010). Application of water quality index for groundwater quality assessment: Thirumanimuttar sub-basin, Tamilnadu, India. *Environmental monitoring and assessment, 171*(1-4), 595-609.

Environmental Management Systems of Dairy Industry Effluents

Vasam Vinila[1], Ankita Tamta[1] and Ambika S[2]*
[1]*Environmental engineering, Department of Civil engineering, NIT Warangal, Telangana, India.*

[2]*Assistant professor, Department of Civil engineering, NIT Warangal*

**corresponding author e-mail:ambika@nitw.ac.in*

Introduction

Environmental management systems is a tool to conserve input resources, water and wastewater conservation and to implement better technologies for sustainable usage of resources in industrial applications. Environment management models offer schemes to move forward for integrated management models and environmental policies according to the company's strategies and policies (Mause, 1998) Increasing environmental responsibilities by adopting environmental policies and strategies to comply with the standards and legislations on dairy industry is the focus of this study. The main problem of dairy industry identified and reported by researchers worldwide is treatment of wastewater produced during the production of dairy products, cleaning and storage. Implementing different technologies for water conservation adopted from various industries (which are being implemented) for better results of conservation of water and wastewater has been evaluated for its application in dairy industry. In order to treat the wastewater its characteristics, possible methods of treatment, availability of resources and planning are very important (J. W. Barnett, 2010). The water mass balance states that dairy industry produces 0.2-10 litres of wastewater per litre of milk processed. Wastewater from dairy industry is mainly organic so when discharged without treatment causes serious environmental pollution and affects aquatic life. It has been reported that by adopting various advanced technologies and their up-gradation in the dairy industry reduces load of the wastewater multiple times on the wastewater treatment plants. The implementation of such techniques in dairy industry can significantly mitigate the impacts on the natural resources, ecology and environment, by reducing the pollution of surface water streams to which effluents are discharged and increasing the life span of water in dairy industry.

Materials and Methods

To study the various aspects and understand the problems "Dodla dairy", a medium scale dairy industry nearby Hyderabad is considered as study industry and analyzed the various aspects to have an effective environmental management system. This paper follows ISO 14001, 2015 for conducting EMS. The dairy plant draws about 500m^3 of water from bore wells every day for processing, cleaning and sanitary uses within the plant. The industry on an average produces 350 m^3 of wastewater per day. About 70% of the water drawn is used for cleaning purposes. Wastewater produced consists of high amounts of BOD, COD, pH, TSS and TDS. Water from the effluent treatment plant is used for gardening purposes.

Techniques to be adopted aiming EMS in dairy industry:

The following are the few strategies suggested to minimise the utilization of natural resources and to reduce the wastewater load which will certainly lead to EMS, effectively.

- **Reduction of water consumption for cleaning:** Using high pressure nozzles for cleaning boilers, milk cans and milk trucks reduces the volume of water to be used and thereby reduces the wastewater production.
- **Minimizing the wastewater load:** At the end of every batch of milk processing, the surplus milk is discharged into the drains. This will increase the wastewater volume. This surplus milk can be added to the next batch after storing them properly.
- **Alternative treatment unit for adequate treatment and recovery options:** Wastewater from cheese production consists of large amounts of cheese solids that can be removed by simple

sedimentation. Discharging the wastewater directly to effluent treatment plant will be a burden on the treatment units. By this cheese solids can be recovered and used back in the production.

- **Active biogas production:** Secondary treatment unit in the effluent treatment plant is a UASB reactor which produces methane gas on the top of it. This gas can be used for burning of wood and the produced heat is used for the boilers.
- **Recovering whey:** Whey protein is the by product from cheese production which is discharged as waste. If this whey concentrate plant is expanded and all the sweet whey is processed to recover permeate for lactose, organic load on the wastewater treatment plant can be reduced.
- **Zero Liquid Discharge focusing water reuse:** The advanced treatment technique used to conserve water. By this technique all the wastewater that is produced can be treated and reused in the industry without any waste age thereby reduces the consumption of fresh water.

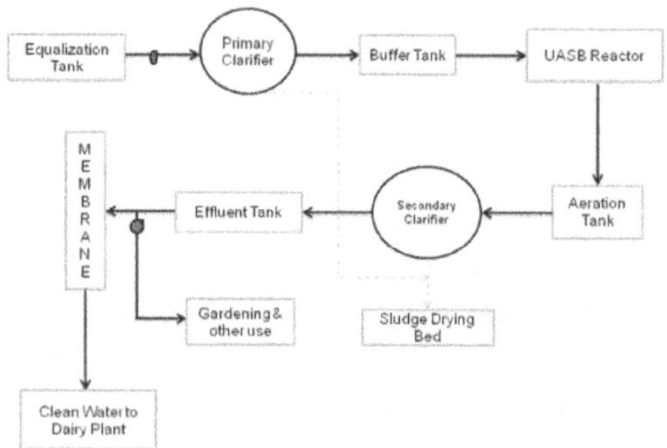

Figure1. Layout of effluent treatment plant in dairy industry

Results and Concluding Remarks

The application of selective strategies learned from different industries reduces the consumption of water in the production processes thereby amount of wastewater generated is reduced thereby reducing the load on effluent treatment plant. All the strategies have shown the best results in terms of profits to the dairy plants. By the usage of high pressure nozzles, an Estonian dairy processing plant reduced water consumption by 30,000m^3/year. A New Zealand dairy company expanded its whey treatment from 1400m^3/day to 2200m^3/day, this expansion of the plant gained a profit of US $ 3 million a year (COWI Consulting Engineers and Planners AS). Recovering whey protein alleviates the organic (lactose in whey) load on wastewater treatment plant. By implementing zero liquid discharge (ZLD) large amount of water can be reused in operating dairy processing units. Permeate from reverse osmosis is used into the production processes and concentrate is used as the cooling water within the industry without discharging it into the surface water streams. Setting up of additional treatment units may increase the capital cost but these has a huge scope on overall and long-term economic benefits along with the apparent environmental benefits. These two upshots make sure the concrete execution of EMS in dairy industry.

References

B.V. Raghunath, A. P. (2016). Impact of Dairy Effluent on Environment—A Review. In A. P. B.V. Raghunath, *Integrated Waste Management in India: Status and Future Prospects for Environmental Sustainability* (pp. 239-249). Tamil Nadu: Springer.

Central Pollution Control Board. (2015). *Guidelines on Techno Economic Feasibility of Implementation of Zero Liquid Discharge (ZLD) for Water Polluting Industries.* Delhi: Ministry of Environment, Forest & Climate Change.

COWI Consulting Engineers and Planners AS, D. (n.d.). Cleaner Production Assessment in Dairy Processing. *United Nations Environment Programme Division of Technology, Industry and Economics* .

J. W. Barnett, S. L. (2010). Environmental Issues in Dairy Processing. *New Zealand Dairy Research Institute, Private Bag 11029.* Palmerston.

Mause, A. (1998). Environmental management in the Dutch dairy industry. *Seventh International Conference of Greening of Industry Network Rome,* (pp. 1-16). Rome.

Fuzzy hyper graph analysis of Solar still with Influence of nano-coated (CuO) in basin to expenditure of polypropylene Honeycomb Material

Y. Ramadevi[a] S. Shanmugan[b]
[a]Research Scholar & [b]Research Centre of Physics, Koneru Lakshmaiah Education Foundation (KLEF - KLU), Green Fields, Guntur District, Vaddeswaram, Andhra Pradesh 522502, India. Phone number: 0863 239 9999
E-mail: mrdphy2016@gmail.com, s.shanmugam1982@gmail.com.
* E-mail: corresponding Author: s.shanmugam1982@gmail.com

Abstract
The solar still has been conducted in nano-coated (CuO) in basin analysis of the expenditure of polypropylene honeycomb material. The solar still is structured as a fuzzy graph with fuzzy vertex set and fuzzy edge set, another two types of fuzzy graph: (i) 6 vertices and 15 edges (ii) 6 vertices and 6 edges. It is the degree of membership values of the vertices for following within high, medium and low level of temperature in the system. Fuzzy incidence matrix and fuzzy cutset matrix are obtained from the two types of fuzzy graphs mentioned as in row sum incidence matrix. This fuzzy set will support to analyse which part of the still are coated with nanoparticle (CuO) to maximize the collection of pure water from the still with more reflections from each vertex carefully. The solar still is improved with nanoparticle coated in the four sides of the basin to produce a high amount of purified water within the support to the analysis of complete fuzzy graph and fuzzy hyper graph. These achieved for production rate during (24h) the investigation range are 5.52Liter/day and is the enactment of 43.51%, respectively.
Keywords: Fuzzy hyper graph; fuzzy incidence matrix; fuzzy cut set matrix; CuO-Nanoparticles, Honeycomb structure, basin solar still.

Introduction

Now a day, water requirement is shortage in many places in our world. The solar still is used as an apparatus, to get purified water from the dirty water or used water. Solar still is manufactured with different shapes and different components are used as a part of the still. Using Nanomaterials are in various parts of the solar still to improve the output of the purified water by many researchers. In [1] modified still is analysed with CaO nano particles are added at water depth of 5 cm and 10 cm. In [2], nano particle is impregnated in phase change material. Nano particles (Al_2O_3) are mixed with mad black paint in the coil inside the still in [3]. In [4] fuzzy logic method can be used to pretend the suggested arrangement of any climatic circumstances for big scale industries, domestic purpose and agriculture. This will helps to decide where the nano particles are required exactly to improve the solar Nano still to maximize the output of the purified water.

Materials and Methods

Structure of the solar still:

Our solar still is prepared with the basin surrounded by 27×27 inches and 1 inch thickness. Copper plate fixed in the bottom with 6mm thickness. Thermopolis with 0.75 inches is used as the insulator and plywood with 0.25 cm is used to cover the outside of the wall. Copper tube with 10mm diameter and wall thickness 0.5 mm is used to increase the temperature of the still which contain few amount of PCM is added. In east direction the back wall is placed with 22cm height and in the west direction the front wall is placed with 40cm height which is the structure of the solar still considers for our research work is shown in Fig.1.
From the deep observations we decided to consider, low temperature of the still is measured from 8.00 am to 10 a.m. and 4.00 p.m. to 5 p.m. Medium temperature of the still is measured from 10.00 am to 11.00 am and 2.00 pm to 4.00 p.m. High temperature of the still is measured from 11.00 am to 2.00 pm as by [5].

Complete fuzzy graph of the solar still:

Now the solar still is structured as a complete fuzzy graph with 5 vertices and 15 edges. The vertices of the complete fuzzy graph is considered as $S_1, S_2, S_3, S_4, S_5, S_6$ which are the parts of the solar still such as Glass cover, four sides of the basin inside the still, bottom of the basin, contaminant water inside the still, copper tube, phase change material inside the copper tube respectively.. The edges of the complete fuzzy graph are $g_1, g_2, g_3, g_4, g_5, g_6, g_7, g_8, g_9, g_{10}, g_{11}, g_{12}, g_{13}, g_{14}, g_{15}$ respectively which is shown in Fig.2.

Fig. 1 Experimental analysis of hybrid solar still bond adsorption for clean water process

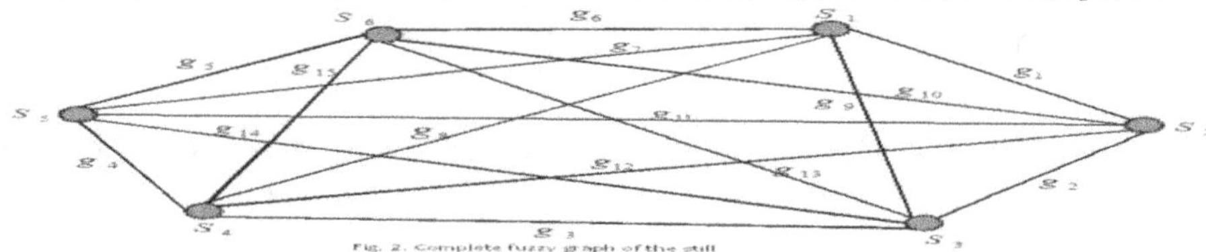

Fig. 2. Complete fuzzy graph of the still

Results and Concluding Remarks

The new fuzzy set defined above will explicitly show the exact transmission amount of solar rays in kW/m^2 by its maximum boiling temperature of the water in every parts of the solar still which is located in Table.7 corresponding to the fuzzy hyper graph.

Table.1. Fuzzy set defined on every member of the solar still by the buttress of complete fuzzy graph

Total transmitted rays totalling by its fuzzy set	S_1	S_2	S_3	S_4	S_5	S_6
High temperature of the still	1	1	1	1	1	1
Medium temperature of the still	1	0	1	1	1	1
Low temperature of the still	1	0	1	1	1	1

$$\alpha(RS_i) = \begin{cases} 1 & \text{if row sum is min} \\ 0 & \text{if row sum is max} \\ \dfrac{RS_i}{No.of\ edges\ incident\ on\ S_i} & \text{if row sum is neither max nor min} \end{cases}$$

This fuzzy set will conclude equal amount (= 0) of rays are transmitted in each removal of the parts of the still. If any vertex is removed, then the incidence edges also removed. If removals of any vertex will states there is no transmission in the solar still.

Conclusion: In our research paper, the output of the purified water collection from the solar still is analysed deeply using a complete fuzzy graph and fuzzy hyper graph are the tools. In this analysis, both graphs will suggest four sides of the single-basin solar still is getting lower transmitted rays. If the four sides of the basin are coated with nanoparticle (Al_2O_3), then it will increase the high amount of solar transmitted rays, which consequently increase the amount of pure water collection in the single basin nano-solar still.

Reference.

1. Bhupendra Guptaa, Prem Shankara , Raghvendra Sharmaa , Prashant Baredarb, (**2016**) 'Performance Enhancement using Nano Particles in Modified Passive Solar Still', Procedia Technology 25 1209 – 1216, 2212-0173 © 2016. Published by Elsevier Ltd.
2. Dsilva Winfred Rufuss, S. Iniyan, L. Suganthi, Davies PA, (**2017**) 'Nanoparticles Enhanced Phase Change Material (NPCM) as Heat Storage in Solar Still Application for Productivity Enhancement', Energy Procedia 141 45–49, 1876-6102 © 2017 , Published by Elsevier Ltd.
3. Selvaraju P., Shanmugan S., (**2018**), "Formation and speciation of sullage water natural conduct analysis of fuzzy formal system application by solar distillation", IJET 7(2.24), PP 444 – 447.
4. Shanmugan S., Krishnamoorthi G., (**2013**) 'Fuzzy logic modeling of single slope single basin solar still', *International Journal of Fuzzy Mathematics and Systems* 3 (2), 125-134.
5. Shanmugan S., Palani S., Janarthanan B., (2018) 'Productivity enhancement of solar still by PCM and Nano particles miscellaneous basin absorbing materials', *Desalination*, Volume 433, Pages 186-198.

Sullage water used in solar still with Permanence of waste contamination water makes bio Ash by Evaluation of sustainable environments

C. Suresh[a] S. Shanmugan[b] V. Chithambaram[c] A. Gangadurai[d]
[a]Research scholar, [c]Research Centre of Physics, Dhanalakshmi College of Engineering, Manimangalam Post, Chennai - 601 301, Chennai, Tamil Nadu 601301.
[a&d]Department of Civil Engineering, Vel Tech Multitech Dr.Rangarajan Dr.Sakunthala Engineering College, Avadi, Chennai, Tamil Nadu, India – 600062.
[b]Research Center of Physics, Koneru Lakshmaiah Education Foundation (KLEF - KLU), Green Fields, Guntur District, Vaddeswaram, Andhra Pradesh 522502, India.
Email ID: sureshc.struct@gmail.com, s.shanmugam1982@gmail.com, chithambaramv@gmail.com, gangadurai15@gmail.com.
** E-mail: corresponding Author: s.shanmugam1982@gmail.com*

Introduction:

Abstract
In this work have been made to the analysis of waste contamination water based on make Bio- ash, which is very favourable stuff in manufacturing an improvement population proclivity India propagate Bio-ash which creates an imperious to the near. A single slope single basin solar still has been offered by the analysis of flowing water, glass cover, water and basin individually and numerical results are confirmed with the experimental for one of the classic days. It is a quotidian output of still around Energy =45.92m^{-2} beneath by the organization. One month of collected waste contamination water by solar still after makes the stone. It is an analysis of stone in mixed nanoparticles and to produce in Bio-ash. Finally, the expenditure of Bio-ash may be a substitute for conventional concrete or cement to accomplish further sustainable, environment-friendly concretes and the ingesting of partial usual properties.

Keywords: Thermal efficiency, Mass flow rate; Basin type solar still, Bio-Ash.

Water is lost daily through various bodily processes. Water is needed by every living cell and almost every process that takes place within the body is dependent on water. Numerical modeling of water flowing over the glass cover in a solar still is very trivial to strategy a frugally prime still and to enhance the manufacture enactment for a given cost. Prakash and Kavatherkar [1] have shown the performance of the regenerative still and its daily yield is about 7.5 l/m^2 compared to conventional solar stills. Dhiman and Tiwari [2-4] have studied a multi-wick solar still and predicted that the output is 10% higher due to the water flow over the glass cover in a thin layer form, as it increased the difference between the temperatures of the water and the glass cover.

Materials and Methods

Structure of the solar still:

The shoot of the new single slope single basin type solar still with water flowing over the glass cover and its sectional opinion are offered in Figs.1. It comprises of outer and inner inclusion completed of plywood with measurement of 1.3 x 1.3 m and 1.25 x 1.25m as follow [5 & 6]. It has a glass wool in thermal conductivity of 0.0038 W/mK. The tallness of the back wall is 0.03m and obverse wall of 0.10m. The glass cover of thickness 4 mm is secondhand as the condensing surface and the slope of the glass cover are fixed as 11º which is equal to the latitude of the location (Chennai) and is completed vapor tight with the help of metal putty. The j-shaped drainage channel is stationary near the obverse wall to amass the distillate yield and the output dropped down to the measuring jar. It is made of basin area in Galvanized iron sheet and a thin copper sheet is pasted in the basin and painted black to absorb more solar radiation. The basin temperature, saline water temperature and condensing cover temperature has been measured by fixing copper-constantan thermocouples which has been regulated initially. Solar radiation intensity and ambient temperature have been measured with solar radiation monitor and digital thermometer.

Figure 1 Photograph of the experimental analysis of solar still.

Results and Concluding Remarks

The instantaneous efficiency of the anticipated system has been found for summer days waste contamination form of bio ash shown in the Figure 2. Instantaneous energy efficiency with and without drip varies from 39 to 59% and 35 to 50%, It has clearly reflected that the instantaneous distillate yield of waste water by solar still is found to be 25% higher form of Bio-ash make the waste water perform of the solar still.

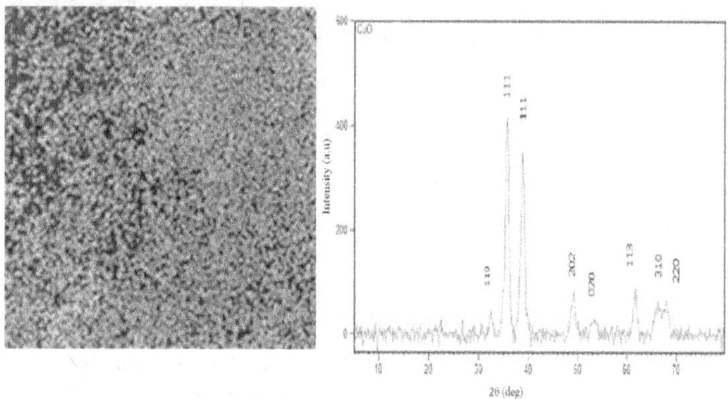

Figure 2 Sem analysis of bio-ash produce by waste contamination water.

Conclusion

The thermal modeling and analytical expressions for the instantaneous energy efficiency is done and substantiated. There is a good settlement between theoretical and an experimental observation of the glass cover has an intriguing role in the performance of the system. The daily distillate manufacture rate of with drip a single basin still is 45% higher than that of and without drip single basin still. The waste contamination is a form of stone after to makes in bio ash performance of good stabiles occupied in cement to control of sustainable environments

Reference

1. Prakash, J., Kavatherkar, A.K., (1986) Performance prediction of a regenerative solar still. Solar & wind Technology 3(2), 119-125.
2. Dhiman, N.K., Tiwari, G.N., (1990) Effect of water flowing over the glass cover of a multi-wick solar still. Energy Conversion Management 30(3), 245-250.
3. Singh, A. K., Tiwari, G. N., (1992) Experimental validation of passive regenerative solar still. International Journal of Energy Research 16(6), 497–506.
4. Selvaraju P., Shanmugan S., (**2018**), Formation and speciation of sullage water natural conduct analysis of fuzzy formal system application by solar distillation, IJET 7(2.24), PP 444 – 447.
5. Shanmugan S., Krishnamoorthi G., (**2013**) Fuzzy logic modeling of single slope single basin solar still, *International Journal of Fuzzy Mathematics and Systems* 3 (2), 125-134.
6. Shanmugan S., Palani S., Janarthanan B., (2018) 'Productivity enhancement of solar still by PCM and Nano particles miscellaneous basin absorbing materials', *Desalination*, Volume 433, Pages 186-198.

FEASIBILITY OF APPLICATION OF NATURAL COAGULANTS TO TREAT BLACKWATER

Kiran Patil G S[1], Udaya Simha L[2], Rohini Pradeep[3]

[1] *Post-graduatet student, Department of Civil Engineering, BMS College of Engineering, Bangalore*

[2] Department of Civil Engineering, BMS College of Engineering, Bangalore, India

[3] Subject coordinator -R&D, CDD Society (Consortium for DEWATS Dissemination Society), Bangalore, India

[*]lusimha.civ@bmsce.ac.in

Introduction

In the areas which are not provided with underground drainage (UGD) facility, the generated blackwater is really a menace. It is a significant public health hazard.

In Bengaluru, the areas around Devanahalli are not provided with the facility of UGD. The Consortium of DEWATs Dissemination Society (CDD Society) has established treatment unit to treat blackwaters delivered to the unit through trucks from pits and septic tanks present in the town. The problem faced is the high turbid effluent from the treatment system.

The coagulation process would be one of the options to improve the quality. The aluminium part in the most widely used coagulant, aluminium sulphate (alum), is reported to be neurotoxin and may lead to the development of neurological disorder (Wang and Cui, 2004, Virginie Rondeau et. al, 2009). Thus, the researchers are investigating the alternate material to the chemical coagulant. The materials of some plant origin are reportedly known to treat turbid water, which are classified as natural coagulants (NC).

The seeds extract of *Moringa Oliefera* (M.O) and other natural plant seeds, like bean, neem, nirmali, soyabean, fenugreek, etc., has been widely investigated as a natural coagulants by many investigators (A.B. Olayami and R.o. Alabi, 1994, Ndabigengesere et. al.,1995, Ndabigengesere and Narasiah, 1998, Subramanium et. al., 2011, Udaya Simha et. al, 2013). The extract of M.O. has shown to be effective in reducing turbidity of water to a greater extent. The overall result of the investigation revealed the potential of NCs to reduce turbidity form water/wastewater. These bio-materials contain some of the hydrophilic amino acids, which would help in the process of coagulation (Sebastien, et al., 2012). The utilisation of natural coagulants is not only limited to reduce the turbidity but also reported to reduce other pollutant indices, namely, BOD and COD (Udaya Simha et. al, 2013).

In the present study, the bio materials, such as extract of M.O and *Arachis Hypogea* (neem) cake have been investigated as natural coagulants to treat blackwater.

Materials and Methods

The seeds of *Moringa Oliefera* and oil extracted *Arachis Hypogea* (neem) cake were obtained from Hiriyur, Chitradurga, Karnataka, India. The bio materials were dried for a day in an oven for 24hrs at 40^0 C. The dried materials were blended and sieved through a sieve size 212μ. The extraction of active agents from the natural coagulants can be obtained from either by distilled water or spring water or tap water or salt solutions (Marobheb et. al, 2007). The extraction of active agents from the materials was obtained with distilled water in the present study. A fixed quantity of prepared bio-material was blended with distilled water and made up to required concentration.

The source of blackwater was collected from the faecal sludge treatment plant located at Devanahalli, Bengaluru. The point of collection was subsequent to settling. The blackwater collect was made to flow through Anaerobic Baffled up-flow Reactor (ABR) of laboratory scale of the field reactor. The effluent collected was analysed for pH, turbidity, BOD, COD, total nitrogen and faecal coliform after running ABR to 52 days. The effluent of ABR was treated with the extracts of bio-coagulants and batch studies were carried out to find out the feasibility of the bio-coagulants.

Results and Concluding Remarks

The first and second rows in table 1 indicate the values of the parameters corresponding to source blackwater and 52nd day effluent from ABR. The quality of effluent from ABR was a matter of concern. The further treatment was hence, planned with NCs. The optimum experimental conditions were investigated by considering the variables like concentration of extract of NCs, dosage of NCs, contact time and mixing speed. Under the optimum conditions, the turbidity was reduced to 71% (MO) and 50% (neem). To increase the efficiency of treatment, alum was considered as coagulant aid with MO and neem extracts in the ratio of 80% (NC) and 20% (alum). The third and fourth rows of table 1 depict the outcome of the investigation. The values within the bracket show the efficiency of percentage removal. The combination of neem+alum was observed to be not fruitful. On the other hand the extract of neem alone was found to be better in reducing the turbidity. However, the combination of extracts of MO +alum was observed to be yielding very good outcome. It can be observed that along with the turbidity, the other parameters also reduced, especially, the faecal coliform (99%). Among the two NCs, the MO in combination with alum was found to be feasible in treating blackwater.

Table 1 Results of Experimental Investigations

	Turbidity, NTU	**pH**	**Temperature, ^{0}C**	**COD, mg/l**	**BOD, mg/l**	**Total N, mg/l**	**Faecal coliform, /100ml**
ABR Infuent	321	7.54	24.8	818	195	339.9	$24x10^6$
ABR Effluent	141(56%)	7.38	24.3	434(47%)	113(42%)	304.3(11%)	$24x10^6$
Neem+alum	164	7.98	24.8	298(31%)	128	224(26%)	$20x10^6$(16%)
MO+alum	17.84(87%)	7.72	25.2	161(63%)	64(43%)	274.1(10%)	1610(99%)

From the study it can thus be concluded that natural coagulants can be adopted for the blackwater treatment. Among the NCs investigated, Moringa *Oliefera* was found to be feasible in combination with alum as a coagulant aid.

Acknowledgement

The authors acknowledge TEQIP III for the financial support extended to participate actively in the international conference.

References

A.B. Olayami and R.o. Alabi, (1994), "Studies on Traditional Water Purification using Moringa *oleifera* seed", African Studt Monographs,15(3), pp. 135-142.

Marobheb N.J., Dalhammar G., and Gunaratna K.R,. 2007, Simple and rapid method for purification and characterization of active coagulants from the seeds of Vigna unguiculata and Parkinsonia aculeata. Environmental technology, 28, pp 671-681

Ndabigengesere A., Narasiah K.S., Talbot B.G., (1995), Active agents and mechanism of coagulation of turbid waters using *Moringa oleifera,* Water Research, 29, 703- 710.

Ndabigengesere, A., and Narasiah, K.S., (1998), Quality of water treated by coagulation using *Moringa oleifera* seeds, Wat. Res., Vol. 32, No. 3, pp. 781- 791,.

Sebastien Tindo Djenontin, Félicien Avlessi, Dominique K C Sohounhloue, Daniel Pioch, (2012), Composition of Azadirachta indica and Carapa procera (Meliaceae) seed oils and cakes obtained after oil extraction. Industrial Crops and Products 38, pp 39– 45

Subramanium Sotheeswaran, Vikashni Nand, Maata Matakite and Koshy Kanayathu, 2011, Moringa oliefera and other local seeds in water purification in developing countries, Research Journal of Chemistry and Environment Vol.15 (2), pp. 135-138

Udaya Simha L, Roopa and Puttaswamy, (2013), Investigation on natural coagulants on water quality parameters, International journal for Earth Science and Engineering, volume 06,No06 (01), pg no: 1637-1645

Virginie Rondeau, He´ le`ne Jacqmin-Gadda, Daniel Commenges, Catherine Helmer, and Jean-Francxois Dartigues (2009), Aluminum and Silica in Drinking Water and the Risk of Alzheimer's Disease or Cognitive Decline: Findings From 15-Year Follow-up of the PAQUID Cohort, American Journal of Epidemiology, Vol. 169, No. 4, pp. 489–496.

Wang. Z, Cui. F, (2004), Decreasing residual aluminiumlevel in drinking water, Trans. Nonferrous Met. Soc. Chaina, Vol 14, No. 5, 1033-40.

Remote Sensing & GIS

Mulberry cultivation in Manipur: An alternative for sustainable livelihood

Deepali Gaikwad[1*], Thiyam Tamphasana Devi[2]
[1]M.Tech Student, [2]Assistant Professor
[1,2]Department of Civil Engineering, National institute of Technology, Manipur, Langol-795004,India
*[*1] e-mail: deepali.dg65@gmail.com (corresponding author), [2]thiyam85@gmail.com*

Introduction

In India, sericulture is one of the prominent agro and forest based cottage industry sectors of economy and plays an important role in programmes ofpoverty alleviation. Literature suggests that compared to agricultural crops, sericulture provide more vibrant employment option all-roundthe year and fetches higher income for rural farm families.Commercialization and wide production of silk will support to the theme of sustainable development as an alternate resource to the rural people. Natural silkworm food plant like castor, mulberry, etc. are the perennial crops which protects the soil from erosion, so it is also an eco-friendly farm activity. India continues to be the second largest producer of silkin the world and has the distinction of producing all the four varieties of silk, viz., Mulberry, Eri, Tasar, and Muga.The major mulberry silk producing states are Karnataka, Andhra Pradesh, West Bengal, Tamil Nadu and Jammu and Kashmir which together account for 92 percent of country's total mulberry raw silk production.In annual report of Ministry of Textiles, it is reported that in 2010-2011, total silk production in the country is 20,410 MT, out of which Mulberry accounted for 80.2%, but total estimated annual consumption is 29,300 MT. And by the year 2025, it is also estimated that domestic demand will increase upto 45,000MT (Patil et al., 2009). Eventually, there will be a huge gap between supply and demand in the coming years which encourages to conduct such study that could help the wide production of silk in future. Due to agro-climate condition of the state (Manipur which is the study area), it is suitable for wide cultivation of all four type of silk - Mulberry, Muga, Tasar Oak and Eri. The major income of the state is agriculture, hence favors the sericulture plantation as an alternative opportunity for additional livelihood for the rural community in the state.

Therefore, the main objective of this study is to identify the potential sites for Mullberry cultivation for silk production in the North-East State of Manipur usinggeographic information system tools and remote sensing. Based on parameters viz., cultivable land, physio-graphy(elevation, aspect, slope), soil characteristics suitability(texture, drainage, depth, erosion) and climate conditions(average temperature, average rainfall, average relative humidity), site suitability were identified.For the quantitative analysis RUSLE method is also used to observe soil erosion. All input layers overlaid in GIS environment and site suitability was identified based on multi-criteria analysis. Four different classes were given to the potential sites viz. highly suitable, moderately suitable, marginally suitable and not suitable.

Materials and Methods

Assessment of suitability of land for sericulture has been done by involving different multi-thematic layer viz, climate conditions(temperature, rainfall and humidity), topographic conditions (slope and elevation), soil properties parameters (drainage, depth, erosion and texture) and cultivable land.

For the climate condition AWS data were used for rainfall map and for Land Surface Temperature map Landsat 8 satellite imagery was used. Soil properties parameters weregenerated by NBSS-LUP soil map of Manipur on 1:500000 scale. Landsat 8 satellite imageryof year 2018was used to compute cultivable land in study area.Topographic mapswere generated from SRTM DEM (30m resolution).

Land use/land cover was classified in six categories: forest, agriculture, wastelands, shifting cultivation, built-up areas and water bodies.For quantitative assessment of soil erosion in Manipur for the year 2018, soil erodibility factor (*K* factor), rainfall erosivity factor (*R* factor), cover management factor(*C* factor), conversation practicefactor (*P* factor) andslope length gradient (*LS* factor) were also calculated by using RUSLE model which can be written as:

$$A = R \ x \ K \ x \ L \ x \ S \ x \ C \ x \ P \ (1)$$

Here, *A*represents the annual average soil loss in tonnes/hectare/year

By using all the information mentioned above, site suitability for mulberry cultivation has mapped by weight overlay method in ArcGIS. Site suitability map has categorized into highly suitable, moderately suitable, marginally suitable and not suitable.

Results and Concluding Remarks

Results are categorized as:

(i) **Annual average soil loss:** Itis estimated by RUSLE model and found that 85.16% area of Manipurcomes under very slight soil erosion range which is 0-5 t/ha/year. Very sever soil erosion classified in 40-80 t/ha/year, range, which occurred in 2.67 km^2 (0.012%)area of Manipur. Where sever soil loss classified in 20-40 t/ha/year, is 46.73km^2 (0.21%) area of Manipur. Moderate soil erosion classified in 10-20 t/ha/year, is 827.81km^2 area of Manipur. Slight soil erosion classified in 5-10 t/ha/year range, is 50173.82 km^2 (23.25%)area of Manipur. By this observation it has concluded that Manipur is not soil erosion hazardous.

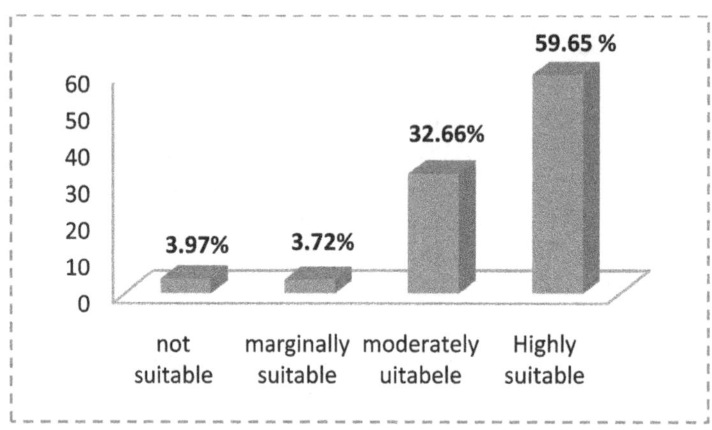

Figure 1: Site suitability chart for mulberry cultivation in Manipur

(ii) Site suitability for Mulberry Cultivation: In Figure 1, it has been observed that 96.03% (21581.42km^2) are the total suitable area in Manipur for mulberry cultivation in which 59.65% (13405.60km^2) were highly suitable, 32.66% (7339.86km^2) were moderately suitable, 3.71% (835.96 km^2) were marginally suitable. However, 3.96% (891.576km^2) were observed to be not suitable for mulberry cultivation. Therefore, whole Manipur fertile land is highly suitable for Mulberry cultivation.

References

Deka A,SharmahBC, Sharmah D (2017)GIS Based study on Development of Sericulture Resources in Assam: ACase study of Goalpara&Sibsagar District. Assam Remote Sensing Application Centre. 4(5):208-215.

Kalita A, Goswami B (2018) Identification of Potential Sites for Mulberry Cultivation in West Garo Hills of Meghalaya using Geospatial Techniques.North Eastern Space Applications Centre (NESAC), Umiam.

Patil BR, Singh KK, Pawar SE, Maarse L, Otte J (2009) A Living from Livestock Research Report. RR No: 09-03.

Sharma A,JeyaseelanBC, Kumar D (2012) Geospatial Technology for Sericulture Development in Jharkhand. Jharkhand Space Applications Center. 3(2):34-45.

Uniyal A, Kandwal BC, Kimothi D (2014) Geospatial technology for identification of suitable sites forMulberry plantation: a case study of uttarakhand state. Uttarkhand space research centre (USAC).

Ministry of Textiles- annual report 2011-2012 (2012) Government of India. Udhyog Bhavan, NewDelhi : 23-38.

Soil Salinity modelling and mapping of Honnali region using Remote sensing and GIS

Varsha Chandran V.[1*], K.P. Indulekha[2]
[1]*Student, Department of Civil Engineering, College of Engineering Trivandrum*
[2] *Assistant Professor,Department of Civil Engineering, College of Engineering Trivandrum*
* *e-mail: varshachandran123@gmail.com*

Introduction

Soil salinization, which is one of the most common threats to the land in arid and semi-arid areas, occurs due to the accumulation of soluble salts in irrigated soil. The continuous use of irrigation water containing high quantities of dissolved salts often degrades the land through salinization. As high levels of soil salinity negatively affect crop growth and productivity, it is vital that proper monitoring and mapping of soil salinity is done at an early stage, to implement soil reclamation program that helps to reduce or prevent future increase in soil salinity. An integrated approach using remote sensing in addition to various statistical methods has shown success for modelling and mapping soil salinity. Numerous studies were done in past years for monitoring and mapping of soil salinity.Engdawork et al. (2018) mapped and modelled the soil salinity of Wonji sugar cane irrigation farm, Ethiopia using remote sensing and geographic information systems.Similarly, Shegena et al. (2017) assessed the level of salinity in Sego irrigated farm, Ethiopia and mapped the temporal and spatial distribution of salt affected soils to support management programmes.This study deals with soil salinity modelling and mapping of Honnalitaluk in Karnataka state, using remote sensing technology.Four villages in Honnalitaluk,which are seriously affected by soil salinization problems, are considered in this study.

Materials and Methods

A total of 40 soil samples were collected from the fieldalong with GPS coordinates. As Electrical conductivity (EC) value is a measure of soil salinity, EC values of the samples were measured using Conductivitymeter. Sentinel 2A satellite imagery having 10 m spatial resolution is used for salinity mapping of the study area. Several remote sensing indices such as NDSI (Normalized Difference Salinity Index), VSSI (Vegetation Soil Salinity Index), BI (Brightness Index), SI-1 (Salinity Index-1) and NDVI (Normalized Difference Vegetation Index) were generated from the satellite imagery. Regression analysis was done to assessthe spatial distribution of EC_e and to predict salinity level at different locations.The salinity index value corresponding to the EC_epoint values were plotted on a scatter diagram and the best fit line, regression equation and correlation were determined. The salinity index showing highest correlation with Field EC_e value was selected and used for model generation. Soil salinity of the entire study areawas mapped using the regression analysis model.The soil salinity map generated using soil salinity model was validated with the help of field data. Correlation between EC_evalue and raster value of the model has been derived by plotting these on a scattered diagram, for validation.

Results and Concluding Remarks

Statistical correlation between field measurements of electrical conductivity (EC_e) and remote sensing spectral indices showed that Salinity index -1 (SI-1) had the highest correlation (R^2=0.882)with EC_e(Figure 1).Hence Soil salinity index -1 model was used to create the soil salinity map of the study area.The map of Soil salinity derived from regression modelrevealed five classes of salinity levels (Figure2), viz. non-saline, slightly saline, moderately saline, strongly saline and very saline soils. Out of the total area, 39.60% were identified underNon-saline class, which is the largestin extent (1162.57 ha).Strongly saline andVery saline soil covered 3.83% and 0.015 % respectively, which was found mainlyin the western edges of the study area. Slightly saline andmoderately saline soils covered 27.34% and 29.21%, respectively,and found scattered throughout the study area.The model validation was done by plotting the field values against SI-1 values from SI-1 model. Scatter plot between field EC_evalues and raster value from salinity index model yielded a coefficient of determination (R^2) value of 0.805 as shown in Figure 3.The sequence ofthe model from detection, site observations, correlation and model validation has proved to be applicablefor mapping and modelling salinity using geospatial techniques.

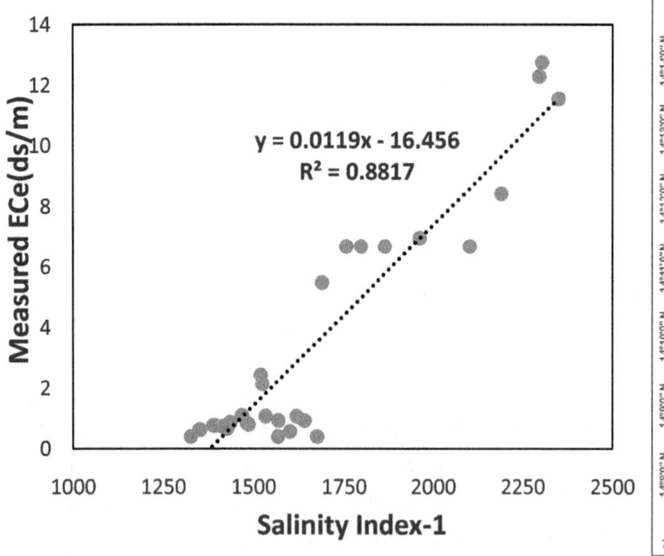

Figure 1 Regression analysis between EC_e and SI-1

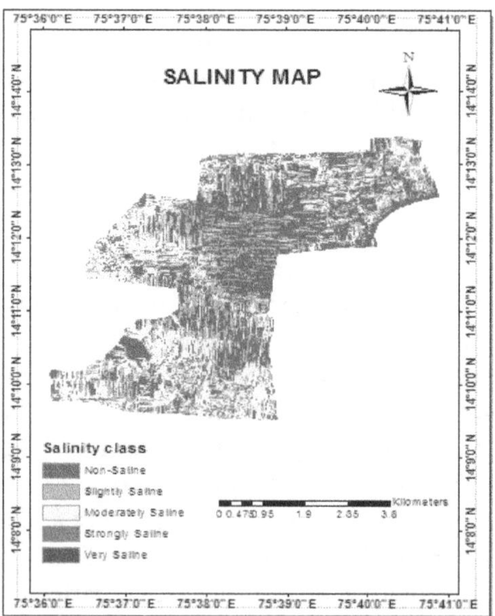

Figure2 Soil Salinity map

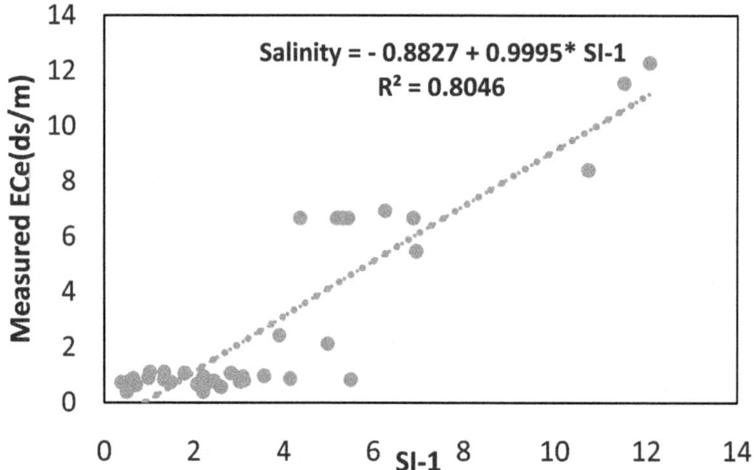

Figure 3 Scatter plot between field ECe values and SI-1 value from salinity model

References

- Asfaw E., Suryabhagavan K.V. andArgaw M. (2018) Soil salinity modeling and mapping using remote sensing and GIS: The case of Wonji sugar cane irrigation farm, Ethiopia, Journal of the Saudi Society of Agricultural Sciences(17) : 250–258
- Nawar S. , Buddenbaum H. , Hill J. and Kozak J.(2014) , Modeling and Mapping of Soil Salinity with Reflectance Spectroscopy and Landsat Data Using Two Quantitative Methods (PLSR and MARS),Remote Sens (6) : 10813-10834
- Zewdu S. , Suryabhagavan K.V., Balakrishnan M. (2017) Geo-spatial approach for soil salinity mapping in Sego Irrigation Farm, South Ethiopia,Journal of the Saudi Society of Agricultural Sciences (16) : 16–24
- Dehni A. and Lounis M. (2012) Remote Sensing Techniques for Salt Affected Soil Mapping:Application to the Oran Region of Algeria ,Procedia Engineering(33) :188 – 198

Drone Remote Sensing Technology for the Sustainable water resources management of small scale water bodies

Manavalan
Centre for Developmet of Advanced Computing (C-DAC), Bangalore, India
rmanavalan@cdac.in

Introduction

Remote Sensing technology become an essential and unavoidable component of water resource management operations of a country without which it is impossible to effectively manage any natural resources. As the technology keep changing its phase and mode, the use of remote sensing techniques as on date currently moving from space borne sensing systems to airborne drone modelling as the first one make available the data of required region only during the satellite pass periods whereas the second one make available the critical information as and when required. At the same time both these technology never be an alternate or can't be swapped each other, as weather and climatic conditions play vital role in deciding the flight planning of drone survey but in general the space bound imaging systems do not have any such constraints other than the optical sensors which can't penetrate the cloud and smoke where RADAR sensors performs well. In line to this, the significance of managing the small scale water bodies through Drone Remote Sensing survey techniques is brought out in this article. Emphasis has been given in understanding the usage nature of small sized low altitude flying drones than large size Unmanned Aircraft Systems (UAS) as small size drones can be handled by an individual of an local institution whereas operational usage of high altitude flying Unmanned Aerial Vehicles (UAV) has many restrictions as well as large UAV should also have many in build equipment's as part of it which are generally out of the scope of individual organization.

Materials and Methods

Materials

Directorate General of Civil Aviation (DGCA), Government of India categorized the Remotely Piloted Aircraft System (RPAS) as Nano Drones (Less than or equal to 250 grams); Micro (Greater than 250 grams and less than or equal to 2 kg); Small (Greater than 2 kg and less than or equal to 25 kg); Medium (Greater than 25 kg and less than or equal to 150 kg) and Large (Greater than 150 kg). The aim and objective of this study is mapping, monitoring and managing the small scale water resources which falls within the jurisdiction of local water resource management institution. Hence, this has to be supported preferably with the help of either Nano Drones or at maximum through Micro Drones. Flying Nano drones over uncontrolled airspace within the height of 50ft and micro drone within the height of 200ft is permitted of which other than the Nano drones rest of the drone survey need to get multiple permissions from different institutions of Government of India. At the same time, either nano or micro drone it should have an efficient video camera as part of it without which the purpose will not meet. Once again flying any customized version of drone falling less than 2kg also need to follow DGCA permission guidelines. Overall for an effective done survey of small scale water bodies a Nano or micro drone, its control units integrated with the mobile or laptop or both are required hardware components.

The drone data processing software in general will be primarily having flight path planning module, photogrammetry functions and other related data processing module which is mostly controlled using a field laptop. All most all leading commercial photogrammetry software's have dedicated modules to support Drone data processing. Open source software's such as Precision Mapper, Web ODM, DJI Ground Station, RAPID and Drone Deploy are also used by small organizations. While selecting drone software it is preferable to decide for a complete solution where both hardware units and software's will be delivered together. It is important to ensure the software should support Digital Elevation Model (DEM), Digital Terrain Model (DTM) for 3D visualization of the survey area. This enhances the visualization and accuracy of the modelling.

Drone Flight Path Planning and Data Processing

During the flight planning operations of mapping and monitoring the small scale water bodies, it is advisable to plan the survey in parallel to the wind direction of the region as this helps in avoiding drastic shift from the planned path. Maintaining a constant altitude both from the water body and surrounding Geomorphological features will ensures a precise 3D mapping of the study area. While flying the smaller drones over the water bodies the following factors has to be taken care with suitable control mechanism which in general achieved through training and experience

- Defining efficient flight planning which can cover the entire water body as the same is generally repeated during temporal surveys
- Managing the effect of wind blow which shakes and drift the drone out of the line of flight plan
- Maintaining right distance between drone and its control unit which is mostly movable in case of Nano drones and less movable when the higher versions of drone is used
- Maintaining a constant flight speed and altitude that can be repeated in subsequent temporal surveys

Use of Nano Drones over the Dam environment requires the support of a manned floating control unit, to maintain and control the flight path. The same is not required while using larger drones where the drone control unit range has considerable aerial coverage than the Nano Drones. Drone survey over the Cannel environment is always planned in parallel to the Cannel and at a constant altitude over the central spatial axis of the cannel over which the camera Nadir tend to follow. However, in practical situations the same is not possible while using the Nano Drones as its camera swath will not be able to reach out both the left and right banks of the cannels. By increasing the altitude of the drones the same can be achieved but miner structural faults or damages of the banks will be missed. In such case the micro and other larger drones has to be used which in general has wider swath. Hence while planning a Cannel survey it is must to ensure that the selected drone will be having suitable camera swath that can cover both the banks.

In case of Drone data processing the captured image frames of a Drone video were initially extracted and a mosaic operation is followed. This requires in having a considerable overlap of each image frames of the video which is taken care during the flight planning operations. Many commercial Drone Processing software performs an automated Drone Mosaic operation based on the control points collected over the overlap regions of the image frames. Though the same can be achieved in open source software, as on date there is considerable shift in the accuracy of the mosaic as the open source software performs the mosaic operation based on the overall histogram intensity of the image. Once the mosaic of image frames are completed the output can be visualized over the DEM (Digital Elevation Model) of the terrain which enhances the interpretation level of the object of the interest.

Conclusion

In the recent times, it has been experienced that during waster scarcity situations the resources from small water bodies were exploited to meet the drinking water demands of larger population. To be specific during recent summer many metro water management boards were looking for the water sources at local levels such as exploiting the water from stone quarries, pumping from nearby cannel beds where availability of storage and groundwater is within a reasonable depth. Due to increasing nature of drought happening in summer seasons this is going to be a future trend where the small water bodies can feed considerable population. Hence, scientifically rejuvenating these smaller water bodies is need of the hour which in turn requires micro level monitoring of respective resources. During summer this has to be done on day today basis. This can be only achieved with the help of Drone remote sensing survey which can map, quantify and make available the day today status of respective water resources and its availability details over a 3D environment. However, in India currently usage of Drone Technology is mostly limited to defence sector and to use other application sectors permissions from nine different institutions has to be obtained. Hence, till the restrictions are further relaxed this article falls in Conceptual Phase but the same is not the case in many developed countries.

Use of simple NDVI for assessment of urban green cover in Vellore using very high resolution Worldview-4 satellite data

Sangeetha Gaikadi [1*], S. Vasantha Kumar[2]

[1] *Ph.D Scholar, School of Civil Engineering, VIT University, Vellore – 632014, Tamilnadu, India.*

[2] *Associate Professor, School of Civil Engineering, VIT University, Vellore – 632014, Tamilnadu, India.*

* *e-mail: sangeetharao888@gmail.com*

Introduction

Migration of people from rural to urban areas in recent decades leads to tremendous pressure on most of the Indian cities and it is said that people who live in cities will increase from 50% at present to 66% in another 30 years. One of the major impacts of urbanization is loss of beautiful green cover in a city apart from other ill effects like water scarcity, polluted air, traffic jams, etc. The green cover in any city includes vegetated areas, trees, parks and gardens and it is very much important for a city as it helps to reduce the temperature, maintain the ecology and biodiversity, provide recreational facilities for the city inhabitants and many more. As green cover is vital for a healthy city, World Health Organization (WHO) recommended minimum 9m^2 of green cover per person in any given city (WHO, 2010). In order to check this criterion, it is essential to first prepare the green cover map using satellite data. Existing studies on green cover mapping have used mostly the low and medium resolution satellite data such as Landsat TM, ETM, OLI, Sentinel, etc. (Kopecka et al., 2017; Texier et al., 2018). Use of very high resolution images like Worldview-4 has not attempted before and the present study tries to explore it for urban green cover assessment. One of the popular indices in the field of remote sensing for extracting vegetated areas from satellite data namely, normalized differenced vegetation index (NDVI) has been used in the present study for extracting green cover from Worldview-4 data. A comparison of the results with supervised classification was also attempted.

Materials and Methods

The present study used satellite image from Worldview-4, which is a multispectral very high resolution commercial satellite launched by Digital Globe, USA on November 11, 2016. It provides satellite imagery in the panchromatic band of 31 cm resolution and 4 multispectral bands of 1.23 m resolution (highest in the world at present compared to other satellites). WorldView-4 has an average revisit time of less than a day and is capable of covering 6,80,000 sq km per day. Worldview-4 data covering Vellore in Tamil Nadu was purchased from Apollo Mapping, USA. Vellore corporation covering an area of 95 sq.km was used as the boundary to clip the image within the corporation area. Vellore is one of the fast-developing cities in the state of Tamil Nadu as it lies in the Chennai – Bengaluru highway and home for many notable educational institutions like VIT University and Christian Medical College (CMC). The urbanization in recent years resulted in loss of green cover in the city and hence necessitates the need to study its extent using high resolution satellite data like Worldview-4. The radiometric, geometric correction and ortho rectification of the satellite data was done by Apollo Imaging and was spatially registered using WGS84 datum and projected using Universal Transverse Mercator (UTM) Projection.

For extracting green cover from Worldview-4, normalized differenced vegetation index (NDVI) has been used in the present study, which is calculated by taking difference between the near infrared (NIR) and red bands and then dividing by their sum. Unlike other indices, the advantage of NDVI is that it has a fixed range of -1 to 1. Areas covered with vegetation will yield positive values due to high NIR reflectance and low red reflectance, whereas rocks, bare soil exhibit NDVI values close to 0. Water, clouds, and snow usually will show negative NDVIs. In the present study, NIR and red bands of Worldview-4 were used to calculate the NDVI and then NDVI output was analyzed with satellite image on the background to find the boundary value above which green cover could be extracted. The pixels above the boundary value were classified as "urban green cover (UGC)" and the remaining pixels were termed as "non-urban green cover (NUGC)". The resulted map was then checked for its accuracy using 500 random points. For each of the points, the classification result (UGC or NUGC) was compared with the actual ground truth information obtained by visually seeing the satellite image. The overall accuracy and Kappa statistic were then calculated and reported. The results were compared with that of the maximum likelihood supervised classification to check whether NDVI or supervised classification is performing well.

Results and Concluding Remarks

The results of NDVI was compared with the original satellite image and it was found that the green covered areas in Vellore has NDVI values of above 0.45. Thus, by keeping 0.45 as the cut-off, the map of UGC and NUGC was prepared (Fig.1). The accuracy assessment showed that the overall accuracy is 98% and the Kappa statistic is 0.94. The supervised classification was also performed using maximum likelihood method by giving training pixels for UGC and NUGC. An overall accuracy of 98% and Kappa coefficient of 0.96 were obtained. The results were almost comparable with that of NDVI method and hence it can be said that both were performing equally well. Generally, an overall accuracy of 85% and above is acceptable. In the present study, an overall accuracy of 98% is obtained which indicates that both NDVI and supervised classification are highly suitable for extracting the green covered areas using Worldview-4 data. The area covered by UGC in Vellore was found to be 72 sq.km and NUGC was 23 sq.km. As the population of Vellore is about 5 lakhs, the UGC required is only 4.5 sq.km as per WHO standards of $9m^2$ per person. However, the proportion of UGC available is more than what has been recommended by WHO. It is important to note here that the calculated UGC of 72 sq.km includes agricultural areas also in addition to trees, parks and gardens. According to Contesse et al. (2018), urban agriculture cannot be a substitute for parks but it can be treated as a complementary form of green space with its own distinctive value. One of the challenges of the present study is to segregate agriculture areas from trees, parks and gardens as all of them exhibit similar NDVI values. The future scope of the work is to separate them and see whether trees, parks and gardens alone fulfil the WHO criteria.

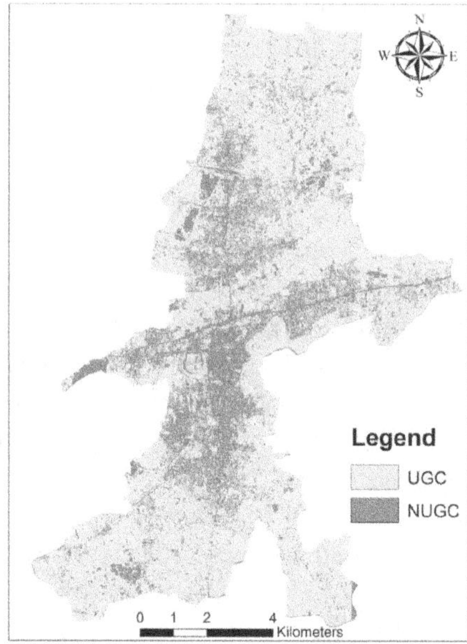

Fig.1 Map showing urban green cover (UGC) and non-urban green cover (NUGC) in Vellore

References

Contesse M, Vliet BJM, Lenhart J (2018) Is urban agriculture urban green space? A comparison of policy arrangements for urban green space and urban agriculture in Santiago de Chile. Land Use Policy 71: 566-577.

Kopecka M, Szatmari D, Rosina K (2017) Analysis of urban green spaces based on Sentinel-2A: Case Studies from Slovakia. Land 6(25):1-17.

Texier M, Schiel K, Caruso G (2018) The provision of urban green space and its accessibility: Spatial data effects in Brussels. PLOS ONE 13(10):1-17.

World Health Organization (WHO) (2010) Urban Planning, Environment and Health: From Evidence to Policy Action 2010. WHO, Denmark.

Construction Technology a Management

Review on Reducing the Construction Time using different techniques utilizing waste materials

Udita Gupta[1,*], Dr.Shobha Ram[1,2], Gaurav Chand [1]
[1]Department of Civil Engineering, Gautam Buddha University, Greater Noida, India[2] Assistant Professor
uditagupta99@gmail.com

Introduction

Many techniques have been discovered keeping in mind the importance of time and sustainable development. These techniques include 3D-printing in construction industry, Precast and Prefabrication, Structural Stay IN Place Structures and Building Information Modelling (BIM). The objective is to reduce the construction time while using waste material or reusing materials wherever possible. This paper reviews various researches done on constructing various structural components with these techniques.

In 3D printing, the printer takes instruction from CAD model to construct a 3D structure in considerably less time while performing miscellaneous functions automatically like reinforcements, aesthetic work like tiling of floors and walls and painting, plumbing, electrical and communications line wiring, etc.Precast and Prefabrication is a technique in which various components of a building are constructed individually in a controlled environment and then placed together on site to obtain the building. Structural Stay-In-Place structure (SIP) is a formwork which is not removed even after concrete has been cast and eliminates the need of reinforcement.Building Information Modelling, BIM is a technology which looks after the building construction management and takes account of every bit of information starting from construction planning to the post construction management. BIM also contains information regarding material, equipment, resource, and manufacturing data. Besides taking information and processing it, it can also track and monitor the construction site (Tay & Panda, 2017). All the mentioned technique aims to decrease construction time.

Literature Review

There has been an often mention of integrating BIM and 3D printing for optimise results (Tay & Panda, 2017).Tumminiastudied(Tumminia et al. 2018)environmental impacts of precast and also mentioned that the various components of precast can be dismantled and reused. RadanTomek focused on various advantages of precast technique especially in bridge construction (Tomek, 2017) stating that it will result in less maintenance cost. Wozniak performed experiments and thus suggested Strain Hardening Cement Composite (SHCC) SiP is better than traditional Steel SiP , also SHCC is mainly made up of glass fibres(that can be used recycled from the waste) (Wozniak et al., 2017) . Another similar research on bonding between concrete and fibre reinforced polymers (FRP) SiP has been carried out (Goyal, Mukherjee, & Goyal, 2016). It is possible to create SiPelements using PVC like panels, connectors and bracing. NoranWahab studied the flexural behaviour of concrete walls made using PVC SiP (Wahab & Soudki,2013) and proved that PVC has improved ductility of wall system. Review papers written on BIM majorly focus on integrating it with other mentioned techniques.

Review Analysis

Although it is mentioned in every paper that these techniques are implemented to accelerate construction time,still there was no experiment conducted to determine how much time these techniques (precast and SIP) will save if implemented. Similarly, even though China constructed two-storey villa using 3D printing claiming that it can withstand an earthquake up to 8-Richter Scale, there has been no

Table 1: Contribution of various techniques towards sustainable growth

Technique	Paper	Contribution / Deductions
3 D printing	(Khoshnevis, 2003)	**Speed** : <2days for 200m^2 double storey house
		Waste : no emission as equipment is electric, even consume waste generated
	(Perkins,2015)	**Speed** :Sq foot of wall will take 2 hours, whole room in a day, 200 sq m single storey building in a day
Precast	(Tomek, 2017),	**Speed** : increases speed by 2 to 3 times when compared to cast-in-situ
	(Tumminia, et al., 2018)	**Environmental impact**: less primary energy consumption during the Use-Stage of prefabricated house module compared on the basis of global energy requirement (GER)
Stay-In-Place	(Wozniak 2017)	**Waste usage**: glass waste can be used in preparing SHCC (Strain Hardening Cement Composite)
	(Wahab ,2013)	**Waste usage:** PVC waste can be utilized.
Building Information Modelling (BIM)	(Tay & Panda, 2017)	**Waste management** : since it can foresee various factors, thus can prevent waste of extra raw material and can save time

experimental research done on seismic properties of 3D printed structures. The table no.1 shows the contribution of various researches towards sustainable growth. The techniques also follow the 4Rs strategy. Prefabricated elements can easily be reused. Recycled glass waste and plastic waste can be adopted at various steps while implementing SIP. 3D printing reduces the wastage of raw material and it also provides the option to recover the wasted raw material.

Conclusion

As discussed, these techniques are capable of reducing construction time and it also utilizes waste upto some extent, increasing sustainable growth. Techniques like 3D printing producevery less negative impact on environment. Prefabricated module has proved to use comparatively less primary energy in the Use-Stage and has the potential to be used again; innovative Stay-In-Place structures made of SHCC and PVC can be used to utilize tons of waste whilst increasing speed and BIM can also save wastage of material and can speed up the process due to its decision-making property.

References

1. Goyal, R., Mukherjee, A., & Goyal, S. (2016). An investigation on bond between FRP stay-in-place formwork and concrete. *Construction and Building Materials*, *113*, 741-751.https://doi.org/10.1016/j.conbuildmat.2016.03.124
2. Khoshnevis, B. (2003). Toward total automation of on-site construction an integrated approach based on contour crafting. *Proceedings of ISARC*, 61-66.https://doi.org/10.22260/ISARC2003/0001
3. Tay, Y. W. D., Panda, B., Paul, S. C., Noor Mohamed, N. A., Tan, M. J., & Leong, K. F. (2017). 3D printing trends in building and construction industry: a review. *Virtual and Physical Prototyping*, *12*(3), 261-276.https://doi.org/10.1080/17452759.2017.1326724
4. Tomek, R. (2017). Advantages of precast concrete in highway infrastructure construction. *Procedia engineering*, *196*, 176-180.https://doi.org/10.1016/j.proeng.2017.07.188
5. Tumminia, G., Guarino, F., Longo, S., Ferraro, M., Cellura, M., & Antonucci, V. (2018). Life Cycle Energy Performances and environmental Impacts of a Prefabricated Building Module. *Renewable and Sustainable Energy Reviews* , 272-283.https://doi.org/10.1016/j.rser.2018.04.059
6. Wahab, N., &Soudki, K. A. (2013). Flexural behavior of PVC stay-in-place formed RC walls. *Construction and Building Materials*, *48*, 830-839.https://doi.org/10.1016/j.conbuildmat.2013.07.073

Hyperreality as a tool for resources optimisation in civil engineering for postmodernist era

Subhadarshini Khatua
IIM Shillong,India
subha.fpm16@iimshillong.ac.in

Introduction

As consumer culture globalizes, there is also the sense, then, of the planetary limits to consumption: that we are literally consuming the planet and our human future at an unsustainable rate (an argument made by James Lovelock (2006) in The Revenge of Gaia). To understand the complexity of marketplace it is necessary to understand postmodern consumers especially in civil engineering related market place. According to Firat & Venkatesh (1993, p. 220), the postmodernism was inspired by a desire to become detached from all metanarratives that require conformity to a single way of perceiving reality. Postmodernists call for a diversity or multiplicity of narratives, a liberation from all conformity, and a freedom to experience as many ways of being as desired.

Hyperreality will help the civil engineering industry with an experience like we see in science fictions. Engineers can use the interactive experience and create an ambiance using sophisticated viewers. It will give a virtual experience of stepping into planned structures much before the execution. This paper would focus on hyperrealities as a tool for persuasion for consumers rather than using traditional tools which uses actual resources.

Postmodern world of hyperreality

According to Firat and Shultz (2001) postmodern consumer has these characteristic traits. First is multiphrenic self in which the individual is encouraged to change the image frequently and therefore, he trying to adapt himself to new roles and new identities. Second is decentred subject where openness and tolerance to experience different lifestyle goals is encouraged. Objects would play a role in guiding the desires of the consumer (Baudrillard 1981). And lastly the need for Hyperreality as consumers would more willing to the idea that reality is a constructed singularity.

Hyperreality and its use

Postmodern consumers adore being individual, they adore continually reinventing themselves through their consumption. They love to develop highly individualised identities, through continually fresh and exciting consumption experiences. In postmodernity, consumers seek different and local experience and they desire to belong to processes and experience immersion in thematic settings rather than merely encounter finished products and images. Therefore, marketing in civil engineering designs has to involve the consumer by considering him as a producer of experiences (Cova, 1996).

In postmodernity, the consumer is not a passive target for image marketing but an active link in the continual production of meanings. He calls for an experience-based designing that emphasizes interactivity, connectivity and creativity. To consume becomes an act of production of experiences, identities or images of itself and not a fact of destruction. (Cinotti, 2007; Filser, 2002). The consumer is an emotional being, centred on the realization of pleasant experiences.

Methodology

The traits identified in postmodernists were assimilated in the questionnaire and a check for minimalistic behaviour was done first. This was done through Belks's model for materialism. The customer segment who showed a minimalistic behaviour were further interviewed in a focus group for identifying postmodernist behaviour. These consumers exposed to traditional as well as virtual walk through of different home designs and then were asked about their perception of hyper reality viz a viz the traditional designs.

Results and Concluding Remarks

The likelihood of purchasing a home where the respondents were given a walk through was observed t be more as compared to the designs which had traditional designs. Satisfaction level of consumers increased with a greater number of walk throughs through. They could remember the details more in case of hyperreality and suggest changes precisely. The probability of referral also increased in case of walk throughs.

Hyperreality interweaves the real-world and visuals to give field workers and civil engineers valuable data such as design specifications, consumer needs etc., making their job on field easy. As changes can be made at conceptual level, therefore the overall cost of manhours and resources reduces significantly.

References

Furat Firat A., Clifford J. Shultz (2001) Preliminary Metric Investigation into the Nature of the "Postmodern Consumer". Marketing Letters 12:2, 189-203, 2001, Netherlands

Russell Belk, Materialsm: Trait aspects of Living in the Material World, Journal of consumer research- December 1985, doi:10.1096/208515

A. Furat Firat, John F. Sherry, Jr and Alladi Venkatesh, Postmodernism, marketing and the consumer, Elsevier Science-1994, 0167-8116/94

Zita Sampaio, The use of Virtual Reality Models in Civil Engineering Training, International Journal of Simulation Modelling 6(2):124-134, Doi 10.2507/IJSIMM06(2)S.07

Title of the paper: Experimental approach towards building services by embedding efficient technologies

A. Karthik M[1], B. Varsha S Danavandi[2*], C. V G Ajey Kumar[3], D. Dilip Raju[4]
[1] *Construction Technology & Management/Dayananda Sagar College of Engineering/Bangalore*
[2] *Construction Technology & Management/Dayananda Sagar College of Engineering/Bangalore*
[3] *Construction Technology & Management/BMS College of Engineering/Bangalore*
[4] *Construction Technology & Management/Dayananda Sagar College of Engineering/Bangalore*
* *e-mail: karthik-ctm@dayandasagar.edu*
varsha-ctm@dayanandasagar.edu
ajeykumar.cct17@gmail.com
regaldsr@gmail.com

Introduction

The building sector represents about 35-40% energy consumption. Keeping this situation, there is a need to adopt sustainable efficient building design approach which is the ultimate solution to reduce the energy demand of the building. Over usage of these conventional building materials, not only cause global warming but also affects the natural environment. Green or sustainable building use key resources like energy, solar, water, materials, and land more efficiently than buildings that are just built in conventional methods. This approach has been undertaken for newly constructed and existing buildings to assess its potential and energy saving capacity.

Isaksson & Karlsson's (2006) investigation of the houses in Lindås, the interviews showed that knowledge about the heating system was an important issue for the residents. Internal heat gain from different light sources in the building lighting systems 23 October 2017, EU directives and the Construction Law have for some time required investors to report the energy consumption of buildings, and this has indeed caused low energy consumption buildings to proliferate. Of particular interest, internal heat gains from installed lighting affect the final energy consumption for heating of both public and residential buildings.

The main object of the work deals with the various energy saving techniques which can be incorporated during the time of planning, designing, construction and execution stage to have energy efficiency in buildings, keeping cost perspective in mind.

Materials and Methods

Ducts were introduced into the false ceiling or ceiling panels. Interior part of Ducts was Coated with Gloss Dark Color, in which the light was not absorbed by the interior surface of the ducts, which indeed enhanced the transfer of light. Various designs of Ducts were incorporated and arrived, depending on the orientation and shape of the building. Mirrors were placed inside the ducts at suitable angle (45 degree) to Transfer the light from the Source. Openings were given in the false ceiling or panels for the light to fall in the occupant area. Acrylic Sheets were used to diffuse the light into Occupant area, which indeed increased the light intensity.

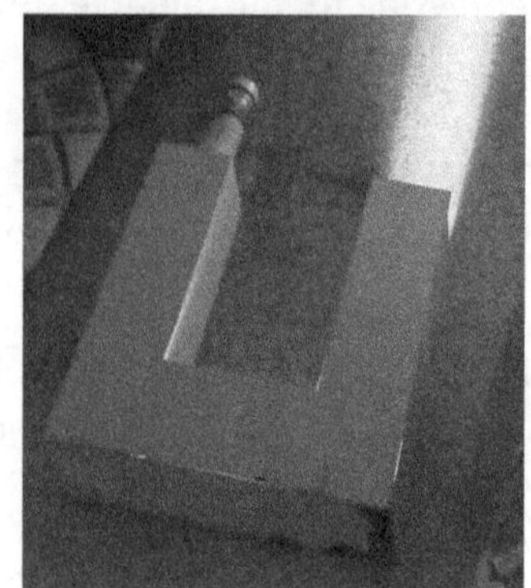

Fig: Model representing light transfer through ducts and different shape of ducts

Results and Concluding Remarks

It was observed that light source had impact on intensity of illumination. Operation cost was reduced.

The various standards were studied and imparted to this project, which enable to understand the behaviour of materials to the surrounding situation.

References

[1]. Tri Endangsih and Hakim,Lecturer in Architectural Studies Program, Faculty of Engineering of Universitas Budi LuhurJl. Ciledug Raya, Petukangan Utara, Jakarta Selatan-Indonesia Volume 116 No. 24 2017, 487-494.

[2]. Barlow, S., Fiala, D., 2007, Occupant comfort in UK offices - How adaptive comfort theories might influence future low energy office refurbishment strategies, Energy and Buildings, vol 39, p. 837-846.

[3]. Isaksson, C., Karlsson, F., 2006, Indoor climate in low-energy houses - an interdisciplinary investigation, Building and Environment, vol 41, p. 1678-1690.

[4]. December 2014 Conference: International Conference on Sustainable Built Environment, At Kandy, Sri Lanka, Volume: 5th Development of Sustainable Roofing material from Waste.

[5]. Internal heat gain from different light sources in the building lighting systems. 23 October 2017, E3S Web of Conferences 19, 01024 (2017) Dariusz Suszanowicz* University of Opole, Faculty of Natural Sciences and Technology, ul. R. Dmowskiego 7-9, 45-365 Opole, Poland.

Sustainable Energy-Efficient Building

Shivam[1], S. Dubey[1], U. A. Shah[1*], K. N. Vishwanath[2], S. Murthy[3]
[1] *Department of Civil Engineering, Dayananda Sagar Academy of Technology and Management, Bangalore*
[2] *Academic Guide, and Head of Department of Civil Engineering*
[3] *External Guide, and Assistant Manager, Tata Consulting Engineers*
[*] *e-mail: udit.shah@outlook.in*

Introduction

Growing populations and affluence around the globe have put increasing pressure on natural resources, including air and water, arable land, and raw materials. Concern over the ability of natural resources and environmental systems to support the needs of global populations, is part of an emerging awareness of the concept of sustainability. Incorporating sustainability into products, processes, technology systems, and services requires assimilating environmental, economic, and social elements in the evaluation of design. Sustainable construction (Iwaro and Mwasha 2014) is a movement towards developing sustainable, energy-efficient and non-toxic buildings. This movement is integrating elements of environmental performance, social responsibility and economics, through design quality, technical innovation and transferability.

The aspects in designing of a building include spatial suitability, spatial functionality, material property, and the interaction among them. Design decisions being made based on each of these aspects have a significant impact on the energy consumption of the building. The location and purpose of requirement define the steps involved in deciding the design. Decisions have to be made on subjects such as form, orientation, function and services that are to be provided. These decisions resolve into selection of, firstly (Perlova et al. 2015), optimal form, orientation, spaces inside and in the surroundings of the building; secondly (Kamal 2012), openings for lighting, circulation and ventilation; thirdly (Kurkinen and Karlsson 2012), efficient filtration approaches to reduce negative effects of outdoor climate on the thermal balance of the building and the quality of indoor environment; fourthly, sources and provisions for, based on availability of, water, energy and materials; and so on. The objective is, therefore, to develop a residential building in a layout with improved outcomes in the stages of construction and operation using

- *Orientation* for efficient lighting and ventilation,
- *Spatial planning* for effective interior and exterior spaces,
- *Passive building design* to maintain comfortable indoor temperatures, and
- *Low energy* and *high thermal mass materials* to reduce environmental impact,

Thereby making it cost-effective, energy-efficient and environment-friendly

Methodology

i. *Theory*: Studying literature and cases to prepare an abstraction of requirements and objectives; the abstraction resulting in the synthesis of a design brief.
ii. *Site Investigation*: The site considered had a flat terrain. Surrounding buildings were not any taller than three storeys. A short unconventional survey of wind directions was conducted.
iii. *Modelling*: Commencing development of design with hand-drawn sketches as ideas. The plan is drafted in Autodesk AutoCAD based on those sketches. 3D models are developed in Autodesk Revit Architecture.
iv. *Optimization*: Optimizing designs by choosing suitable materials and developing appropriate spaces for better indoor environment in terms of thermal comfort, ventilation and lighting. Analysing based on materials and energy, as well as cost function in a brief view, with aid of an open source software, EDGE, developed by the International Finance Corporation.
v. *Application*: The developed design shall have great scope in application, repetition and reiteration with minimal changes to the design.

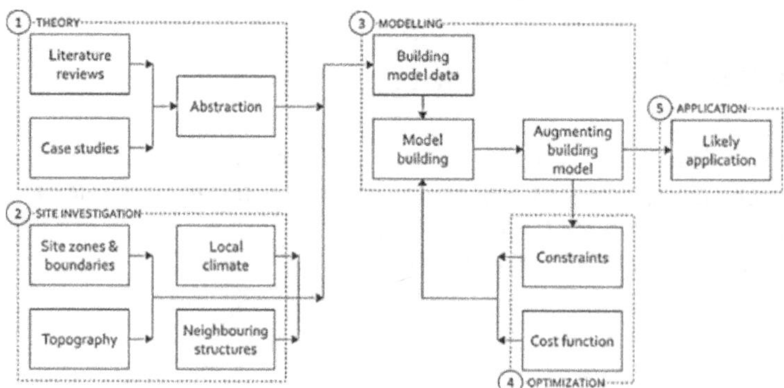

Figure 1: Flowchart showing methodology adopted.

Results and Concluding Remarks

Sustainability is evaluated based on many criteria, and in case of a residential building, the criteria are defined by the physical, material, environmental, economic and social characteristics of the surroundings, the site and the building itself. This work is relying on the concepts of embodied energy, low energy emission and high thermal mass in case of materials, and on orientation, spatial planning and passive building design in case of design.

The designs developed by using these concepts, by themselves did not seem credible, until they were analysed against a unique base case developed by EDGE specifically for this project using empirical data from actual buildings reflecting current practices around the world. Comparing the designs that we developed and the materials that we are proposing, to that of the base case, the project is attaining commendable *savings of 46.55% in energy efficiency* and *73.18% in material efficiency*—both values being appreciably *greater than the minimum standard of 20%* set by EDGE.

This project is a successful example that can be reiterated with or without any modifications on sites of similar conditions, all over the city of Bengaluru, KA, India. The layout in consideration, has multiple sites at the same orientation adjacent to one another, and this interactive design can be used in all cases. However, design is subjective and not constant for everybody. For the same requirements, there may be multiple designs giving similar outcomes in energy consumption. Parallelly, there may also be many materials that may have comparable life-cycle cost. Therefore, several perspectives of development can result in differing yet efficient designs.

Acknowledgments: The authors would like to thank Ar. Amod Shah for assisting them in developing the designs of the building.

References

International Finance Corporation, World Bank Group (2019), accessed 23 May 2019, Excellence in Design for Greater Efficiency <https://www.edgebuildings.com/>

Iwaro J, Mwasha A (2014), The Impact of Sustainable Building Envelope Design on Building Sustainability Using Integrated Performance Model. International Journal of Sustainable Built Environment: 153-171

Kamal M A (2012), An Overview of Passive Cooling Techniques in Buildings: Design Concepts and Architectural Interventions; Acta Technica Napocensis. Civil Engineering & Architecture Vol. 55, No. 1

Kurkinen E L W, Karlsson J (2012), Different Materials with High Thermal Mass and its Influence on a Buildings Heat Loss – An Analysis based on the Theory of Dynamic Thermal Networks. 5th International Battery Production Conference

Perlova E, Platonova M, Gorshkov A, Rakova X (2014), Concept Project of Zero Energy Building. DAAAM 2014: 1505-1514

Construction Waste Optimization – A Lean Six Sigma Technique

HA. Nishaant[1*], J. Sudhakumar[2]
[1]*Research Scholar, Department of Civil Engineering,National Institute of Technology,Calicut,Kerala,India*
[2]*Professor and Head,Department of Civil Engineering,National Institute of Technology,Calicut,Kerala,India*
* *e-mail: nishaant_p180003ce@nitc.ac.in*

Introduction

One of the major bottlenecks in anyconstruction project is the issue of construction waste management. The prime reasons for a large generation of construction wastes are improper planning and execution, unskilled labours, errors in design andthe use substandard materials. Owing to this, there is a fall in the quality of construction as well as an increased cost. So, it becomes imperative for all construction projects to optimize the wastes.

Six sigma, a highly effective quality improvement technique has been extensively used in the recorded history. A similar technique of lean methodology can be used to minimize or nullify the waste in various processes.A combination of Lean and Six Sigma concepts enable the achievement of multi-dimensional quality improvement as well as waste minimization. This study implements the DMAIC (Define, Measure, Analyse, Improve and Control) methodology, a prominent feature of the six sigma ideology,in which the controlling causes of waste generation are identified during the define phase, quantified during the measure phase, examined subsequently in the analyse phase. Post the analysis, methods to overcome the problem and the techniques for improvement are discussed during the improve phase. Recommendation for an optimized execution of the project with the expected quality and least wastage are derived at the control phase.

Materials and Methods

This research work is limited to the incorporation ofLean and Six Sigma concepts, to study the nature and causes of increased generation of wastes in a construction project. DMAIC technique is used as a key factor for framing sequential steps, with which lean concepts are integrated at each phase.A framework is established and suggestions for waste optimization are presented.

Results and Concluding Remarks

A framework to minimizethe construction wastes is presented and recommendations are suggested. The advantages of implementing Lean Six Sigma are also discussed.

References

L. S. Pheng and M. S. Hui(2004) Implementing and Applying Six Sigma in Construction. Journal of Construction Engineering and Management 130(4): 482-489

A. Banawi and M. M. Bilec (2014) Aframework to improve construction processes: Integrating lean, green and six sigma. International Journal of Construction Management 14(1): 45-55

Evaluation of alternative building materials for passive cooling to achieve thermal comfort and energy efficiency – A systematic review

Mohammed Shibin N[1*], Dr.Deepthi Bendi[2],

[1] *Research scholar, Department of Architecture and Planning, National institute of technology, Calicut*

[2] *Assistant Professor, Department of Architecture and Planning, National institute of technology, Calicut*

* *e-mail: ar.mohammedshibin@gmail.com*

Introduction

Built environment is known as intense consumer of energy. Heating and cooling are the major functions that require multiple energy resources. In 2016-17, households accounted for about 24% of the total units of electricity consumed in India. The electricity consumption in residential sector has increased 7.93% that has grown at a rapid pace compared to other sectors during 2007-08 to 2016-17 (Energy statistics 2018). Several researchers have highlighted the shortcomings of the conventional building materials to address the energy consumption and enhance thermal efficiency in built environment. This highlights the need for innovation in the research and invention of alternative building materials that enables optimal thermal comfort in the built environment.

Alternative thermal efficient materials keep the building comfortable and reduce the load on mechanical air conditioning systems. Olgyay (1963) reports materials that reflect rather than absorb radiation and that more readily release the absorbed quantity as thermal radiation will cause lower temperature within the structure. He further adds that white materials may reflect 90 % or more, black material 15 % or less, of radiation received. Givoni (1976) points out that when the indoor thermal conditions are not controlled by mechanical means, the materials affect the temperatures of both the indoor air and surfaces and thus have a very pronounced effect on the occupant' comfort. Even when control is used, in the form of heating or air-conditioning for instance, the thermo physical properties of the materials used determine the amount of heating or cooling which is provided and also the temperature of the internal surfaces (radiant temperature). Therefore, even in these circumstances, the materials have an effect on the comfort of the occupants, as well as on the economic efficiency of the control systems.

The primary objective of the study is to investigate the performance of passive cooling building materials for thermal efficiency considering factors including thermal transmittance value, thermal mass, insulation properties etc. This paper is focusing particularly on the innovation done in the last few years in the field of passive cooling building materials. The emerging trends in passive cooling materials are aimed to achieve energy efficiency as well as to enhance the occupant's' comfort with respect to thermal environment. The research proposes to evaluate selected alternative building materials for passive cooling using Computer aided simulation models.

Materials and Methods

The flow chart below presents the proposed structure of the research. It involves five major stages. In stage 1 –Topic introduction describes the scope and purpose of the research topic which leads from a general research area into a refined topic for the study.

Stage 2, the literature review comprises a detailed literature study to understand the passive cooling building materials. The literature study will provide the fundamental knowledge for the research.

The next stage, consists of two phases. The first phase involves the collection of properties of selected thermal efficient materials. The second phase is a detailed computer simulated data on material evaluation incorporating thermal properties of materials. Stage 4 – deals with the analysis – In this stage, energy based software(s) will be used to analyse the thermal performance of materials in built environment.

The final stage (Stage 5) explains the findings and inferences from the findings. It will also discuss the relevance of the findings and direct towards the future research in this area.

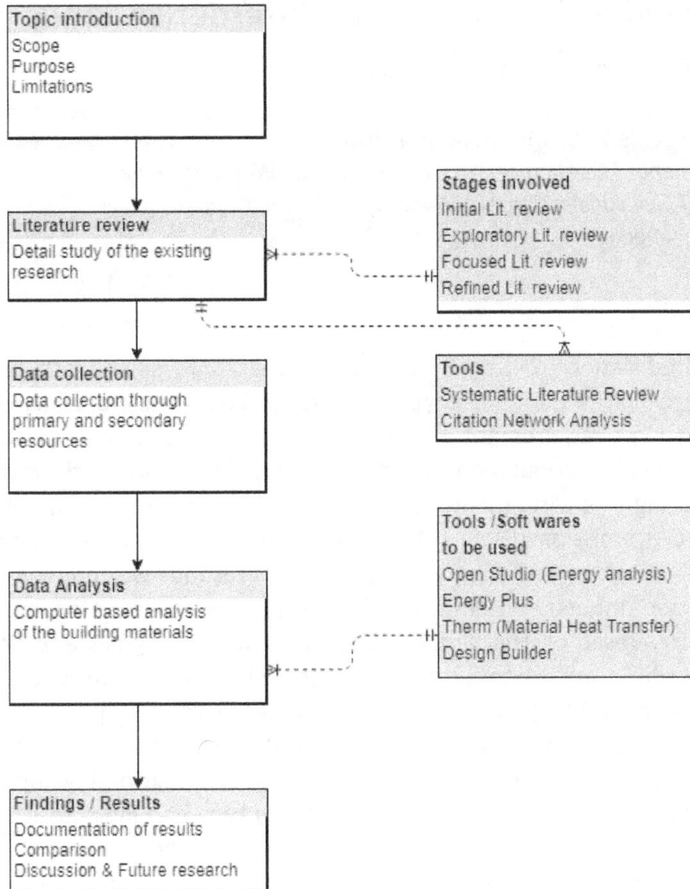

Figure 1. Research stages

Results and Concluding Remarks

Building materials have a significant role in reducing cooling and heating loads thereby achieving energy efficiency. The appropriate selection of building materials can reduce the energy load required on air conditioners. This paper is limited to evaluate building materials used to achieve thermal efficiency, other sustainable parameters are not been addressed. The quality of passive cooling materials in built environment is connected with several different aspects. The research also covers the factors to be considered for the selection of building materials which influence the energy requirements of building for heating and cooling load.

References

Olgyay, Victor. (1963). Design with Climate: Bioclimatic Approach to Architectural Regionalism. New Jersey: Princeton University Press, pp. 14-114

Givoni, Baruch. (1976). Man, Climate and Architecture.

Energy statistics 2018, 25th issue, Central statistics office, Ministry of statistics and programme implementation, Government of India.

A fuzzy AHP model in evaluating interdependency of sustainable indicators and criteria- Case of India

A.Suchith Reddy[1*], P. Rathish Kumar[2], P. Anand Raj[3]
[1]*Research Scholar, Department of Civil Engineering, NIT Warangal, Warangal, India*
[2,3]*Professor, Department of Civil Engineering, NIT Warangal, Warangal, India*
* *e-mail: asr.nitwarangal@gmail.com*

Introduction

The According to United Nations(UN), by the end of the year2050, globally around 6.3 billion people are expected to live in the cities (Mahmoud, Zayed, & Fahmy, 2019) due to which the urban inhabitants increase rapidly with the huge requirement of infrastructural facilities for transportation, housing, health, and education. This unintended population growth will have to sufferfromthe availability of resources, energy,and pollution leading to environmental degradation. Today's cities energy consumption accounting to 70% of Greenhouse Gas emissions(GHG) (McCormick et al, 2013). Infrastructure buildings consume a huge quantity of natural resources and energy and produce hazardous waste along with pollution. To take control of the consumption (inputs) and emissions (outputs), it is necessary to implement and adopt the principles of sustainability. Based on the directions of UN Sustainable goals and principles, localized and specific indicators and criteria are essential in achieving sustainable construction(Reddy, Raj, & Kumar, 2018; Zhang, Zhan, Wang, & Li, 2019). The present study investigates the specific sustainable criteria with respect to Social, Environmental, Economic and Technological (SEET) indicators keeping in view the regional variations, culture, heritage, climatic, geographical, and regional context of developing countries like India. The objective of the study is to quantify the interdependency between indicators and criteria using Fuzzy Analytical Hierarchy Process (FAHP). The findings of the study simplify the decisions to be taken by the stakeholders in improving the sustainable performance of buildings and further facilitates in developing sustainable building assessment tool.

Research Methodology

The approach to evaluating the performance of the building can meet the design objective of the building. The scientific evidence proposes that the assessment of the significant performance of indicator can be performed by a consensus-based process which best suits the comprehensive analysis (Vyas, Jha, & Patel, 2019). The hierarchical model facilitates to provide a common platform for assessing the building in all disciplines of sustainability and also serves for the development of design and technical solution for achieving sustainability. Keeping in view, varied local and regional context, climate changes, construction procedures, topographical and cultural changes, the study identified four sustainable indicators Social, Environmental, Economic, and Technological and eight sustainable criteria including Water Efficiency (WE), Materials & Waste Management (MW), Human Well-being (HW), Energy Efficiency (EE), Sustainable Sites (SS), Social Welfare (SW), Transportation (T) and Management (M). Further, based on Content Analysis (CA) and Delhi Technique (DT) the criteria are refined and tailored to suit the Indian context. To evaluate the relative weights and quantify the interrelationship, a questionnaire survey was drawn out with 7-point Likert scale and experts from public and private organizations (including academicians, designers, architects, consultants, clients, contractors, and others) to analyze their significance towards sustainability. The data extracted was observed to be consistent using Cronbach's alpha coefficient for four dimensions of sustainability are found to be above 0.85. They are then processed, analyzed and interpreted using statistical techniques to extract the required information. This information is then analyzed using Fuzzy

Analytical Hierarchy Process (FAHP) a Multi-Criteria Decision-Making (MCDM) method to establish interrelationship among criteria and indicators.

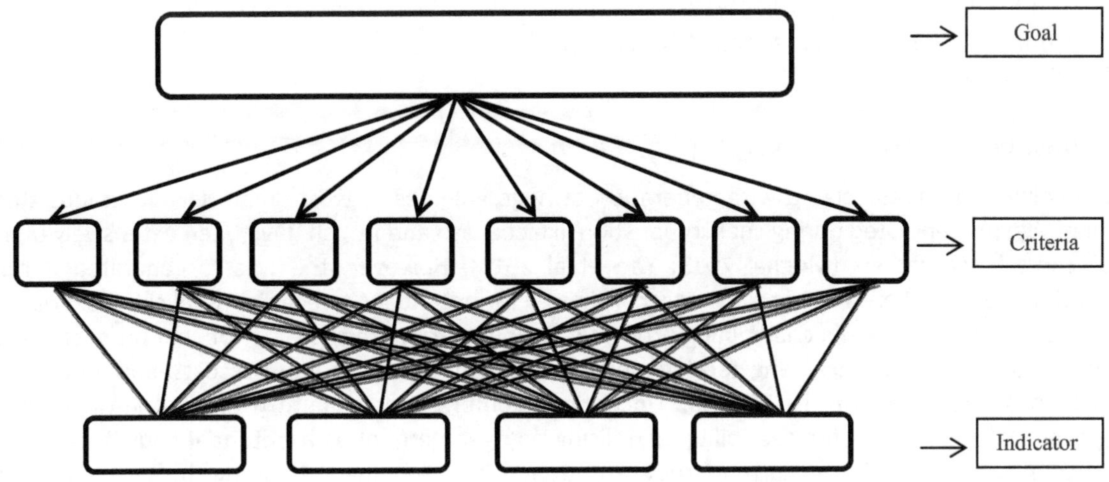

Fig.1 Hierarchy of the model to assess the interdependency of indicators and criteria

Results and Concluding Remarks

The study found that development of the nation involved with the adoption of sustainable principles in construction industry promotes the overall growth without disturbing the eco-system and avoids adverse impacts caused by the conventional principles and practices in India.

- The study brought out the significance of the proposed 'Technological' indicator and encouraged Quadra-Bottom Line approach in implementing and achieving sustainable construction. This facilitates to incorporate innovative ideas and then implement the concept of Reduce, Recycle and Reuse (3R's) into design principles. Among SEET indicators, Environmental indicator has secured the highest weight of 30.15% and Technological indicators of 28.52%.
- The normalized interdependency of Technological indicator has attained thehighest weight of 28.40% prior to Environmental indicator of 27.01%, by this, the study derives the integration of Technological indicator with a triple-bottom-line approach to form quadra-bottom-line approach in achieving sustainable construction.
- The Material and Waste Management criterion has attained the highest relative weight and interdependency 13.96% and 15.56% respectively.Based on the interdependency, the criteria WE, MW, SS, and S are categorized under Social indicator, MW, EE, T, and M are categorized under Environmental aspect. Similarly, WE, HW, EE, and SS are grouped under Economic indicator and HW, SS, SW and M criteria are categorized under the Technological indicator.

References

1. Mahmoud, S., Zayed, T., & Fahmy, M. (2019). Development of sustainability assessment tool for existing buildings. *Sustainable Cities and Society*, *44*(May 2017), 99–119. https://doi.org/10.1016/j.scs.2018.09.024

McCormick, K., Anderberg, S., Coenen, L., & Neij, L. (2013). Advancing sustainable urban transformation. *Journal of Cleaner Production*, *50*, 1–11. https://doi.org/10.1016/j.jclepro.2013.01.003

Reddy, A. S., Raj, P. A., & Kumar, P. R. (2018). Developing a Sustainable Building Assessment Tool (SBAT) for Developing Countries—Case of India. In *Urbanization Challenges in Emerging Economies*. https://doi.org/10.1061/9780784482032.015

Vyas, G. S., Jha, K. N., & Patel, D. A. (2019). Development of Green Building Rating System Using AHP and Fuzzy Integrals: A Case of India, *25*(2), 1–12. https://doi.org/10.1061/(ASCE)AE.1943-5568.0000346.

Zhang, X., Zhan, C., Wang, X., & Li, G. (2019). Asian green building rating tools: A comparative study on scoring methods of quantitative evaluation systems. *Journal of Cleaner Production*, *218*, 880–895. https://doi.org/10.1016/j.jclepro.2019.01.192

Strength behavior of coal gangue stabilised soils

Mohammed Ashfaq[1*] M Heeralal[2] Arif Ali Baig Moghal[3]
[1, 2, 3] *Department of civil engineering, NIT Warangal.*
* e-mail: mhl@nitw.ac.in

Introduction

Coal mining is often associated with generation of huge volumes of waste materials at various stages of mining. Wastes generated during coal combustion like coal ash and fly ash have been extensively utilised in geotechnical applications (Moghal 2013; Yao et al 2015). However, the wastes generated during the mineral processing phase has been sparsely considered for potential geotechnical applications. Coal gangue is a solid waste generated in coal mining process contributes to more than 15% of coal produced. Utilising coal gangue in geotechnical engineering applications can evade economic costs and environmental implications associated with its disposal. Unconfined compressive strength (UCS) is one of the principal property that determines the feasibility of utilising any geomaterial as a potential substitute of existing soils. With this in view, in the current study, an attempt has been made to study the UCS of coal gangue stabilised soils. Further, the effect of stabiliser on 'UCS' of coal gangue was studied by considering lime as the potential stabiliser

Materials and Methods

Soil considered for the study is locally available BC soil. Analytical grade hydrated lime is used as a stabiliser. Coal Gangue was procured from Kakatiya Coal mines, Bhupalpally, Telangana. The sourced samples were immediately sealed in plastic bags for 30 days to reach radioactive equilibrium. Unconfined compressive strength of specimens was measured as per ASTM D2166.

Results and Concluding Remarks

Firstly, the coal gangue addition to BC soil up to 50 % (by dry weight). The optimum coal gangue content was arrived at 40 %. Subsequently, for 40: 60 mixture (coal gangue: BC soil), stabiliser (lime) content is varied. The results presented in Figure 1 and Figure 2 are for a curing period of 28 days.

Effect of lime on strength behaviour of stabilised soils

The results obtained from the graph it is observed that the coal gangue stabilization of soil increases the strength significantly and the maximum strength increases with increase in coal gangue content. It is observed that after addition of coal gangue only, the improvement in strength is very marginal. The strength does not improve with curing period as well. In fact, the marginal increase in strength might be due to flocculation of soil particles. The optimum strength was obtained at a specific water content, which was close to the optimum water content. Increasing additions of coal gangue the matrix of the BC soil is disturbed more and more, leading to a decrease in strength. After curing, the strength of BC soil increased with an increase in coal gangue addition. Generally the trend between the coal gangue content and strength of soil is that the higher the coal gangue content, the higher the strength.

Effect of lime on strength behaviour of stabilised soils

The results observed from the graph, it can be observed that all coal gangue contents the strength of BC soil with lime is higher than without lime. This is because of an enhanced pozzolanic reaction due to the increased lime content. The strength of given coal gangue-soil mixture varies with lime content. Generally increase in strength with an increase in lime content up to a certain percentage of lime, beyond which it may reduce or the rate of increase in strength is negligible. The very similar immediate unconfined compressive strength of BC soil on addition of low percentage of coal gangue shows that the decrease has been compensated for the pozzolanic reaction that takes place during this lapsed time. The significant increase in strength is observed with addition of 6 % lime to soil-coal gangue mix.

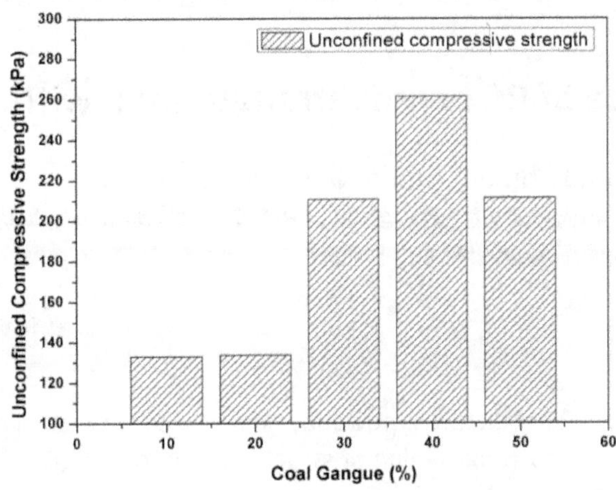

Figure 1. Variation in UCS of BC soil with Coal gangue addition

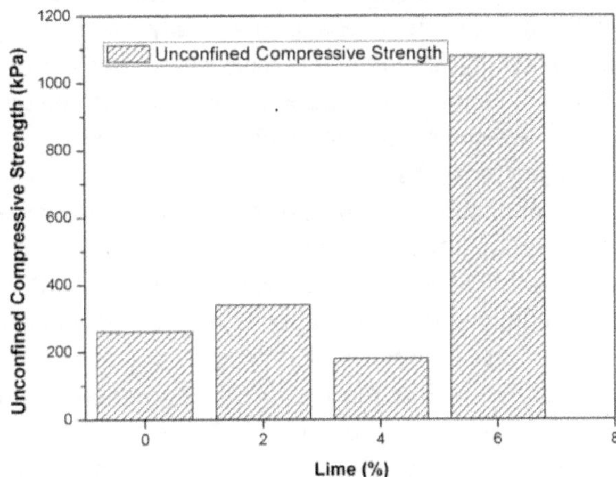

Figure 2. Variation in UCS of Coal gangue stabilised BC soil with Lime

From the results it was observed that UCS behaviour of untreated BC soil increased with coal gangue addition. The addition of lime has contributed to further increase in UCS of BC soil. The effect is more pronounced for specimens with higher curing period. This behaviour can be attributed to the change in pore size distribution of BC due to the formation of secondary hydration compounds. Due to its ability to fill the pore spaces, the dosage of lime has also played vital role for the increase in UCS (Prakash et al 1989; Bell 1996; Rao and Shiavananda 2005). Thus, from the results it can be concluded that coal gangue can be utilised in geotechnical applications as a substitute/supplementary material to natural soils.

References

Journal article example:

ASTM D2166 / D2166M-16, Standard Test Method for Unconfined Compressive Strength of Cohesive Soil, ASTM International, West Conshohocken, PA, 2016

Bell FG (1996) Lime stabilization of clay minerals and soils. Engineering Geology 42(4):223–37.

Moghal, A.A.B. (2013). "Geotechnical and Physico-chemical characterization of low lime fly ashes." Advance Material Science and Engineering. Article ID: 674306.

Prakash K, Sridharan A and Rao SM (1989) Lime addition and curing effects on the index and compaction characteristics of a montmorillonitic soil. Geotechnical Engineering 1989; 20: 39-47.

Rao SM and Shivananda P (2005) Compressibility behaviour of lime-stabilized clay. Geotech Geol Eng. 2005;23:309–19.

Yao, Z.T., Ji, X.S., Sarker, P.K., Tang, J.H., Ge, L.Q., Xia, M.S., and Xi, Y.Q. (2015). "A comprehensive review on the applications of coal fly ash." Earth Science Reviews 141,105–121

Flexural performance of RC beams strengthened with FRP techniques

Toufeeq Anwar[1*],Syed Jawwad Ahmed[2],Mohd Nadeem Uddin[3]
[1] Associate professor, [2] Assistant professor and [3] PG scholar, Dept. Of Civil Engg., Muffakham Jah College of Engineering and Technology, Hyderabad, India
** e-mail: toufeeqanwar@mjcollege.ac.in*

Introduction

Strengthening of inadequate RC structures, nowadays have become a burning need in the construction industry world-wide. Necessity for strengthening of structures emerges due to inadequacy of structures to resist the loads acting on it which can be caused mainly due to changing the usage type of the structure, deterioration of concrete mainly due to corrosion or committing errors in designing the structure. Strengthening of structures with the help of traditional methods such as Grouting, Shotcreting, Jacketing etc., may not always offer a significant solution. In many cases, using fibre reinforced polymers (FRP) for Strengthening purposes provides a cost-effective and practically superior option to the conventional methods. Presently, use of FRP for strengthening works have been imposed to strengthen structures such as slabs, columns, beams, chimneys, tunnels etc. The FRPs have sound ductility and durability, excellent heat resistance, higher strength-to-weight ratios than traditional materials used for retrofitting purposes and requires less workforce and appurtenances.

J. Gopi Krishna et al. 2015 has conducted test on beams retrofitted with externally bonded carbon fiber reinforced polymer (CFRP) with various types of resins and also the load displacement response of strengthened and control beams were investigated. Narayan Tiadi et al. 2018 investigated the behavior RC continuous beams strengthened in flexure with glass fiber reinforced polymer (GFRP) externally bonded to the tension side of the beams. The test results showed an increase of 61.92%, 59.61% and 63% in load carrying capacity of strengthened beam compared with unstrengthened beams. A.N. Nayak et al. 2018 investigated the effect of glass fiber reinforced polymer (GFRP) in flexural strengthening of beams using externally bonded technique. A.S.D. Salama et al. 2018 investigated the flexural behaviour of reinforced concrete beams strengthened with carbon fiber reinforced polymer (CFRP) externally bonded to the side faces of beams Test results showed an increase of 39.7-93.4% for side bonded beams and 62-92% for bottom bonded strengthened beams. An experimental study is presented on the flexural performance of RC beams strengthened with Basalt FRP using different FRP techniques. In the previous studies extensive research has been done using Carbon, Glass and Aramid FRP's. Whereas, limited researches has been found for strengthening of structures using Basalt FRP.

Materials and Methods

In order to study the flexural performance of RC beams strengthened with different FRP techniques, series of tests were conducted. Table 1 shows details of 15 RC beams that were cast to carry out the study.

Table 1. Details of RC beams that were cast in the present study.

RC beams	No. of Specimen	Size of specimen	Reinforcement provided
Set -1	3	100mm x 175mm x 1000mm	2# 8mmϕ on top & 2#12mmϕ at bottom
Set -2	3	100mm x 175mm x 1000mm	2# 8mmϕ on top & 2#12mmϕ at bottom
Set -3	3	100mm x 175mm x 1000mm	2# 8mmϕ on top & 2#12mmϕ at bottom
Set -4	3	100mm x 175mm x 1000mm	2# 8mmϕ on top & 2#12mmϕ at bottom
Set -5	3	100mm x 175mm x 1000mm	2# 8mmϕ on top & 2#12mmϕ at bottom

Specimens mentioned in the table 1 were cast and cured for 28 days. Set 1 being control specimen was tested for their flexural capacities. Other sets were strengthened with different FRP techniques, in which set-2 beams were strengthened with Near Surface Mounted(NSM) technique using 2#6mmϕ Basalt rebar provided on tension face by making shallow grooves of size 10mm x 10mm along the length of the beam and filled with epoxy mortar, set-3 beams were strengthened using Basalt FRP fabric externally bonded to the tension face throughout the length of the beams using epoxy resin as adhesive, set-4 beams were strengthened with Hybrid technique of NSM and External bonding of Basalt FRP fabric on tension face and set-5 beams were strengthened with Hybrid technique of NSM and U-wrapping of Basalt FRP fabric throughout the length of beams. All the strengthened beams were tested in UTM for their flexural capacities under four point bending system.

According to the data provided by the manufacturer, the orientation of Basalt FRP fabrics used for external bonding is uni-directional, thickness is 0.158mm and density is 200 gsm. For casting of beams ordinary portland cement used was of grade 53 having specific gravity of 2.9 and the maximum size of coarse aggregates was about 20 mm. Mix design proportion used for concreting was 1:1.45:2.62 by weight. Cubes and cylinders were cast and tested along with beams in order to determine concrete compressive strength and split tensile strength. Results showed an average cube compressive strength of 40.65 N/mm^2 and an average split tensile strength of 3.85 N/mm^2 for 28 days of curing.

Results and Concluding Remarks

Using FRP techniques for strengthening of RC beams shows a noteworthy improvement in their flexural capacities. Table 2 shows the ultimate load carrying capacities of control and strengthened RC beams and the percentage increment in the load capacities of strengthened beams.

Table 2. Load at failure of control and strengthened beams and percentage increment of strengthened beams.

Beams	Strengthening technique	Avg. Ultimate Load	% Increment
Set-1	Control specimen	91.43 kN	-
Set-2	Near surface mounted(NSM)	110.08 kN	21.8%
Set-3	External bonding of BFRP on tension face	114.45 kN	25.1%
Set-4	Hybrid (NSM + bonding of BFRP on tension face)	116.45 kN	26.87%
Set-5	Hybrid (NSM + U-wrapping of BFRP)	124.8 kN	36.5%

The flexural capacity of strengthened beams using various FRP techniques were found to be greater than that of control beams. Hence, it can be concluded that Basalt FRP fabrics are noteworthy for increasing the shear strength of beams and punching strength of slabs.

Acknowledgments: The R&D cell MJCET provided financial assistance for this research, to which writers are grateful.

References

A.N. Nayak, A. Kumara, R.B. Swain Hawileh, J.A. Abdalla rahman, Zhangjian Wu, Lee S.Cunningham (2017) Experimental and numerical investigation i into strengthening flat slabs at corner columns with externally bonded CFRP. Journal of Construction and building materials 139(2017): 132-147. https://doi.org/10.1016/j.conbuildmat.2017.02.056

J.H. Gonzalez-Libreros, L.H. Sneed, T. D'Antino, C. Pellegrino (2017) Behavior of RC beams strengthened in shear with FRP and FRCM composites. Journal of Engineering Structures 150(2017): 830-842. https://doi.org/10.1016/j.engstruct.2017.07.084

M. Gherdaoui, M. Guenfoud, R. Madi (2018) Punching behavior of strengthened and repaired RC slabs with CFRP. Journal of Construction and building materials 170(2018): 272-278. https://doi.org/10.1016/j.conbuildmat.2018.03.093

Riza Secer Orkun Keskin, Guray Arslan, Kadir Sengun (2017) Influence of CFRP on the shear strength of RC and SFRC beams. Journal of Construction and building materials 153(2017): 16-24. https://doi.org/10.1016/j.conbuildmat.2017.06.170

Shape Memory Alloy Bars as Internal Reinforcement in Structural Elements – A Review Paper

Gisha George[1*], Bindhu K.R[2]
[1] *Research Scholar, Civil Engineering Department/College of Engineering Trivandrum, Kerala, India*
[2] *Professor, Civil Engineering Department/College of Engineering Trivandrum, Kerala, India*
** e-mail: geo.gisha@gmail.com*

Introduction

Shape memory alloy bars having super elasticity and shape memory effect is used in construction industry in many forms. The first type of shape memory alloy (SMA) used in structural elements are NiTi due to its self-centring property and high corrosion resistance. Shape memory alloy have the ability to undergo large deformation and it can come back to initial position with the removal of load and upon heating. This effect of SMA is known as super elasticity and shape memory effect (SME) respectively. Superelasticity exists in the alloy when it is above the austenite finish temperature. In this phase SMA can sustain significant deformation cycles upto their maximum recoverable strains without any permanent deformation. SMA bars were provided at plastic hinge region of structural elements to improve seismic performance, which can be connected to steel reinforcement using mechanical screw type couplers. Several study on structural elements reveals that elements reinforced with SMA bars in plastic hinge region show reduced residual displacement and crack width. It also shows high ductility ratio and adequate energy dissipation capacity. Use of SMA bars along with engineered cement composite (ECC) can effectively increase the energy dissipation capacity and crack closing with the removal of load.

Materials and Methods

The present study concentrates on the effectiveness of using SMA bars as main reinforcement in plastic regions of structural elements . For better performance of structures during severe seismic event it should have high ductility, energy dissipation, less residual displacement and crack width. Saiidi and Wang (2006) conducted shake table test on circular column reinforced with NiTi SMA bars in plastic hinge length connected to steel bars using steel couplers. The maximum displacement of 66 mm with a drift ratio of 4.8% after 11 th run and the residual displacement were 1.3 mm. Then column is repaired with ECC using poly vinyl alcohol fibers along a height of 254mm from top of footing sustain 15 run with maximum drift of 135 mm with 9.8% drift ratio and the residual displacement were 3.3 mm. The results of two columns compared with normal reinforced column with residual displacement of 45 mm during 7 th run. This experimental study shows that SMA-RC column can withstand large amplitude of earthquake with less residual displacement. Alam et al., (2008) conducted experimental study on the seismic performance in terms of load -displacement and energy dissipation in exterior beam column joint (BCJ). An exterior BCJ isolated from eight story RC building with moment resisting frame designed for maximum moments and shear forces according to Canadian standards. The BCJ was scaled down by 3/4 th and subjected to cyclic loading under constant axial load of 350 kN. It was designed with superelastic NiTi SMA in plastic region and other regions with GFRP bars. The plastic hinge length calculated using Paulay and Priestly equation and screw lock adhesive couplers are used to connect SMA and GFRP bars. At 4.4 % drift the flexural crack width was 5.5 mm and it closes to 1.5 mm. SMA-FRP beam column joint dissipate energy of 6.29 kN m and normal BCJ dissipate 6.76 kN m of energy. It takes 89 % of full load carrying capacity at 4 % drift and shifts the plastic hinge region away from column face to one-quarter beam depth.

Billah and Alam (2012) conducted study on square hybrid column with shape memory alloy in critical region and FRP in other region, shape memory alloy in critical region and other region with stainless steel, stainless steel in critical region and FRP bars in other region and stainless steel throughout the structure. SMA and SS is provided in plastic hinge region. Dynamic time history analysis conducted using 20 selected ground motion shows displacement more than 100 mm. In the selected ground motion Loma Prieta in 1989 with peak ground acceleration of 0.605 g develops maximum drift of 3.8%. The residual drift for the same case is less than 0.2% for SMA-FRP and SMA-SS. But SS-FRP and SS column have residual displacement around 0.9%. So this study shows that for columns reinforced with SMA in plastic hinge region have less residual displacement. Ductility is the ability to undergo large displacement maintaining post yield load level. It is the ratio of ultimate displacement corresponding to first yielding of tension reinforcement to the yield displacement. Stainless steel reinforced column have ductility of 9%, SS-FRP have 6%, SMA-SS have 5% .These shows columns reinforced with SMA in critical region show less ductility compared to others.

Nickel- Titanium shape memory alloy in the form of cable was used along with ECC in beams to study the effectiveness under dynamic loading. Li et al., (2015) tested four cantilever beam with two point reverse cyclic loading. ECC beam reinforced with SMA (SMA-ECC), SMA reinforced normal concrete (SMA-RC), plain ECC, and normal concrete (RC) beam. Among 4 beams SMA-RC beam have maximum deflection of 9.6 mm, residual deflection of 1.5 mm and residual crack of 1.5 micrometer. ECC beam have maximum deflection of 7.2 mm, residual deflection of 3.6 mm and residual crack of 2.614 micrometre. SMA-ECC have maximum deflection of 7.9 mm, residual deflection of 1.2 mm and residual crack of 0 micrometre. Under reverse cyclic loading SMA-ECC undergo large deformation with less residual displacement and full recovery upon unloading. SMA-ECC beam have higher energy dissipation capacity compared to ECC and SMA-RC beam. Rashedul et al., (2017) numerically analysed BCJ using SeismoStruct V7.0.4 with NiTi, FeNCATB, CuAlMn and FeMnAlNi SMA in plastic hinge region. Among these different SMA iron based carries higher beam tip load and takes drift more than 4%. BCJ reinforced with FeMnAlNi having higher super elastic strain length show energy dissipation capacity of 34.2 kN-m with storey drift of 5.27%. Iron based FeNCATB SMA with 13.5 superelastic strain in plastic region of BCJ produce less residual displacement of 1.22%.

Results and Concluding Remarks

Seismic analysis of structural elements reinforced with shape memory alloy in plastic region produce less residual displacement and can sustain large amplitude of earthquake compared to steel reinforced concrete elements. Use of shape memory alloy bars in critical region along with engineered cement composite exhibits full crack recovery and high energy dissipation capacity. Iron based SMA with higher modulus of elasticity and superelastic strain have better seismic performance compared to Nickel based SMA. Hence it can be effectively used in plastic hinge regions of structural elements along with ECC in seismic region to make the structure serviceable after strong earthquake.

References

M Saiidi Saiidi and Hongyu Wang (2006) Exploratory study of seismic response of concrete column with shape memory alloy reinforcement. ACI Structural Journal, Vol 103, 436-443.

Billah A H M, Alam M S (2012) Seismic performance of concrete columns reinforced with hybrid shape memory alloy and fibre reinforced polymer bars. Construction and Building Materials 28(2012) 730-742.

Xiaopeng Li, Mo Li and Gangbing Song (2015) Energy dissipating and self repairing SMA-ECC composite material system. Smart Materials and Structures 24,025024.

Rashedul K M, Sosan R, Shahria A M (2017) Seismic performance of beam- column joint reinforced with different shape memory alloy alternatives. Conference on Leadership in Sustainable Infrastructure, Vancouver, Canada, May 31- June3.

Mix Proportioning of High Strength Concrete Using VSI Sand

R. R. Khartode[1*], D. D. Ahiwale[1], G. N. Narule[1] and D. A. Nikam[2]
[1]*Assistant Professor, Department of Civil Engineering, VPKBIET, Baramati, Pune.*
[2]*P.G.Student, Department of Civil Engineering, VPKBIET, Baramati, Pune.*
[*] *e-mail: rushikesh.khartode@vpkbiet.org*

Introduction:

High strength concrete plays an important role in present constructional activities such as high rise buildings, offshore structures, long-span bridges and structures at marine environment. These structures required high strength concrete for its stability and durability for a lifetime. Mineral admixtures such as silica fume are used to fill microscopic voids to get required strengths. To overcome the deficiency of river sand and to meet the requirement of the high strength concrete, it is required to formulate the high strength concrete using VSI sand.

The concrete mix was designed as per IS 10262:2019 to achieve a grade of M75. Nine different mix proportions were casted with 8%, 10% & 12% of silica fume with partial replacement of cement respectively whereas super plasticizer is used at 1%, 1.2% & 1.4%.

Materials used

1 Cement: Ordinary Portland cement of 53 grade confirming to IS 12269-1989, (Locally available brand Birla super) is used for this present study. The 28 days compressive strength of cement obtained is 63.50 MPa.
2 Coarse Aggregates: Crushed stone aggregates having a maximum size of 20mm size is used. Sieve analysis has been carried out by blending the aggregates of size 20mm to 12.50mm to get the good fineness modulus. The coarse aggregate having specific gravity 2.96 and fineness modulus 5.67 is used.
3 Fine Aggregates (VSI sand): VSI/Artificial sand is manufactured in vertical shaft impact crusher was collected from VSI crusher, Vasunde, dist. Pune. It is used as an alternative for river sand in the fully replacement of fine aggregates. The fine aggregate having specific gravity 2.94 and fineness modulus 4.54 is used.
4 Silica Fume: It is obtained from the Sudha enterprises, Pune. The specific gravity of Silica Fume is 2.2.
5 Superplasticizers: MasterEase 5801 as high range water reducing admixture was used provided by BASF Pvt. Ltd. Bhosari, Pune. Superplasticizer based on Polycarboxylate ether was used to impart additional desired properties to the high strength concrete.
6 Water: Potable tap water is used for mixing and curing which is free from deleterious materials.

Figure 1: VSI Sand.

Figure 2: Testing of Specimens.

Methodology

1) Collection of materials.
2) Testing of materials for their physical &chemical properties.
3) Proportioning of aggregates on the maximum density approach.
4) Mix design calculations for given cementitious content by blending of Silica fume in different percentages.
5) Finding water content for a given mix.
6) Carry out trial mixes to get required slump & homogenous mix without honeycombing & segregation.
7) The casting of samples for various cementitious contents with different percentages of silica fume and superplasticizer.
8) Testing of samples at 7 & 28 days age.

Results and discussions

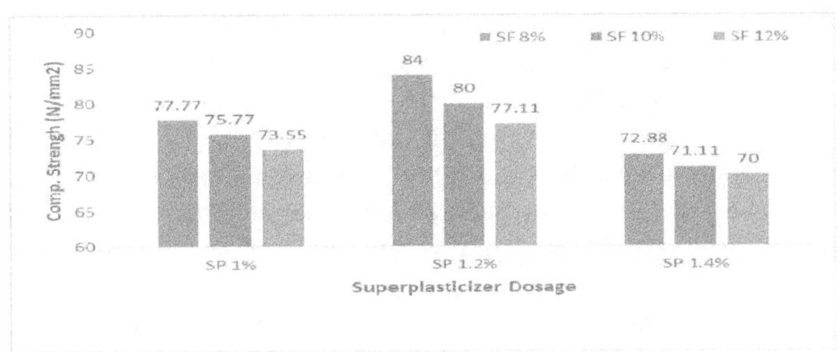

Figure 3: 28 Days Compressive strength results for w/b ratio 0.27

From the graph, it is observed that the compressive strength is maximum for a proportion of 8% silica fume and 1.2% superplasticizer with w/b ratio of 0.27. The compressive strength obtained is 77.77 N/mm^2 for an 8% slilica fume with a 1% superplasticizer. For a further increment of superplasticizer by 0.2%, the compressive strength is increased by 8.01%. But for the next increment of superplasticizer by 0.2%, the reduction in compressive strength observed is about 6.28% of its initial proportion.

Concluding Remarks

1. The addition of Silica fume up to 8%, leads to a significant increase in the characteristic strength of concrete.
2. The workability of concrete decreases with the addition of Silica fume as compared to the conventional mix.
3. Results shows that, the river sand can be fully replaced by VSI sand as per the experimental results.

References

[1] Adams Joe, A. Maria Rajesh, P. Brightson, M. Prem Anand (2013), "Experimental investigation on the effect of M-sand in high performance concrete" American Journal of Engineering Research, Volume-02, Issue-12, pp-46-51.

[2] Boopathi and J. Doraikkannan (2016) "Study on M-sand as a partial replacement of fine aggregate in concrete" International Journal for Research in Applied Science & Engineering Technology, Vol. 3, Special Issue 2, Page 746 – 749.

[3] Kulkarni D. B. and Khartode R. R. (2016) "Mix proportioning of HSC using manufactured sand" International Journal of Science and Research Volume 5 Issue 7, pp-423-429.

[4] Nimitha Vijayaraghavan, and A. S. Wayal (2013) "Effects of manufactured sand on compressive strength and workability of concrete" International Journal of Structural and Civil Engineering Research Vol. 2, No. 4.

[5] Prakash. P (2013) "Strength of concrete by partially replacing the fine aggregate using M-sand" Sch. J. Eng. Tech., 1(4):238-246.

[6] Priyanka A. and Dilip K. Kulkarni (2012) "An experimental investigation on the properties of concrete containing manufactured sand", International Journal of Engineering Research & Technology, Vol. III, Issue II, pp-101-104.

Thermal Energy Storage Building Materials using PCM– The State of art Review

Prathik Kulkarni[1], A.Muthadhi[2*]
[1] *Research Scholar,Department of Civil Emgineering,Pondicherry Engineering College,Pondicherry*
[2] *Assistant Professor, Department of Civil Emgineering,Pondicherry Engineering College,Pondicherry.*
** e-mail: prathik.kulkarni89@pec.edu*

Introduction

India, home to 18% of the total world's population, uses solely six percentage of world's Pre-eminent energy. Energy consumption in India has virtually increased from 2000 and therefore it is likely for additional zoom is gigantic. India is about to contribute over the other country to the projected rise in international energy demand, around one-quarter of the total: however, energy demand per capita in 2040 continues to be four-hundredth below the global average. The two parts of energy use within the buildings sector (the residential and services sectors) have terribly completely different patterns of consumption. In India these days the residential sector depends in the main on solid biomass, with oil an overseas second (LPG for preparation, coal oil for preparation and lighting) followed by electricity. The services sector, that tends to be focused in urban areas, is already mostly keen about electricity. Future will increase in energy consumption within the services sector – together with a jump in demand for area cooling in buildings.

Thermal energy has a key role in conservation of energy, which might be greatly achieved by incorporation of heat storage materials in several merchandise. In residential parts, that stored heat throughout peak power operation will cut back at constant time area acquisition energy consumption. PCMs are one of the most trusted thermal mass parts in buildings for the past 60 years. A key goal on low energy building analysis is to search out different methods that to manage variations in time between consumption of energy and sources of energy.

When any substance is heated up, heat will transfers mainly in two ways: Heat of transformation and Sensible heat. The construct of heat and heat energy were discovered by Scottish someone chemist within the mid-eighteen century. Ice is the best-noted PCM utilized by a human for preparation of food, acquisition of food, soft drinks, and area cooling. Igloos are the primary noted that the application of the phase-change heat in building structures. In step with Espada et.al 2001; inside igloo internal temperature falls between 9 and 15°C, once it is occupied and even throughout winters in harsh arctic region the outside temperature will fall to – 45°C.

PCM works on the principle of Heat of transformation Regin et.al;2008. An activity material PCM is a substance with high heat of transformation that melts and solidifies at a precise temperature range, is capable of storing and cathartic giant amounts of energy. Heat is absorbed or discharged once the fabric changes from solid to liquid and the other way around. Heat permits PCM to manage the temperature. PCM employed in construction modification from Liquid to Solid and Vice-versa at 23ºC – 26ºC. They absorb heat from area in day timings and melt at the desired temperature and then is unbroken constant till the modification of state is complete. Due, to the reduction in temperature at night time PCM changes its state to solid by evening time ventilation / mechanical suggests that. The same cycle repeats.

Materials and Methods

There as many methods to incorporate PCM in building materials but the three predominant methods are.

1) Direct mixing
2) Immersion
3) Impregnation

1) **Direct mixing:-** on the spot mixing of powdery kind, liquid kind, liquid cake kind or encapsulation will be done at the time of blending the concrete or cement mortar. Norvell et.al; 2013 these days encapsulation of PCM is enjoying a crucial role in building materials. A method by that each and every particle is coated with a continual film of compound material to make it in the form of capsules known as microcapsules. These microcapsules will vary from Micrometer to millimeter, in powder form.

2) **Immersion Technique:-**Hawes et.al; was the person to use this technique and introduce it for the first time in 1991 in this method light weight aggregate of any porous materials are soaked in liquid PCM. Kemal Cellat et.al; 2017 the time required for the PCM to be fully soaked in aggregate was supported by mainly criteria: The capability of absorption of aggregates, the temperature range required, and PCM used.

3) **Impregnation:-**This method is principally concerned within the removal of air or water from the aggregates victimisation vacuum impregnation or kitchen appliance drying at high temperature. Then the light weight aggregates are immersed within the liquid PCM in desired environmental conditions.

Results and Concluding Remarks

Organic PCMs are one of the most suitable materials for storage of latent heat and in particular paraffin wax PCM, has been used by most of the researchers in construction materials. PCM incorporation by direct mixing resulted in reduction in mechanical properties and there was significant improvement in thermal properties of cement mortar/concrete depending on type of PCM used. When PCM incorporated/impregnated into various light weight aggregates such as expanded perlite, shale, clay there was loss in mechanical strength was reduced and there was slight change in the volumetric behavior. Addition/replacement of microencapsulated PCM resulted in reduction in workability, delay in setting time & significant reduction in strength properties. Incorporation of PCM up to 5% by mass of cement at the end of mixing process will result in improvement in workability and heat capacity of concreteAlkali in concrete can attack some of the PCM in concrete and can be reduced using some of the pozzolonic materials like Fly ash, Silica Fume, GGBS needed regarding the use of PCM in building materials. Micro-Encapsulated PCM if used in construction then it has to be added at the end of mixing to reduce the negative effect of workability and to reduce the breakage.

Acknowledgments: There was a financial support from UGC Minor Project [MRP-6460/16(SERO/UGC)] in the year 2017 and Research work carried out at Pondicherry Engineering College was highly appreciated.

References

1) Chad Norvell, David J.sailor, and Peter Dusicka (2013) The Effect of Microencapsulated Phase-Change Materials on the Compressive strength of Structural Concrete: Journal of Green Buildings, 8 (2013), 116-124.
2) D.W.Hawes, D.Banu, and D.Feldman (1993) Latent heat storage in building materials: Energy and Buildings, 20 (1993) 77-96.
3) Hawes DW, Feldman D (1992) Absorption of phase change material in concrete. Sol Energy Mater Sol Cells 1992;27:91-101.
4) Kemal Cellat, Beyza Beyhan, Berk Kazanchi, Yeliz Konuklu and Halime Paksoy (2017) Direct Incorporation of Butyl Stearate as Phase Change Material in Concrete for Energy saving in Buildings: Journal of Clean Energy Technology, Vol. 5, No. 1, January 2017.
5) Regin AF, Solanki SC (2008) Heat transfer characteristics of thermal energy storage system using PCM capsules: A Review. Renew Sust Energy Rev 2008;12:2438-58.

Sponsors

Sponsors

Manglam Consultancy Services was founded in 2001 at Vadodara, Gujarat for giving quality of work in the field of Civil engineering material testing and subsoil exploration work. MCS provides services in many fields such as

- Highway engineering & Road-testing.
- Third party inspection for Quality Assurance.
- Building Material Testing Laboratory.
- NDT Testing.
- Soil investigation.
- Surveying.

Manglam Consultancy Services Hyderabad was Established in 2017 and this organization is equipped with all the latest equipment for soil, Building material, highway material and survey activities. All these equipment and machineries are duly calibrated and maintained standards of NABL. MCS is recognized by our clients for quality work & services. MCS Hyderabad is having ISO 9001:2015 Certification, **NABL** Accreditation along with major Central govt., State govt., & private sector approval.

www.ingramcontent.com/pod-product-compliance
Lightning Source LLC
LaVergne TN
LVHW080850240726
843527LV00052B/299
9789389354577